MAGNETISM AND METALLURGY *OF* SOFT MAGNETIC MATERIALS

by

CHIH-WEN CHEN

Lawrence Livermore National Laboratory
Livermore, California
Formerly Professor of Metallurgy
Iowa State University

DOVER PUBLICATIONS, INC.
NEW YORK

Published in Canada by General Publishing Company, Ltd., 30 Lesmill Road, Don Mills, Toronto, Ontario.
Published in the United Kingdom by Constable and Company, Ltd., 10 Orange Street, London WC2H 7EG.

This Dover edition, first published in 1986, is an unabridged and corrected republication of the work originally published by North-Holland Publishing Company, Amsterdam, in 1977, as Vol. 15 of their "Series of Monographs on Selected Topics in Solid State Physics," edited by E. P. Wohlfarth.

Manufactured in the United States of America
Dover Publications, Inc., 31 East 2nd Street, Mineola, N.Y. 11501

Library of Congress Cataloging in Publication Data

Chen, Chih-wen.
Magnetism and metallurgy of soft magnetic materials.

Reprint. Originally published: Amsterdam ; New York : North-Holland Pub. Co., 1977. (Series of monographs on selected topics in solid-state physics ; 15)
Bibliography: p.
Includes indexes.
1. Magnetic materials. 2. Magnetism. 3. Physical metallurgy. I. Title.
QC765.C47 1986 538′.4 85-10393
ISBN 0-486-64997-0 (pbk.)

To the Memory of My Father

CHEN RUENN-LIN (1879–1946)

A Dedicated Educator

PREFACE

The development of new materials and novel applications for the computer and telecommunications industries during the past two decades has immensely broadened the scope and altered the nature of soft magnetic materials. No longer is this field dominated by metallic substances in terms of technological importance. Nonmetallic compounds and amorphous thin films share the spotlight, if not steal the show. Consequently, a basic knowledge of this large group of materials involves two major components: (1) traditional magnetism, which imparts magnetization and spin-dependent properties, and (2) metallurgy, which governs the preparation of the products and the attainment of their structure-sensitive properties. The present book has been written in this context and is therefore divided into two parts to cover these two components. The word "metallurgy" in the title is used in a broad sense. It would have been replaced by "materials science" had the word "materials" not already appeared at the end of the title.

Combining magnetism and materials science, the aim of the book is to present, with coherency and consistency and in logical sequence, the physical principles that underlie the intrinsic as well as the applied properties of soft magnets. Discussions on some of the metallurgical phenomena are rather brief in order to hold down the volume of the book. Further information on these phenomena can be found in several texts on physical metallurgy cited in Chapter 5. In discussing many topics of great fundamental importance in Chapters 2, 3 and 4, I have tried to make the presentations as complete as possible without burdening the reader with tedious mathematics and trivial details. Strong emphasis is placed on their physical implications and effects on applications, however. Throughout the book, a policy is adopted and practised as much as possible to refer the theories and experimental studies originated before 1948 to later pertinent publications. Thus the reader will not be distracted by the antiquity of the original work, while the later publications are generally more accessible and understandable.

The book is intended for researchers and development workers who wish to obtain a comprehensive working knowledge of this field. Also, I have twice used it with good results as a textbook for an interdisciplinary graduate course on magnetic materials at the Iowa State University. Therefore I should think, and hope that the book can be used likewise at other universities. For the latter purpose, I should point out that although the book deals exclusively with soft magnetic materials, all principles thus discussed herein apply equally well to hard materials. The only item missing for the permanent magnets is the description of their design, preparation and applications. The missing information can be found in the review articles published in IEEE Trans. Magnetics, Vol. 4, pp. 84–99 and 221–8 (1968).

I am indebted to the Editor of the Series, Professor E. P. Wohlfarth, without whose encouragement and advice I would not have undertaken this time-consuming task. For the preparation of the manuscript, grateful acknowledgment is made to Mrs. Carol Greiner for her editorial assistance, to Miss Verna Thompson and Mrs. Erma Miller for their elegant typing and to D. W. Sailsbury, who did all the photographic reproduction of numerous figures. I am deeply grateful to W. J. Carr, Jr. of the Westinghouse Laboratories, W. H. Hu of the US Steel Research Center, R. R. Coltman of the Oak Ridge National Laboratory and many of my colleagues, M. F. Berard, O. N. Carlson, K. A. Gschneidner, Jr., F. X. Kayser and M. L. Covault for their kindness in reading portions of the manuscript and suggesting improvements. I also wish to express my appreciation to S. Legvold and to S. H. Liu for helpful discussions on ferromagnetism in lanthanides. I must thank many authors and publishers for permitting me to reproduce many figures and tables from their papers or books. In this connection, I am especially gratefui to Dr. R. M. Bozorth, Professor S. Chikazumi and Professor F. Brailsford and their publishers (Litton Educational Publishing International, John Wiley & Sons, and D. Van Nostrand Company Ltd, London, respectively), who granted me their kind permission for reprinting numerous figures from their renowned books. Thanks are due to the Ames Laboratory, US Energy Research and Development Administration, for generous secretarial and graphic services.

Chih-Wen CHEN

ACKNOWLEDGEMENTS

I am indebted to the authors and publishers for permission to reproduce the following figures appearing in this book:

4.12, Anderson, J. C., 1968, Magnetism and Magnetic Materials (Chapman and Hall, London). 2.13, Argyle, B. E., S. Charap and E. W. Pugh, 1963, Phys. Rev. **132**, 2501. 5.37, 5.52, 5.68, Barrett, C. S. and T. B. Massalski, 1966, Structure of Metals, 3rd ed., by permission, copyright McGraw-Hill Book Company. 5.45, Bates, L. F., D. F. Craik and E. D. Issac, 1962, Proc. Phys. Soc. **79**, 970. 5.51, 6.61, Bechtold, J. H. and G. W. Wiener, 1965, The Sorby Centennial Symposium on the History of Metallurgy, ed. C. S. Smith (Gordon and Breach, New York). 7.17, 7.19, 7.21, Bobeck, A. H., 1967, Bell Sys. Tech. J. **46**, 1901, by permission, copyright 1967, the American Telephone and Telegraph Company. 7.18, 7.22–7.26, 7.28, 7.29, 7.30, 7.33, 7.34, 7.36, 7.37, Bobeck, A. H. and E. Della Torre, 1975, Magnetic Bubbles (North-Holland, Amsterdam). 1.1, 3.7, 4.5, 4.6, 4.8, 4.10, 4.22, 5.5, 5.6, 5.25, 5.29, 5.30, 5.31, 5.32, 5.35, 5.57, 5.61, 5.63, 5.64, 5.65, 5.78, Bozorth, R. M., 1951, Ferromagnetism (van Nostrand Co., New York), reproduced with the kind permission of the author and publisher. 1.3, 2.2, 2.3, 2.8, 3.10, 3.11, 4.1, 4.2, 4.24, 4.25, 4.30, 4.32, 4.33, 5.23, 5.24, 5.26, 5.27, 5.28, Brailsford, F., 1966, Physical Principles of Magnetism (van Nostrand, New York), reproduced with the kind permission of the author and publisher. 7.8, Brinkman, J. A., 1962, Societa Italian Di Fisica, XVIII Corso (Academic Press, New York). 3.2, Carr, W. J., 1958, Magnetic Properties of Metals and Alloys (Am. Soc. for Metals, Metals Park, Ohio, USA). 7.1, Chadderton, L. T., 1965, Radiation Damage in Crystals (Methuen, London). 5.34, Chen, C. W., 1963, Physics Letters **7**, 16. 5.4, Chen, C. W., 1964, Phys. Rev. **129**, 121. 5.58, Chen, C. W., 1961, J. Appl. Phys. **32**, 348S; 3.1, 4.11, Chen, C. W., 1958, J. Appl. Phys. **29**, 1337. 2.4, 3.14, 3.17, 4.4, 4.21, 4.23, 4.35, 5.3, 5.46, 5.69, 5.70, 5.72, 5.73, 5.75, 5.76, 6.11, A.3.1, Chikazumi, S., 1964, Physics of Magnetism (Wiley, New York) with the kind permission of the author and publisher. 5.74, Chikazumi, S., J. Appl. Phys. **29**, 346. 5.67, Chikazumi, S. and T. Oomura, 1955, J. Phys. Soc. Japan, **10**, 842. 5.77, Chikazumi, S., K. Suzuki and H. Iwata, 1957, J. Phys. Soc. Japan **12**, 1259. 3.16, 3.18, 3.19, 6.18, 6.21, Craik, D. J. and R. S. Tebble, 1965, Ferromagnetism and Ferromagnetic Domains (North-Holland, Amsterdam). 4.9, De Barr, A. E., 1953, Soft Magnetic Materials Used in

Industry (Inst. Physics, London). 4.28, 4.29, Dillon, J. F., Jr. and H. E. Earl, Jr., 1959, J. Appl. Phys. **30**, 202. 5.39, Dykstra, L. J., 1969, Magnetism and Metallurgy (Academic Press, New York). 5.54, Fisher, R. M. and M. J. Marcinkowski, 1960, Phil. Mag. **6**, 1385. 7.13, 7.14, 7.15, Gordon, D. I., 1960, Proc. Inst. Environmental Sci. Annual Tech. Meeting, 205. 7.12, 7.16, Gordon, D. I. and R. S. Sery, 1964, J. Appl. Phys. **35**, 879. 5.48, 5.49, 5.50, Graham, A., 1969, by permission, Magnetism and Metallurgy (Academic Press, New York). 5.55, 5.56, Guttman, L., 1956, Solid State Physics, Vol. 3 (Academic Press, New York). 5.60, Hall, R. C., 1960, J. Appl. Phys. Suppl. **31**, 157S. 7.9, 7.10, Hall, R. C., W. S. Byrnes and R. G. Crawford, 1959, J. Appl. Phys. **30**, 288S. 5.47, Hibbard, W. R., Jr. and C. G. Dunn, 1956, Acta Met. **4**, 306. 3.3, Honda, K. and S. Kaya, 1926, Sci. Rep. Tohoku Univ. **15**, 721. 3.4, Kaya, S., 1928, Sci. Rep. Tohoku Univ. **17**, 639. 3.5, Kaya, S., 1928, Sci. Rep. Tohoku Univ. **17**, 1157. 1.5, 2.5, 5.53, Kittel, C., 1976, Int. to Solid State Physics, 5th ed. (Wiley, New York). 3.20, 4.15, 4.17, 4.19, 4.20, Kittel, C., 1949, Phys. Rev. **80**, 918. 2.14, Kittel, C., 1971, Introduction to Solid State Physics, 4th ed. (Wiley, New York). 5.59, Kouvel, J. S., 1969, Magnetism and Metallurgy, Vol. 2, Berkowitz, A. E. and E. Kneller, eds. (Academic Press, New York). 1.4, 6.2, Lee, E. W., 1963, Magnetism (Penguin, London). 4.26, Lee, E. W., 1958, Proc. Inst. Electrical Engineers, **105C**, 337. 3.8, 3.9, Lee, E. W., 1955, Rept. Prog. Phys. **18**, 184. 5.2, Legvold, S., 1973, in: Magnetic Properties of Rare Earth Metals, ed. R. J. Elliott (Plenum Press, London). 7.5, Leibfried, G., 1959, J. Appl. Phys. **30**, 1388. 2.7, Lomer, W. M., 1967, Theory of Magnetism in Transition Metals, ed. W. Marshall (Academic Press, New York). 5.33, 6.3, 6.4, 6.6, 6.7, 6.8, by permission, Metals Handbook, Vol. 1, 1961, copyright American Society for Metals. 5.17, McGuire, T. R. and W. M. Shafer, 1964, J. Appl. Phys. **35**, 984. 6.17, 6.18, 6.20, 6.23, Middelhoek, A., 1971, Magnetic Properties of Materials, ed. J. Smit, by permission, copyright McGraw-Hill, New York. 6.19, Moon, R. M., 1959, J. Appl. Phys. **30**, 83S. 2.11, 4.16, 4.18, 4.27, Morrish, A. H., 1965, The Physical Principles of Magnetism (Wiley, New York). 7.2, Nelson, R. S., 1968, The Observation of Atomic Collisions in Crystalline Solids (North-Holland, Amsterdam). 7.39, O'Dell, T. H., 1974, Magnetic Bubbles (MacMillan, London). 6.27, Olmen, R. W., 1963, Intermag. **12**, 1. 5.21, Parr, N. L., 1960, Zone Refining and Allied Techniques (George Newnes, London) reproduced by permission of the Hamlyn Publishing Group Ltd. 5.18, 5.20, Pfann, W. G., Zone Refining, 2nd ed. (Wiley, New York). 2.7, Phillips, J. C., 1967, Theory of Magnetism in Transition Metals, Marshall, W., ed. (Academic Press, New York). 6.16, 6.25, Pugh, E. W. and T. O. Mohr, 1964, Thin Films, by permission, copyright American Society for Metals, Ohio, USA. 4.31, Rado, G. T. et al., 1950, Phys. Rev. **80**, 273. 5.40, Reed-Hill, R. E., 1973, Physical Metallurgy Principles, 2nd ed. (van Nostrand Co., New York). 6.26, 6.28, 6.29, Renwick, W. and A. J. Cole, Digital Storage Systems (Chapman and Hall, London). 6.22, Rushton, S., 1963, Thesis, Univ. Nottingham. 5.41, Seeger, A., H. Kronmüller, H. Rieger and H. Traüble, 1964, J.

Appl. Phys. **35**, 740. 2.12, Sinclair, R. N. and B. N. Brockhouse, 1960, Phys. Rev. **120**, 1638. 5.37, 5.62, Smallman, R. E., 1970, Modern Physical Metallurgy (Butterworths, London). 4.34, 5.1, 6.13, 6.14, Standley, K. J., 1972, Oxide Magnetic Materials, 2nd ed. (Oxford University Press). 6.9, Smit, J. and H. P. J. Wijn, 1954, Adv. Electronics, **6**, 69. 5.22, 6.5, Stanley, J. K., 1963, Electrical and Magnetic Properties of Metals (Am. Soc. Metals, Ohio), by permission, copyright American Society for Metals. 7.7, Thompson, M. W., 1969, Defects and Radiation Damage in Metals (Cambridge University Press, London). 5.42, 5.44, Traüble, H., 1969, Magnetism and Metallurgy, Berkowitz, A. E. and E. Kneller, eds. (Academic Press, New York). 5.43, Traüble, H., O. Boser, H. Kronmüller and A. Seeger, 1965, Phys. Stat. Solidi, **10**, 283. 6.10, Verweel, A., 1971, Magnetic Properties of Materials, ed. J. Smit, by permission, copyright McGraw-Hill, New York). 6.15, Weiss, R. S., 1971, Magnetic Properties of Materials, ed. J. Smit (McGraw-Hill, New York). 5.7, Went, J. J. and E. W. Gorter, 1952, Philips Tech. Rev. **13**, 181. 5.50, Wiener, G., P. A. Albert, R. F. Trapp and M. F. Littman, 1958, J. Appl. Phys. **29**, 366. 6.24, Williams, H. J. and R. C. Sherwood, 1957, J. Appl. Phys. **28**, 548.

LIST OF MOST IMPORTANT SYMBOLS

a	Lattice parameter
A	Area
A_{ex}	Exchange stiffness
$\boldsymbol{B}$, B	Magnetic induction
B_r	Remanence
$B_{r,\infty}$	Retentivity
C	Curie constant
d_w	Thickness of domain wall
E_e	Magnetoelastic energy
E_s	Magnetostatic energy
f	Frequency
$\boldsymbol{F}$	Force
F	Helmholtz free energy
g	Landé factor
G	Gibbs free energy
h	Planck's constant
$\boldsymbol{H}$, H	Applied magnetic field
H_a, H_k	Anisotropy field
H_c	Coercive force
$H_{c,\infty}$	Coercivity
H_d	Demagnetizing field
H_W	Weiss molecular field
$\boldsymbol{I}$, I	Electric current
j	Total angular momentum quantum number of electron
J	Total angular momentum quantum number of atom
J_{ex}	Isotropic exchange integral
k	Boltzmann's constant
K_1, K_2	First and second magnetocrystalline anisotropy constants
l	Orbital angular momentum quantum number of electron
L	Orbital angular momentum quantum number of atom
m	Total magnetic quantum number, magnetic pole strength

m_l	Orbital magnetic quantum number of electron
m_0	Rest mass of electron
m_s	Spin magnetic quantum number of electron
$\mathbf{M}$, M	Intensity of magnetization, or simply magnetization
M	Total magnetic quantum number of atom
M_A	Magnetization for a mass of the substance corresponding to the atomic weight
M_s	Saturation magnetization
M_0	Saturation magnetization at 0°K
n	Total or principal quantum number of electron
n_{eff}	Effective Bohr magneton number per atom
N_0	Avogadro's number
N	Number of atoms per unit volume
N_d	Demagnetizing factor
$\mathbf{p}$, p	Angular momentum
s	Spin quantum number of electron
S	Spin angular momentum quantum number of atom, entropy
T	Temperature, torque
T	Tesla
U	Internal energy
V	Volume, electric potential
W_h	Hysteresis loss
z	Coordination number or number of nearest neighbors
Z	Atomic number
Z_n	Number of outer electrons per atom
γ	Magnetomechanical or gyromagnetic ratio
γ_{ex}	Exchange energy of a domain wall per unit area
γ_w	Surface energy density of domain wall
θ_C	Curie temperature
θ_F	Ferrimagnetic Néel temperature
θ_N	Néel temperature
λ	Weiss molecular field constant, characteristic length of a bubble material
λ_s	Saturation longitudinal magnetostriction
μ	Absolute normal permeability, magnetic dipole moment
μ_i	Initial permeability
μ_{max}	Maximum permeability
μ_r	Relative normal permeability
μ_0	Permeability of vacuum
μ_B	Bohr magneton

ξ	Mobility of domain wall
ρ	Density, electrical resistivity
ϕ	Magnetic flux
χ, χ'	Absolute and relative magnetic susceptibility per unit volume
χ_m, χ'_m	Absolute and relative mass susceptibility
χ_A, χ'_A	Absolute and relative atomic susceptibility
χ_M, χ'_M	Absolute and relative molar susceptibility
ω	Angular velocity, angular frequency
ω_R	Resonance angular frequency

CONTENTS

MAGNETISM AND METALLURGY *OF* SOFT MAGNETIC MATERIALS

CHAPTER 1

INTRODUCTION

§1. HISTORICAL

The history of the development of magnetic materials is undeniably long and splendid. According to the history of Asia Minor, a cradle of civilization, lodestone was discovered as a natural magnet in Magnesia more than 3500 years ago. The stone was called magnes lapis, which means Magnesian stone. "Magnet" and "magnetism" were derived from the word "Magnesian." Thales of Miletus, the Greek philosopher, stated that the magnetic interaction between magnetite and iron was known before 600 BC. Contemporary Chinese historians have shown evidence that the world's first compass—a south-point spoon of lodestone balanced on a round, bronze heaven plate within a square earth plate—was invented in China about 85 AD (see figure at left).

Magnetic materials also played a prominent role in the discovery of the New World and in the development of modern technology. Without the compass Christopher Columbus would not have made his voyages and discoveries. The outstanding magnetic properties and low cost of iron have made it possible to generate inexpensive electricity on a massive scale since 1886 when Westinghouse Electric Company built the first commercial AC generating station at Buffalo, New York. The use of magnetic materials to perform vital functions is not limited to the utility industry, other industries consuming appreciable amounts of magnetic materials include communications, computer, audio-visual, and home appliances, to name just a few.

From the scientific viewpoint, it is understandable that magnetic materials have contributed vitally to the history of civilization and technology. In the first place, all substances, whether solid, liquid or gas, display certain magnetic characteristics at all temperatures. Hence magnetism is one of the basic properties of materials. Secondly, although only three elements (Fe, Co, and Ni) out of a total of 106 are ferromagnetic at room temperature, we are fortunate that iron constitutes approximately 3.3% of the earth's crust, and hence, it is abundant in natural resources and relatively inexpensive. Thirdly, metallic meteorites are known to consist largely of pure iron. It is logical to assume that, by the time ancient man appeared on the earth, pure iron had already existed for millions of years. The natural occurrence of Fe_3O_4 as lodestone is just as old as meteorites and probably more widespread than the latter. Thus it is not surprising that the magic stone and pure iron were used so early in our history.

Magnetism appears in various forms, as will be discussed in §4; but the kinds that have practical value, and therefore interest us here, are known as ferromagnetism and ferrimagnetism. Traditionally, only those elements that exhibit ferromagnetic properties are called magnetic. Besides Fe, Co and Ni, six elements in the lanthanide series (Gd, Tb, Dy, Ho, Er and Tm) become ferromagnetic at ≈20°C (Gd) and subzero temperatures. The list of magnetic materials, nevertheless, is infinitely long because an unlimited number of magnetic solid solutions and compounds can be formed of magnetic and certain nonmagnetic elements.

Soft magnetic material is more a technical term than a scientific one. The word "soft" means temporary in the sense that the ferromagnetism emerges only when a magnetic field is applied. In contrast, hard or permanent magnets display ferromagnetism in the absence of an external field*. All magnetic elements in the pure form are soft; whereas magnetic solid solutions and compounds can be either soft or hard, depending on the composition and metallurgical treatments. In this book, we are concerned mainly with soft magnetic materials. However, the physical principles discussed in this book are also applicable to hard magnets.

§2. UNITS

In accordance with the 1960 adoption by the Conférence Générale de Poids et Mesures (Chiswell and Grigg [1971]), the Système International (SI) d'Unités is used throughout the book. This system is based on the metre–kilogram–second–ampere (MKSA) system. It includes such practical units as ampere, volt and ohm; hence, it is particularly convenient when eddy currents and other phenomena relating magnetism to elec-

*Although the earth produces a magnetic field, its strength (from $32\ A \cdot m^{-1}$ or 0.4 Oe at the equator to $64\ A \cdot m^{-1}$ or 0.8 Oe at the poles) is too small to be considered as an effective external field.

tricity are discussed. The system offers another advantage in that the cumbersome factor 4π is eliminated in (1.6), which links magnetic induction $\boldsymbol{B}$ with magnetization $\boldsymbol{M}$. The various SI units for magnetic quantities will be discussed in the next section.

The adoption of the SI units may cause inconvenience to some readers who are more accustomed to the CGS units. Furthermore, the CGS system has thus far been the choice of most scientists in their publications. For these reasons, the equivalent values in CGS units are given in parentheses following many of the quantities expressed in SI units. A table of two-way conversion between major units of the two systems is given in appendix 1.

§3. FUNDAMENTAL QUANTITIES OF MAGNETISM

The quantities discussed in this section are considered fundamental because they deal with elemental magnetic properties covering all kinds of magnetism.

§3.1. *Magnetic poles*

Experiment shows that a natural magnet of lodestone in the shape of a long and thin bar sets up a magnetic field at each end. This field is radially outward at one end (N or positive) and radially inward at the other end (S or negative). The magnet may then be considered as possessing two equal and opposite poles at the ends. The strength of each pole m is defined in the Coulomb interaction that the force $\boldsymbol{F}$ exerted on a magnetic pole of strength m_2 by another pole of strength m_1 at a distance r is

$$\boldsymbol{F} = K\frac{m_1 m_2}{r^2}\boldsymbol{r}_0, \tag{1.1}$$

where $\boldsymbol{r}_0$ is a unit vector along r. In the SI, the magnitude of $\boldsymbol{F}$ is expressed in newton ($1\ \mathrm{N} = 1\ \mathrm{kg\cdot m\cdot s^{-2}}$) and m, a scalar quantity, in $\mathrm{A\cdot m}$ when the constant of proportionality K takes the rationalized value of $1/4\pi\mu_0$ where $\mu_0 \equiv 4\pi \times 10^{-7}$ henry/metre ($\mathrm{H\cdot m^{-1}}$) is the magnetic permeability of free space (vacuum).

§3.2. *Magnetic dipole and magnetic moment* $\boldsymbol{\mu}$

Magnetic poles have never been observed in isolated form, but occur in pairs. Such a pair is called a magnetic dipole. The magnetic moment of a dipole is defined as

$$\boldsymbol{\mu} = m\boldsymbol{l}, \tag{1.2}$$

where $\boldsymbol{l}$ is the vector pointing from the S pole to the N pole and μ is expressed in $A \cdot m^2$.

§3.3. *Magnetic field* ***H***

The strength of the magnetic field H induced by either pole of a bar magnet of pole strength m_1, at a distance r, is defined as the magnetizing force acting on a unit test pole ($m_2 = 1$) placed at that position. Accordingly,

$$\boldsymbol{H} = \frac{\boldsymbol{F}}{m_2} = K\frac{m_1}{r^2}\boldsymbol{r}_0. \tag{1.3}$$

The popular oersted (Oe) is the unit of $\boldsymbol{H}$ in the CGS system. In the latter system, since K has the rationalized value of unity, one oersted corresponds to a repulsive force of one dyne acting on a unit test pole by another unit pole of like kind at a distance of one centimeter. The SI unit of H is based on the fact that a magnetic field is often produced by a current of electricity. For a long solenoid having N turns per metre and carrying a current of I amperes, the magnetic field produced inside the solenoid is given by

$$H = NI. \tag{1.4}$$

Hence the SI unit of H is the ampere-turn per metre, or simply, the ampere per metre ($A \cdot m^{-1}$). One ampere per metre is equivalent to 0.0126 oersted and one oersted is roughly 80 $A \cdot m^{-1}$.

§3.4. *Magnetization* ***M*** *and magnetic induction* ***B***

Magnetization, or more completely, the intensity of magnetization $\boldsymbol{M}$, is the total magnetic moment of dipoles per unit volume in units of $A \cdot m^2$ per m^3 or $A \cdot m^{-1}$. Considering a bar magnet of pole strength m, length l and cross-sectional area A, the vector quantity $\boldsymbol{M}$ has its direction pointing from the S pole to the N pole and its magnitude given by

$$M = \mu / V = m / A. \tag{1.5}$$

So magnetization is also the pole strength per unit area in units of $A \cdot m^{-1}$, which is equivalent to 10^{-3} gauss (G) in the CGS system.

Magnetic induction or magnetic flux density $\boldsymbol{B}$ is the flux per unit area and expressed in units of $Wb \cdot m^{-2}$ or tesla (T). By flux is meant the number of lines of induction crossing a given area at a right angle. The weber is the magnetic flux that, linking a circuit of one turn, produces in the circuit an electromotive force of one volt as the flux is reduced to zero at a uniform rate in one second. The idea of using endless flow lines between the opposite poles of a magnet to represent magnetic induction was first conceived by M. Faraday. Indeed when a magnet is brought underneath a piece of cardboard on which iron filings are sprinkled, the magnet will force the filings to form endless flow lines emanating from the N pole and converging at the S pole, thus depicting the pattern of the magnetic induction, as shown in fig. 1.1. Hence $\boldsymbol{B}$ also is called the magnetic flux density.

In free space a magnetic field produces a magnetic induction given by $\boldsymbol{B} = \mu_0 \boldsymbol{H}$. If the space filled with any magnetic substance in which the induced magnetization is $\mu_0 \boldsymbol{M}$, the total induction now becomes

$$\boldsymbol{B} = \mu_0(\boldsymbol{H} + \boldsymbol{M}). \tag{1.6}$$

Thus both the magnetizing field and the magnetization contribute to the induction. For ferromagnetic materials, however, the contribution of $\boldsymbol{M}$ usually dominates $\boldsymbol{B}$.

In the CGS system, since $\mu_0 = 1$, (1.6) becomes

$$\boldsymbol{B} = \boldsymbol{H} + 4\pi \boldsymbol{M}. \tag{1.7}$$

The factor 4π arises from the fact that a unit pole produces a unit field everywhere on the surface of a sphere of unit radius (1 cm) enclosing the pole, and the surface area of this sphere is 4π cm^2. The unit of $\boldsymbol{B}$ in

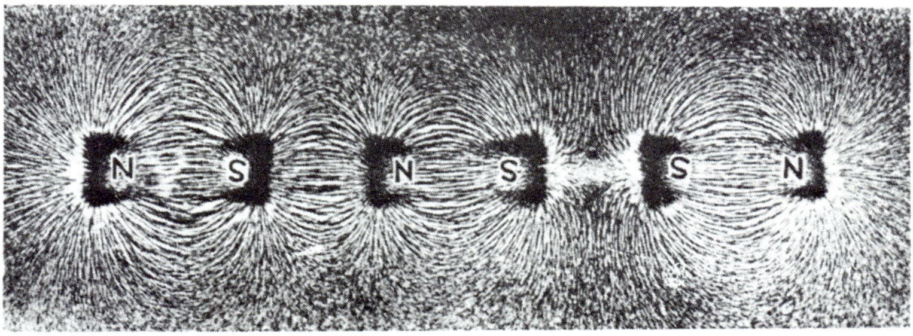

Fig. 1.1. Lines of magnetic induction, formed by iron filings, depicting the magnetizing force of a field near a group of magnets (Bozorth [1951]).

the CGS system is the popular gauss. One gauss is equal to one line of induction per cm^2. In the SI, as indicated by (1.6), $\boldsymbol{H}$ and $\boldsymbol{M}$ have the same unit in $A \cdot m^{-1}$. In the CGS system, since $\boldsymbol{B} = \boldsymbol{H}$ in free space, gauss frequently is used to express the magnitude of $\boldsymbol{H}$ in place of oersted. For conversion it is well to remember that one tesla is equivalent to 10^4 G.

§3.5. *Susceptibility χ and permeability μ*

To compare the magnetic response of various materials, we define the volume susceptibility, or simply susceptibility, χ, and the absolute permeability μ by

$$\chi = M/H \quad \text{and} \quad \mu = B/H, \tag{1.8}$$

where H is the applied magnetic field, and M and B are the induced magnetization and induction in the direction of H, respectively. Thus in SI, since $\mu_0 = 4\pi \times 10^{-7}\ H \cdot m^{-1}$, not unity, the absolute value of permeability divided by μ_0 gives the relative permeability, $\mu' = \mu/\mu_0$. While χ and μ' are dimensionless, μ is expressed in henry per metre ($H \cdot m^{-1}$), or appropriate SI units per m^3. (In the CGS system, since $\mu_0 = 1$, the absolute and relative values of μ are identical, and they are numerics as χ.)

In the literature, the susceptibility has two other forms: (A) mass susceptibility, χ_m, which is related to χ by

$$\chi_m = \chi/\rho, \tag{1.9}$$

where ρ is the density expressed in $kg \cdot m^{-3}$ in SI and χ_m is based per kg; and (B) atomic susceptibility, χ_A, which is related to χ_m by

$$\chi_A = \chi_m \cdot w, \tag{1.10}$$

where w is the atomic or molecular weight and χ_A is based per gram-mole. Whenever the material is monatomic, the atomic susceptibility is also the molar susceptibility, χ_M, which refers to one molecular gram-weight (one mole) of the material dissolved in sufficient solvent to make one cubic decimeter ($= 10^{-3}\ m^3$) of solution. Experimenters favor the volume susceptibility because it can be measured directly, whereas theoreticians prefer the mass and the atomic or molar susceptibilities in their analysis. For copper at 18°C, the volume susceptibility is $\chi = -9.24 \times 10^{-6}$ per m^3. With $\rho = 8.96 \times 10^3\ kg \cdot m^{-3}$ and $w = 63.54 \times 10^{-3}\ kg \cdot \text{g-mole}^{-1}$ for Cu, we have $\chi_m = -9.24 \times 10^{-6}/8.96 \times 10^3 = -1.03 \times 10^{-9}$ per kg and $\chi_A = \chi_M = -1.03 \times 10^{-9} \times 63.54 \times 10^{-3} = -6.55 \times 10^{-11}$ per g-mole.

From (1.6) and the definition of μ', we obtain

$$\mu' = 1 + \chi. \tag{1.11}$$

If other forms of χ are used in (1.11), there will be corresponding changes in μ'.

§4. VARIOUS KINDS OF MAGNETISM

We have stated at the outset that all substances display certain magnetic properties at all temperatures, regardless of their composition and state. Now we shall review the various kinds of magnetism that have been observed in solids. Broadly speaking, there are five basic kinds of magnetism; namely, (1) diamagnetism, (2) paramagnetism, (3) ferromagnetism, (4) ferrimagnetism and (5) antiferromagnetism. In this listing, metamagnetism is not included because it is merely a phenomenon in which antiferromagnetism is replaced by ferromagnetism upon the application of a strong magnetic field or a decrease in temperature. Nor are superparamagnetism, parasitic ferromagnetism and mictamagnetism included, which may be viewed as variations of the ones listed.

In simple terms, a solid is said to be diamagnetic if it is repelled by a permanent magnet and to be paramagnetic, ferromagnetic or ferrimagnetic if it is attracted. The classification of magnetism can be made more precisely, of course. A convenient way to define the first four kinds of magnetism is to use the susceptibility or relative permeability as follows:

Criterion for classification	Diamagnetism	Paramagnetism	Ferro- or ferri-magnetism
Susceptibility, χ	<0	$\gtrsim 0$	$\gg 0$
Relative permeability, μ'	<1	$\gtrsim 1$	$\gg 1$

Table 1.1 lists some outstanding examples of each kind and typical values of their susceptibility or relative permeability at room temperature. In the Periodic Table, various groups of elements are shown to display certain kinds of magnetism. For instance, inert gases and noble metals (Cu, Ag and Au) are diamagnetic, alkali metals (Li, Na, K and Rb) and all transition metals except Fe, Co and Ni are paramagnetic, iron group transition metals and heavy lanthanide metals are ferromagnetic, etc. Ferrimagnetism exists only in compounds and alloys

TABLE 1.1
Examples of diamagnetism, paramagnetism, ferromagnetism and ferrimagnetism and their susceptibility or relative permeability values* at 18°C

Kind of magnetism and examples	Susceptibility, χ
(I) Diamagnetism	
Copper	-1.0×10^{-5}
Zinc	-1.4×10^{-5}
Gold	-3.6×10^{-5}
Mercury	-3.2×10^{-5}
Water	-0.9×10^{-5}
Hydrogen (n.t.p.)**	-0.2×10^{-8}
(II) Paramagnetism	
Lithium	$+4.4\times 10^{-5}$
Sodium	$+0.62\times 10^{-5}$
Aluminum	$+2.2\times 10^{-5}$
Vanadium	$+38\times 10^{-5}$
Palladium	$+79\times 10^{-5}$
Neodymium	$+34\times 10^{-5}$
Air† (n.t.p.)	$+36\times 10^{-8}$
	Approximate highest value of relative permeability, $\mu'(=1+\chi)$
(III) Ferromagnetism	
Iron crystals	1.4×10^{6}
Cobalt crystals	$\approx 10^{3}$
Nickel crystals	$\approx 10^{6}$
3.5% Si–Fe (transformer sheets)	7×10^{4}
(single crystals)	3.8×10^{6}
(IV) Ferrimagnetism	
Fe_3O_4	100
Ferrites	5000

*References: Brailsford [1966a] and Bates [1963a].
**Normal temperature (18°C) and pressure (1 atmosphere).
†Oxygen is paramagnetic and nitrogen is diamagnetic.

because it involves at least two magnetic moments, μ_A and μ_B (the origin of which is discussed in ch. 2, §1) carried by ions of different species. While all moments of ions of the same species may be coupled

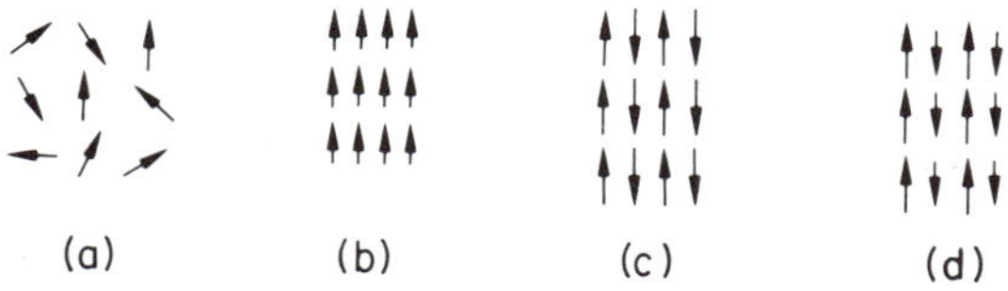

Fig. 1.2. Spin arrangement in a crystal or domain illustrating (a) paramagnetism, (b) ferromagnetism, (c) antiferromagnetism and (d) ferrimagnetism.

together ferromagnetically, the coupling between μ_A and μ_B is antiferromagnetic (see fig. 1.2d). The alignment of magnetic moments in a simple antiferromagnet is shown in fig. 1.2c in comparison with those of paramagnetic and ferromagnetic materials. Several metals show antiferromagnetism, notably Cr, Mn and some of the light lanthanides.

Next we shall examine the magnetic susceptibility for diamagnetism and paramagnetism. A prominent group of diamagnetic substances is the inert gases. These gas atoms have their outermost subshell completely filled. Therefore, their ground states all have the common term $^1S_0$*, of which the magnetic moment is zero because the spin, orbital and total angular momentum quantum numbers are all zero. (See ch. 2, §1 for detailed explanation.)

The observed diamagnetic susceptibility may be explained, however, by drawing an analogy to the Lenz law in electrodynamic theory. This law states that, when the flux through an electrical circuit is changed, an induced current is produced in such a direction as to oppose the flux change. The application of a magnetic field H to an inert gas would then cause the orbital motion of each electron in the atom to precess about H, thus inducing a magnetic moment μ_a for the collection of Z electrons in each atom. The volume susceptibility of a rare gas and other diamagnetic substances can be shown (Kittel [1976a]) as

$$\chi = \frac{N\mu_a}{H} = -\frac{NZ\mu_0 e^2}{6m_0}\langle r^2\rangle, \tag{1.12}$$

where N is the number of atoms per unit volume, m_0 is the rest mass of electron, e is the electronic charge and $\langle r^2\rangle$ is the mean square radius of

*For metals, the $^1S_0$ ground state term is neither a necessary condition (e.g. Cu with $^2S_{1/2}$, In with $^2P_{1/2}$ and Pb with $^3P_0$ are diamagnetic) nor a sufficient condition (e.g. Mg, Ca, Ba and Yb are paramagnetic) for the occurrence of diamagnetism.

the electron charge cloud about each atom. This is the formula originally derived from the classical treatment, but also arrived at in the quantum-mechanics analysis. It worked well for the inert gases. For example, an atomic susceptibility of -2.39×10^{-11} in SI units per g-mole was obtained for helium as compared with measured value of -2.40×10^{-11}.

From (1.12) we conclude that diamagnetic susceptibility is due entirely to the orbital motion of electrons associated with Larmor precession. For metals, it is important to remember that their atomic cores, having completely filled subshells, are normally associated with a small, negative susceptibility. Hence the negative diamagnetic effect is always present in metallic materials even though it may be overwhelmed by the positive paramagnetic or ferromagnetic effect. This explains why many metals other than the noble ones, notably Be, Zn, Cd, In, Hg, Pb and Bi also are diamagnetic.

There are three forms of paramagnetism. Traditionally, however, paramagnetism refers to a system in which each atom carried a magnetic moment, but the atomic moments point in all directions, thereby generating no net magnetization, as shown in fig. 1.2a. The complete disorder of atomic moments comes about because there is no magnetic interaction between neighboring atoms on one hand and because the thermal agitation causes the moments to be aligned randomly on the other. This form of paramagnetism was first treated classically by Langevin (see Bates [1963a]) who considered a system of atoms in which each atom has a magnetic moment μ in Bohr magnetons, but the moments are not coupled. When a magnetic field H is applied, the moments will tend to align as close to the field as the thermal effect at temperature T permits. Applying the Maxwell–Boltzman law, Langevin arrived at the expression for the average magnetic moment $\bar{\mu}$ of the atom in the direction of H

$$\bar{\mu}/\mu = \coth a - 1/a = L(a), \tag{1.13}$$

where $a = \mu_0\mu H/kT$ and $L(a)$ is known as the Langevin function.

Usually a is very small. A typical value for μ is $1\,\mu_B = 9.2732\times10^{-24}\ \mathrm{A\cdot m^2}$. For $H = 8\times10^5\ \mathrm{A\cdot m^{-1}}$ ($\sim10^4$ Oe) and at room temperature ($T\sim300$°K), $\mu_0\mu H/kT \simeq 2\times10^{-3}$, thus $\mu_0\mu H \ll kT$. Under these conditions,

$$L(a) = \frac{a}{3} - \frac{a^3}{45} + \frac{2a^5}{945} - \cdots \approx \frac{a}{3}. \tag{1.14}$$

Hence, in weak to moderate fields and at high enough temperatures,

$$\bar{\mu}/\mu \simeq \mu_0\mu H/3kT. \tag{1.15}$$

The magnetization per atomic mass (i.e. the mass corresponding to the atomic weight) is given by

$$M_A = N_0\bar{\mu} = \mu_0 N_0 \mu^2 H/3kT, \tag{1.16}$$

where N_0 is the Avogadro number, and the atomic susceptibility is given by

$$\chi_A = M_A/H = \mu_0 N_0 \mu^2/3kT = C/T, \tag{1.17}$$

where $C = \mu_0 N_0 \mu^2/3k$ is known as the Curie constant. The relation $\chi_A = C/T$ is known as the Curie law. This is the essential result of the classical theory of paramagnetism of the Langevin or spin type. In this theory, the magnetic axis was assumed to take any angular position with respect to the field direction.

Another type of paramagnetism was proposed by Pauli (see Kittel [1976b]) to account for the magnetic behavior of the conduction electrons. Take copper as a typical example. This metal may be envisaged as consisting of ionic cores with closed orbits up to the 3d subshell and the conduction electrons, one per atom, in the 4s band. Hence, this is a system in which the diamagnetic ionic cores are immersed in an electron gas. Set the ions aside and consider only the conduction electrons. In the absence of a magnetic field, the band structure of the 4s electrons is shown by a density-of-states diagram in fig. 1.3a. When a field is applied and remembering that the inherent magnetic moment per

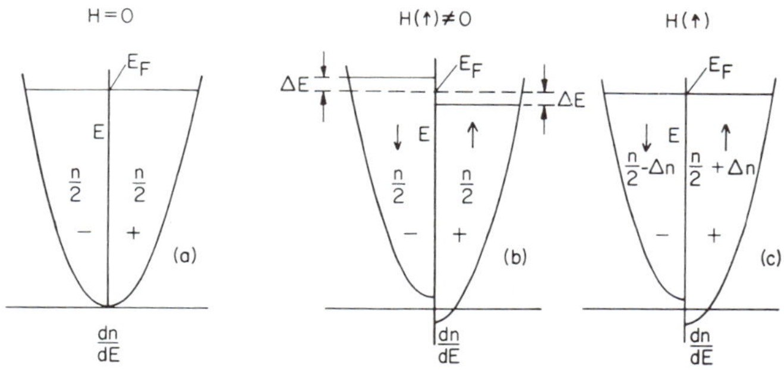

Fig. 1.3. Pauli paramagnetism of the conduction electrons: (a) $H = 0$, (b) and (c) $H \neq 0$ applied in the upward vertical direction (Brailsford [1966]).

electron is one Bohr magneton, μ_B (see discussion in ch. 2, §1), electrons with positive and negative spins will gain potential energies of the magnitudes $\Delta E = \mp \mu_0 H \mu_B$. This means that electrons with positive spins will have lower potential energy than those with negative spins in the field by the amount $2\mu_0 H \mu_B$. This energy difference is equivalent to shifting the two half s bands as shown in fig. 1.3b. Furthermore, there will be a transfer of Δn electrons from the negative spin half band to the other half band so as to attain an equilibrium distribution of the conduction electrons as shown in fig. 1.3c. On the basis of fig. 1.3c, the volume susceptibility of the Pauli paramagnetism is given by

$$\chi_P = \Delta M / H = \mu_0 m_0 \mu_B^2 (3n)^{1/3} / \hbar^2 \pi^{4/3}, \tag{1.18}$$

where n is the total number of conduction electrons per unit volume and $\hbar = h/2\pi$ with h being Planck's constant. Note that χ_P given by this expression is always positive and independent of temperature. Landau later showed that the presence of a magnetic field will force the conduction electrons to move in helical paths, thus resulting in an additional diamagnetism. If this effect is included, the induced Pauli susceptibility will be reduced by one-third of the value given in (1.18).

For the entire system of Cu, the total volume susceptibility is given by combining (1.12) and (1.18),

$$\begin{aligned} \chi_{\text{total}} &= \chi_{\text{core}} + \chi_{\text{cond. elect.}} \\ &= \frac{-N(Z-1)\mu_0 e^2}{6m_0}\langle r^2 \rangle + \frac{2\mu_0 m_0 \mu_B^2 (3N)^{1/3}}{3\hbar^2 \pi^{4/3}}, \end{aligned} \tag{1.19}$$

remembering that $N = n$ in the case of noble metals. These two terms are of the same order of magnitude for nontransition metals. Hence most of these metals are either diamagnetic or paramagnetic, depending on which of the two terms is larger. For Cu, the diamagnetism outweighs the Pauli paramagnetism; thus negative susceptibility prevails. The reverse is true for the alkali metals and for Mg, Al and Ca. Magnetic susceptibilities of the elements up to $Z = 83$ are shown in fig. 1.4.

Van Vleck [1932] uncovered a third type of paramagnetism in a rigorous quantum-mechanics analysis of the diamagnetic susceptibility of free molecules. Before his analysis, the direction of the applied field had been assumed to be an axis of symmetry of the molecular system. Van Vleck pointed out that such an assumption would rarely be valid for most molecules; thus there exists a nondiagonal matrix element

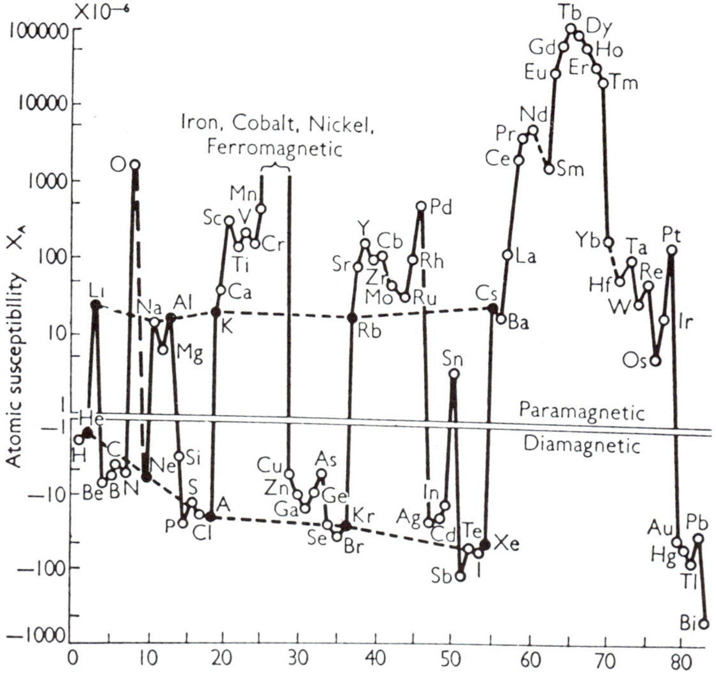

Fig. 1.4. The magnetic susceptibilities of the elements (Lee [1963]).

$\langle s|\mu_z|0\rangle$ of the z component of the orbital magnetic moment linking the ground state 0 with the excited state s. This matrix element gives rise to an additional susceptibility term in (1.12). For a system of polyatomic molecules with zero spin quantum number the total volume susceptibility should be

$$\chi_{\text{total}} = \frac{-NZ\mu_0 e^2}{6m_0}\langle r^2\rangle + 2N\sum_s \frac{|\langle s|\mu_z|0\rangle|^2}{E_s - E_0}, \tag{1.20}$$

where N is the number of molecules per unit volume and $E_s - E_0$ is the energy difference between the ground and the excited states. The second term represents the van Vleck paramagnetism. Again, as with nontransition metals, the molecular systems will exhibit either diamagnetism or paramagnetism, depending on which of the two terms is larger.

The van Vleck paramagnetism also is important for certain ions, such

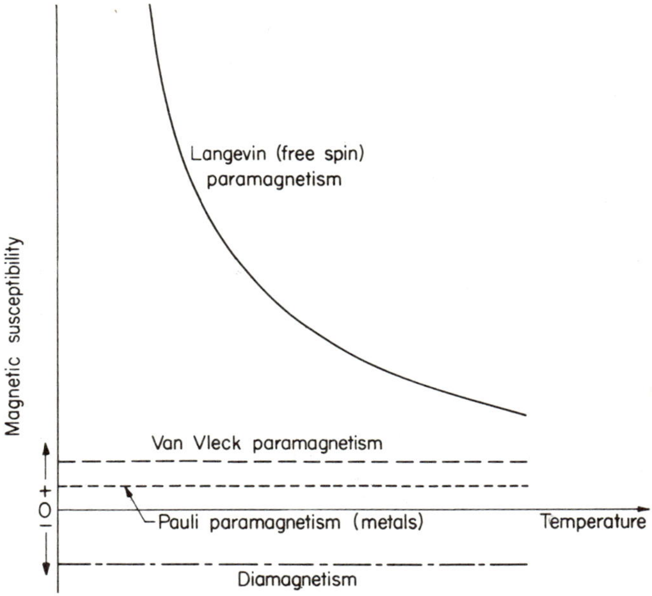

Fig. 1.5. Relative values and temperature dependence (or independence) of magnetic susceptibility in diamagnetism and three kinds of paramagnetism (Kittel [1976]).

as Eu^{++}, for which the *LS* coupling is very weak. Like the Pauli paramagnetism, the van Vleck paramagnetism also is independent of temperature. Fig. 1.5 shows diagrammatically the relative magnitudes and temperature dependence of susceptibilities of the three types of paramagnetism together with that of diamagnetism. Only the Langevin paramagnetism displays a temperature dependence, and consequently, the Pauli and van Vleck paramagnetisms become important at rising temperatures when the Langevin paramagnetism weakens.

Suggestions for further reading

Bates, L. F., 1963, *Modern Magnetism*, 4th ed. (Cambridge University Press, London). Chapter 1 on §4.

Chiswell, B. and E. C. M. Grigg, 1971, *SI Units* (Wiley, Sydney).

Lee, E. W., 1963, *Magnetism* (Penguin, London).

Morrish, A. H., 1965, *The Physical Principles of Magnetism* (Wiley, New York). Chapters 1, 2 and 5 on §3 and §4.

CHAPTER 2

FERROMAGNETISM AND FERRIMAGNETISM

§1. MAGNETIC MOMENTS OF ELECTRONS AND ATOMS

A solid will display ferromagnetism if it meets two conditions: (1) part or all of its atoms each carry a magnetic moment and (2) the moments are aligned overwhelmingly in one direction. An understanding of ferromagnetism thus necessitates the answering of two basic questions:

(1) What is the origin of the magnetic moment of the atoms?
(2) How are the atomic moments coupled?

The first question will be discussed in this section and the second question in the next section. To answer the first question it is important to realize that atoms derive their magnetic moments from electrons, not from the nuclei, because the magnetic moments of the latter are too small (by roughly three orders of magnitude) in comparison with the moments of the electrons. There are two sources from which the electron can obtain its magnetic moment. One source involves the orbital motion of the electrons around the nuclei. From the electromagnetic theory, for an electron with rest mass m_0 and charge $-e$ moving in a circular orbit of radius r at a velocity $\boldsymbol{v}$ (see fig. 2.1), this orbital motion produces a magnetic moment $\boldsymbol{\mu}_l$, which is given in $\mathrm{A \cdot m^2}$ units by

$$\begin{aligned}\boldsymbol{\mu}_l &= \boldsymbol{n}(\text{current}) \cdot (\text{orbit area}) \\ &= \boldsymbol{n}\left(-e\frac{|\boldsymbol{v}|}{2\pi r}\right) \cdot (\pi r^2) = -\frac{1}{2}evr\boldsymbol{n},\end{aligned} \tag{2.1}$$

where $\boldsymbol{n}$ is a unit vector of the normal to the orbital plane. According to classical mechanics, the orbital angular momentum of the electron is $\boldsymbol{p}_l = m_0\boldsymbol{v} \times \boldsymbol{r}$, indicating that $\boldsymbol{u}_l$ and $\boldsymbol{p}_l$ are collinear but opposite to each other as shown in fig. 2.1, and their magnitudes have the ratio

$$\mu_l/p_l = e/2m_0. \tag{2.2}$$

Quantum mechanics also yields the same ratio, but the magnitude of the

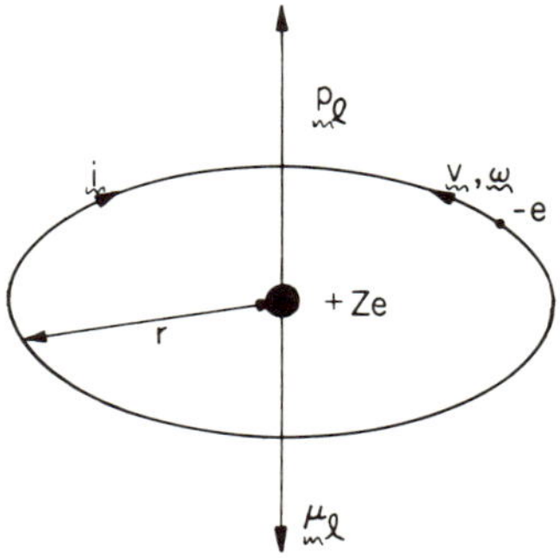

Fig. 2.1. The angular momentum $\boldsymbol{p}_l$ and the magnetic moment $\boldsymbol{\mu}_l$ produced by the orbital motion of an electron are collinear, but opposite to each other.

orbital angular momentum is given in units of $\hbar$ by

$$p_l = [l(l+1)]^{1/2}\hbar = l^*\hbar, \tag{2.3}$$

where l is called the orbital angular momentum quantum number. Substituting (2.3) into (2.2) gives

$$\mu_l = l^*(e\hbar/2m_0) = l^*\mu_B. \tag{2.4}$$

Here we introduce an important unit of the magnetic moment, namely, the Bohr magneton μ_B, which has the fixed value $\mu_B \equiv e\hbar/2m_0 = 9.2732\times10^{-24}\ \text{A}\cdot\text{m}^2$ ($\mu_B \equiv e\hbar/2m_0c = 9.2732\times10^{-21}\ \text{erg}\cdot\text{G}^{-1}$ in the CGS system). Since l has the values $0, 1, 2, 3, \ldots$, corresponding to the s, p, d, f, ... states of electrons, the orbital magnetic moment of an electron is thus $0, \sqrt{2}, \sqrt{6}, \sqrt{12}, \ldots$ Bohr magnetons, respectively.

The second source of the magnetic moment of electrons is based on the spin of the electrons. Unlike the orbital motion, electron spin is strictly a quantum-mechanical effect. In analogy with (2.3), the spin angular momentum is given by

$$p_s = [s(s+1)]^{1/2}\hbar = s^*\hbar, \tag{2.5}$$

where s is called the spin momentum quantum number. Since s always has the value 1/2, s^* has the invariant value $\sqrt{3}/2$. This spin angular momentum also gives rise to a spin magnetic moment $\boldsymbol{\mu}_s$. Experimental evidence, however, requires the ratio μ_s/p_s to be double the value for the orbital motion given in (2.2). Hence the spin magnetic moment i given by

$$\mu_s = 2s^* \mu_B = \sqrt{3} \mu_B. \tag{2.6}$$

Before answering the second question concerning the coupling of atomic moments, we must know how the orbital and spin magnetic moments are combined into a resultant magnetic moment of an electron. The combination can be conveniently effected by a Larmor vector model shown in fig. 2.2. The scheme is known as the spin–orbital interaction, or simply, the *ls* coupling. An electrostatic interaction takes place between the orbital motion and the spin of the electron because each motion produces a magnetic field that affects the other motion.

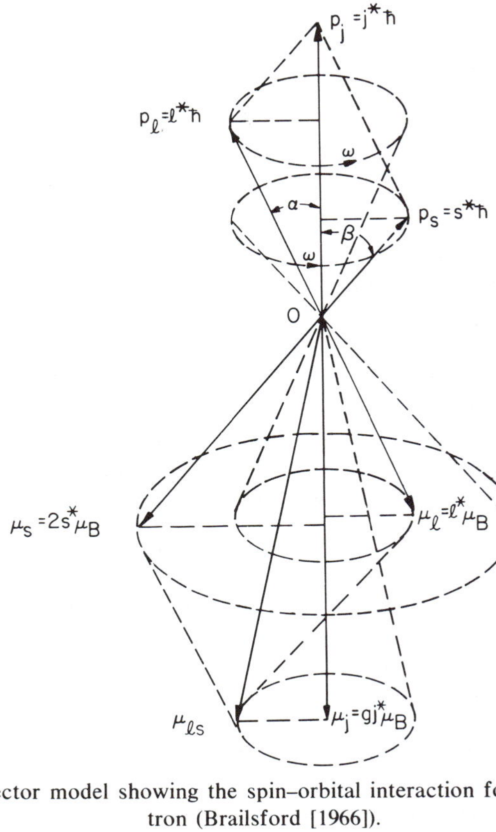

Fig. 2.2. Vector model showing the spin–orbital interaction for a bound electron (Brailsford [1966]).

The result is that both the spin and orbital axes will precess with the same angular velocity, $\omega = H(\mu_s/p_s)$ and in the same direction about a common axis along which lies the resultant angular momentum $\boldsymbol{p}_j$, as shown in the upper part of fig. 2.2. The magnitude of $\boldsymbol{p}_j$ is given by

$$p_j = [j(j+1)]^{\frac{1}{2}}\hbar = j^*\hbar, \tag{2.7}$$

where $j = l \pm s$ is called the total angular momentum quantum number. But if $l = 0$, $j = +\frac{1}{2}$ only. The lower part of fig. 2.2 indicates that the vector sum of $\boldsymbol{\mu}_l$ and $\boldsymbol{\mu}_s$ is $\boldsymbol{\mu}_{ls}$, which is not the resultant magnetic moment $\boldsymbol{\mu}_j$ of the electron because of the factor of 2 introduced in (2.6) in comparison with (2.5). To obtain $\boldsymbol{\mu}_j$, one needs to project $\boldsymbol{\mu}_{ls}$ onto the common axis of the precession. The magnitude of $\boldsymbol{\mu}_j$ is shown to be

$$\mu_j = j^* g\mu_{\mathrm{B}}, \tag{2.8}$$

where g is called the Landé factor of the value

$$g = 1 + \frac{(j^*)^2 + (s^*)^2 - (l^*)^2}{2(j^*)^2}. \tag{2.9}$$

This expression of g is derived readily from fig. 2.2 and the derivation is left as an exercise to the reader. As is the case of $\boldsymbol{\mu}_l$ and $\boldsymbol{\mu}_s$, $\boldsymbol{\mu}_j$ also is collinear with, but opposite to, $\boldsymbol{p}_j$.

We next consider the effect of an external magnetic field H on the ls coupling. The applied field is considered to be weak when it is much smaller than the magnetic field H_i induced by the orbital motion, where $H_i = l^*\hbar Ze/4\pi m_0 r^3$ with Z being the nuclear charge number of the atom to which the electron belongs. Under a weak H, the spin–orbital interaction will prevail. The applied field, however, will cause $\boldsymbol{p}_j$ and $\boldsymbol{\mu}_j$ to precess about H as shown in fig. 2.3a. The resolved components of $\boldsymbol{p}_j$ and $\boldsymbol{\mu}_j$ in the direction of H will take discrete values given by

$$p_H = m\hbar \quad \text{and} \quad \mu_H = mg\mu_{\mathrm{B}}, \tag{2.10}$$

where m is a magnetic quantum number and has the allowable values $\pm\frac{1}{2}, \pm\frac{3}{2}, \ldots, \pm j$. As an example, let us consider an electron in the s state. Since $l = 0$ and $s = \frac{1}{2}$, we then have $j = \frac{1}{2}$ and $m = \pm\frac{1}{2}$. Substituting these values in (2.9) and (2.10) gives $g = 2$ and $\mu_H = \pm 1\ \mu_{\mathrm{B}}$. Thus an electron with zero average orbital motion has a resolved component of magnetic moment of one Bohr magneton in the direction of a weak external field.

The well-known Zeeman effect refers to the splitting of energy levels

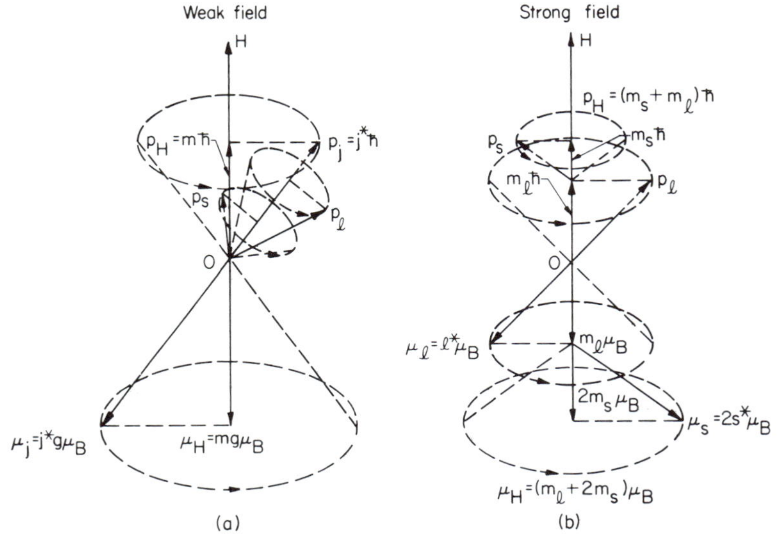

Fig. 2.3. Precession of angular momenta and magnetic moments in a weak (a) and strong (b) applied field (Brailsford [1966]).

caused by a weak applied field. For instance, the energy level of an s electron is split into two sublevels corresponding to $\mu_H = \pm 1\ \mu_B$. For the electrons in the p state, we have $l = 1$, $j = 1 \pm \frac{1}{2}$ and two values $\frac{2}{3}$ and $\frac{4}{3}$, for g. Consequently m has six values, which in turn, result in six magnetic moments ($\pm\frac{1}{3}$, $\pm\frac{2}{3}$ and $\pm 2\ \mu_B$) in the direction of the external field. Thus the p energy level is split into six sublevels because of the Zeeman effect. Similar analysis will show that the d, f, . . . energy levels are split into 10, 18, . . . sublevels, respectively.

When the external field is very strong in comparison with H_i, the ls coupling can no longer hold, but will break down. Then the orbital and spin axes will precess independently about the applied field, as shown in fig. 2.3b. This phenomenon is known as the Paschen–Back effect.

Now we need to introduce two new magnetic quantum numbers (MQN), namely, m_l, the orbital MQN and m_s, the spin MQN, to indicate the quantization of the resolved components of the angular momenta and magnetic moments associated with the orbital and spin motions of an electron in the direction of H. The situation is rather complex and is best summarized in table 2.1. The reason for $m = m_l + m_s$ is that, when

TABLE 2.1
Three magnetic quantum numbers for each bound electron in an atom

	Magnetic quantum number	Resolved components along H: Angular momentum	Magnetic moment
Orbital	$m_l = 0, \pm 1, \pm 2, \dots, \pm l$	$p_{l,H} = m_l \hbar$	$\mu_{l,H} = m_l \mu_B$
Spin	$m_s = \pm \frac{1}{2}$	$p_{s,H} = m_s \hbar$	$\mu_{s,H} = 2\, m_s \mu_B$
Total	$m = m_l + m_s$	$p_H = (m_l + m_s)\hbar$	$\mu_H = (m_l + 2\, m_s)\mu_B$

the strength of the external field is reduced from the Paschen–Back effect to the Zeeman effect, there should be a continuity in the fine structure of the energy state of an electron.

Now we are in a position to describe the quantum state of various bound electrons in a free atom. According to the Pauli exclusion principle, each electron can be completely specified by a set of four quantum numbers. Though there are several choices, the set most suitable for our purposes consists of n, l, m_l and m_s, where n is the total or principal quantum number and has the values $1, 2, 3, 4, \dots$ corresponding to the K, L, M, N, ... shells in X-ray spectroscopy. For a free hydrogen atom ($Z = 1$) in the ground state, the bound electron has these quantum numbers: $n = 1$, $l = 0$, $m_l = 0$ and $m_s = +\frac{1}{2}$ or $-\frac{1}{2}$. For a free helium atom ($Z = 2$), there are two bound electrons, which have the characteristic values in the ground state $1, 0, 0, +\frac{1}{2}$ and $1, 0, 0, -\frac{1}{2}$ for the four quantum numbers n, l, m_l and m_s, respectively.

The same procedure may be used to determine the electronic configuration for other free atoms of larger Z values. The problem is greatly simplified, however, if we remember the following rules governing the atomic structure:

(A) The maximum number of electrons in the K, L, M, N, ... shells is $2, 8, 18, 32, \dots$, respectively, or simply, there is a maximum of $2n^2$ electrons in each shell.

(B) For each l subshell, the allowable maximum number of electrons is $2(2l + 1)$. Thus there are $2, 6, 10, 14, \dots$ electrons in the s, p, d, f, ... subshells.

(C) As far as the resultant angular momentum, $\boldsymbol{p}_J$, and resultant magnetic moment, $\boldsymbol{\mu}_J$, of the atom are concerned, we only need to consider the electrons in the incompletely filled subshell. This is so because all electrons in the inner subshells result in zero p_J and μ_J. For example, there are 26 bound electrons in a free atom of iron, but only 8 are in the subshells outside the completed 3p subshell. Further, two of the eight electrons are known to fill up the 4s subshell. Hence, only the remaining six electrons in the 3d subshell will concern us, when discussing free atoms.

It is generally true that the total energy of an electron in a particular state depends upon the value of n and l. Thus, the relative positions of the various energy levels start with 1s at the bottom of the ladder and progressively move up to the states 2s, 2p, 3s and 3p. A reversal takes place, however, in the energy-level sequence between the 3d and 4s states. This reversal arises because the wave pattern of an s electron has no nodes through the nucleus; thus the s electrons spend relatively more time near the nucleus. The wave function of a d electron, on the other hand, has either two nodal planes or a conical nodal surface through the nucleus (fig. 2.4), and thus the d electrons spend far less time near the nucleus. The less time an electron in a state spends near the nucleus, the higher is its energy. This reversal between the nd and $(n+1)$s states prevails in all three series of transition elements. In the lanthanides, an nf electron is raised to a higher energy level

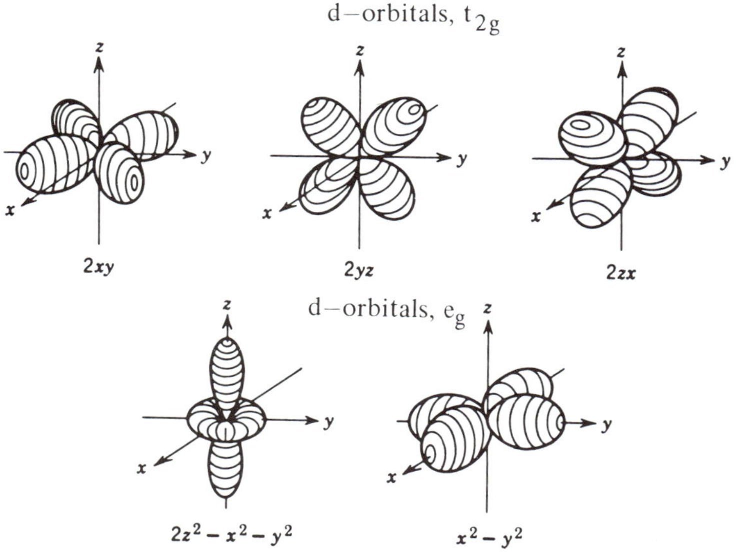

Fig. 2.4. Spatial distribution of the t_{2g} and e_g orbits (Chikazumi [1964]).

close to those of $(n+2)$s and $(n+1)$d electrons. This explains why the electronic configuration of a gadolinium free atom $(Z=64)$ is $K(1s^2)\ L(2s^22p^6)\ M(3s^23p^63d^{10})\ 4s^24p^64d^{10}\ 5s^25p^64f^75d^16s^2$. The electronic configurations of all magnetic elements in the ground or normal state of their free atoms are listed in table 2.2.

A scheme has been devised under which the angular momenta and the magnetic moments of electrons in the outermost subshell(s) of an atom can be combined. The scheme is known as the Russell–Saunders coupling or the *LS* coupling. For a total of *i* outer electrons, the coupling may be represented diagramatically as follows:

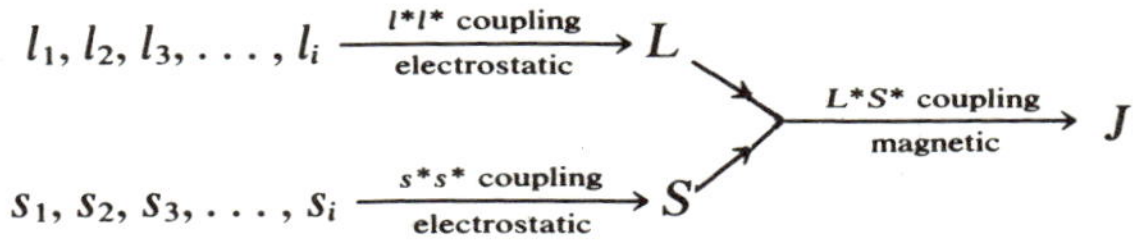

where *S*, *L* and *J* are for the atom the counterparts to *s*, *l* and *j* for the individual electrons. Note that the *ll*, *ss* and *LS* couplings actually involve l^*, s^* and L^* and S^*. Certain quantum rules must be observed in forming *J*. Consequently, *L* may assume only the values $0, 1, 2, 3, \ldots$; *S* may assume only the values $0, \frac{1}{2}, 1, \frac{3}{2}, 2, \ldots$ and *J* may assume only the values $0, 1, 2, \ldots$ when *S* is an integer and $\frac{1}{2}, \frac{3}{2}, \frac{5}{2}, \ldots$ when *S* is a half-integer. The relations of S^*, L^* and J^* to the respective *S*, *L* and *J* are the same as for the electrons. In other words,

$S^* = [S(S+1)]^{\frac{1}{2}}$, from which the spin magnetic moment of the atom is given by $\mu_S = 2S^*\mu_B$;

$L^* = [L(L+1)]^{\frac{1}{2}}$, from which the orbital magnetic moment of the atom is given by $\mu_L = L^*\mu_B$ and

$J^* = [J(J+1)]^{\frac{1}{2}}$, from which the total magnetic moment of the atom is given by $\mu_J = J^*g\mu_B$

where the Landé factor has the same form as expressed in (2.9) with S^*, L^* and J^* replacing s^*, l^* and j^* respectively.

The electronic configuration of an atom thus is characterized by *S*, *L* and *J*. These three quantum numbers also determine the energy level of an atom, which is customarily expressed by a term of the form

$$^{2S+1}L_J.$$

For each particular term, *L* actually is replaced by a capital letter S, P, D, F, ... for $L = 0, 1, 2, 3, \ldots$, respectively. For example, $^1S_0$ cor-

TABLE 2.2
Electronic configuration of magnetic elements in the ground state of free atoms

Shell		K	L		M			N				O			P	Normal
n		1	2		3			4				5			6	state
Subshell		1s	2s	2p	3s	3p	3d	4s	4p	4d	4f	5s	5p	5d	6s	term
l		0	0	1	0	1	2	0	1	2	3	0	1	2	0	
Z	Element															
26	Fe	2	2	6	2	6	6	2								$^5D_4$
27	Co	2	2	6	2	6	7	2								$^4F_{9/2}$
28	Ni	2	2	6	2	6	8	2								$^3F_4$
64	Gd	2	2	6	2	6	10	2	6	10	7	2	6	1	2	$^9D_2$
65	Tb	2	2	6	2	6	10	2	6	10	8	2	6	1	2	
66	Dy	2	2	6	2	6	10	2	6	10	9	2	6	1	2	
67	Ho	2	2	6	2	6	10	2	6	10	10	2	6	1	2	
68	Er	2	2	6	2	6	10	2	6	10	11	2	6	1	2	
69	Tm	2	2	6	2	6	10	2	6	10	12+1	2	6	(1)	2	$^2F_{7/2}$

responds to $S = 0$, $L = 0$, and $J = 0$ for an atom having a full outermost subshell. Also, the term $^4F_{3/2}$ signifies an atomic state with an incompletely filled outermost subshell, as is reflected by $S = \frac{3}{2}$, $L = 3$ and $J = \frac{3}{2}$.

To determine which of the allowable terms will represent the lowest energy level of an atom, we rely on Hund's rules, which are summarized as follows:

(1) First select the terms of highest $S (S = \Sigma\, m_s)$ consistent with the Pauli principle.
(2) Among the terms having the same highest value of S, select the one with the highest L ($L = \Sigma\, m_l$).
(3) For the determination of J, two distinct cases exist:
 (A) If the subshell is less than half full, set $J = |L - S|$.
 (B) If the subshell is half full or more, set $J = L + S$.

The application of Hund's rule is clearly seen in table 2.3, which gives a detailed analysis of the normal state terms for the transition and rare earth ions. The physical significance of the first Hund's rule is that the spins of electrons in a subshell tend to align in one direction until the maximum multiplicity is attained. After that, the spins of electrons will align in the opposite direction. Thus in the 3d subshell, a maximum of five electrons can have their spins aligned in one direction, giving $S = \frac{5}{2}$ as in the case of Mn^{2+}. For lanthanide elements, which have their magnetic electrons in the 4f subshell, a maximum of 7 electrons will align in one direction with $S = \frac{7}{2}$, as in the case of Gd^{3+}.

The magnetic moment of an ion, or atom, is given by $\mu = J^* g \mu_B$. Since J^* is the vector sum of L^* and S^*, in principle the magnetic moment is derived from both the orbital and spin motions. If, for some reason to be discussed later, $L^* = 0$, then $J^* = S^*$ and $g = 2$. Whence $\mu = \mu_s = 2S^* \mu_B$, signifying that the magnetic moment is due entirely to spin, not to the orbital motion at all. The orbital motion and therefore its associated angular momentum and magnetic moment are said to be quenched.

This phenomenon is seen to occur in most of the transitional metal ions and atoms, but hardly in lanthanides. The experimental evidence has long been furnished by the susceptibility data compiled in table 2.3. To see the evidence, let us refer back to the expression (1.17) for the atomic susceptibility of a paramagnet given by the classical theory of Langevin. The quantum theory of Langevin-type paramagnetism is discussed in the next section, which defines the magnetic moment μ in

(1.17) as $\mu = n_{\mathrm{eff}}\mu_{\mathrm{B}}$, where $n_{\mathrm{eff}} = g[J(J+1)]^{\frac{1}{2}}$, if the moment comes from both the orbital and spin motions, or $n_{\mathrm{eff}} = 2[S(S+1)]^{\frac{1}{2}}$, if the spin is the only source of the moment. Calculated values of the effective number of magnetic moment n_{eff} for ions of the transitional metals and lanthanides are compared with the deduced values from the susceptibility data in table 2.3. Clearly, quenching is seen to take place in practically all transitional metal ions (except two cases where the 3d subshell is either just half-filled as in Mn^{2+} and Fe^{3+} or completely full as in Cu^{+} and Zn^{2+}, for which $gJ^* = 2S^*$, thus the evidence for quenching is masked); but none is seen in the lanthanide ions.

In quantum theory, an orbital angular momentum is said to be quenched when its component L_z in the polar axis of the spherical polar-coordinate system is no longer a constant of the motion, even though L^2 may still be a constant. The quantum-mechanical treatment of the quenching effect is beyond the scope of the book and can be found in Morrish [1965a] and Kittel [1976c]. In the present discussion, we are interested in the physical significance and implications to various magnetic systems. Physically quenching occurs when the electric fields in which the electrons move about the nucleus are noncentral. In a central field directed toward a fixed nucleus, the plane of a classical orbit is fixed in space; thus all components of the orbital angular momentum L_x, L_y and L_z are constant. In a noncentral field, however, the plane of the orbit will move around, resulting in a zero L_z on the average. The noncentral fields responsible for the quenching effect on transitional metal ions in paramagnetic salts are identified with the crystalline fields produced by neighboring ions. These electric crystal fields are not only intense in comparison with any applied magnetic field, but also inhomogeneous as evidenced by their noncubic (e.g., rhombic) symmetry, even though the crystal itself may have the cubic symmetry. The interaction of these crystal fields with the orbital motions of the 3d electrons produces two major effects: the breakdown of the Russell–Saunders or LS coupling and the splitting of an energy level characterized by L to $2L+1$ degenerate sublevels. Furthermore, the splittings caused by the crystal fields are about $1 \times 10^6\ \mathrm{m}^{-1}$ (~ 1.24 eV), which are much larger than those produced by the spin–orbit coupling. Thus the crystal field splitting diminishes the contribution of the orbital motions of the 3d electrons to the magnetic moment.

The basic reason for the different magnetic behavior between transitional metal ions and lanthanide ions can be explained readily in relation to their electronic configurations. The transitional metal ions

TABLE 2.3

A detailed analysis of the normal states of magnetic d and f subshells and values* for n_{eff}, the effective number of magnetic moment per ion.

Subshell	Number of elect.	Typical ion	Normal state S	Normal state L	Normal state J	Normal state Term	n_{eff} Calc. from $g[J(J+1)]^{\frac{1}{2}}$	n_{eff} Obs.	n_{eff} Calc. from $2[S(S+1)]^{\frac{1}{2}}$
	1	Ti^{3+}, V^{4+}	$\frac{1}{2}$	2	$\frac{3}{2}$	$^2D_{\frac{3}{2}}$	1.55	1.70	1.73
	2	V^{3+}	1	3	2	$^3F_2$	1.63	2.61	2.83
	3	V^{2+}, Cr^{3+}	$\frac{3}{2}$	3	$\frac{3}{2}$	$^4F_{\frac{3}{2}}$	0.77	3.85	3.87
	4	Cr^{2+}, Mn^{3+}	2	2	0	$^5D_0$	0	4.82	4.90
d	5	Mn^{2+}, Fe^{3+}	$\frac{5}{2}$	0	$\frac{5}{2}$	$^6S_{\frac{5}{2}}$	5.92	5.82	5.92
	6	Fe^{2+}	2	2	4	$^5D_4$	6.70	5.36	4.90
	7	Co^{2+}	$\frac{3}{2}$	3	$\frac{9}{2}$	$^4F_{\frac{9}{2}}$	6.63	4.90	3.87
	8	Ni^{2+}	1	3	4	$^3F_4$	5.59	3.12	2.83
	9	Cu^{2+}	$\frac{1}{2}$	2	$\frac{5}{2}$	$^2D_{\frac{5}{2}}$	3.55	1.83	1.73
	10	Cu^{+}, Zn^{2+}	0	0	0	$^1S_0$	0	0	0

	1	Ce^{3+}	$\frac{1}{2}$	3	$\frac{5}{2}$	$^2F_{\frac{5}{2}}$	2.54	2.51
	2	Pr^{3+}	1	5	4	$^3H_4$	3.58	3.56
	3	Nd^{3+}	$\frac{3}{2}$	6	$\frac{9}{2}$	$^4I_{\frac{9}{2}}$	3.62	3.3–3.71
	4	Pm^{3+}	2	6	4	$^5I_4$	2.68	–
	5	Sm^{3+}	$\frac{5}{2}$	5	$\frac{5}{2}$	$^6H_{\frac{5}{2}}$	0.85(1.55)**	1.74
	6	Eu^{3+}	3	3	0	$^7F_0$	0(3.40)**	3.4
f	7	Gd^{3+}, Eu^{3+}	$\frac{7}{2}$	0	$\frac{7}{2}$	$^8S_{\frac{7}{2}}$	7.94	7.98
	8	Tb^{3+}	3	3	6	$^7F_6$	9.72	9.77
	9	Dy^{3+}	$\frac{5}{2}$	5	$\frac{15}{2}$	$^6H_{\frac{15}{2}}$	10.63	10.63
	10	Ho^{3+}	2	6	8	$^5I_8$	10.60	10.4
	11	Er^{3+}	$\frac{3}{2}$	6	$\frac{15}{2}$	$^4I_{\frac{15}{2}}$	9.59	9.5
	12	Tm^{3+}	1	5	6	$^3H_6$	7.57	7.61
	13	Yb^{3+}	$\frac{1}{2}$	3	$\frac{7}{2}$	$^2F_{\frac{7}{2}}$	4.53	4.5
	14	Lu^{3+}, Yb^{2+}	0	0	0	$^1S_0$	0	–

* References: Kittel [1976d] and Brailsford [1966d].

** Values given by van Vleck (The Theory of Electric and Magnetic Susceptibilities, Oxford Univ. Press, 1932, p. 243).

derive their magnetic moments from electrons in the outermost 3d subshell, which are not shielded, especially in the ionic crystals. On the other hand, the lanthanide ions owe their magnetic moments to the 4f electrons, which are deep-seated inside the complete 5s and 5p subshells. Hence the transitional metal ions suffer from the crystal fields, leading to the quenching of the orbital motions of the 3d electrons. Meanwhile, the lanthanide ions with their magnetic electrons so well screened from the crystal fields are not subject to the quenching effect. Thus the magnetic moments of the lanthanide ions react to the applied magnetic field in the manner depicted by the Brillouin function in the next section.

The quenching effect deserves two further comments. First, the observed values of n_{eff} for some of the transitional metal ions, notably Fe^{2+}, Co^{2+} and Ni^{2+} fall between the calculated values for $n_{\text{eff}} = g[J(J+1)]^{\frac{1}{2}}$ and $n_{\text{eff}} = 2[S(S+1)]^{\frac{1}{2}}$, indicating that quenching is actually incomplete. The second comment concerns the occurrence of the quenching effect in metals. For atoms of the iron group and lanthanide elements in their metallic states, again quenching is seen to occur only in the metals of Fe, Co and Ni, not in metallic lanthanides. The evidence for quenching in the iron-group metals is shown by the experimental values deduced for the Landé g-factor, which are 1.92, 1.85 and 1.84 for Fe, Co and Ni, respectively (Scott [1962]). These values are indeed close to $g = 2$, which corresponds to a complete quenching. The deduced g values thus indicate the preponderant occurrence of quenching in the transitional metals, but the contribution from orbital motions to the magnetic moment is not entirely negligible.

§2. THE WEISS MOLECULAR FIELD AND THEORIES OF FERROMAGNETISM

In the previous section, we have seen that a free atom can acquire a magnetic moment from the orbital and spin motions of its electrons. When atoms are condensed into a crystalline solid, magnetization may or may not occur, depending on how the atomic moments are coupled in the solid. Hence the second question posed at the outset of this chapter deals with a basic requirement of all magnetic materials.

At this point, it is useful to discuss the quantum theory of Langevin-type paramagnetism. Brillouin (see Bates [1963b]) showed that the component of the magnetic moment μ of atoms must be quantized in the direction of the magnetic field H. Hence for each atom, there are these characteristic quantum numbers: L, S, J and M_a, where $M_a = J, J-1, \ldots, -(J-1), -J$, and the components of the magnetic moment in H are $M_a g \mu_B$. Applying the Maxwell–Boltzmann law gives

$$M_A = N_0 \bar{\mu} = \frac{N_0 \sum_{M_a=-J}^{M_a=J} M_a g \mu_B \exp(M_a g \mu_0 \mu_B H/kT)}{\sum_{M_a=-J}^{M_a=J} \exp(M_a g \mu_0 \mu_B H/kT)} \tag{2.11}$$

and for $M_a g \mu_0 \mu_B H/kT \ll 1$,

$$\begin{aligned} \chi_A &= (N_0 \mu_0/3kT) J(J+1) g^2 \mu_B^2 \\ &= (N_0 \mu_0/3kT)(n_{\text{eff}} \mu_B)^2 \\ &= C/T, \end{aligned} \tag{2.12}$$

where the Curie constant

$$C = (N_0 \mu_0/3k)(n_{\text{eff}} \mu_B)^2, \tag{2.13}$$

and

$$n_{\text{eff}} = g[J(J+1)]^{\frac{1}{2}}, \tag{2.14}$$

is known as the effective number of Bohr magnetons per atom. Note that the spin paramagnetism in quantum theory also follows the Curie law except that the Curie constant is changed from $C = N_0 \mu_0 \mu^2/3k$ to $C = N_0 \mu_0 J(J+1) g^2 \mu_B^2/3k$.

When the applied field is strong or the temperature is sufficiently low, $g \mu_0 J \mu_B H/kT$ is not small compared with unity. In the latter case, we obtain

$$M_A = N_0 g J \mu_B \left[\frac{2J+1}{2J} \coth\left(\frac{2J+1}{2J}\right) a - \frac{1}{2J} \coth\left(\frac{a}{2J}\right)\right], \tag{2.15}$$

where

$$a = Jx = g \mu_0 J \mu_B H/kT. \tag{2.16}$$

Putting

$$B_J(a) = \frac{2J+1}{2J} \coth\left(\frac{2J+1}{2J}\right) a - \frac{1}{2J} \coth\left(\frac{a}{2J}\right), \tag{2.17}$$

we finally have

$$M_A = N_0 J g \mu_B B_J(a), \tag{2.18}$$

which shows that the magnetization of a Langevin-type paramagnet varies with temperature through the expression $B_J(a)$, which is known as the Brillouin function.

If an infinitely strong field is applied or, alternatively, if the temperature is close to absolute zero, M_A will approach a saturation value. Under these conditions, coth $a \to 1$ and $B_J(a) \to 1$. Whence

$$M_A = N_0 J g \mu_B. \tag{2.19}$$

This corresponds to the limiting case that all atomic moments take the allowed angular position nearest to the field; i.e. $M_a = J$. Thus $M_A \to M_0$, which is the saturation magnetization. Combining (2.18) and (2.19) gives

$$M_A(T)/M_0 = B_J(a). \tag{2.20}$$

Assuming that a paramagnet with $J = \frac{3}{2}$ and $g = \frac{4}{3}$ attains a magnetization value $M_A(T) \sim 0.1\, M_0$ at room temperature, then from (2.17) and (2.20), we obtain $a \sim 0.18$, which, upon substituting in (2.16), gives $H \sim 3.1 \times 10^7\ \text{A} \cdot \text{m}^{-1}$ (3.9×10^5 Oe). Such a high field is barely attainable in the pulse form in a few laboratories. Therefore, under a static field of typical strength of $1.6 \times 10^6\ \text{A} \cdot \text{m}^{-1}$ (20 000 Oe), only a small percentage of M_A can be induced in most paramagnetic materials at room temperature.

The situation with ferromagnets is entirely different. In an iron crystal, for instance, not only do the atoms carry a magnetic moment, but also the metal is broken up into small regions known as the magnetic domains. In each domain all the atomic moments are coupled together in a preferential crystallographic direction, and the material reaches the saturation value of magnetization even though no field is applied. The domains are thus said to be spontaneously magnetized.

To account for the onset of the spontaneous magnetization, Weiss [1907] postulated that an intense internal field, now known as the Weiss molecular field, H_W, exists in a ferromagnet. (The word molecular unfortunately is retained because Weiss's theory was proposed originally for magnetic systems of molecules.) He further assumed that

$$H_W = \lambda M, \tag{2.21}$$

where the constant of proportionality λ is called the molecular field constant. The molecular field theory of Weiss will be further discussed later. Here we shall see how his theory determines the susceptibility of ferromagnets.

Because of the internal field, application of an external field will cause the atomic moments in a ferromagnet to be acted on by a total field,

$$H_{\text{total}} = H + H_{\text{W}}. \tag{2.22}$$

Since a ferromagnet becomes paramagnetic above the Curie temperature θ_{C}, it is reasonable to assume that the Curie law still applies, but the law must be modified in that the field acting on the moments will be H_{total}, not just H. Hence, following (1.16) and (1.17), we have for the magnetization per unit volume

$$M = C'(H + H_{\text{W}})/T = C'(H + \lambda M)/T, \tag{2.23a}$$

where $C' = Ng^2\mu_0\mu_{\text{B}}^2 J(J+1)/3k$ according to (2.13) and (2.14) with N being the number of atoms per unit volume. Another expression for M is obtained by first converting (1.16) to

$$M = N\bar{\mu}, \tag{1.16.1}$$

where $\bar{\mu}$ is calculated from the quantum-mechanical counterpart to (1.13) by replacing the Langevin function with the Brillouin function,

$$\bar{\mu}/\mu = B_J(a), \tag{1.13.1}$$

where a and $B_J(a)$ are defined in (2.16) and (2.17), respectively and $\mu = gJ\mu_{\text{B}}$. For $J = \frac{1}{2}$, it is easy to show that

$$B_{\frac{1}{2}}(a) = \tanh(a). \tag{2.17.1}$$

Substituting H by $H + \lambda M$ in (2.16) so as to define a in the Weiss theory and combining the last three equations yields

$$M = N\mu_{\text{B}} \tanh\left[\mu_0\mu_{\text{B}}(H + \lambda M)/kT\right]. \tag{2.23b}$$

This alternative expression of M is used in the discussion of the Ising model and several other aspects of magnetization.

From (2.23a), we obtain

$$\chi = C'/(T - \theta_{\text{C}}) \tag{2.24}$$

and

$$\theta_{\text{C}} = C'\lambda = [\mu_0 NJ(J+1)g^2\mu_{\text{B}}^2/3k]\lambda. \tag{2.25a}$$

Equation (2.24) is known as the Curie–Weiss law. As $T \to \theta_{\text{C}}$, $\chi \to \infty$; we may state that below θ_{C}, the magnetic susceptibility of a ferromagnet displays complex behavior as shown in fig. 2.5.

In the plot of reciprocal susceptibility versus temperature shown in fig. 2.5, the intercept produced by extrapolation on the temperature axis

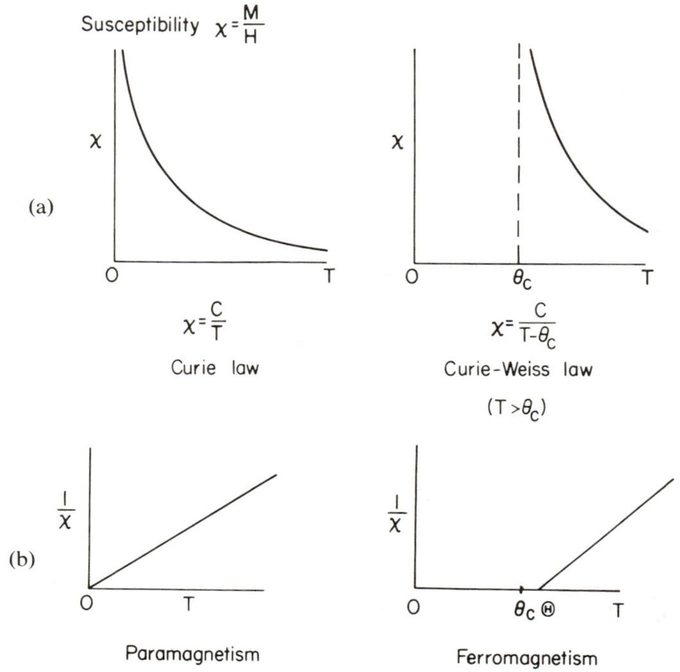

Fig. 2.5. Temperature dependence of (a) the magnetic susceptibility and (b) the inverse magnetic susceptibility in paramagnets and ferromagnets (Kittel [1976]).

is known as the paramagnetic Curie temperature Θ. This temperature should not be confused with the true Curie temperature θ_C. Usually θ_C lies slightly below Θ. The determination of θ_C requires a series of magnetization measurements to be conducted at various temperatures below θ_C. Several methods are devised to use the data obtained from these measurements to deduce the value of θ_C. All methods are based on the relation,

$$a(T - \theta_C)M + bTM^3 = H, \tag{2.26}$$

where H is the effective magnetic field, that is, the applied field corrected for the demagnetization effect and a and b are constants. Readers who are interested in detailed information on the determination of θ_C are referred to a review article by McGuire and Flanders [1969].

The Weiss theory of ferromagnetism postulates the existence of an internal field H_W in the ferromagnetic materials. This so-called molecular field is responsible for the parallel coupling of atomic moments in each domain. We now want to see the strength of the molecular field and its origin.

To calculate H_W, we must first evaluate the molecular field constant λ. From (2.25a), we obtain an expression for the dimensionless constant λ

$$\lambda = \theta_C / C' = 3k\theta_C / \mu_0 N J(J+1) g^2 \mu_B^2. \tag{2.25b}$$

For iron, $\theta_C = 770°C = 1043°K$, $N = N_0 \times (\text{density/atomic weight}) = 8.49 \times 10^{28}\ \text{m}^{-3}$. For an estimate of λ, we set $[J(J+1)]^{1/2} g\mu_B = n_{\text{eff}}\mu_B = 2.2 \times 9.2732 \times 10^{-24} = 2.04 \times 10^{-23}\ \text{A} \cdot \text{m}^2$. Substituting these values in (2.25b) gives

$$\lambda = \frac{3 \times 1.38 \times 10^{-23} \times 1043}{8.49 \times 10^{28} \times (2.04 \times 10^{-23})^2 \times 4\pi \times 10^{-7}} = 9.73 \times 10^2.$$

The magnetization of iron may be calculated from

$$M_S = n_{\text{eff}} \times \mu_B \times N = 1.73 \times 10^6\ \text{A} \cdot \text{m}^{-1} \tag{2.27}$$

The internal field in Fe is $H_W = \lambda M_S \simeq 9.73 \times 10^2 \times 1.73 \times 10^6 \simeq 1.7 \times 10^9\ \text{A} \cdot \text{m}^{-1} (\sim 2.1 \times 10^7\ \text{Oe})$. A very intense field, indeed! Such a field is much stronger than the magnetic field ($\sim 8 \times 10^4\ \text{A} \cdot \text{m}^{-1}$) produced in a crystal by magnetic ions at a lattice point and cannot be matched even by the most powerful pulse magnetic fields. Because of the intense H_W, all ferromagnetic materials are spontaneously magnetized to saturation within each domain below the Curie temperature.

The origin of the internal field was first explained by Heisenberg [1928] in terms of a positive exchange interaction between spins in neighboring atoms. This interaction is strictly a quantum mechanical effect and is of an electrostatic nature. It is best explained with reference to a hydrogen molecule. (In fact, Heisenberg's original analysis was based on the Heitler–London treatment of the hydrogen molecule.) The main features of the Heisenberg theory are presented here; detailed treatment may be found in the excellent review articles by Herring [1962, 1966]. Fig. 2.6 shows the two atomic systems in a hydrogen molecule: nucleus a-electron 1 and nucleus b-electron 2. The Hamiltonian operator is resolved into two parts:

$$\mathcal{H} = \underbrace{-\left[\frac{\hbar^2}{2m_0}(\nabla_1^2 + \nabla_2^2) + \frac{e^2}{r_{a1}} + \frac{e^2}{r_{b2}}\right]}_{\mathcal{H}_{\text{noninter}}} + \underbrace{\left[\frac{e^2}{r_{ab}} + \frac{e^2}{r_{12}} - \frac{e^2}{r_{a2}} - \frac{e^2}{r_{b1}}\right]}_{\mathcal{H}_{\text{inter}}}, \tag{2.28}$$

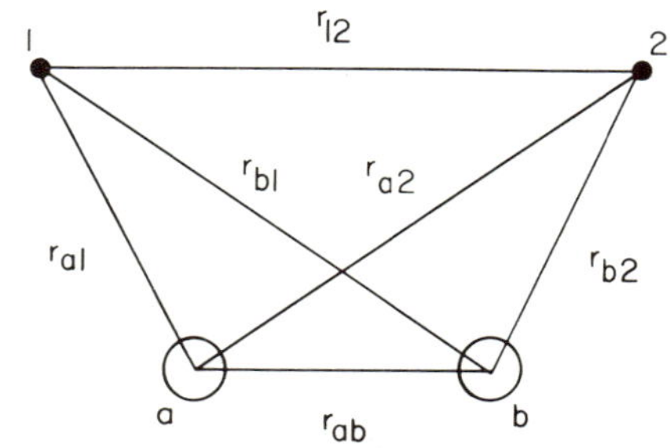

Fig. 2.6. Interactions between electrons and nuclei in a hydrogen molecule.

where $\nabla_i^2 = \partial^2/\partial x_i^2 + \partial^2/\partial y_i^2 + \partial^2/\partial z_i^2$. The four terms in the first brackets represent the kinetic and potential energies of the two electrons within the two atoms, thus constituting the noninteracting part of $\mathscr{H}$. The four terms in the second brackets account for the interactions between the electrons and the nuclei shown in fig. 2.6, thus constituting the interacting part of $\mathscr{H}$. The expectation value of the energy of interactions is given by

$$E = \int \phi^* \mathscr{H}_{\text{inter}} \phi \, \mathrm{d}\tau, \tag{2.29}$$

where ϕ is an appropriate wave function and ϕ^* is the complex conjugate of ϕ. To formulate ϕ, let us denote $\psi_a(1)$ to be the normalized one-electron wave function for electron 1 in state a. In quantum mechanics, $\psi_a(1)$ is a solution of the one-electron Schrödinger equation,

$$\left[-\frac{\hbar^2}{2m_0}\nabla_1^2 + V_{a1}(\boldsymbol{r})\right]\psi = E_a\psi, \tag{2.30}$$

where $V_{a1}(\boldsymbol{r})$ is the potential energy of the electron. Similar definitions are for $\psi_a(2)$, $\psi_b(1)$ and $\psi_b(2)$.

From the noninteracting part of the Hamiltonian, we obtain two forms for the orbital wave functions of the molecule

$$\psi_{\substack{\text{sym,}\\ \text{antisym}}}(1,2) = \frac{1}{\sqrt{2}}[\psi_a(1)\psi_b(2) \pm \psi_a(2)\psi_b(1)], \tag{2.31}$$

where the factor $1/\sqrt{2}$ is the normalizing constant. Both symmetric and antisymmetric wave functions are acceptable in (2.31) as far as the Pauli exclusion principle is concerned. This is because the complete time-independent wave function ϕ is composed of an orbital part ψ and a spin part χ; that is,

$$\phi = \psi(\boldsymbol{r})\chi. \tag{2.32}$$

For the spin wave function χ, we recall (ch. 2, §1 after (2.10)) that when a magnetic field $\boldsymbol{H}$ is applied in the z-direction, the component of the spin angular momentum of an s electron will be quantized in that direction by assuming only $+\hbar/2$ or $-\hbar/2$ value, thereby acquiring a magnetic moment μ_z with magnitude of one Bohr magneton lying either parallel or antiparallel to $\boldsymbol{H}$. A similar situation prevails for the d electrons in transition metals because of the effect of quenching previously discussed. We denote these two spin states by $\chi_\alpha(i)$ and $\chi_\beta(i)$ for μ_z of the ith electron parallel and antiparallel to $\boldsymbol{H}$, respectively. Various spin wave functions then can be generated of either symmetric or antisymmetric form

$$\chi_{\text{sym}}(1,2)_{\text{I}} = \chi_\alpha(1)\chi_\alpha(2), \tag{2.33.1}$$

$$\chi_{\text{sym}}(1,2)_{\text{II}} = \chi_\alpha(1)\chi_\beta(2) + \chi_\alpha(2)\chi_\beta(1), \tag{2.33.2}$$

$$\chi_{\text{sym}}(1,2)_{\text{III}} = \chi_\beta(1)\chi_\beta(2), \tag{2.33.3}$$

$$\chi_{\text{antisym}}(1,2) = \chi_\alpha(1)\chi_\beta(2) - \chi_\alpha(2)\chi_\beta(1). \tag{2.33.4}$$

In compliance with the Pauli exclusion principle, the acceptable wave functions in (2.29) must be antisymmetric. This condition is satisfied if we choose $\psi_{\text{sym}}(1,2)\chi_{\text{antisym}}(1,2)$ or $\psi_{\text{antisym}}(1,2)\chi_{\text{sym}}(1,2)$. Explicitly, the acceptable wave functions for the hydrogen molecule are

$$\phi_{\text{I}} = A[\psi_a(1)\psi_b(2) + \psi_a(2)\psi_b(1)][\chi_\alpha(1)\chi_\beta(2) - \chi_\alpha(2)\chi_\beta(1)], \tag{2.34.1}$$

$$\phi_{\text{II}} = B[\psi_a(1)\psi_b(2) - \psi_a(2)\psi_b(1)]\begin{bmatrix} \chi_\alpha(1)\chi_\alpha(2) \\ \chi_\alpha(1)\chi_\beta(2) + \chi_\alpha(2)\chi_\beta(1) \\ \chi_\beta(1)\chi_\beta(2) \end{bmatrix}, \tag{2.34.2}$$

where A and B are normalizing factors.

By using ψ_{sym} or ψ_{antisym}, (2.29) yields

$$E_{\text{I}} = A^2(K_{12} + J_{12}) \quad \text{or} \quad E_{\text{II}} = B^2(K_{12} - J_{12}), \tag{2.35}$$

where

$$K_{12} = \int \psi_a^*(1)\psi_b^*(2)\mathcal{H}_{\text{inter}}\psi_a(1)\psi_b(2)\,\mathrm{d}\tau, \tag{2.36}$$

is called the Coulomb integral, and

$$J_{12} = \int \psi_a^*(1)\psi_b^*(2)\mathcal{H}_{\text{inter}}\psi_a(2)\psi_b(1)\,d\tau, \tag{2.37}$$

is called the exchange integral. The latter integral is so named because J_{12} differs from K_{12} only in the exchange in position between the electrons in the wave functions $\psi_a\psi_b$ in the integrand. This position exchange is required in quantum theory because electrons are indistinguishable.

The exchange integral J_{12} may be positive or negative, depending on the relative magnitudes of the positive and negative terms in $\mathcal{H}_{\text{inter}}$. For the hydrogen molecule, J_{12} is negative; hence the spins of the two electrons are antiparallel in the 1s ground state. Heisenberg postulated that J_{12} is positive in Fe, Co and Ni; therefore these metals are ferromagnetic. His postulate was later substantiated in a semi-empirical manner by Slater and Bethe, as we shall see in ch. 5, §1. In terms of the complete wave function ϕ, the exchange energy as a part of E_{I} or E_{II} has a spin-dependent term of the form

$$E_{\text{ex}} = -2J_{ij}\boldsymbol{S}_i \cdot \boldsymbol{S}_j, \tag{2.38}$$

where $\boldsymbol{S}_i = \Sigma\, s_i$ and $\boldsymbol{S}_j = \Sigma\, s_j$ are the total spins of atoms i and j, respectively. For crystals in which the exchange integral is isotropic, we have for the total exchange energy

$$E_{\text{ex}} = -J_{\text{ex}} \sum_i \sum_j \boldsymbol{S}_i \cdot \boldsymbol{S}_j, \tag{2.39}$$

where the first sum is over all the atoms in the crystal and the second sum is over the nearest neighbors of atom i. Note that in the last expression, the factor two is dropped to prevent counting the atoms twice.

To see the connection of the molecular field with the exchange interaction, let us examine a ferromagnetic system of pure metal. For simplicity, we assume that the magnetic moment per atom, μ, in the system is due to spins only, with no contribution from the orbital motion of electrons, as in Fe, Co and Ni. According to (2.38) and (2.39), the energy of the exchange interaction between spins of the ith atom and its nearest neighbors at sites j can be expressed by

$$E_{\text{ex}}^{ij} = -J_{\text{ex}} \sum^{z} \boldsymbol{S}_i \cdot \boldsymbol{S}_j = -zJ_{\text{ex}}\boldsymbol{S}_i \cdot \boldsymbol{S}_j, \tag{2.40}$$

where z is the coordination number (8 for Fe and 12 for Ni and approximately 12 for Co). The exchange energy also may be considered as the potential energy of atom i in a field H produced by the lattice of atoms j; that is

$$E_{\mathrm{ex}}^{ij} = -\boldsymbol{\mu}_i \cdot \boldsymbol{H}_{\mathrm{W}}\mu_0, \tag{2.41}$$

where

$$\boldsymbol{\mu}_i = \boldsymbol{S}_i g\mu_{\mathrm{B}}, \tag{2.42}$$

is the magnetic moment of atom i. By definition, the magnetization $\boldsymbol{M}$ of the system is the total magnetic moment per unit volume of the lattice comprising atoms j. Thus

$$\boldsymbol{M} = N\boldsymbol{S}_j g\mu_{\mathrm{B}}, \tag{2.43}$$

where N is the number of atoms per unit volume. From (2.40)–(2.43), we obtain

$$\boldsymbol{H}_{\mathrm{W}} = \frac{zJ_{\mathrm{ex}}}{Ng^2\mu_{\mathrm{B}}^2\mu_0}\boldsymbol{M}. \tag{2.44}$$

Turning to the Weiss theory of ferromagnetism, because $\boldsymbol{H}_{\mathrm{W}} = \lambda\boldsymbol{M}$, we further obtain the relation

$$\lambda = zJ_{\mathrm{ex}}/Ng^2\mu_{\mathrm{B}}^2\mu_0. \tag{2.45}$$

Both (2.44) and (2.45) show the relationships between the exchange integral J_{ex} and the molecular field H_{w} or its constant λ.

The Weiss–Heisenberg theory of ferromagnetism, as viewed from (2.39) and (2.44), is based on the exchange interaction between localized electrons in the "magnetic" subshell at the atoms. Hence, this theory represents the localized electron model. It is doubtless applicable to lanthanide metals, in which the "magnetic" electrons are deeply seated in the 4f subshell of the atomic core. The theory also works well with most ferrimagnetic oxides and compounds because of the ionic nature of these materials.

The localized electron model has not been so successful with the 3d transition metals. One main difficulty is caused by the nonintegral values of the effective Bohr magneton number, n_{eff}, per atom observed in Fe, Co and Ni. The difficulty is seen in the following argument. In terms of the Heisenberg theory, if we assume one electron per atom in the 4s subshell for Fe, Co and Ni, then the number of electrons in the 3d subshell are 7, 8 and 9, giving S the values of $\frac{3}{2}$, 1 and $\frac{1}{2}$, respectively.

If the contribution to the moment from the orbital motion of the 3d electrons is considered negligibly small as a result of the quenching effect on the spin–orbital coupling previously discussed, the magnetic moment per atom for Fe, Co and Ni should have integral values of 3, 2 and 1 in units of μ_B, respectively. From saturation measurements and neutron diffraction experiments, however, the deduced values of the magnetic moment are 2.216, 1.716 and 0.616 μ_B per atom at 0°K, instead, (Danan et al. [1968]). These nonintegral values of the apparent Bohr magneton number per atom for Fe, Co and Ni thus present a stumbling block to the application of the localized electron model to the transition metals.

Several alternative theories have been proposed to account for the nonintegral values of n_{eff}. Among them, the collective electron, or the band, theory initiated by Stoner (see Lomer [1967]) has drawn the most attention and is particularly attractive for Ni and Co. In this theory, the 3d electrons are considered to form a band just as the 4s electrons form the conduction band. Hence, the 3d electrons belong to the whole crystal, not attached to individual atoms. Because of the spin effect, the 3d band is split into halves with opposite spins, each capable of accommodating a maximum of five electrons, as shown in fig. 2.7b.

The band structures of Fe and Ni have been calculated by many workers and re-examined recently by Lomer [1973]. The new calculations revealed that Ni and even the bcc Fe have band structures quite similar to that of Cu calculated by Phillips [1967]. The similarities are clearly seen in fig. 2.7. Also noteworthy in this connection is the overlap between the 3d and 4s bands. This overlap, coupled with the fact that both bands must be filled with electrons to the same energy level, would permit the average number of electrons per atom in both bands to have nonintegral values. For Ni, for instance, the split of the 10 electrons outside the 3p subshell is 0.6 in the 4s band and 9.4 in the 3d band. If the polarization effect on the conduction electrons is ignored, the two half 4s bands will be equally occupied by 0.3 electron per atom in both the paramagnetic and ferromagnetic states. The two half 3d bands also will be equally occupied by 4.7 electrons per atom in the paramagnetic state, but unequally occupied in the ferromagnetic state.

Stoner's theory also assumes the existence of a molecular field in the ferromagnetic transition metals. As a result, the half d band with negative spins will be displaced upward by an amount $H_W n_{eff} g \mu_B \mu_0$, whereas the other half d band with positive spins will be displaced

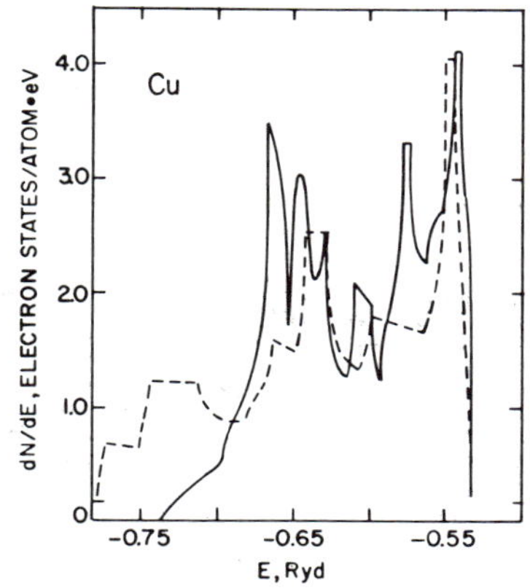

Fig. 2.7a. The 3d energy bands in copper, (Phillips [1967]).

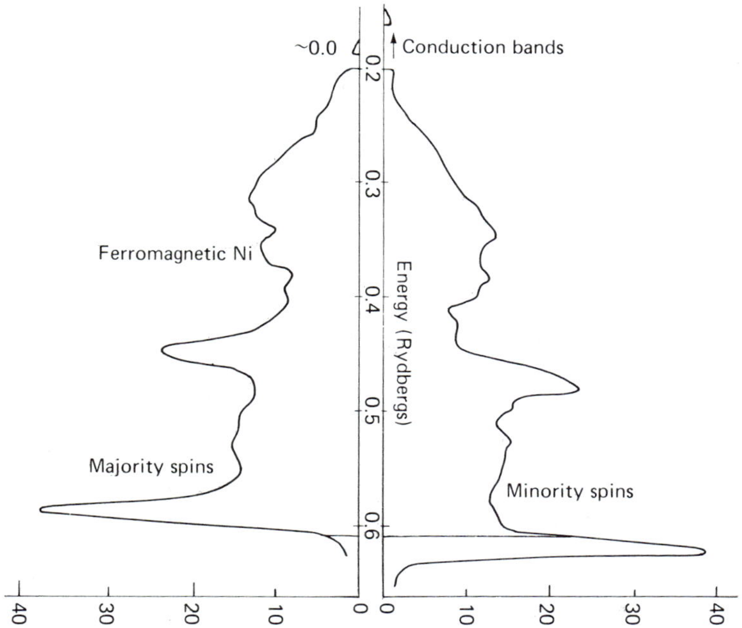

Fig. 2.7b. The 3d energy bands in nickel, (Lomer [1967]).

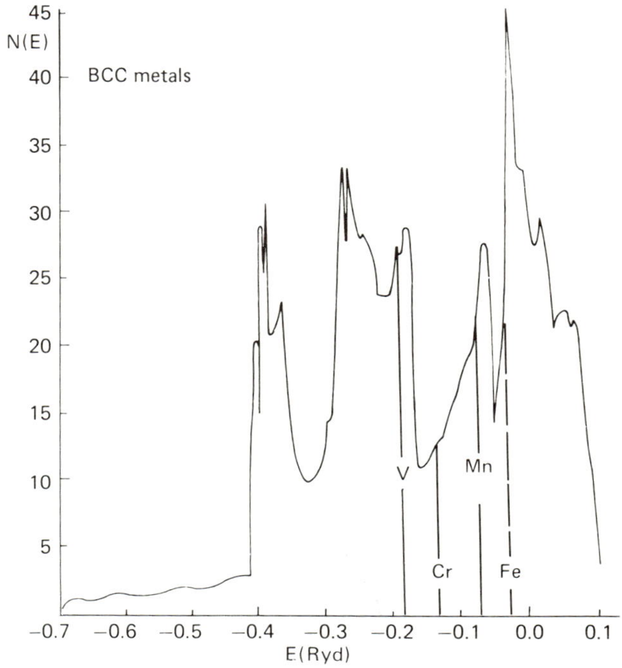

Fig. 2.7c. The 3d energy bands in the bcc transition metals, (Wohlfarth and Cornwell [1962]).

downward by the same amount. The occupancy of the two half d bands thus becomes unequal. Calculation shows that, in ferromagnetic Ni, although the half d band with positive spins is completely filled with 5 electrons per atom, the other half d band is filled with 4.38 electrons per atom. Hence $n_{\text{eff}} \simeq 0.62\mu_B$ for Ni which is in good agreement with the experimental results. Fig. 2.8 indicates diagrammatically the electron distribution in the 3d and 4s bands in Ni, Co and Fe. Note that, although Co has the same full occupancy of the half d band with positive spins as Ni, Fe has holes in both half d bands. The band theory provides a natural explanation of the nonintegral values of n_{eff} for the transition metals. Detailed comparison with experimental results of magnetic and thermal measurements and neutron-diffraction studies shows that, for Ni, Co and their alloys, the collective electron model works particularly well, whereas for Fe, lanthanide metals, and ferri-

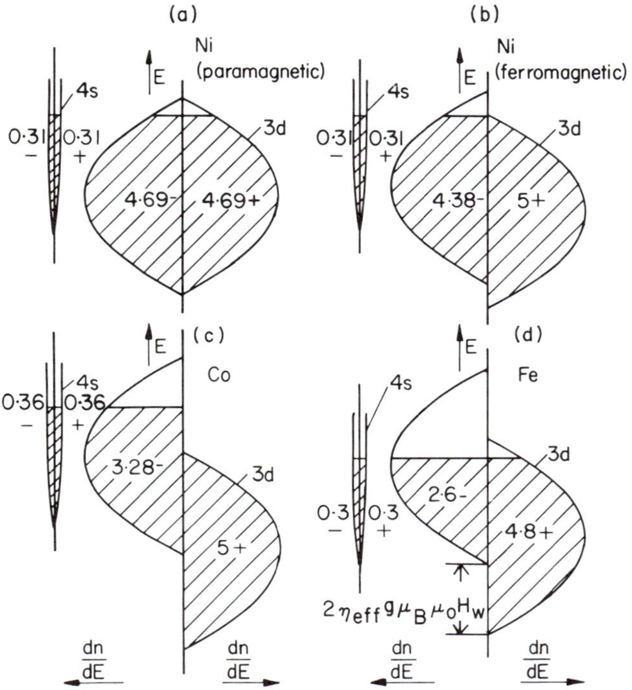

Fig. 2.8. Distribution of electrons in the 3d and 4s energy bands (a) in paramagnetic nickel and (b)–(d) in ferromagnetic nickel, cobalt and iron (Brailsford [1966]).

magnetic materials, the localized electron model is favored.

Although Heisenberg's idea of using the sign of the exchange integral J_{ex} in (2.37) to determine the mode of spin coupling proved to be useful, the theory has an inherent drawback in that the spin coupling thus defined can be accurately evaluated only in the limit of very low and very high temperatures. The evaluation becomes very difficult at intermediate temperatures, including the temperature of transition from ferro- to paramagnetism, where our interest is greatest. Moreover, an important enterprise of the theory of ferromagnetism is to calculate the thermodynamic properties as a function of temperature or magnetic field in terms of the eigenvalues of the proper spin Hamiltonian. This is a formidable task because, in the first place, it is virtually impossible to write the exact spin Hamiltonian, which would have to include fluctuations and correlations between the spins on neighboring sites. Secondly, even if a good approximation can be found for the spin Hamiltonian, the mathematical problems in the calculation of the thermodynamic properties are still enormous.

In lieu of an exact spin Hamiltonian, following Heisenberg [1928] and Dirac [1929], we represent every spin except the one we are dealing with by an average expectation value for the internal field. In this way, a crude spin Hamiltonian is obtained for a magnetic system in an applied field $\boldsymbol{H}$ in the form:

$$\mathscr{H} = -\sum_{n}\sum_{n'} J_{nn'}\boldsymbol{S}_n \cdot \boldsymbol{S}_{n'} - 2\mu_0\mu_{\mathrm{B}}\boldsymbol{H} \cdot \sum_{n} \boldsymbol{S}_n, \tag{2.46}$$

where the summation Σ_n runs over all sites and $\Sigma_{n'}$ runs over all but the nth sites. Based on this Hamiltonian, the theory has proliferated in two directions, leading to two models known as the Ising model and the theory of spin waves.

The Ising model. In his model, Ising made two important assumptions: (1) For a first approximation, it suffices to consider only the interactions between direct pairs. Consequently, the second sum in the first term of $\mathscr{H}$ in (2.46) will only run over the z nearest neighbors. (2) As far as the spins are concerned, the system is quantized along a certain direction – usually the direction of the applied field $\boldsymbol{H}$. Whence,

$$\boldsymbol{S}_i \cdot \boldsymbol{S}_j = S_{xi}S_{xj} + S_{yi}S_{yj} + S_{zi}S_{zj} \simeq S_{zi}S_{zj}, \tag{2.47}$$

because spin components in the x and y directions are zero. The correspondingly simplified $\mathscr{H}$ is called a truncated Hamiltonian, which has the attraction that all its variables commute, thus leaving out quantum mechanics in the subsequent treatment. The energy levels associated with the truncated $\mathscr{H}$ are given by

$$E = -\sum_{i,j} J_{ij}S_{zi}S_{zj} - 2\mu_0\mu_{\mathrm{B}}H \sum_{i} S_{zi}, \tag{2.48}$$

where $\Sigma_{i,j}$ runs over all pairs of nearest neighbors. In (2.47) and (2.48), we suppose that the spin quantum numbers S_z can only be $+1$ or -1. In this context, the magnetic system is like a binary alloy in which $S_{zi} = +1$ corresponds to an A atom on the ith site, and $S_{zi} = -1$, to a B atom on that site. The application of the Ising model to the order–disorder transition will be discussed in ch. 5, §2.5.

The principal advantage of the Ising model is that, by avoiding the difficulties of finding the exact eigenfunctions of a spin system, it reduces the calculation of the thermodynamic properties of the system to a combinatorial problem of the statistical distribution of the states of the system. To illustrate this point, first we write the partition function

$$Z = \sum_{\nu} \exp(-E_\nu/kT) = \sum_{S_{zi}=\pm 1} \exp\left[\frac{1}{kT}\left(\sum_{i,j} J_{ij}S_{zi}S_{zj} + 2\mu_{\mathrm{B}}\mu_0 H \sum_{i} S_{zi}\right)\right], \tag{2.49}$$

where the summation $\Sigma_{S_{zi}=\pm 1}$ goes over all states, or the 2^N combinations of + spins and − spins on the N sites of the lattice. From Z, we can calculate the internal energy of the system from the relation

$$U = kT^2(\partial \ln Z/\partial T), \tag{2.50}$$

the Helmholtz free energy

$$F = -kT \ln Z, \tag{2.51}$$

and the total magnetic moment

$$M = kT(\partial \ln Z/\partial H). \tag{2.52}$$

However, the evaluation of Z for cases of physical interest is a difficult problem. The nature of, and recourse to, the problem are examined in a review article by Newell and Montroll [1953].

To demonstrate the usefulness of the Ising model, we shall show how the Weiss theory of the internal field expressed in (2.23b) can be substantiated with a physical interpretation of the molecular field constant. Let N_+ be the number of + spins and N_- be the number of − spins, and define their concentrations by

$$c_+ = \frac{N_+}{N_+ + N_-} = \frac{N_+}{N}, \qquad c_- = \frac{N_-}{N} = 1 - c_+. \tag{2.53}$$

By using the Boltzmann relation for the entropy, $S = k \ln p$, where p is the number of possible independent arrangements of the ± spins on the N sites ($p = N!/N_+!N_-!$) and Stirling's approximation, the entropy of mixture is given by

$$S_{\mathrm{mix}} = -Nk(c_+ \ln c_+ + c_- \ln c_-). \tag{2.54}$$

For the calculation of the internal energy of the system, following

(2.46), we write

$$U = J_{ex}(N_{+-} - N_{++} - N_{--}) - 2\mu_B\mu_0 H(N_+ - N_-), \tag{2.55}$$

where N_{++} is the number of $++$ pairs, etc. In figuring out these numbers, we realize that: (1) there are zN_+ and zN_- nearest neighbors of the $+$ spins and $-$ spins, respectively; (2) each parallel ($++$ or $--$) pair will be counted twice and each antiparallel pair only once (thus, $2N_{++} + N_{+-} = zN_+$ and $2N_{--} + N_{+-} = zN_-$) and (3) $N_{++} = \frac{1}{2}(c_+N)\cdot(zc_+) = \frac{1}{2}Nzc_+^2$, $N_{--} = \frac{1}{2}Nzc_-^2$ and $N_{+-} = Nzc_+c_-$. Substituting these relations in (2.55) yields

$$U = \tfrac{1}{2}zNJ_{ex}(c_+^2 + c_-^2 - 2c_+c_-) - 2\mu_B\mu_0 HN(c_+ - c_-). \tag{2.56}$$

To simplify further, consider M to be a variable through the relations

$$c_+ = \tfrac{1}{2}(1 + (M/N\mu_B)), \qquad c_- = \tfrac{1}{2}(1 - (M/N\mu_B)). \tag{2.57}$$

We then obtain for U

$$U = -\tfrac{1}{2}(zJ_{ex}M^2/n\mu_B^2) - 2MH\mu_0 \tag{2.58}$$

and for S, from (2.54),

$$S = Nk[\ln 2 - \tfrac{1}{2}(1 - (M/N\mu_B))\ln(1 - (M/N\mu_B)) \\ - \tfrac{1}{2}(1 + (M/N\mu_B))\ln(1 + (M/N\mu_B))]. \tag{2.59}$$

The condition for minimum free energy is obtained by setting $\partial F/\partial M = 0$. This yields, after some rearrangement of terms,

$$M = N\mu_B \tanh\left(\frac{\mu_0\mu_B H}{kT} + \frac{zJ_{ex}M}{4N\mu_B kT}\right). \tag{2.60}$$

This expression is identical with (2.23b) with the molecular field constant taking the physical interpretation

$$\lambda = zJ_{ex}/4\mu_B^2\mu_0 N. \tag{2.61}$$

In both expressions $N \sim 1/r^3$, where r is the distance between nearest neighbors. Thus $\mu_0\mu_B^2 N \sim \mu_0\mu_B^2/r^3$ is of the order of the magnetic interaction energy between two neighboring spins, which is $\sim 10^{-23}$ J (10^{-16} ergs). The exchange integral in the numerator is more like an electrostatic quantity, which may approach 10^{-19} J (10^{-12} ergs). Again, this explains why λ would have a large value as has been deduced experimentally.

Further applications of the Ising model will be discussed in connection with another cooperative phenomenon (namely, the order–disorder transition discussed in ch. 5, §2.5). As far as ferromagnetism is concerned, analyses based on the model give the following conclusions:

(1) A one-dimensional lattice or a linear chain cannot show ferromagnetism.
(2) A two-dimensional square lattice exhibits ferromagnetism, for which $J_{ex}/kT_c = 0.441$, where T_c is the transition temperature (McCoy and Wu [1973]).
(3) For three-dimensional lattices, there are no exact solutions of the Ising problem. They probably will show long-range order because the topological conditions are even more favorable to cooperative effects than they are in two dimensions (Ziman [1964a]).

Spin waves. Besides the limited applicability to magnetism stated above, the Ising model has another drawback in that the suppression of the x and y components of spins, as expressed by (2.47), also produces some undesirable consequences. For instance, the model deprives magnetism of its angular momentum aspect, thus tending to give false information about the dynamic behavior of spin systems. Another consequence of this deficiency of the Ising model is seen as follows. For a simple ferromagnet at absolute zero, the third law of thermodynamics requires that all spins be perfectly ordered; that is, the spins are parallel in the classical picture. When the temperature is slightly raised, how would the spin system accommodate to the small thermal disturbance? The Ising model would predict a reversal of some spins, instead of the creation of spin waves, which turns out to be the correct answer. Hence, the Ising model is incapable of handling the effect of thermal agitation on the spin alignment.

Spin waves are the elementary excitation of a spin system in a wavelike form. To see how they come about, let us assume that the simple ferromagnet previously cited contains N atoms per unit volume, with each atom having a spin number S. At absolute zero, the system is in the ground state with all spins* aligned parallel in zero magnetic field.

*Strictly speaking, the spin alignment discussed here is confined to a domain, because the sample contains a number of domains in the demagnetized state. To simplify the discussion, however, we may assume that the size of crystal is sufficiently large and the domain structure is such as to permit the unit volume to fall within one domain.

If we treat the spins as classical vectors, then in the ground state, $\boldsymbol{S}_i \cdot \boldsymbol{S}_j = S^2$ and, according to (2.39), the total exchange energy of the system is given by

$$E_{ex}(0°\mathrm{K}) = E_{ex}^0 = -Nz\, J_{ex} S^2. \tag{2.62}$$

In the Ising model, the first excited state of the system would be one with a particular spin reversed. Such a spin reversal will create two pairs of antiparallel spins in a linear chain, as shown in fig. 2.9a, and according to (2.38) it will increase E_{ex} by $8J_{ex}S^2$. Whence $E_{ex}^1 = E_{ex}^0 + 8J_{ex}S^2$. The same excitation can be effected with much lower energy increase if all the spins equally share the reversal, as shown in fig. 2.9b. There is another reason that the spin wave formation is favored. The excited state envisaged in the Ising model is unstable because each of the atoms has an equal probability to reverse the spin. Therefore, the reversed spin will not stay at one atom; rather, the reversed spin tends to travel from atom to atom. The propagation of the reversed spin throughout the domain or crystal actually constitutes a spin wave. Formal proof of the existence of spin waves in ferromagnets at finite temperatures by quantum theory can be found in textbooks on solids by Ziman [1964b] and Kittel [1963a].

In the theory of spin waves, it is shown in fig. 2.10 that the spin vectors are assumed to precess on the surfaces of cones with successive spins advanced in phase by a constant phase angle qa, where $q = |\boldsymbol{q}|$, $\boldsymbol{q}$ being the wavevector of the spin wave and a the lattice constant. The radius of the circle of precession u is the amplitude of the spin wave, which is seen in the figure to have the relation with S,

$$u \sin \tfrac{1}{2}qa = S \sin \tfrac{1}{2}\theta, \tag{2.63}$$

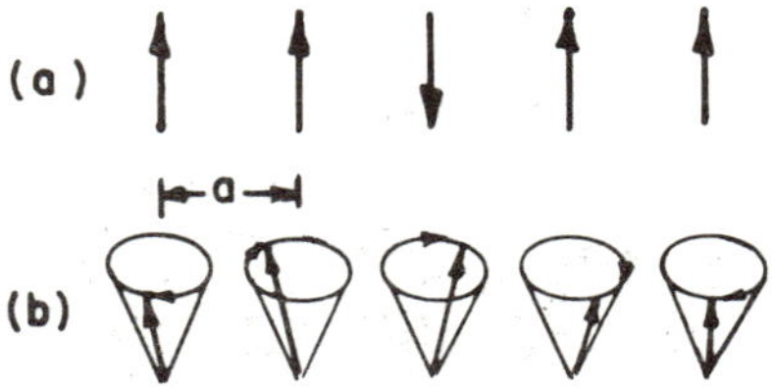

Fig. 2.9. Two possible excitations of spins in a simple ferromagnet: (a) One spin is reversed and (b) each spin undergoes an elementary excitation to form spin waves.

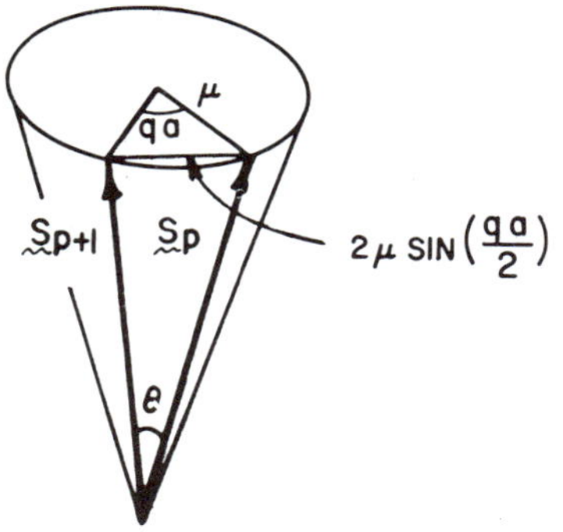

Fig. 2.10. Diagram showing the angle θ betwen two successive spin vectors relating to the spin wave magnitude μ and the phase angle qa.

where θ is the angle between two successive spin vectors. A spin wave in a linear chain is shown in fig. 2.11, of which (a) delineates the change in position of the precessing spins at a given time in perspective and (b) shows the spins as viewed from above. A wave emerges in one wave length in fig. 2.11b, when the ends of the spin vectors are connected.

From fig. 2.10, the z component of a spin is given by

$$S_z = (S^2 - u^2)^{\frac{1}{2}} \simeq S - u^2/2S, \tag{2.64}$$

when $u/S \ll 1$. Quantum theory allows only integral values for $S - S_z$. Hence spin waves are quantized by a quantum number n_q for each wave-vector $\boldsymbol{q}$. From (2.64), we obtain the semiclassical quantization condition

$$n_q \simeq Nn_q^2/2S, \tag{2.65}$$

where n_q must be an integer $(1, 2, 3, \ldots)$. In our system, $NS - n_q$ gives the total spin quantum number when a spin wave of wave vector $\boldsymbol{q}$ is excited. Quantized spin waves are called magnons to reflect their analogy with photons and phonons. Because a spin wave may be treated as a particle associated with a momentum $\boldsymbol{p}_q$ and an energy E_q $(= p_q^2/2m^*$, where m^* is an effective mass) and also because a wave for a particular vibrational mode of wave vector $\boldsymbol{q}$ is the same as that of a simple harmonic oscillator of the same frequency, we have by invoking the de Broglie relation,

$$E_q = n_q \hbar\omega_q = n_q \hbar^2 q^2/2m^*. \tag{2.66}$$

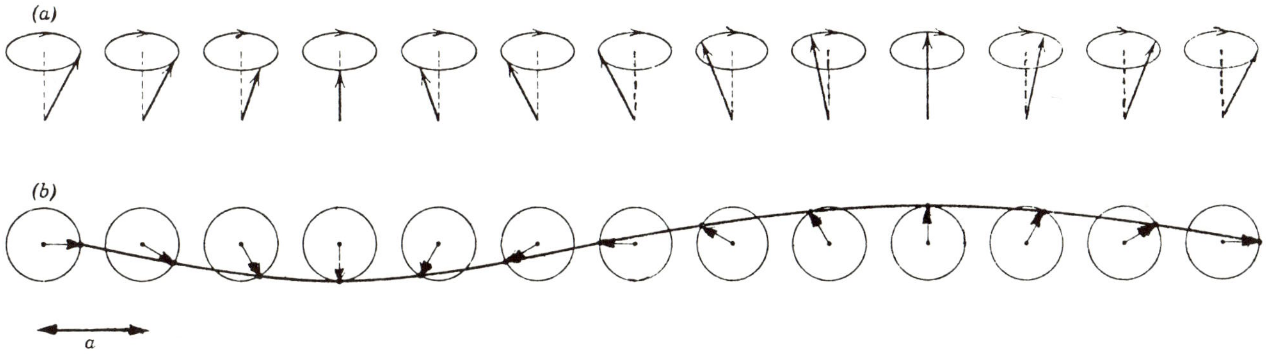

Fig. 2.11. A spin wave on a line of spins. The spins are viewed (a) in perspective and (b) from above, showing one wavelength. The wave is drawn through the ends of the spin vectors (Morrish [1965]).

For a typical ferromagnet with a Curie temperature of a few hundred degrees K, m^* is about 10 times the mass of a free electron.

An important result of the quantum theory of spin waves is that the energy in the $\boldsymbol{q}$th spin-wave mode is quantized in units

$$\hbar\omega_q = \sum_{\boldsymbol{r}} 2SJ_{\text{ex}}(\boldsymbol{r})(1 - e^{-i\boldsymbol{q}\cdot\boldsymbol{r}}), \tag{2.67}$$

where $\boldsymbol{r} = \boldsymbol{r}_i - \boldsymbol{r}_j$ is the position vector between two nearest neighbors and the sum is over the z nearest neighbors. For small values of q, since $|\boldsymbol{q}\cdot\boldsymbol{r}| \ll 1$ and $e^{i\boldsymbol{q}\cdot\boldsymbol{r}} \simeq \cos(\boldsymbol{q}\cdot\boldsymbol{r})$, (2.67) reduces to

$$\hbar\omega_q = 2SJ_{\text{ex}}\left[z - \sum_{\boldsymbol{r}} \cos(\boldsymbol{q}\cdot\boldsymbol{r})\right]. \tag{2.68}$$

For all three cubic lattices (sc, bcc and fcc), the leading terms in the expansion of (2.68) give

$$\hbar\omega_q \simeq 2SJ_{\text{ex}}q^2a^2. \tag{2.69}$$

This is known as the dispersion relation for spin waves with long wavelengths ($qr \ll 1$). Neutron inelastic scattering experiments conducted on a fcc Co alloy with 8% Fe by Sinclair and Brockhouse [1960] gave a partial dispersion curve shown in fig. 2.12. An excellent agreement is seen between the experimental results and the theoretical prediction. Their results are considered as strong evidence for the existence of spin waves in a ferromagnet.

An interesting aspect of the magnons is their thermal excitation, with which saturation magnetization of the system varies as a function of temperature. Because the magnons obey Bose–Einstein statistics, the average value of the quantum number n_q in thermal equilibrium at the temperature T is given by

$$\bar{n}_q = [\exp(\hbar\omega_q/kT) - 1]^{-1}. \tag{2.70}$$

The total number of magnons in all modes excited at this temperature is

$$\begin{aligned}\bar{n}_T = \sum_q \bar{n}_q &= \frac{1}{8\pi^3}\int_0^\infty \frac{4\pi q^2\,\mathrm{d}q}{\exp(2SJ_{\text{ex}}q^2a^2/kT) - 1}\\ &= \frac{1}{4\pi^2}\left(\frac{kT}{2SJ_{\text{ex}}a^2}\right)^{\frac{3}{2}}\int_0^\infty \frac{x^{\frac{1}{2}}\,\mathrm{d}x}{e^x - 1},\end{aligned} \tag{2.71}$$

where $x = E/kT$. The integral has the value $(0.0587)4\pi^2$, thus $\bar{n}_T = 0.0587(kT/2SJ_{\text{ex}}a^2)^{\frac{3}{2}}$, which represents the number of spins reduced

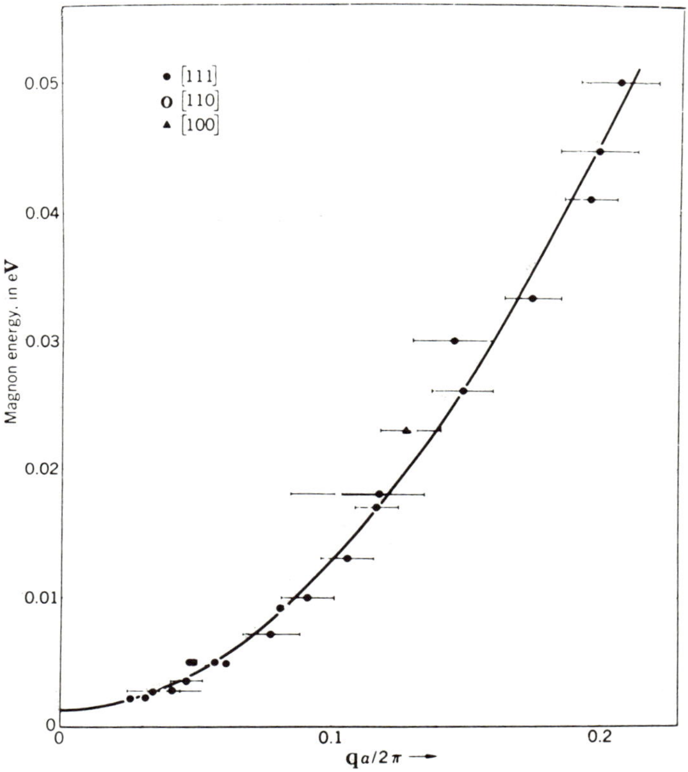

Fig. 2.12. The energy spectrum of magnons in a fcc alloy ($Co_{0.92}Fe_{0.08}$) at room temperature (Sinclair and Brockhouse [1960]). The solid line is the theoretical curve adjusted to best fit the data points.

from NS as a result of thermal excitation of magnons at T. Hence $\bar{n}_T/NS$ represents the fractional change in magnetization, $M_s(T)/M_o$, where M_o is the saturation magnetization at 0°K and

$$M = M_o\left[1 - \frac{0.0587}{\alpha S}\left(\frac{kT}{2J_{ex}S}\right)^{\frac{3}{2}}\right], \tag{2.72}$$

where $\alpha = 1$, 2 and 4 for sc, bcc and fcc lattices, and it is introduced for the fact that $N = \alpha/a^3$. Equation (2.71) is known as the Bloch $T^{\frac{3}{2}}$ law. The theory of spin waves provides a natural explanation of the observation on the temperature dependence of saturation magnetization of a ferromagnet expressed by the empirical formula,

$$\Delta M_s(T)/M_0 = \beta T^{\frac{3}{2}}, \tag{2.73}$$

where $\beta = (7.5 \pm 0.2) \times 10^{-6} \deg^{-\frac{3}{2}}$ for Ni deduced from the experimental data shown in fig. 2.13, and $\beta = (3.4 \pm 0.2) \times 10^{-6} \deg^{-\frac{3}{2}}$ for Fe. Substituting these values for β and $S = 1$ in (2.72) gives $J_{ex} = 230\, k$ for Ni and 205 k for Fe.

Spin waves in ferromagnets have several other interesting properties. One is their contribution to the specific heat. Another is the interaction

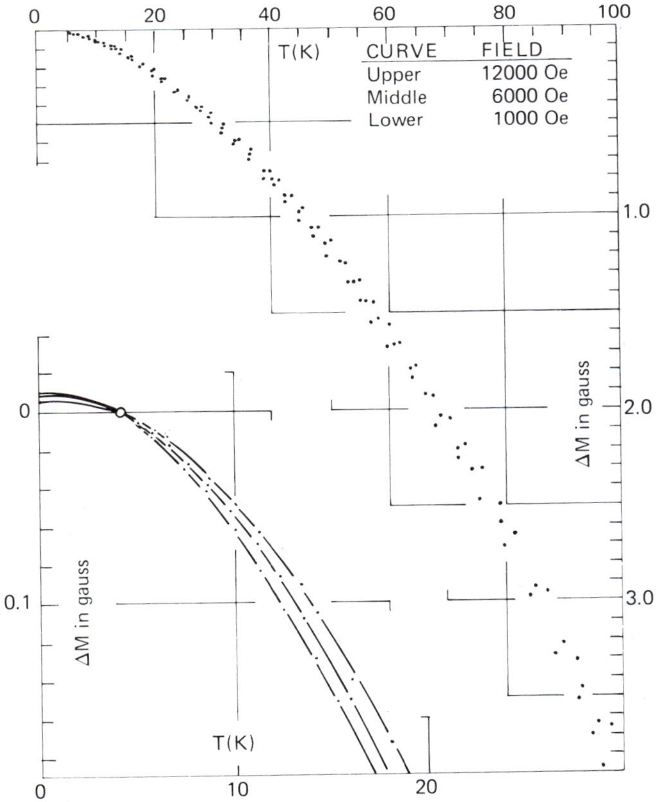

Fig. 2.13. Experimental results showing the decrease in magnetization of nickel with temperature above 4.2°K (Argyle et al. [1963]).

between magnons and phonons, which can give rise to inelastic scattering of neutrons (or photons) leading to the results shown in fig. 2.12. Spin waves also play an important role in resonance phenomena in ferrites and other ferrimagnetic systems at microwave frequencies.

§3. FERRIMAGNETISM

Thus far we have considered only simple magnetic systems in which the magnetic atoms or ions are of the same species. Such systems can be either ferromagnetic with a nonzero magnetization or antiferromagnetic with a zero magnetization, depending on whether the exchange integral J_{ex} is positive or negative. There is a group of complex systems, however, that features

(1) two or more species of magnetic ions,
(2) ordering of all constituent ions,
(3) a mixed scheme of ferromagnetic and antiferromagnetic spin coupling of the magnetic ions, and
(4) a net magnetization.

These materials are known as ferrimagnets, and their magnetic behavior is referred to as ferrimagnetism. Because they have antiferromagnetic spin coupling among some of the magnetic ions, the ferrimagnets resemble antiferromagnets; otherwise, they are more like ferromagnets by virtue of their net magnetization. Many ferrimagnets offer attractive properties for magnetic applications, especially at high frequencies. A more detailed description of the material characterization of ferrimagnets will be given in ch. 5, §4. Here, we shall only consider their fundamental magnetic characteristics.

A well-known example of ferrimagnets is the ancient magic stone, or the magnetite. Its chemical composition is Fe_3O_4, which implies the presence of two magnetic ions, Fe^{2+} and Fe^{3+}, in the compound. The ferric (Fe^{3+}) ions are in a state with zero orbital moment and spin $\frac{5}{2}$, whereas the ferrous (Fe^{2+}) ions are in a state with $S = \frac{4}{2}$ and a nonzero orbital moment. On the basis of spins only, each Fe^{2+} ion carries a moment of $4\,\mu_B$ and each Fe^{3+} ion carries $5\,\mu_B$. If the spins of the ferrous ions are coupled ferromagnetically with those of the ferric ions, obviously, based on the approximate formula $Fe_3O_4 = Fe^{II}Fe_2^{III}O_4$, the effective Bohr magneton number per formula unit of Fe_3O_4 would be $n_{eff}/\text{f.u.} = (2 \times 5) + 4 = 14$. Alternatively, if the spins are coupled antiferromagnetically, $n_{eff}/\text{f.u.} = (2 \times 5) - 4 = 6$. The observed value is

4.1 μ_B per formula unit, implying that neither of the simple coupling schemes is true. The observed n_{eff} value necessitates a configuration in which the spins of the Fe^{3+} ions are evenly split and coupled antiferromagnetically, whereas the spins of the Fe^{2+} ions are coupled ferromagnetically, as shown in fig. 2.14. The ferrous ions are the sole contributors to the net magnetization. Such a complex spin configuration is possible because the Fe^{3+} ions are divided into two equal parts, each occupying the tetrahedral and half of the octahedral sites in the so-called inverse spinel lattice. Neutron diffraction experiments have confirmed the complex spin alignment in the compound. A detailed description of the spin configuration consistent with the crystal structure of Fe_3O_4 is given in ch. 5, §1.3.1.

A large number of ferrimagnets are known as ferrites, which are expressed by the chemical formula $MO{\cdot}Fe_2O_3$, where M is a divalent cation of Zn, Cd, Fe, Ni, Cu or Mg. The crystal structure of ferrites having the normal spinel structure comprises two types of lattice sites, tetrahedral A and octahedral B sites, in which the occupants are M^{2+} and Fe^{3+} ions, respectively. An interesting feature of these ferrites is

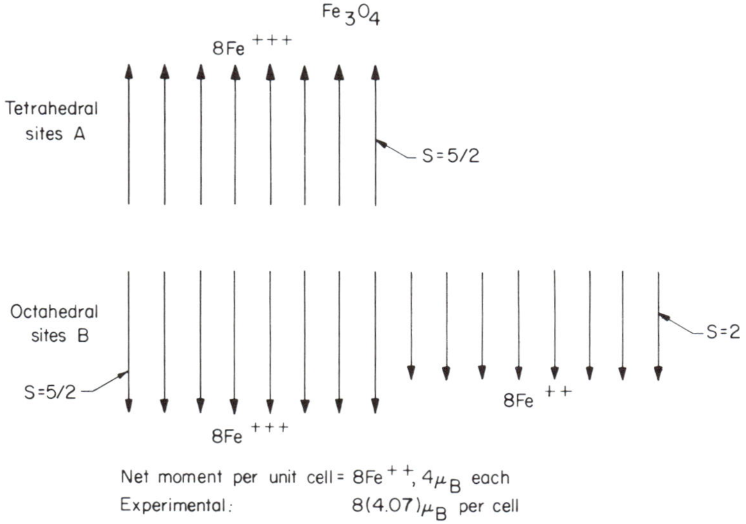

Fig. 2.14. Schematic diagram showing the alignment of spins associated with the iron ions in a unit cell of magnetite (Kittel [1971]).

that, although the spins within each sublattice are aligned parallel to each other, the calculated exchange integrals, J_{AA} and J_{BB}, are negative, not positive. This situation seems contradictory to the Heisenberg theory and is thought possible because J_{AB} is also negative and, moreover, $J_{AB} \gg J_{AA}$ and J_{BB}. Consequently, the weak antiferromagnetic intrasublattice exchange interactions are overruled by the much stronger intersublattice interaction. The spin alignment within each sublattice is thus parallel, opposite to what the sign of the exchange integral would indicate in the Heisenberg theory.

The susceptibility of ferrimagnets is readily formulated if we assign separate Curie constants, C'_A and C'_B, to the two sublattices. Since J_{AA} and J_{BB} are generally much weaker than J_{AB}, we may neglect the intrasublattice interaction. Analogous to the derivation of eqs. (2.21) and (2.23a), we have for the ferrimagnets:

Sublattice A		Sublattice B	
$H_{W,A} = -\lambda M_B,$	(2.74a)	$H_{W,B} = -\lambda M_A;$	(2.74b)
$M_A T = C'_A(H - \lambda M_B),$	(2.75a)	$M_B T = C'_B(H - \lambda M_A).$	(2.75b)

The last two equations will have a nonzero solution for M_A and M_B in the absence of an applied field if

$$\begin{vmatrix} T & \lambda C'_A \\ \lambda C'_B & T \end{vmatrix} = 0 \tag{2.76}$$

from which the ferrimagnetic characteristic temperature is given by

$$\theta_F = \lambda (C'_A C'_B)^{\frac{1}{2}}. \tag{2.77}$$

Solving (2.75a) and (2.75b) for M_A and M_B, we obtain an expression for the ferrimagnetic susceptibility at temperatures above the ferrimagnetic Néel temperature θ_F

$$\chi_F = \frac{M_A + M_B}{H} = \frac{(C'_A + C'_B)T - 2\lambda C'_A C'_B}{T^2 - \theta_F^2}, \tag{2.78}$$

which is more complicated than the Curie–Weiss law expressed in (2.24) for ferromagnets at $T > \theta_C$. The plot of $1/\chi_F$ against T is shown in fig. 2.15. It should be compared with similar plots for paramagnetic and ferromagnetic susceptibilities in fig. 2.5. Note that the reciprocal

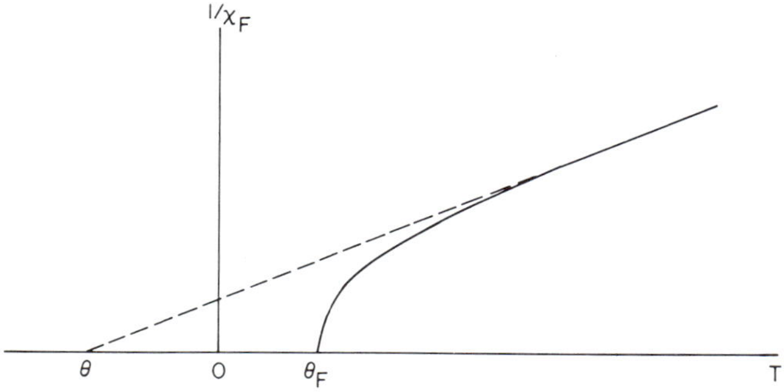

Fig. 2.15. Temperature dependence of the inverse magnetic susceptibility in ferrimagnets.

ferrimagnetic susceptibility shows considerable curvature as the temperature approaches θ_F and becomes infinite at the characteristic temperature.

Although antiferromagnetism per se does not concern us here, the analysis on the spin structure of ferrites involves antiferromagnetic configurations. Therefore it seems necessary to discuss briefly the basic features of antiferromagnetism. A simple antiferromagnet could be a pure metal (Cr, α-Mn and some light lanthanide metals such as Sm), a compound (FeO, CoO, NiO, MnS, etc.), or an alloy (Fe_3Mn, CrPt, ordered Pt_3Fe, etc.), in which the spins are ordered in an antiparallel fashion with zero net moment at temperatures below the ordering point, known as the Néel temperature, θ_N. Susceptibility of an antiferromagnet can be formulated easily by treating the material as a special case of a ferrimagnet for which both sublattices A and B have equal saturation magnetization. Thus $C'_A = C'_B = C'$, $\theta_N = \lambda C'$ from (2.77) and by setting $T - \lambda C' = T - \theta_N$ in (2.78),

$$\chi_{AF} = \frac{2C'T - 2\lambda C'^2}{T^2 - \theta_N^2} = \frac{2C'}{T + \theta_N}. \tag{2.79}$$

At $T = \theta_N$, χ_{AF} is not infinite as in the case of ferromagnets (fig. 2.5). The temperature dependence of χ_{AF} is shown in fig. 2.16. The χ_{AF} versus T plot displays a cusp, whereas the linear plot of $1/\chi_{AF}$ versus T intercepts, by extrapolation, the temperature axis at a negative value as in the case of ferrimagnets (fig. 2.15).

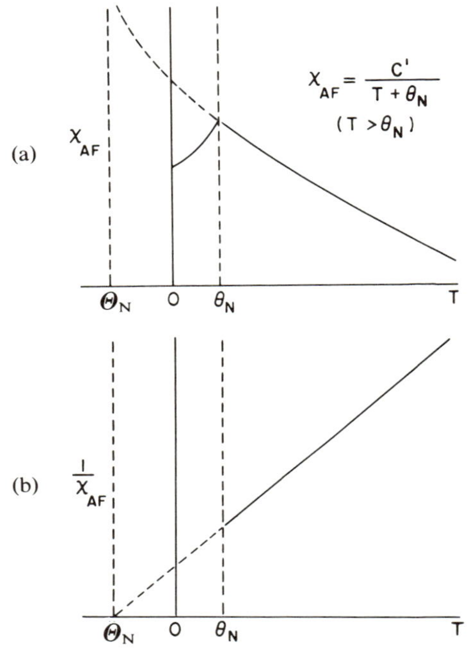

Fig. 2.16. Temperature dependence of (a) the magnetic susceptibility and (b) the inverse magnetic susceptibility in antiferromagnets (Kittel [1976]).

Experimental results for many antiferromagnetic compounds indeed show a linear relationship in the $1/\chi$ versus T plot in conformity with (2.79). But the negative intercept Θ_N differs substantially from θ_N in practically all compounds. In other words, the value for the ratio Θ_N/θ_N always deviates from unity and falls in a range between 0.4 for $FeCO_3$ and 5.3 for MnO. In analogy with the ferromagnetism, Θ_N is called the paramagnetic Néel temperature. The discrepancy between θ_N and Θ_N, however, can be resolved when interactions between magnetic ions in the next nearest neighbors are taken into account. The effect of the latter interactions is represented by a second molecular field constant λ'. Accordingly, as Néel [1948a] and Luttinger [1951] have shown, (2.74a) and (2.74b) are expanded into the form

$$H_{W,A} = -\lambda M_B - \lambda' M_A; \qquad H_{W,B} = -\lambda M_A - \lambda' M_B, \tag{2.80}$$

and consequently,

$$\frac{\Theta_N}{\theta_N} = \frac{(\lambda + \lambda')}{(\lambda - \lambda')}. \tag{2.81}$$

In this way, Θ_N/θ_N is not expected to be unity unless λ' is negligibly small.

Another interesting aspect of the antiferromagnetism observed in compounds is the mechanism leading to the antiparallel alignment of spins. In the previous analysis on magnetic susceptibilities of ferrimagnets and antiferromagnets, we have assumed that the Weiss molecular field theory applies. For the present compounds, we may further assume that the Heisenberg theory and, therefore, (2.39) also apply except here $J_{ex} < 0$; hence (2.39) becomes

$$E_{ex} = |J_{ex}| \sum_i \sum_j \boldsymbol{S}_i \cdot \boldsymbol{S}_j. \tag{2.82}$$

However, the origin of the antiferromagnetic spin coupling in compounds such as MO, where M = Mn, Fe, Co or Ni, cannot be the same as that envisaged for the ferromagnetic metals because the bonding in the oxides and many similar compounds (MS, MSe, etc.) is ionic in nature, definitely not metallic. Thus the 3d electrons would not form energy bands, nor would the polarization of the conduction electrons occur in these materials. The negative J_{ex} must then arise from a different mechanism.

The new mechanism is known as the superexchange interaction, which is best described by reference to the crystal and magnetic structures of MnO. This simple oxide has the NaCl crystal structure, wherein the Mn^{2+} ions form a face-centered cubic lattice with a = 4.4345 Å at room temperature (Nagamiya et al. [1955]). Neutron scattering experiments (Shull et al. [1951]) determined the magnetic structure with the following features: (1) The magnetic unit cell is eight times the chemical unit cell. (2) The magnetic moments are localized on the Mn^{2+} ions. And (3) the spins are aligned in the (111) planes to form a layer structure. Fig. 2.17 shows the spin arrangement, in which all moments are coupled ferromagnetically in the same (111) plane, but the spin direction alternates between [001] and [00$\bar{1}$] in successive (111) layers.

If the effect of the oxygen anions is neglected, we may make the following observations from fig. 2.17 on the mode of spin coupling in

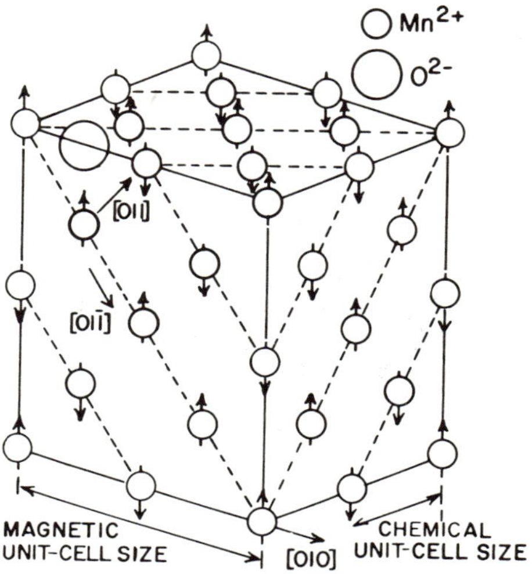

Fig. 2.17. Ordered arrangement of spins of the Mn^{2+} ions in antiferromagnetic MnO, as determined by neutron diffraction. The O^{2-} ions are not shown (Shull et al. [1951]).

three important directions:

Crystal direction	Interaction distance D (Å)	Spin coupling mode
[010]	4.435	Antiferromagnetic
[01$\bar{1}$]	3.136	Ferromagnetic
[011]	3.136	Antiferromagnetic

These observations clearly indicate the invalidity of the Bethe–Slater (B&S) criterion for the determination of the sign of the exchange integral J_{ex} in the Heisenberg theory in this type of material. This is because the B&S criterion is based on the ratio D/d, where d is the diameter of the 3d subshell of the magnetic ions. Now assuming $d = 1.71$ Å (table 5.1) for the Mn ions, the ratio has values of 1.83 in [01$\bar{1}$] and [011] and 2.59 in [010]. According to the criterion, these values imply positive exchange interactions in all three directions. Such implications are obviously erroneous in view of the observed modes of spin coupling listed.

The foregoing discussion points to the important fact that the presence of oxygen ions plays a dominant role in spin coupling in MO and similar compounds, leading to the proposal of the superexchange interaction by Kramer [1934] and later analysis by Anderson [1950, 1963]. The interaction invokes the transfer of electrons from the oxygen ions to neighboring metal ions, thereby permitting the cations to interact indirectly via the intervening anions. To see how the superexchange comes about, let us examine the ground state and a possible excited state of a Mn^{2+}–O^{2-}–Mn^{2+} complex in the [010] direction shown in fig. 2.18. In the ground state, the electronic configuration of the Mn^{2+} and O^{2-} ions are $(3d)^5$ and $(2p)^6$, respectively. The six p electrons of the oxygen ion are coupled in three pairs, each pair forming a ∞-shaped orbital, as shown in fig. 2.18a. Having a full 2p subshell, the oxygen ion cannot interact with the iron ions on either side; nor would the interaction between two Mn^{2+} ions be significant because of the great distance. There is, however, a certain probability of transferring one of the p electrons of oxygen to a Mn^{2+} ion. Specifically, by virtue of the dumbbell shape of the p-orbitals, it is reasonable to identify the p_x' electron as the one being transferred, as shown in fig. 2.18b. Now the complex is in an excited state in which momentarily the left Mn^{2+} ion becomes Mn^{1+} with $(3d)^6$ and the O^{2-} ion becomes O^{1-} with $(2p)^5$. Now the unpaired p_x electron on the O^{1-} ion can interact with the Mn^{2+} ion on the right on account of the overlap of the p_x-orbital with the d-orbitals of the Mn^{2+} ion. The latter interaction leads to an indirect interaction between the two Mn ions through an intricate mechanism

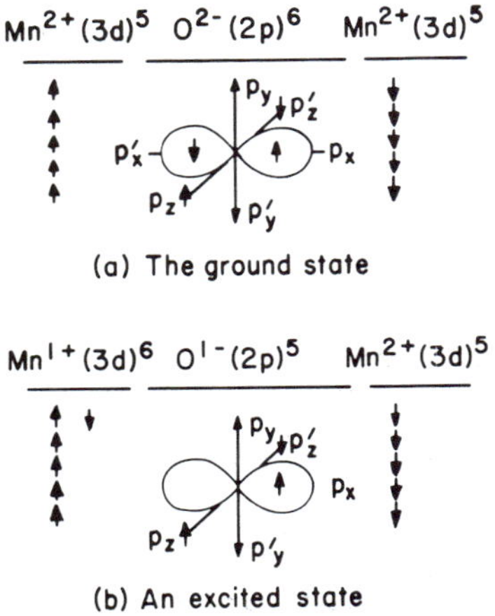

Fig. 2.18. Illustrating the superexchange interaction in MnO. (a) The ground state and (b) an excited state.

thus described, which Kramer called a superexchange interaction.

Anderson's calculations indicate a negative interaction of the right order of magnitude in MnO and other MO compounds between the M ions in the M–O–M complexes. The theory further shows that if the 3d subshells of the metal ions are less than half full, the superexchange should favor a positive J_{ex}, whereas for 3d subshells half filled or over, a negative J_{ex} is expected. Meanwhile, the strength of the superexchange interaction depends on the distance between ions in the M–O–M complex as well as the angular positions of the M-ions with respect to the O–ion. Because of the dumbbell shape of the p-orbitals, it is easy to see that the interaction is the strongest when ions in the M–O–M complex lie collinearly and very weak when the complex assumes a $\mathrm{M{-}}\underset{\substack{|\\ \mathrm{M}}}{\mathrm{O}}$ configuration. The angular dependence of the superexchange interaction accounts very well for the occurrence of antiferromagnetism in MnO. The collinear M–O–M complexes in the [010] direction produce the dominant, antiparallel spin coupling. While the superexchange between the Mn ions in the $[01\bar{1}]$ direction is so greatly weakened by the 90° position of the two Mn^{2+} ions with respect to the intervening O-ion that direct exchange interaction between the Mn^{2+} ions dominates to yield a net ferromagnetic coupling in that direction.

The superexchange interaction not only gives a satisfactory account for the antiferromagnetism in MO and many other compounds, but also inspired other workers to propose similar mechanisms, such as the double exchange (Zener [1951]. Its significance in connection with the ferrimagnetic materials lies in the fact that many ferrites owe their magnetic properties to the superexchange interactions, which occur in more than one form of the M–O–M complexes at varying distances and angles (Gorter [1955]). The applicability of the Kramer–Anderson theory to different ferrites has been examined by several workers, notably Gorter [1954–5] and Keffer [1966].

Suggestions for further reading

Brailsford, F., 1966, *Physical Principles of Magnetism* (van Nostrand, London). Chapters 2 and 5 on §1.

Morrish, A. H., 1965, *The Physical Principles of Magnetism* (Wiley, New York). Chapters 6 and 9 on ferromagnetism and ferrimagnetism.

Standley, J. K., 1972, *Oxide Magnetic Materials*, 2nd ed. (Oxford University Press, London). Chapters 4 and 5 on ferrimagnetism.

Marshall, W., ed., 1967, *Theory of Magnetism in Transition Metals* (Academic Press, New York).

CHAPTER 3

MAGNETIZATION AND DOMAIN STRUCTURE

§1. THE MAGNETIZATION CURVE

The behavior of a ferromagnet in a magnetic field is considered the primary factor in the practical evaluation of the material. For this evaluation, the magnetization M or the induction B is plotted against the magnetizing field H. Such magnetization curves are not only useful for technical purposes, but also indispensable in elucidating the process of magnetization under different conditions.

A typical B–H magnetization curve is shown in fig. 3.1 for 3% Si–Fe that is polycrystalline, in sheet form, well annealed, grain-oriented and commercially pure (>99.5%). With the sample initially unmagnetized, the curve will start out from the origin with a finite slope, which approaches the initial permeability at zero field, $\mu_i \simeq (\mathrm{d}B/\mathrm{d}H)_{H\to 0}$. The curve is usually divided into three parts. The first part extends from the toe (O) to the instep (A) and is the shortest of the three. The shape of the first part often follows the Rayleigh relation

$$B = \mu_i H + \nu H^2, \tag{3.1}$$

where $\nu = \mathrm{d}\mu/\mathrm{d}H$ and has a constant value. This part is reversible in the sense that the segment of the curve is retraceable upon the reduction of the field. The second part covers the steepest segment of the curve between the instep and the knee (B). The slope rises very rapidly to a maximum value, near which the maximum permeability μ_{max} is reached; then drops off gradually. The third part lies above the knee and ultimately reaches the saturation value of induction B_s, where $B_s = \mu_0(H + M_s)$. Both the second and third parts are irreversible, or more appropriately irretraceable.

Upon close examination, the second part is seen to consist of many steps, as shown in the inset of fig. 3.1. In a noted magneto-audio experiment, Barkhausen [1919] demonstrated that each step consists of an irreversible jump in induction from point a to b, c to d, etc., and a reversible, much slower rise in induction from point b to c, d to e, etc.

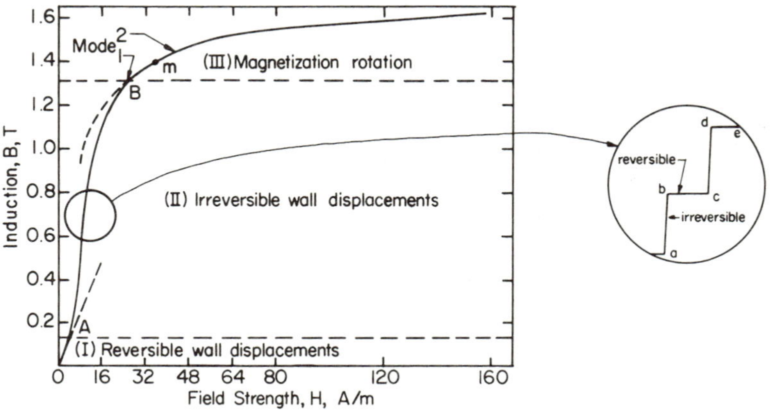

Fig. 3.1. Magnetization curve for 3% Si–Fe at 27°C. Inset shows the Barkhausen effect in the second part of the curve (Chen [1958]).

The physical meaning of the Barkhausen effect will be discussed in §2.

If the magnetization curve is stopped at m in fig. 3.1 and the field is slowly reduced to zero, the curve is not retraced; instead, the induction decreases from B_m at slower rates, and a residual induction B_r remains at $H = 0$, as shown in fig. 3.2. To remove the remanent induction B_r, a reverse field $-H_c$ must be applied. When the reverse field is continually increased until $-B_m$ is attained, the curve between B_m and $-B_m$ constitutes half of a symmetric loop known as the hysteresis loop. Fig. 3.2 shows a stretched elliptical loop. The shape and size of the hysteresis loop vary widely according to the composition and metallurgical conditions of the material. A detailed discussion of the loop variation will be given in ch. 4, §2.1.3. At present, it suffices to state that a hysteresis loop is characterized by B_r, the remanence, and by H_c, the coercive force. These quantities together with μ_i, μ_{max}, and B_s measure the applicability of a ferromagnetic material.

Like the hysteresis loop, the magnetization curve can have different shapes according to (1) whether the specimen is a single crystal or polycrystalline, (2) the predominant orientation of the crystals and (3) other factors, such as the state of stress, temperature, etc. Shown in figs. 3.3–3.5 are the M–H curves for single crystal specimens of Fe and Ni with the magnetizing field applied in the [100], [110] and [111] directions and curves for Co single crystals with the field applied in the

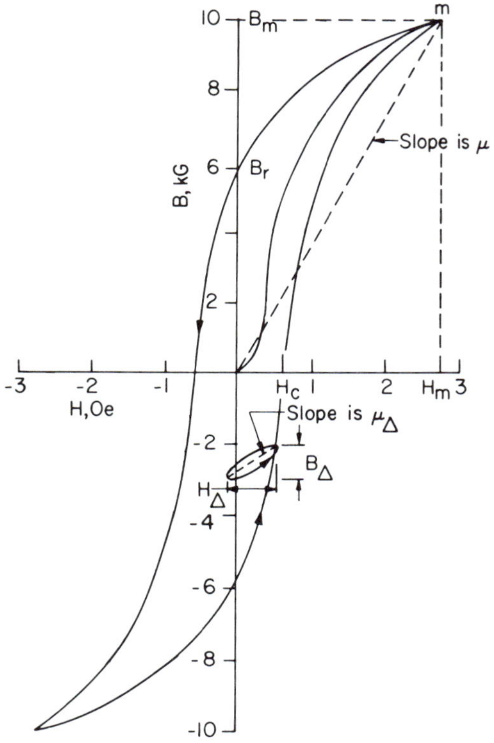

Fig. 3.2. Hysteresis loop for 3% Si–Fe featuring the remanence, B_r, coercive force, H_c and a minor loop (Carr [1959]).

[0001]* and [*uvt*0]* directions. Note the vast differences in the shape of the magnetization curves shown. For each metal, there are crystal directions in which the magnetization is attained at all levels with ease, or more specifically, in which both the knee and μ_{max} emerge at the

*Directions in the hexagonal crystals can be expressed by [*uvtw*], where *u*, *v*, *t* and *w* are integers, including zero, corresponding to the components of the vector representing the direction; that is

$$\boldsymbol{r}_{uvtw} = u\boldsymbol{a}_1 + v\boldsymbol{a}_2 + t\boldsymbol{a}_3 + w\boldsymbol{c}.$$

where $\boldsymbol{a}_1$, $\boldsymbol{a}_2$, $\boldsymbol{a}_3$ and $\boldsymbol{c}$ are the unit vectors along the four axes. Thus the [0001] direction is along the *c*-axis, and the [*uvt*0] directions are in the basal plane.

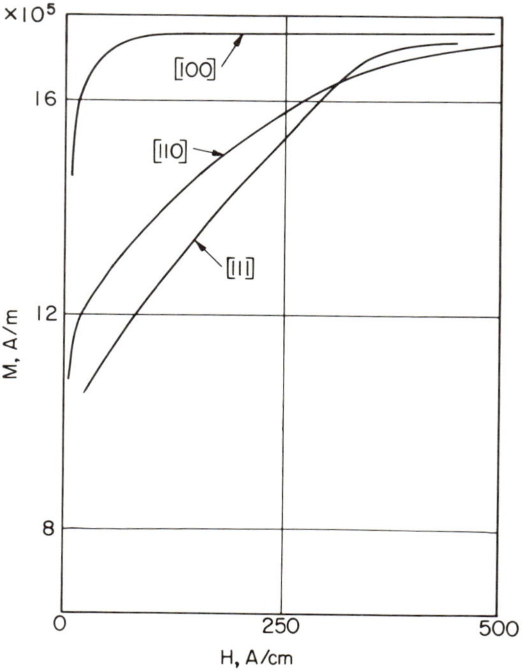

Fig. 3.3. Magnetization curves for single crystals of iron in the [100] easy, [110] intermediate and [111] hard directions, based on experimental results of K. Honda and S. Kaya, Sci. Rep. Tohoku Univ. **15**, 721 (1926).

lowest field strengths. Such directions are called the axes or directions of easy magnetization, or simply the easy axes or directions. For Fe, Ni and Co, the easy directions are ⟨100⟩*, ⟨111⟩* and [0001], respectively.

As might be expected, there are directions of hard magnetization, or simply the hard directions, in which the process of magnetization encounters the greatest difficulty. The hard directions for Fe, Ni and Co are ⟨111⟩, ⟨100⟩ and ⟨*uvt*0⟩, respectively. This phenomenon of showing

*Square brackets [ ] signify a specific direction. Thus, [*uvw*] represents the direction of the vector $\boldsymbol{r}_{uvw} = u\boldsymbol{a}_1 + v\boldsymbol{a}_2 + w\boldsymbol{a}_3$. Carets ⟨ ⟩ indicate a family of crystallographically equivalent directions. For example, $\langle 100\rangle = [100] + [\bar{1}00] + [010] + [0\bar{1}0] + [001] + [00\bar{1}]$ and $\langle 111\rangle = [111] + [\bar{1}\bar{1}\bar{1}] + [\bar{1}11] + [1\bar{1}\bar{1}] + [1\bar{1}1] + [\bar{1}1\bar{1}] + [\bar{1}\bar{1}1] + [11\bar{1}]$.

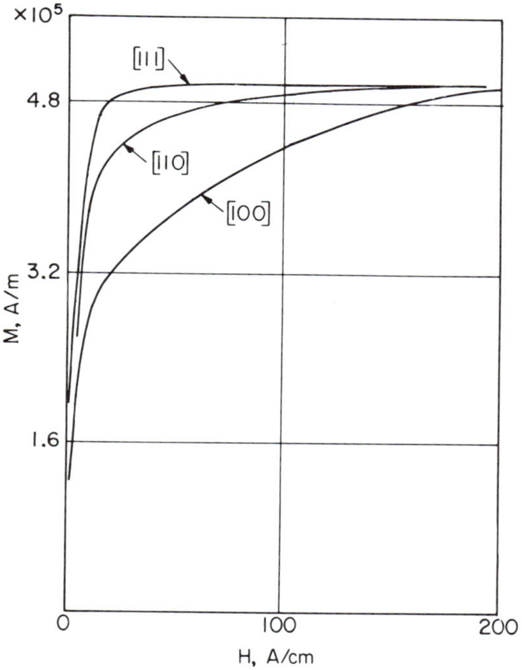

Fig. 3.4. Magnetization curves for single crystals of nickel in the [111] easy, [110] intermediate and [100] hard directions based on experimental data of S. Kaya, Sci. Rep. Tohoku Univ. **17**, 639 (1928).

crystallographic preference for magnetization is known as the magnetocrystalline, or simply magnetic, anisotropy, which will be discussed in §3.

§2. FERROMAGNETIC DOMAINS AND MAGNETIZATION PROCESSES

It was cited in ch. 2, §2 that ferromagnets are characterized by spontaneous magnetization; that is, they attain saturation magnetization in each of the domains even though no field is applied. The magnetization curve, on the other hand, asserts that an unmagnetized ferromagnetic sample displays no overall magnetization in zero field. Moreover, it takes an appreciable field to produce magnetization close to the saturation value. Would this imply a serious discrepancy between the theory and the observation?

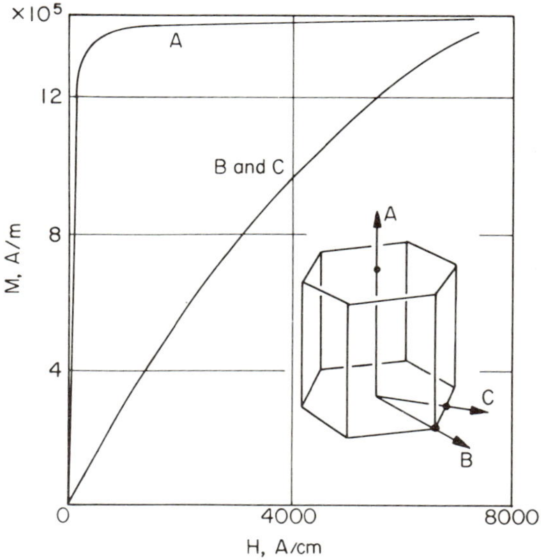

Fig. 3.5. Magnetization curves for single crystals of cobalt in the [0001] easy and two hard directions based on experimental data of S. Kaya, Sci. Rep. Tohoku Univ. **17**, 1157 (1928).

To reconcile this seeming discrepancy, Pierre Weiss [1929] proposed a magnificent theory, which may be stated simply as follows: A ferromagnet of macroscopic size contains numerous regions called magnetic domains in the demagnetized state. Within each domain, all the atomic moments are aligned in one of the easy directions; therefore, spontaneous magnetization does exist. The direction of spontaneous magnetization, however, varies from domain to domain so as to minimize the magnetostatic energy. On a purely statistical basis, all available easy directions will be used equally in the material. For instance, if there are n domains of approximately equal volume in a demagnetized iron specimen, the number of domains spontaneously magnetized in each of the six $\langle 100 \rangle$ easy directions will be $n/6$. Hence the specimen as a whole will not show a net magnetization in the absence of an applied field.

In terms of Weiss's domain theory, the process of magnetization is readily described. Consider a single crystal of iron in the form of a square sheet with two (001) faces (fig. 3.6). In the demagnetized state, the crystal is supposed to have a domain structure like that shown in fig. 3.6a. There are four triangular domains of equal volume, of which two are spontaneously magnetized in the ±[100] directions and the other two magnetized in the ±[010] directions. (Domains magnetized in the ±[001] directions are not expected because the crystal is essentially two dimensional with most of its free surfaces lying in the (001) plane.) Since the magnetization in every direction is counter-balanced, the crystal would not exhibit a net magnetization. Now let a magnetic field be applied at a moderate angle, say 10°, from [010], as shown in fig. 3.6b. Because this field direction is closer to [010] than any other ⟨100⟩ directions, the [010] domain will expand at the expense of the other three domains. Fig. 3.6b shows that the expansion is effected by the

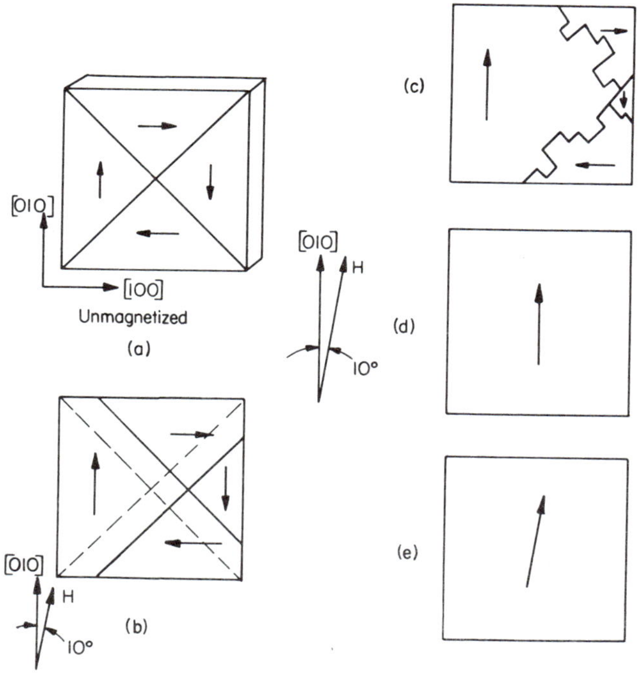

Fig. 3.6. Domain structure of a single crystal slab of iron (a) in the unmagnetized state; (b) after reversible movements of domain walls; (c) after irreversible movements of domain walls; (d) after all magnetization vectors have rotated into [010] and (e) during the rotation of magnetization vector from [010] to the field direction.

movement of the boundaries between domains. In the early stage, the movement of domain boundaries usually is reversible; that is, the boundaries will tend to move back toward the original positions when the field is reduced.

Crystals always contain imperfections, such as impurity atoms, dislocations, etc., which may act as barriers to the movement of domain boundaries. Thus, in the later stage, the movement of domain boundaries becomes nonuniform (see fig. 3.6c) and jerky, because only those segments of boundaries that have acquired enough energy from the field can overcome the barriers and continue the movement until new barriers are encountered. The movements of domain walls become irreversible after the initial stage, and the jerky style of the movements gives rise to the Barkhausen effect.

As the field is increased still further the atomic spins begin to rotate in two modes. One is the sudden 90° turning or 180° reversal of the spins in the ±[100] or [010] domains into the [010] direction (fig. 3.6d). This mode of rotation will eliminate all but the [010] domains. Consequently the entire crystal will have only one domain because all spins are now aligned in the [010] direction. The second mode of rotation involves the gradual turning of the spins from [010] to the direction of the field. The crystal eventually attains saturation magnetization when all spins have rotated into the field direction, as shown in fig. 3.6e. The spin rotation in either mode is reversible.

Several methods have been developed to observe domain structures. The oldest and the most popular method was devised by Bitter [1931], who used fine magnetite powder in suspension to trace the domain boundaries. (Readers interested in the experimental techniques for revealing domains should consult another book of this series, *Ferromagnetism and Ferromagnetic Domains*, by D. J. Craik and R. S. Tebble.) Supplementary examination of both the magnetization curve and the domain structure has asserted that each part of the magnetization curve shown in fig. 3.1 is due predominantly to one of the processes of magnetization just described. Specifically the first part, OA, involves the reversible displacement of domain walls; the second part, AB, involves the irreversible displacement of domain walls; the knee marks the onset of the spin rotation of the first mode and the third part, BC, involves the spin rotation of the second mode. Exceptions to this general scheme arise, however, when the specimens are either single crystals of certain orientations or under stress conditions. Some of the exceptional cases will be discussed later.

§3. MAGNETOCRYSTALLINE ANISOTROPY

The phenomenon of magnetocrystalline anisotropy was introduced in §1. Table 3.1 summarizes the directions of both easy and hard magnetization for Fe, Ni and Co, which also represent three major crystal structures; namely, body-centered cubic (bcc), face-centered cubic (fcc) and close-packed hexagonal (hcp), respectively. In this section, we shall first derive expressions for the magnetic energy as a function of the direction of magnetization with respect to the crystal axes, followed by a brief discussion on the origin of magnetic anisotropy.

When a ferromagnetic body becomes magnetized in a field, its free energy is increased by an amount E_m. To formulate E_m, we must first review the pertinent thermodynamic relationships. Suppose the body has a volume V before magnetization and a suitable shape, such as that of a long rod or a closed ring, so as to avoid the complication caused by the demagnetizing effect. The first law of thermodynamics states that

$$dU = dQ - dW, \tag{3.2}$$

where dU is the change in internal energy, dQ is the heat absorbed and dW is the amount of work done *by* the body. (No effort is made here to differentiate exact from inexact differentials.) A general expression of

TABLE 3.1
The easy and hard axes of magnetization for iron, nickel and cobalt

Metal	Crystal structure	Axes of magnetization: Easy	Axes of magnetization: Hard
Iron	Body-centered cubic	⟨100⟩ (6, cube edge)	⟨111⟩
Nickel	Face-centered cubic	⟨111⟩ (8, cube diagonal)	⟨100⟩
Cobalt*	Close-packed hexagonal	±[0001] (2, c-axis)	All directions in the basal plane

*As shown in fig. 5.82, the sum of the first and second constants of magnetocrystalline anisotropy for Co changes sign from positive to negative upon heating at ~250°C. Consequently at temperatures above 250°C, the easy axes change to $\langle 11\bar{2}0\rangle$ in the basal plane and the c-axis becomes the direction of hard magnetization.

dW is

$$\mathrm{d}W = \sum Q_\mathrm{i}\,\mathrm{d}Q_\mathrm{e}, \tag{3.3}$$

where Q_i is an intensive quantity, such as force, pressure, magnetic field, voltage, etc., and Q_e is a related extensive quantity, such as distance, volume, magnetization, current, etc. An extensive quantity is one which has its value as the sum of the value in each of the subsystems. For a magnetic system, we have

$$\mathrm{d}W = P\,\mathrm{d}V - \mu_0 H\,\mathrm{d}M. \tag{3.3.1}$$

The negative sign of the magnetic term signifies that this work is done *on* the system by the field.

The Helmholtz free energy is defined as

$$F = U - TS, \tag{3.4}$$

where S is the entropy. For reversible processes, $\mathrm{d}Q = T\,\mathrm{d}S$ and

$$\mathrm{d}F = -\mathrm{d}W - S\,\mathrm{d}T. \tag{3.5}$$

Combining (3.3.1) and (3.5) gives

$$\mathrm{d}F = -P\,\mathrm{d}V + \mu_0 H\,\mathrm{d}M - S\,\mathrm{d}T. \tag{3.5.1}$$

The change in free energy at constant temperature can then be attributed to a mechanical work $-P\,\mathrm{d}V$ by the body and a magnetic work $\mu_0 H\,\mathrm{d}M$ on the body by the applied field. The mechanical work is related to the phenomenon of magnetostriction, which will be discussed in the next section. Generally the volume change dV associated with the magnetization is very small; therefore, the magnetic term $\mu_0 H\mathrm{d}M$ dominates dF in (3.5.1). Denoting the change in free energy per unit volume by $(\Delta F)_{M_\mathrm{t}}$ as a result of magnetization of the body to the value M_t, we have the relation $E_\mathrm{m} \sim (\Delta F)_{M_\mathrm{t}}$. From (3.5.1), we further obtain

$$(\Delta F)_{M_\mathrm{t}} = \int (\mathrm{d}F)_{M_\mathrm{t}} = \int_0^{M_\mathrm{t}} \mu_0 H\,\mathrm{d}M. \tag{3.6}$$

The integral $\int_0^{M_\mathrm{t}} \mu_0 H\,\mathrm{d}M$ gives the energy of magnetization, which is always positive because work must be done on the body in order for the field to establish the magnetization. Consequently, the free energy of the body is increased by approximately this amount.

By definition, as shown in fig. 3.7, the integral $\int_0^{M_\mathrm{t}} \mu_0 H\,\mathrm{d}M = F_2$,

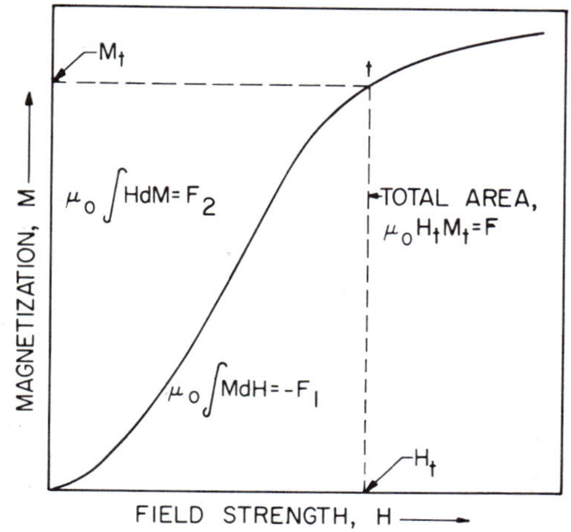

Fig. 3.7. Work and energy involved in magnetization (Bozorth [1951]).

represents an area under a M–H curve with respect to the ordinate axis up to M_t*. A comparison of the various M–H curves for single crystals of Fe, Ni and Co shown in figs. 3.3 to 3.5 will readily reveal the physical meaning of the energy E_a associated with the magnetocrystalline anisotropy. These curves are the M–H plots up to the saturation with the field applied in the easy, hard and intermediate directions. They clearly show that the value of the integral $\int_0^{M_s} \mu_0 H \, dM$ varies from a minimum, when the field is applied in the easy direction, through intermediaries to a maximum, when the field is applied in the hard direction. Thus we may resolve E_m into two components $E_{m,[easy]}$ and E_a, to

*The total area covered by the magnetization curve up to M_t in fig. 3.7 is $F = \mu_0 H_t M_t$ which is known as the mutual magnetostatic energy between the body of unit volume and the field. The difference between F and F_2 is

$$F_1 = F - F_2 = \mu_0 H_t M_t - \int_0^{M_t} \mu_0 H \, dM = \int_0^{H_t} \mu_0 M \, dH. \qquad (3.7)$$

If the zero of energy is chosen for the unmagnetized body in zero field, then $-\int_0^{H_t} \mu_0 M \, dH$ represents the work done in moving the body into the field H_t. The negative sign signifies that the energy associated with the change in both the magnetization of the body and the mutual action between the body and the field has decreased to establish a stable state.

emphasize the fact that magnetization in the directions other than the easy ones will require an extra work E_a to turn the atomic spins from a nearby easy axis into the field direction.

For a single crystal of unit volume magnetized to technical saturation in a noneasy $[uvw]$ direction,

$$E_a = (\Delta F)_{M_{s,[uvw]}} - (\Delta F)_{M_{s,[easy]}} = \int_0^{M_s} \mu_0 (H \, dM)_{[uvw]} - \int_0^{M_s} \mu_0 (H \, dM)_{[easy]}. \tag{3.8}$$

This expression indicates that the magnetic anisotropy energy is given by the area between the M–H curve in the $[uvw]$ direction and the corresponding curve in the easy direction.

As might be expected, the expression of E_a depends on the crystal structure of the material. For this reason, Co and other hcp materials represent the simplest case because of their uniaxial anisotropy largely along the c-axis* and will be discussed first. Since [0001] and $[000\bar{1}]$ are identical as far as magnetization is concerned, E_a must contain terms of only even-power functions of the angle between the magnetization direction and the c-axis. Therefore, E_a has the general expression

$$E_a = K_1 \sin^2\theta + K_2 \sin^4\theta + \cdots + K_n \sin^{2n}\theta = \sum_n K_n \sin^{2n}\theta, \tag{3.9}$$

where $K_1, K_2, \ldots, K_n$ are the magnetocrystalline anisotropy constants independent of θ. Experimental results for Co show that, for a first-order approximation, only the first two terms need to be retained, and the first and second constants have these values at 20°C: $K_1 = 4.3 \times 10^5\ \mathrm{J \cdot m^{-3}}$ ($1\ \mathrm{J \cdot m^{-3}} = 10\ \mathrm{ergs \cdot cm^{-3}}$, so $K_1 = 4.3 \times 10^6\ \mathrm{ergs \cdot cm^{-3}}$) and $K_2 = 1.2 \times 10^5\ \mathrm{J \cdot m^{-3}}$.

For cubic crystals, the expression of E_a has to satisfy two restrictions imposed by the cubic symmetry: (1) the expression must contain only even-power terms of direction cosines $(\alpha_1, \alpha_2, \alpha_3)$ of the angles between the direction of magnetization and the three cubic axes. (2) It must be invariant in any interchange of the α_i among themselves. Remembering the relation $\alpha_1^2 + \alpha_2^2 + \alpha_3^2 = 1$ and that E_a contains only those terms that reflect the variation of the energy with the α's, we

*A small anisotropy may exist in the basal plane.

obtain the general expression of E_a up to sixth order for cubic crystals,

$$\begin{aligned}E_a = {} & A(\alpha_1^4+\alpha_2^4+\alpha_3^4)+B(\alpha_1^2\alpha_2^2+\alpha_2^2\alpha_3^2+\alpha_3^2\alpha_1^2)\\ & +C(\alpha_1^6+\alpha_2^6+\alpha_3^6)+D(\alpha_1^4\alpha_2^2+\alpha_2^4\alpha_3^2+\alpha_3^4\alpha_1^2+\alpha_2^4\alpha_1^2\\ & +\alpha_3^4\alpha_2^2+\alpha_1^4\alpha_3^2)+E(\alpha_1^2\alpha_2^2\alpha_3^2). \qquad (3.10)\end{aligned}$$

Utilizing the following three transformations:

(1) $$\begin{aligned}\alpha_1^4+\alpha_2^4+\alpha_3^4 &= (\alpha_1^2+\alpha_2^2+\alpha_3^2)-2(\alpha_1^2\alpha_2^2+\alpha_2^2\alpha_3^2+\alpha_3^2\alpha_1^2)\\ &= 1-2(\alpha_1^2\alpha_2^2+\alpha_2^2\alpha_3^2+\alpha_3^2\alpha_1^2),\end{aligned}$$

(2) $$\begin{aligned}&\alpha_1^4\alpha_2^2+\alpha_2^4\alpha_3^2+\alpha_3^4\alpha_1^2+\alpha_2^4\alpha_1^2+\alpha_3^4\alpha_2^2+\alpha_1^4\alpha_3^2\\ &\quad= (\alpha_1^2+\alpha_2^2+\alpha_3^2)(\alpha_1^2\alpha_2^2+\alpha_2^2\alpha_3^2+\alpha_3^2\alpha_1^2)-3\alpha_1^2\alpha_2^2\alpha_3^2\\ &\quad= (\alpha_1^2\alpha_2^2+\alpha_2^2\alpha_3^2+\alpha_3^2\alpha_1^2)-3\alpha_1^2\alpha_2^2\alpha_3^2 \quad \text{and}\end{aligned}$$

(3) $$\begin{aligned}\alpha_1^6+\alpha_2^6+\alpha_3^6 &= 1-3(\alpha_1^4\alpha_2^2+\alpha_2^4\alpha_3^2+\alpha_3^4\alpha_1^2+\alpha_2^4\alpha_1^2+\alpha_3^4\alpha_2^2+\alpha_1^4\alpha_3^2)\\ &= 1-3(\alpha_1^2\alpha_2^2+\alpha_2^2\alpha_3^2+\alpha_3^2\alpha_1^2)+9\alpha_1^2\alpha_2^2\alpha_3^2,\end{aligned}$$

we simplify the general expression (3.10) to the form

$$E_a = K_1(\alpha_1^2\alpha_2^2+\alpha_2^2\alpha_3^2+\alpha_3^2\alpha_1^2)+K_2(\alpha_1^2\alpha_2^2\alpha_3^2). \qquad (3.11)$$

Experimental evaluation of the constants at room temperature gave $K_1 = 4.8\times10^4\ \mathrm{J\cdot m^{-3}}$ and $K_2 = 1.5\times10^4\ \mathrm{J\cdot m^{-3}}$ for Fe and $K_1 = -4.5\times10^3\ \mathrm{J\cdot m^{-3}}$ and $K_2 = 2.4\times10^3\ \mathrm{J\cdot m^{-3}}$ for Ni. These anisotropy constants are smaller than those for Co by a factor of 10 or more. This difference is in accord with the theory that the magnitude of E_a depends on the symmetry of the crystal structure. The higher the symmetry, the lower will be E_a and, hence, the anisotropy constants.

The listing of the easy and hard axes of magnetization observed in Fe, Co and Ni in table 3.1 seems to indicate a dependence of these axes on the crystal structure. This dependence, however, is subject to the effect of temperature and composition in a basic manner. It has already been stated in the footnote in table 3.1 that as the temperature is raised above 250°C, the easy and hard axes of magnetization in Co are switched to $\langle 11\bar{2}0\rangle$ and [0001] from [0001] and $\langle uvt0\rangle$, respectively. Another example is seen in the equiatomic alloy NiFe, for which the easy and hard axes are ⟨100⟩ and ⟨110⟩, instead of ⟨111⟩ and ⟨100⟩, respectively, even though the alloy retains its fcc crystal structure by rapid cooling from temperatures above 350°C. The reason for such drastic changes is not precisely known, but the changes are consistent with the reversal of sign of the first and second anisotropy constants for the alloy ($K_1 = 3.3 \times 10^3\ \mathrm{J \cdot m^{-3}}$ and $K_2 = -18.0 \times 10^3\ \mathrm{J \cdot m^{-3}}$ at 20°C). A detailed description of the changes in values of K_1 and K_2 with composition for alloys in the entire binary system of Ni and Fe can be found in Bozorth [1951a].

The anisotropy constants are evaluated by measurements using a torque magnetometer. They can also be evaluated, with less accuracy, from magnetization curves for single crystals in different orientations and from ferromagnetic resonance experiments for ferrites. The various methods are reviewed by Zijlstra [1967].

The energy associated with the exchange interaction between the atomic spins expressed in (2.38) is isotropic because the scalar product $\boldsymbol{S}_i \cdot \boldsymbol{S}_j$ is independent of the crystal structure and the choice of the coordinate system. Thus, the magnetocrystalline anisotropy obviously must not have its origin closely related to the exchange interaction. As a result of several theoretical analyses, notably by van Vleck (see Kanamori [1965]), a qualitative theory has emerged, which attributes the magnetocrystalline anisotropy largely to the spin–orbit coupling. In crystals, because of the interactions between neighboring atoms that arise from the crystalline electric fields and the overlap of the wave functions, the orbital motions of the magnetic electrons are strongly tied to the lattice. Indeed, the orbital–lattice coupling is so strong that (1) the orbital wave functions must reflect the symmetry of the lattice, (2) the orbital motions are quenched in the iron-group metals because of the effect of the crystalline electric field and (3) the application of a magnetic field has little effect on the coupling. Then through the spin–orbit coupling, the atomic spins are also tied to the crystal lattice, although the spin–lattice coupling is comparatively weak. When a field is applied in such a direction as to cause the spins to rotate into it, the orbital motion will tend to rotate with the spins because of the spin–lattice coupling. Both rotations require extra energy, thus leading to the term E_a. The theory becomes quite complicated in the quantitative treatment. Kanamori [1965] has given a comprehensive review of the quantitative theories of magnetic anisotropy.

Magnetic anisotropy can be induced in directions other than the inherent easy axes listed in table 3.1 by a magnetic anneal, by crystal growth and by plastic deformation. The induced anisotropy is usually uniaxial, in contrast to the multiple axes of easy magnetization for cubic crystals. The phenomenon of induced magnetic anisotropy will be discussed in ch. 5, §2.6.

For many material systems, especially those of ferrites and thin films, it is useful to characterize the magnetic anisotropy, either inherent or induced, in terms of an anisotropy field H_a. The expression for H_a varies according to the type or symmetry of anisotropy. For uniaxial anisotropy, we first rewrite (3.9) to the approximate form,

$$E_a \sim K_1 \sin^2 \phi \sim K_1 \phi^2, \tag{3.9.1}$$

when the magnetization is nearly parallel to the easy direction, thus ϕ is very small and $\sin \phi \sim \phi$. If we envisage H_a as a field being applied parallel to the easy direction, the associated magnetic energy is given by

$$E_{H_a} = -M_s H_a \mu_0 \cos \phi \sim \text{const.} + \tfrac{1}{2} M_s H_a \mu_0 \phi^2, \tag{3.12}$$

recalling that $\cos \phi \sim 1 - 2 \sin^2 (\phi/2) \sim 1 - \frac{1}{2}\phi^2$ for very small ϕ. Dropping the constant term in (3.12) and equating E_{H_a} to E_a (more accurately, we should equate the associated torque $\mathrm{d}E_a/\mathrm{d}\phi$ to $\mathrm{d}E_{H_a}/\mathrm{d}\phi$, which eliminates the constant term) yields

$$H_a = 2K_1/M_s \mu_0. \tag{3.13}$$

Similar consideration will show that the anisotropy field has the same expression (3.13) for the $\langle 100 \rangle$ easy axes and

$$H_a = -(4/3 M_s \mu_0)(K_1 + \tfrac{1}{3} K_2), \tag{3.13.1}$$

for the $\langle 111 \rangle$ easy axes in cubic crystals (Brailsford [1966a]). Some applications of the anisotropy field are described in ch. 6, §3 and ch. 7, §1.

§4. MAGNETOSTRICTION

As early as 1842, Joule observed that, during the magnetization of a nickel rod along its axis up to the technical saturation, the length of the rod suffered a steady, though small, decrease. Such a dimensional change accompanying magnetization is known as magnetostriction. Broadly speaking, magnetostriction refers to any change in dimension or volume during magnetization. It includes the following three types:

(1) Longitudinal magnetostriction, $\lambda = (\mathrm{d}l/l)_L$ – the change in length in the direction of magnetization; also known as the Joule effect.
(2) Transverse magnetostriction, $(\mathrm{d}l/l)_T$ – dimensional changes in the directions perpendicular to that of magnetization.
(3) Volume magnetostriction, $\omega = \mathrm{d}V/V$.

As opposed to the volume magnetostriction, the longitudinal and transverse magnetostrictions constitute the linear magnetostriction. Analogous with the elastic behavior of isotropic solids, the transverse magnetostriction is assumed equal to $-\frac{1}{2}\lambda$ in all directions normal to that of magnetization. Consequently, the volume change resulting from (1) and (2) is very small. There is, however, another source of volume magnetostriction that occurs when the applied field is further increased after saturation magnetization has been attained in the field direction. It is called the "forced" magnetostriction. Fig. 3.8 shows the onset of the latter volume magnetostriction along with the longitudinal magnetostriction at various stages of magnetization in iron. In cubic polycrystals, the latter volume magnetostriction is isotropic, which implies that the accompanying linear normal strain is independent of direction and approximately one-third the volume strain. In hexagonal crystals, this volume change may be accompanied by a shape change, signifying that there are shear strains as well as normal strains associated with the rotation of atomic spins.

The longitudinal magnetostriction can be positive (elongation) or negative (contraction). Some typical curves showing the variation of $(\mathrm{d}l/l)_{\mathrm{L}}$ with the applied field are shown in figs. 3.8 and 3.9. Note that in Fe and the 30% Ni–Co alloy, $(\mathrm{d}l/l)_{\mathrm{L}}$ changes sign as the field is increased. Also note the very small magnitudes, of the order of 10^{-5}, equivalent roughly to the length change produced by a temperature change of 1°C, of the longitudinal magnetostriction of most ferromagnets. Likewise, the volume magnetostrictions are small. For Ni at room temperature, $\mathrm{d}\omega/\mathrm{d}H$ is approximately $\sim 1.3\times 10^{-12}$ per $\mathrm{A\cdot m^{-1}}$ ($\sim 1\times 10^{-10}$ per oersted), and the total $\mathrm{d}V/V$ is around 3×10^{-4}.

Despite the overall small magnitudes of magnetostriction in most cases, this phenomenon is of great importance, both practically and scientifically and deserves further exploration. In the following discussion, we shall focus our attention on the linear magnetostriction, however. From the viewpoint of crystal symmetry, Becker and Döring [1939] suggested the following general expression of linear (both longitudinal and transverse) magnetostriction at saturation for Fe and Ni

$$
\begin{aligned}
(\mathrm{d}l/l)_{\mathrm{s}} = {} & h_1(\alpha_1^2\beta_1^2+\alpha_2^2\beta_2^2+\alpha_3^2\beta_3^2-\tfrac{1}{3})+2h_2(\alpha_1\alpha_2\beta_1\beta_2 \\
& +\alpha_2\alpha_3\beta_2\beta_3+\alpha_3\alpha_1\beta_3\beta_1)+h_4(\alpha_1^4\beta_1^2+\alpha_2^4\beta_2^2+\alpha_3^4\beta_3^2+\tfrac{2}{3}A-\tfrac{1}{3}) \\
& +2h_5(\alpha_1\alpha_2\alpha_3^2\beta_1\beta_2+\alpha_2\alpha_3\alpha_1^2\beta_2\beta_3+\alpha_3\alpha_1\alpha_2^2\beta_3\beta_1) \\
& +h_3(A-\tfrac{1}{3}) \quad \text{when } K_1<0 \text{ as in Ni} \\
& +h_3A \qquad\quad \text{when } K_1>0 \text{ as in Fe,}
\end{aligned}
\tag{3.14}
$$

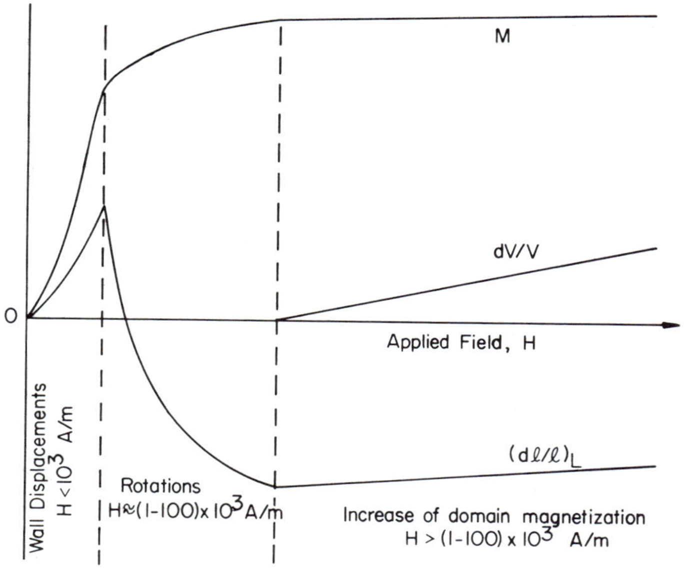

Fig. 3.8. Longitudinal and volume magnetostriction accompanying magnetization of iron to saturation (Lee [1955]).

where, α_i and β_i $(i = 1, 2, 3)$ are the direction cosines, relative to the crystal axes, of the magnetization vector and the direction of observation, respectively, $h_1, \ldots, h_5$ are constants called the magnetostrictive coefficients and $A = \alpha_1^2\alpha_2^2 + \alpha_2^2\alpha_3^2 + \alpha_3^2\alpha_1^2$. The last term in (3.14) contains h_3 and α_i, but no β_i, and represents the volume change caused by the rotation of atomic spins. When the direction of magnetization and the direction of observation are along [100], $\alpha_1 = \beta_1 = 1$ and $\alpha_2 = \alpha_3 = \beta_2 = \beta_3 = 0$; hence

$$\begin{aligned} \lambda_{s,[100]} &= \tfrac{2}{3}h_1 - \tfrac{1}{3}h_3 + \tfrac{2}{3}h_4 \quad \text{for Ni,} \\ &= \tfrac{2}{3}h_1 + \tfrac{2}{3}h_4 \quad \text{for Fe.} \end{aligned} \tag{3.15}$$

Similarly, when both directions are along [110], $\alpha_1 = \alpha_2 = \beta_1 = \beta_2 = 1/\sqrt{2}$ and $\alpha_3 = \beta_3 = 0$; hence

$$\begin{aligned} \lambda_{s,[110]} &= \tfrac{1}{6}h_1 + \tfrac{1}{2}h_2 - \tfrac{1}{12}h_3 + \tfrac{1}{12}h_4 \quad \text{for Ni,} \\ &= \tfrac{1}{6}h_1 + \tfrac{1}{2}h_2 + \tfrac{1}{4}h_3 + \tfrac{1}{12}h_4 \quad \text{for Fe.} \end{aligned} \tag{3.16}$$

When both directions are along [111], $\alpha_1 = \alpha_2 = \alpha_3 = \beta_1 = \beta_2 = \beta_3 =$

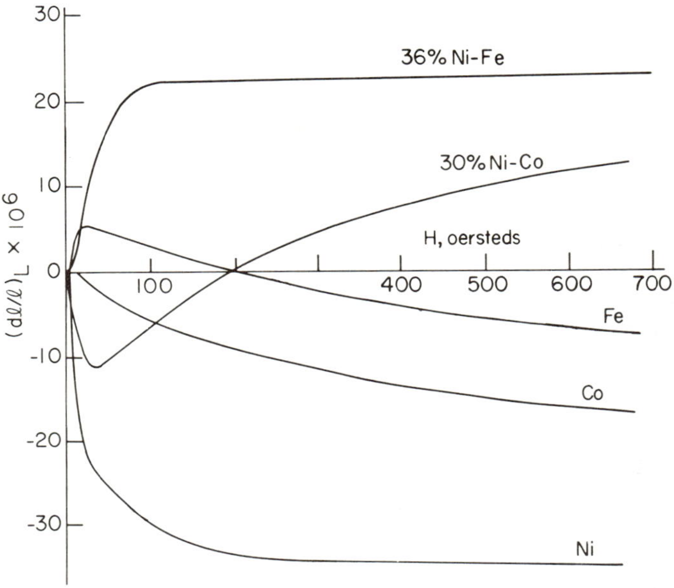

Fig. 3.9. Field dependence of longitudinal magnetostriction in Fe, Co, Ni, 30% Ni–Co and 36% Ni–Fe (Lee [1955]).

$1/\sqrt{3}$; hence

$$\begin{aligned}\lambda_{s,[111]} &= \tfrac{2}{3}h_2 + \tfrac{2}{9}h_5 && \text{for Ni,}\\ &= \tfrac{2}{3}h_2 + \tfrac{1}{3}h_3 + \tfrac{2}{9}h_5 && \text{for Fe.}\end{aligned} \tag{3.17}$$

Many workers simply write $\lambda_{s,[100]}$ as λ_{100}, etc.

The five magnetostrictive coefficients may be experimentally evaluated by measuring the induced strain at saturation for fixed β and varying α. The experimental values for Fe, Ni, 78% Ni–Fe and Fe_3O_4 at room temperature are given in table 3.2. In these and many other materials, the values for h_3, h_4 and h_5 are much smaller than those of h_1 and h_2. It is then sufficient to neglect terms that contain h_3, h_4 and h_5. Furthermore, we can express $(dl/l)_s$ in terms of λ_{100} and λ_{111} by combining (3.14), (3.15) and (3.17),

$$\begin{aligned}(dl/l)_s = {} & \tfrac{3}{2}\lambda_{100}(\alpha_1^2\beta_1^2 + \alpha_2^2\beta_2^2 + \alpha_3^2\beta_3^2 - \tfrac{1}{3})\\ & + 3\lambda_{111}(\alpha_1\alpha_2\beta_1\beta_2 + \alpha_2\alpha_3\beta_2\beta_3 + \alpha_3\alpha_1\beta_3\beta_1).\end{aligned} \tag{3.18}$$

TABLE 3.2

Magnetostriction constants† for some cubic materials at room temperature*

Material	h_1	h_2	h_3	h_4	h_5	λ_{100}	λ_{111}
Fe	32	−32	1.4 ± 0.4	−3.2 ± 0.8	4.7 ± 1	20.7	−21.2
Ni	−68.8 ± 3.8	−36.5 ± 1.9	−2.8 ± 3.1	−7.5 ± 5.2	7.7 ± 3.1	−45.9	−24.3
78% Ni–Fe							
(A) Quenched	13.7 ± 1.0	2.6 ± 0.5	−0.3 ± 0.8	1.1 ± 1.4	−0.1 ± 0.8		
(B) Slowly cooled	20.9 ± 0.7	2.8 ± 0.3	1.7 ± 0.5	−1.4 ± 1.0	−0.2 ± 0.5		
Fe_3O_4							
(A) Natural	−15.3	142.4	−9.43	−9.70	−50.5		
(B) Synthetic	−24.5	123.3	−0.40	−4.76	−20.1		

†All constants must be multiplied by 10^{-6} to attain the absolute magnitudes.
*References: Lee [1955] and Carr [1959].

TABLE 3.3

Magnetostriction constants† for some hexagonal metals

Metal	λ_A	λ_B	λ_C	λ_D	Reference
Co(R.T.)	−0.045	−0.095	0.11	−0.10	Bozorth [1954]
Gd(0°K)	−0.107	−0.217	0.082	0.025	Alstad and Legvold [1964]
Tb(0°K)	6.95	−1.75	−9.0	6.99	Rhyne [1972]

†All constants must be multiplied by 10^{-3} to attain the absolute magnitudes.

This is the expression for linear magnetostriction most often used for cubic crystals. It is simplified to various forms for different values of α_i and β_i. Three cases are particularly important. First, if the crystal is assumed to be magnetostrictively isotropic, that is, $\lambda_{100} = \lambda_{111} = \lambda_{[uvw]} = \lambda_0$, then (3.18) becomes

$$(\mathrm{d}l/l)_s = \tfrac{3}{2}\lambda_0[(\alpha_1\beta_1 + \alpha_2\beta_2 + \alpha_3\beta_3)^2 - \tfrac{1}{3}] = \tfrac{3}{2}\lambda_0(\cos^2\theta - \tfrac{1}{3}), \quad (3.18a)$$

where θ is the angle between the directions of magnetization and observation. Note that for longitudinal and transverse magnetostrictions, since $\theta = 0$ and $\frac{1}{2}\pi$, $\lambda_s = (\mathrm{d}l/l)_{s,L} = \lambda_0$ and $(\mathrm{d}l/l)_{s,T} = -\frac{1}{2}\lambda_0 = -\frac{1}{2}\lambda_s$, respectively. The values usually assigned to λ_0 are -7×10^{-6} for iron and -34×10^{-6} for nickel. The second case deals with the general expression for longitudinal magnetostriction. Here, with $\alpha_i = \beta_i$ and using an appropriate transformation relation, we have

$$(\mathrm{d}l/l)_{L,s} = \lambda_{100} - 3(\lambda_{100} - \lambda_{111})(\alpha_1^2\alpha_2^2 + \alpha_2^2\alpha_3^2 + \alpha_3^2\alpha_1^2). \quad (3.19a)$$

The third case is concerned with longitudinal magnetostrictions in polycrystalline materials. To obtain an average of λ_s, we have to average the $\alpha_i^2\alpha_j^2$ terms in (3.19a). Because $\overline{\alpha_1^2\alpha_2^2 + \alpha_2^2\alpha_3^2 + \alpha_3^2\alpha_1^2} = \frac{1}{5}$, so

$$\bar{\lambda}_s = \tfrac{2}{5}\lambda_{100} + \tfrac{3}{5}\lambda_{111}. \quad (3.19b)$$

This expression often is used to correlate magnetostriction data for polycrystalline samples with those for single crystals. Using the values listed for λ_{100} and λ_{111} for Fe and Ni, (3.19b) gives the values of $\bar{\lambda}_s$ at -4.4×10^{-6} for Fe and -32.9×10^{-6} for Ni, as compared with experimental values of -8.0 and -34×10^{-6} for Fe and Ni, respectively.

For the hexagonal metals, the expression for $\mathrm{d}l/l$ at saturation is

$$\begin{aligned}(\mathrm{d}l/l)_s = {} & \lambda_A[(\alpha_1\beta_1 + \alpha_2\beta_2)^2 - (\alpha_1\beta_1 + \alpha_2\beta_2)\alpha_3\beta_3] + \lambda_B[(1-\alpha_3^2)(1-\beta_3^2) \\ & - (\alpha_1\beta_1 + \alpha_2\beta_2)^2] + \lambda_C[(1-\alpha_3^2)\beta_3^2 - (\alpha_1\beta_1 + \alpha_2\beta_2)\alpha_3\beta_3] \\ & + 4\lambda_D(\alpha_1\beta_1 + \alpha_2\beta_2)\alpha_3\beta_3, \end{aligned} \quad (3.20)$$

where α_i and β_i are the direction cosines defining the directions of magnetization and observation, respectively (Mason [1954]) and $\lambda_A, \ldots, \lambda_D$ are the magnetostrictive coefficients. The experimental values of $\lambda_A, \ldots, \lambda_D$ for Co, Gd and Tb are given in table 3.3. Note the "giant" magnetostriction constants found in Tb, and possibly in Dy, Ho and Er.

Since magnetostriction and magnetocrystalline anisotropy are intrinsic, directional properties of ferromagnets, it is reasonable to assume that both phenomena share a common origin. Indeed, Vonsovsky [1940] showed that calculated magnetostrictions based on the model of spin–orbital coupling agree with the experimental results within an order of magnitude. The theory that magnetostriction, like magnetic anisotropy, also has its origin in the spin–orbital coupling is strengthened by the work of Kittel [1949], who established a close analytical relationship between constants of anisotropy and magnetostriction and concluded that linear magnetostriction would not have occurred if the anisotropy energy had been independent of the state of strain of the crystal. Various mechanisms of magnetostriction are described in the review articles by Lee [1955] and Carr [1966].

An important phenomenon related to magnetostriction deals with the effect of an internal stress on the magnetization in a ferromagnetic body. This is known as the magnetoelastic, or inverse magnetostrictive, effect. The effect of an external force applied, or of an internal stress existing, in a direction other than that of the applied field is to change the direction of magnetization via magnetostriction. Expressions for the energy associated with the magnetoelastic effect E_e, can be derived in terms of magnetostrictive coefficients (see Chikazumi [1966a]) and the general form for cubic crystals:

$$\begin{aligned} E_e = &-\tfrac{3}{2}\lambda_{100}\sigma(\alpha_1^2\delta_1^2+\alpha_2^2\delta_2^2+\alpha_3^2\delta_3^2)\\ &-3\lambda_{111}\sigma(\alpha_1\alpha_2\delta_1\delta_2+\alpha_2\alpha_3\delta_2\delta_3+\alpha_3\alpha_1\delta_3\delta_1), \end{aligned} \tag{3.21}$$

where σ is a tensile stress induced in a ferromagnet by an external force, and α_i and $\delta_i (i = 1, 2, 3)$ are the direction cosines of the magnetization vector and this stress with respect to the cubic axes, respectively. Various simplified forms emerge under special conditions. Three cases are especially of interest. First, when the direction of the internal stress coincides with that of magnetization ($\alpha_i = \delta_i$), we have

$$E_e = -\tfrac{3}{2}\lambda_{100}\sigma + 3\sigma(\lambda_{100}-\lambda_{111})(\alpha_1^2\alpha_2^2+\alpha_2^2\alpha_3^2+\alpha_3^2\alpha_1^2). \tag{3.22}$$

In the second case, if the magnetostriction is isotropic ($\lambda_{100} = \lambda_{111} = \lambda_0$), (3.22) is further reduced to the simplest form

$$E_e = -\tfrac{3}{2}\lambda_0\sigma, \tag{3.23}$$

in the direction of magnetization. It will be shown in the next section that the anisotropy energy per unit volume is represented approximately by the first anisotropy constant K_1. Thus (3.23) implies that the magnetoelastic effect of an internal stress σ on the magnetocrystalline anisotropy is to change the first anisotropy constant by an amount $-\frac{3}{2}\lambda_0\sigma$. Finally, if the stress direction deviates from that of magnetization by an angle θ, the magnetoelastic energy will be given by

$$E_e = -\tfrac{3}{2}\lambda_0\sigma \cos^2\theta. \tag{3.24}$$

This expression is the one most often used to analyze the behavior of polycrystalline cubic ferromagnets under stress.

§5. THE BLOCH WALL AND DOMAIN STRUCTURE

Crystalline substances of macroscopic size rarely appear in single crystal form unless special efforts have been made in the preparation procedure. Likewise, ferromagnetic materials of macroscopic size seldom possess a single domain unless they are magnetized beyond the knee in the magnetization curve. Despite the presence of the boundaries between grains and domains, both structures exist in stable equilibrium with the energy of the system tending to a minimum. Thus domain boundaries are an integral part of the domain structure just as grain boundaries exist inherently in the grain structure.

Domain boundaries are called the Bloch walls since it was Bloch [1932] who first argued from the energy point of view that, in crossing the wall, the atomic spins would prefer to change directions gradually in many steps (fig. 3.10b), rather than abruptly in one step (fig. 3.10a). Therefore, a domain wall must have a finite thickness to accommodate to the gradual change in spin direction. Consider, for example, a 180°

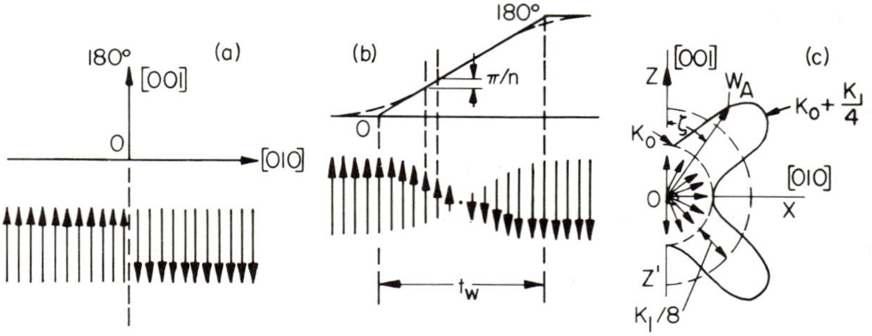

Fig. 3.10. Spin alignment (a) between neighboring domains and (b) inside a 180° Bloch wall in iron. (c) Showing the angular dependence of the anisotropy energy of spins with an average value $\bar{E}_a \sim \frac{1}{8}K_1$. The direction-independent component of the anisotropy energy is represented by K_0, thus $K_0 = \int_0^{M_s} (\mu_0 H \, dM)_{[easy]}$ (Brailsford [1966]).

wall, which separates two domains of opposite magnetization vectors (fig. 3.10a). According to (2.38), the exchange energy between spins aligned ferromagnetically within the domains is given by

$$E_{ex}(\uparrow\uparrow) = -2J_{ex}\boldsymbol{S}_i \cdot \boldsymbol{S}_j = -2J_{ex}S^2 \cos\theta = -2J_{ex}S^2.$$

If the spin flips over in one step across the wall, the associated exchange energy will be

$$E_{ex}(\uparrow\downarrow) = -2J_{ex}S^2 \cos\pi = 2J_{ex}S^2. \tag{3.25}$$

Therefore, the change in E_{ex} due to the spin flip is

$$\Delta E_{ex} = 2J_{ex}S^2 - (-2J_{ex}S^2) = 4J_{ex}S^2. \tag{3.26}$$

If the spin reversal is accomplished in n equal steps, each of which involves a tilting angle θ of $180/n$ degrees, the exchange energy in each step is

$$E_{ex}(\uparrow\nearrow) = -2J_{ex}S^2 \cos\pi/n. \tag{3.27}$$

When n is sufficiently large, $\cos\theta \simeq (1 - \frac{1}{2}\theta^2)$; so (3.27) becomes

$$E_{ex}(\uparrow\nearrow) = -2J_{ex}S^2[1 - \tfrac{1}{2}(\pi/n)^2]$$

and the change in E_{ex} in each step is

$$(\Delta E_{ex})_\theta = 2J_{ex}S^2(\tfrac{1}{2}\theta^2) = J_{ex}S^2(\pi/n)^2. \tag{3.28}$$

Whence the total change in E_{ex} over $n+1$ spins is

$$\Delta E_{ex} = \sum^{n}(\Delta E_{ex})_\theta = J_{ex}S^2\pi^2/n. \tag{3.29}$$

Comparison of (3.26) and (3.29) confirms that the n-step spin-tilting scheme consumes less energy than the one-step spin-flip scheme when $n > \frac{1}{4}\pi^2$, or n has the minimum value of 3.

From the exchange energy consideration alone, it looks as if n might take an infinite value and the wall will be indefinitely thick. This, of course, is not true because, when the spins deviate progressively from the easy direction in the many-step scheme, there will be a corresponding steady increase in magnetocrystalline anisotropy energy. Hence a domain wall is also controlled by anisotropy energy. As a first-order approximation, we may neglect the magnetostrictive effect and calculate the thickness d_w and surface energy density γ_w of a domain wall from the exchange energy γ_{ex} and the anisotropy energy

γ_a. We further assume that the 180° wall lies in the (010) surface. The wall energy per unit area associated with the exchange interaction is

$$\gamma_{ex} = (J_{ex}S^2\pi^2/n)\cdot(1/a^2), \tag{3.30}$$

where a is the lattice constant expressed in metres and $1/a^2$ gives the number of atoms per unit area of a (010) surface in a simple cubic lattice. The anisotropy energy per unit volume is given approximately by the first anisotropy constant K_1^*. The volume of the wall per unit area is na times unit area; thus

$$\gamma_a = naK_1. \tag{3.33}$$

Combining (3.30) and (3.33) and setting $d\gamma_w/dn = 0$, we obtain an optimum value for n,

$$n_0 = (\pi^2 J_{ex}S^2/K_1a^3)^{\frac{1}{2}}. \tag{3.34}$$

The thickness of the wall is given by

$$d_w = n_0a = (\pi^2 J_{ex}S^2/K_1a)^{\frac{1}{2}} = \pi(A_{ex}/K_1)^{\frac{1}{2}}, \tag{3.35}$$

where $A_{ex} = J_{ex}S^2/a$ is called the exchange stiffness or the exchange constant. The corresponding minimum energy of the wall is

$$\begin{aligned}\gamma_w &= \gamma_{ex} + \gamma_a \\ &= (\pi^2K_1J_{ex}S^2/a)^{\frac{1}{2}} + (\pi^2K_1J_{ex}S^2/a)^{\frac{1}{2}} \\ &= 2(\pi^2K_1J_{ex}S^2/a)^{\frac{1}{2}} = 2\pi(A_{ex}K_1)^{\frac{1}{2}}\end{aligned} \tag{3.36}$$

*The following discussion will show that $E_a \approx K_1$ within the order of magnitude. Referring to (3.11), if the magnetization vector is at an angle ζ with [001] and its projection on the (001) plane is at an angle ϕ from [100], fig. 3.11 shows that $\alpha_1 = \sin\zeta\cos\phi$, $\alpha_2 = \sin\zeta\sin\phi$ and $\alpha_3 = \cos\zeta$. Substituting these relations in (3.11) gives

$$\begin{aligned}E_a = (K_1/64)[&(3 - 4\cos 2\zeta + \cos 4\zeta)(1-\cos 4\phi) \\ &+ 8(1-\cos 4\zeta)] + (K_2/256)[(2-\cos 2\zeta - 2\cos 4\zeta \\ &+ \cos 6\zeta)(1-\cos 4\phi)].\end{aligned} \tag{3.31}$$

For a 180° wall lying in the (010) plane, $\phi = 0$; so

$$E_a = \tfrac{1}{8}K_1(1-\cos 4\zeta). \tag{3.32}$$

The angle ζ can have any value from 0 to π. Figure 3.10c shows that the average value of E_a is $\frac{1}{8}K_1$. Similar analyses can be made for other types of walls, and the results also support the $E_a \approx K_1$ approximation.

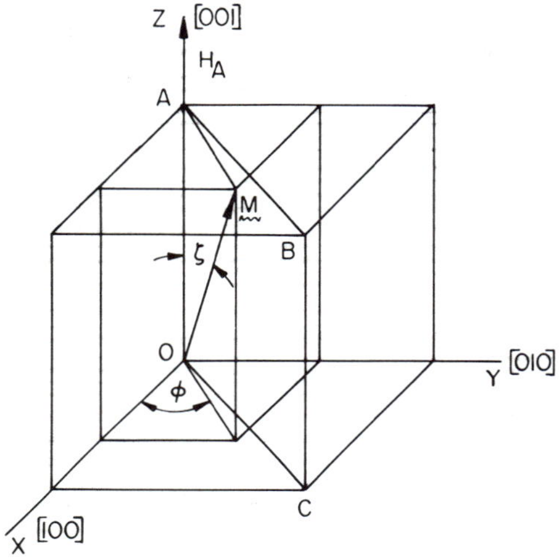

Fig. 3.11. Illustrating the geometry of a randomly oriented magnetization vector $\boldsymbol{M}$ in a cube whose (110) plane is shown by OABC (Brailsford [1966]).

Note that the wall energy is equally divided between the exchange energy and the anisotropy energy and that combining the last two equations gives

$$\gamma_w = 2d_w K_1. \tag{3.37}$$

In a more detailed analysis, Chikazumi [1964b] expressed the exchange stiffness as $A_{ex} = 2J_{ex}S^2/a$ for bcc lattices and $A_{ex} = 4J_{ex}S^2/a$ for fcc lattices; also, taking into account the variation of the anisotropy energy as a function of the spin rotation between the two faces of the wall, he attained the result

$$\gamma_w = 4(A_{ex}K_1)^{\frac{1}{2}}. \tag{3.36.1}$$

This wall energy density is smaller than that given by (3.36) by a factor of 0.64. For iron, $J_{ex} = 0.20\ k\theta_C = 2.83 \times 10^{-21}$ J, $K_1 = 4.8 \times 10^4\ \mathrm{J \cdot m^{-3}}$, $S = 1$, and $a = 2.87 \times 10^{-10}$ m. From these data we obtain $n_0 = 157$, $d_w = n_0 a = 4.50 \times 10^{-8}$ m or 450 Å, and $\gamma_w = 4.32 \times 10^{-3}\ \mathrm{J \cdot m^{-2}}$ (= 4.32 ergs · cm^{-2}). The calculation of n_0, d_w and γ_w can be improved by using a more exact value of E_a and by including the magnetoelastic effect. In a more refined calculation, Lilley [1950] obtained $n_0 \approx 490$ and $\gamma_w = 1.24 \times 10^{-3}\ \mathrm{J \cdot m^{-2}}$ for iron.

In addition to the thickness and surface energy density, the geometry of the Bloch wall also is important. There are two geometric aspects to be considered. First, a wall in iron can separate two domains whose magnetization vectors are either 90° or 180° apart. Thus there are 90° and 180° walls in iron. In nickel the ⟨111⟩ easy axes permit the Bloch walls to take the angular values of 70.53°, 109.47° and 180°, whereas in cobalt the ±[001] easy axes restrict all walls at 180°.

Another geometric aspect is the crystallographic plane in which the wall lies. At first glance, it seems that the wall surface could be in any crystal plane between the magnetization vectors of the neighboring domains. The choice of the wall surface, however, is not so wide and random because it tends to meet the condition that no free poles are created at the wall. This condition is satisfied if the components of magnetization parallel to the wall, denoted by $M_{\parallel}$, are completely cancelled and the components of magnetization normal to the wall, denoted by $M_{\perp}$, are either completely cancelled or continuous. Fig. 3.12

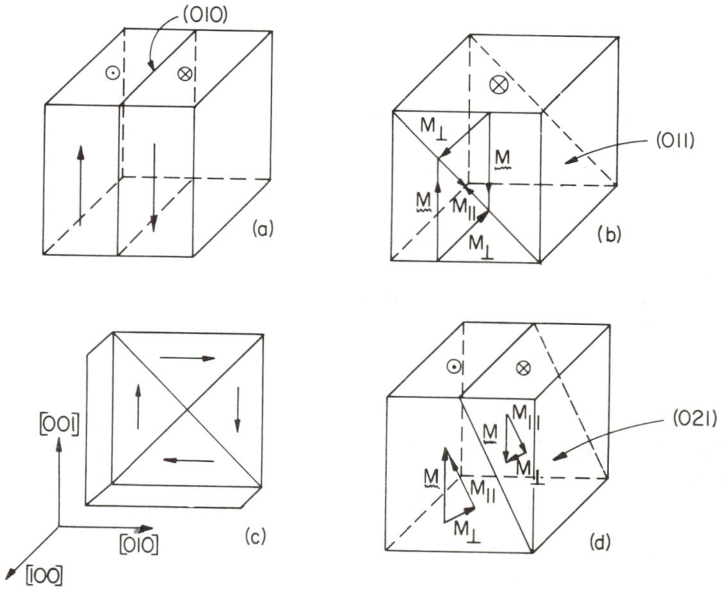

Fig. 3.12. Four types of domain walls in iron: (a) a 180° wall lying in (010); (b) a 180° wall lying in (011); (c) four 90° walls lying in {110} and (d) a 180° wall lying in (021).

shows four types of domain walls in iron: (a) a 180° wall lying in the (010) plane, (b) a 180° wall lying in the (011) plane, (c) four 90° walls lying in the {110} planes, and (d) a 180° wall lying in the (021) plane. The figure also indicates that there is no $M_\perp$ component, and $M_\parallel$ are cancelled in (a); both $M_\perp$ and $M_\parallel$ are cancelled in (b); $M_\parallel$ are cancelled, and $M_\perp$ are continuous in (c); and both $M_\parallel$ and $M_\perp$ are not completely cancelled in (d). Consequently, the first three walls are energetically favored, but the last one is unfavored. Listed below are the important domain walls in

(A) Fe: 90° lying in (100), (110) and (111) planes and 180° lying in (100), (110) and planes in between;
(B) Ni: 70.53° lying in (100) and (110) planes, 109.47° lying in (100), (110) and (111) planes and 180° lying in (110) and (112) planes; and
(C) Co: 180° lying in any (*hki*0) plane.

Lilley [1950] has calculated the thickness and energy density of the prominent walls in Fe, Co and Ni. His results are summarized in table 3.4. The surface energy densities of various types of domain walls range between 0.1 and 8×10^{-3} J · m^{-2} (0.1 and 8 ergs · cm^{-2}), which is very small in comparison with the energy densities of grain boundaries (0.3–0.5 J · m^{-2}) and free surfaces of metals (1–2 J · m^{-2}). Cobalt has the highest density of surface energy in its only type of domain walls, of which the thickness is therefore the smallest.

The Bloch wall is always associated with a positive surface energy, signifying that a finite work is required to create the wall. In justifying a domain structure, however, it should be realized that wall energy is only one of several terms involved. Other energies, such as magnetostatic, anisotropy, and magnetostrictive, must be considered.

The magnetostatic energy E_s arises when magnetic poles appear at the free surface of the specimen. A detailed discussion of this energy is given in appendix 3. As shown in that discussion, the magnitude of E_s per unit volume is given by

$$E_s = \tfrac{1}{2} H_d M_s \mu_0 = \tfrac{1}{2} N_d M_s^2 \mu_0, \tag{3.38}$$

where H_d and N_d are the demagnetizing field and factor, respectively. The magnetostatic energy could be very large in comparison with other energy terms. Hence, frequently it plays a dominant role in determining the domain structure. In fact, the correct concept of domain structure did not emerge until Landau and Lifshitz and Néel explored the

TABLE 3.4
Densities of surface energy and thicknesses of various domain walls calculated by Lilley [1950]

Metal	Domain wall		Density of wall energy, $\gamma_w \times 10^3$ (J · m^{-2})	Wall thickness d_w (Å)
	Plane	Type		
Iron	(001)	90°	0.62	408
	(001)	180°	1.24	1410
	(110)	90°	1.07	516
	(110)	180°	1.71	728
	(111)	90°	0.74	408
Nickel	(001)	70.53°	0.076	1000
	(001)	109.47°	0.152	∞*
	(110)	70.53°	0.0645	1110
	(110)	109.47°	0.192	860
	(110)	180°	0.306	2060
	(111)	109.47°	0.181	1000
Cobalt		180°	8.2	157

*If the magnetostriction term is not included, the (001), 109.47° wall in Ni can be regarded as two successive (001), 70.53° walls; thus d_w is ∞.

significance of the magnetostatic energy (see Kittel and Galt [1956]).

Magnetoelastic energy E_e must be considered whenever magnetization induces a mechanical strain in the lattice and (or) the magnetized specimen is under external forces. Equation (3.24) is used to calculate E_e in isotropic media. For anisotropic magnetostriction in single crystals, E_e may be calculated from the general expression of the elastic strain energy, where, in the absence of shearing stresses and strains,

$$E_e = \tfrac{1}{2}(\sigma_x \epsilon_x + \sigma_y \epsilon_y + \sigma_z \epsilon_z), \tag{3.39}$$

with $\sigma_x, \sigma_y, \sigma_z$ and $\epsilon_x, \epsilon_y, \epsilon_z$ being the normal stress and strain components along the Cartesian coordinates axes, respectively.

As far as domain structure is concerned, the total magnetic energy of a specimen consists of four terms:

$$E_{total} = E_a + E_e + E_s + E_w, \tag{3.40}$$

where the subscripts a, e, s and w stand for anisotropy, magnetostrictive, magnetostatic and wall, respectively. A stable domain structure corresponds to a minimum total energy by minimizing the various terms under the existing conditions. By conditions, we mean the composition, shape, size and crystal orientation of the specimen and whether or not an external field is present. The following examples illustrate the ways in which (3.40) attains the optimum value in various configurations of domains.

As a first example, let us consider a square-shaped single crystal slab of iron. The slab has the dimensions 1 cm × 1 cm × 0.5 cm, and its square faces are in the (001) plane. The simplest domain structure, of course, contains just one domain shown in fig. 3.13a. This structure has zero E_a and zero E_w. Its E_e also is approximately zero because the single domain can adjust to the magnetostriction without causing internal stresses. The only nonzero energy term is E_s because of the appearance of free poles at the (100) faces. Hence,

$$E^{a}_{total} = E^{a}_{s} = (\tfrac{1}{2} N_d M^2_s \mu_0) \cdot V, \tag{3.41}$$

where V is the volume of the specimen. The calculation of the magnetostatic energy is given in appendix 3. According to Rhodes and

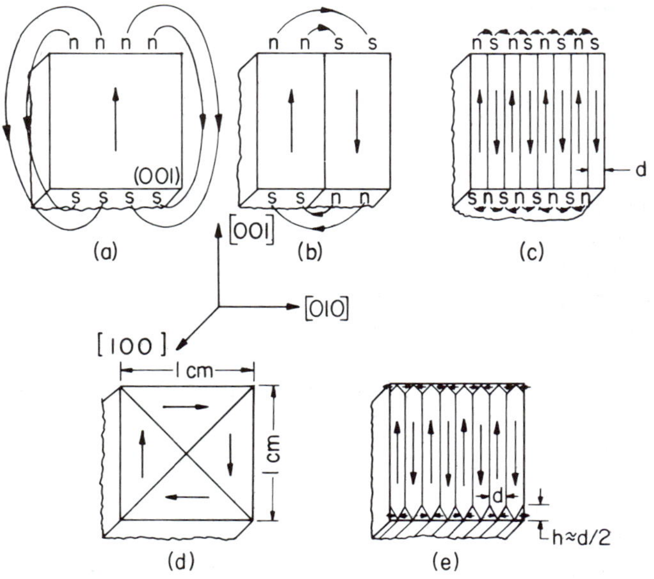

Fig. 3.13. Possible domain structures in a square single crystal slab of iron having the top and bottom faces in the (001) plane.

Rowland [1954], the slab has an N_d value of $\frac{1}{3}$, approximately the same as for spherical specimens. Substituting $M_s = 1.73 \times 10^6$ A · m^{-1} and $V = 5 \times 10^{-7}$ m^3 in (3.41) gives $E^{a}_{\text{total}} = 0.313$ J, which is enormous in comparison with most wall energies.

It is possible to reduce the magnetostatic energy by dividing the crystal into multiple domains. Let us examine this approach first with a bi-domain structure shown in fig. 3.13b. Since an equal number of N and S poles appear on the same (100) face, the spatial extension of the demagnetizing field produced by the free poles is reduced roughly by a factor of two. Consequently, the magnetostatic energy also is reduced by the same amount. We have for the bi-domain structure

$$\begin{aligned} E_{\text{total}} &= E^{b}_{s} + E_{w} \\ &= \tfrac{1}{2}E^{a}_{s} + \gamma_{180^\circ,(010)} \cdot A_b, \end{aligned} \tag{3.42}$$

where A_b is the surface area of the 180° wall. Substituting $\gamma_{180^\circ,(010)} = 1.24 \times 10^{-3}$J · m^{-2} (from table 3.4) and $A_b = 5 \times 10^{-5}$ m^2 in (3.42) gives $E^{b}_{\text{total}} = 0.157 + 0.62 \times 10^{-7} \sim 0.157$ J. The magnetostatic energy, although reduced to half of the original value, is still very large in comparison with the wall energy.

To further reduce the magnetostatic energy, the slab may be divided into n thin lamellar domains as shown in fig. 3.13c. For the latter structure,

$$E^{c}_{\text{total}} = \frac{1}{n}E^{a}_{s} + \gamma_{180^\circ,(010)}(n-1)A_b. \tag{3.43}$$

Assuming arbitrarily $n = 20$, we have $E^{c}_{\text{total}} = 1.57 \times 10^{-2} + 1.62 \times 10^{-6} \approx 0.016$ J. This total energy is 10 times less than E^{b}_{total}, but still is much greater than the wall energy.

Although the crystal dividing method can be made more attractive by adopting a very large value of n, another method is capable of eliminating the magnetostatic energy altogether. This method involves the introduction of closure domains that allow the flux circuit to be completed within the crystal, thereby avoiding the formation of magnetic poles at the free surface. Figure 3.13d shows four triangular prism domains separated by 90° walls lying in the {011} planes. In this configuration, the magnetization components parallel to the walls are completely cancelled, whereas those normal to the walls are continuous. Consequently, no free poles are produced at free surfaces of the crystal, resulting in zero E_s. There is, however, a magnetostrictive energy associated with this structure because the domains impose constraints upon each other, thus they are prevented from changing dimensions to relax the induced magnetostrictive stresses. To calculate E_e here, we can use (3.39) by setting $\epsilon_x = \lambda_{100}$ according to (3.18), $\sigma_x = \lambda_{100} \cdot c_{11}$, where c_{11} is an elastic constant, and other stress and strain components at zero. Thus,

$$E_{\text{total}}^{\text{d}} = E_e + E_w = \tfrac{1}{2}\lambda_{100}^2 \cdot c_{11} \cdot V + \gamma_{90°,(110)} A_d, \tag{3.44}$$

where A_d is the total domain wall area. Substituting $\lambda_{100} = 2.07 \times 10^{-5}$, $c_{11} = 2.41 \times 10^{11}\ \text{N} \cdot \text{m}^{-2}$, $\gamma_{90°,(110)} = 1.07 \times 10^{-3}\ \text{J} \cdot \text{m}^{-2}$ (from table 3.4), and $A_d = 2 \cdot \sqrt{2} \cdot 5 \cdot 10^{-5} = 1.41 \times 10^{-4}\ \text{m}^2$ in (3.44) gives $E_{\text{total}}^{\text{d}} = 5.16 \times 10^{-6} + 0.15 \times 10^{-6} = 5.31 \times 10^{-6}$ J. This energy is much smaller than those calculated for structures with magnetic poles on the free surface. Finally, both the crystal-dividing method and the formation of closure domains can be put together to achieve maximum reduction of magnetic energy. The structure shown in fig. 3.13e is essentially a combination of figs. 3.13c and 3.13d. The triangular domains at the (100) end faces serve as closure domains to prevent the formation of free poles. In this structure, $E_s = E_a = 0$ and

$$E_{\text{total}}^{\text{e}} = E_e + E_w = \tfrac{1}{2}\lambda_{100}^2 \cdot c_{11} V_c + \gamma_{180°,(010)} A_{180°} + \gamma_{90°,(110)} A_{90°}, \tag{3.45}$$

where V_c is the volume of closure domains and $A_{180°}$ and $A_{90°}$ are areas of the 180° and 90° walls, respectively. Assuming eight lamellar domains in fig. 3.13e, we obtain $d = 0.01/n = 1.25 \times 10^{-3}$ m, $V_c = \frac{1}{2}d^2 \cdot (0.005) \cdot n = 3.13 \times 10^{-8}\ \text{m}^3$, $A_{180°} = (n-1) \cdot (l - 2h) \cdot 0.005 = 3.06 \times 10^{-4}\ \text{m}^2$, $A_{90°} = 2n \cdot (d/\cos\frac{1}{4}\pi) \cdot 0.005 = 1.41 \times 10^{-4}\ \text{m}^2$. Substituting these values in (3.45) gives $E_{\text{total}}^{\text{e}} = (1.84 + 0.43 + 0.15) \times 10^{-6} = 2.42 \times 10^{-6}$ J, which is the lowest energy among the five domain structures considered. Indeed, domain structures observed in bcc crystals in the demagnetized state usually feature both closure and lamellar domains. Figure 3.14 shows such a domain structure obtained from a single crystal of Si–Fe, which has an orientation similar to that depicted in fig. 3.13a. When a magnetic field is applied in a direction close to [100], the lamellar domains with the [100] magnetization vector in fig. 3.13e grow at the expense of the $[\bar{1}00]$ domains. As the field strength is increased, the growth of the [100] domains continue, while the closure domains gradually shrink and eventually disappear before the onset of the knee in the magnetization curve. A single domain structure will exist beyond the knee where the

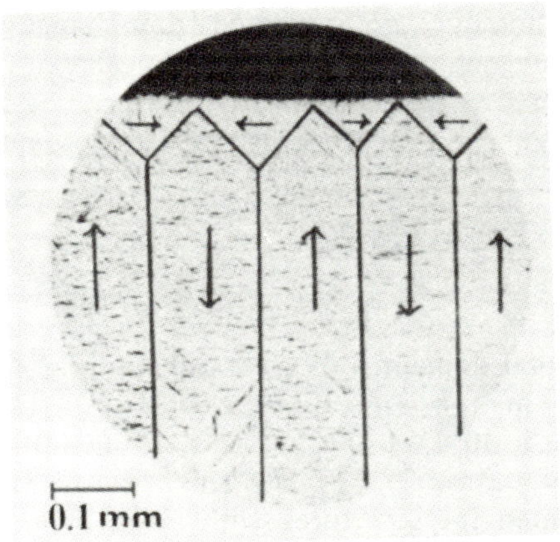

Fig. 3.14. Closure and lamellar domains observed on the (001) surface of silicon–iron (Chikazumi [1964]).

field is strong enough to provide the necessary energy to support free poles at the (100) faces. The single domain structure may also exist in single crystals of minute size if the magnetostatic energy is made negligibly small by virtue of the small volume involved. These crystals are called single domain particles.

As a second example, let us consider a single crystal of nickel shown in fig. 3.15a. This crystal is assumed to have the same size and shape as the iron crystal, but its square faces are assumed to lie in the (110)

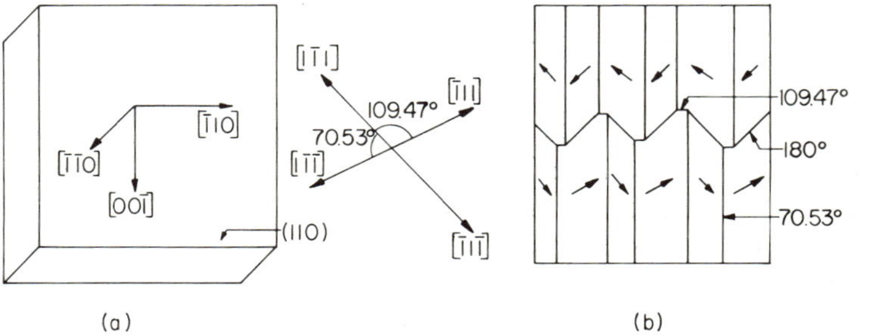

Fig. 3.15. Showing (a) a square slab of nickel single crystal having the top and bottom faces in the (110) plane and (b) an energetically favored domain structure which contains lamellar domains with zigzag walls.

plane so that they will contain four easy axes of magnetization; namely, $\pm[1\bar{1}1]$ and $\pm[\bar{1}11]$. A domain configuration that would minimize the magnetostatic energy and also keep the anisotropy energy at a minimum will contain lamellar domains with the magnetization vector aligned zigzag along the four ⟨111⟩ directions. Figure 3.15b depicts such a structure, in which the domain walls are 70.5°, 109.5° or 180°. If the crystal faces are changed to be in the (211) plane, which contains two easy axes in $\pm[\bar{1}11]$, the ideal domain pattern would consist of $\pm[\bar{1}11]$ lamellae domains separated by 180° walls. When the crystal faces are in the planes of either {100} or {111}, where there are no easy axes, the observed domain patterns contain dense branching lines, as shown in fig. 3.16, which may be interpreted as featuring fine closure domains compactly packed into the crystal.

For cobalt and other ferromagnets of uniaxial anisotropy, the domain structure predominantly consists of lamellae (fig. 3.17a). Such a structure can be improved by introducing closure domains (fig. 3.17b). In the

Fig. 3.16. Complicated domain patterns observed on a nickel crystal surface not containing an easy axis (Craik and Tebble [1965]).

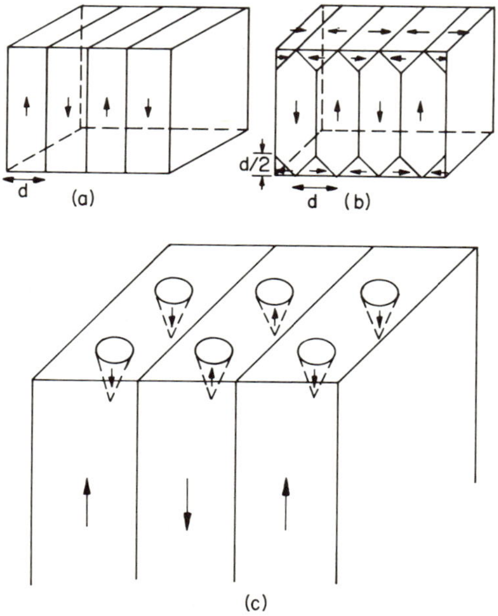

Fig. 3.17. Three possible domain structures in a single crystal of cobalt with the top face in the basal plane. Note the formation of island-like reverse domains in (c) (Chikazumi [1964]).

latter arrangement, the magnetostatic energy is eliminated, but, since the closure domains are magnetized in a hard direction, their presence would introduce a large anisotropy energy as well as a magnetoelastic energy. To reduce these energies, the closure domains can be modified into isolated conical-shaped domains with magnetization vectors in directions opposite to those of the lamellar domains (fig. 3.17c). Such island-like closure domains are called domains of reverse magnetization, or simply reverse domains. Figure 3.18 shows a checkerboard domain pattern observed on the (0001) surface of a cobalt single crystal. Each ring in the pattern represents a reverse domain. Such a domain pattern is commonly found in high anisotropy, uniaxial ferromagnets such as Co and certain feerites (e.g., $BaFe_{12}O_{19}$, and $BaCo_{1.85}Fe_{16.15}O_{27}$).

We have thus far considered the various domain structures under ideal conditions; that is, the crystal surface has exactly the specified orientation, and the specimen is free from strain and impurity particles.

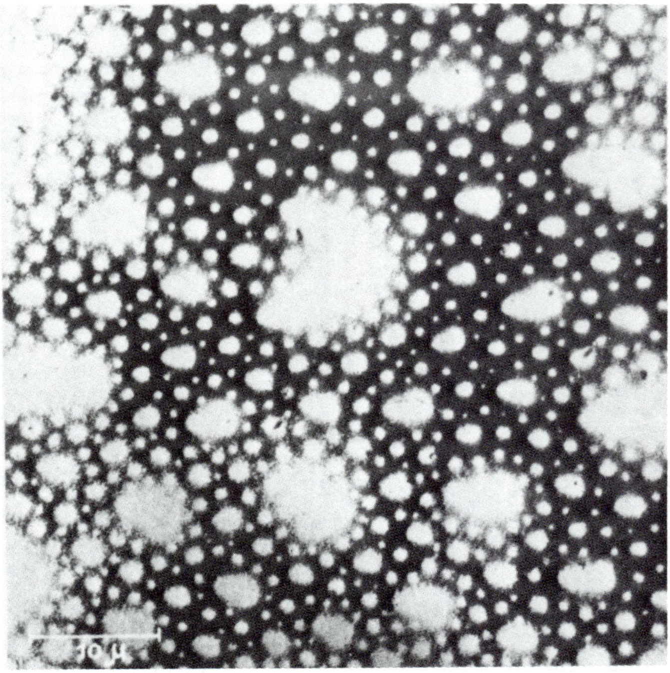

Fig. 3.18. Checkboard domain patterns observed on the (0001) surface of a cobalt single crystal in a high field (Craik and Tebble [1965]).

In practice, these ideal conditions are seldom met. Therefore, the actual domain structures observed are more complicated. A common complication is the attachment of needle-like domains to the lamellar domain walls. This occurs when the crystal surface deviates from the favorable orientation (e.g., the (100) surface for Fe and Si–Fe) by a few degrees. Figure 3.19 shows a lamellar structure decorated with a tree pattern observed on a Si–Fe crystal surface slightly off the (100) plane. The triangular domains are usually superficial. The reason for the superposition of the tree pattern is explained by the arrows, which show that the branch domains serve as closure domains to minimize the magnetostatic energy associated with free poles created on the inexact (100) surface.

Another important complication is the formation of spikes at cavities, inclusion particles, and sometimes grain boundaries. Figure 3.20b shows

Fig. 3.19. Lamellar domains decorated with superficial tree patterns on an inexact (110) surface of a Si–Fe crystal. Arrows indicate the distribution of magnetization vectors (Craik and Tebble [1965]).

typical spikes formed around a cavity. The interpretation of this domain pattern is given in fig. 3.20a. Néel [1944] predicted that the presence of spikes would redistribute the free poles along the elongated curved boundaries, instead of centering at the cavity shown in fig. 3.20a. Néel's calculation showed that the combination of the reduced magnetostatic energy and the wall energy in the spike structure is lower than the original magnetostatic energy associated with the free poles right at the cavity.

The present discussion of the domain structure is far from complete. Further discussion will be given under related subjects. For instance, the unique domain walls and domain structures found in thin films will be described in ch. 6, §3. The bubble-domain technology based on the formation and motion of isolated cylindrical domains in orthoferrites rare-earth garnets and amorphous transition metal-lanthanide alloys will be treated in ch. 7, §2.

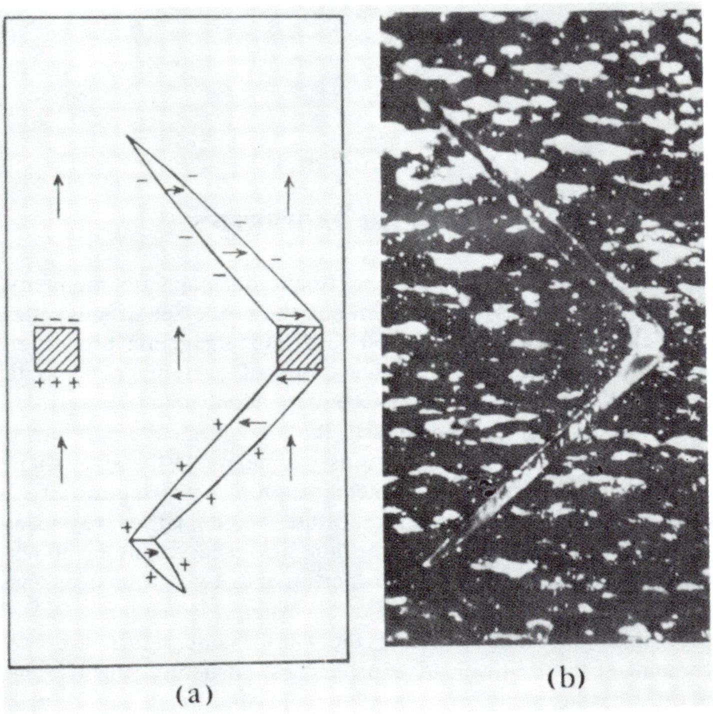

Fig. 3.20. Spike domains (a) predicted by Néel [1944] and (b) observed around a cavity (Kittel [1949]).

Suggestions for further reading

Brailsford, F., 1966, *Physical Principles of Magnetism* (van Nostrand, London). Chapters 6 and 7.

Chikazumi, S., 1964, *Physics of Magnetism* (Wiley, New York). Chapters 6 to 9 and 11.

Craik, D. J. and R. S. Tebble, 1965, *Ferromagnetism and Ferromagnetic Domains* (North-Holland, Amsterdam). On §5.

Lee, E. W., 1955, Rept. Prog. Phys. **18**, 184. On magnetostriction.

Kittel, C. and J. K. Galt, 1956, Solid State Phys. **3**, 437. On ferromagnetic domains.

Morrish, A. H., 1965. *The Physical Principles of Magnetism* (Wiley, New York). Chapter 7.

CHAPTER 4

MAGNETIC PROPERTIES

Chapters 2 and 3 dealt with the origin of magnetic moments, the characterization of ferromagnetism and the processes of magnetization. In this chapter, we shall proceed to discuss magnetic properties that play important roles in the applications of soft magnetic materials. Although all magnetic properties are dependent upon composition and temperature, not all are sensitive to the metallurgical conditions of the material. By metallurgical conditions we mean (A) the size, shape and orientation of grains; (B) the concentration and distribution of various crystal imperfections and (C) the state of the lattice regarding impurities, residual stresses and the atomic arrangement in alloys. Accordingly, we divide magnetic properties into two groups: intrinsic, or structure-insensitive, and extrinsic, or structure-sensitive.

Saturation magnetization and the Curie temperature are the two prominent structure-insensitive properties. Structure-sensitive properties, on the other hand, are numerous and may be classified further as static or dynamic, according to whether or not the property displays a frequency dependence. Induction, permeabilities, hysteresis loop and associated energy loss, coercive force and remanence are structure-sensitive, static properties. Eddy-current loss and resonance of spins and domain walls typify structure-sensitive, dynamic properties.

§1. INTRINSIC PROPERTIES

§1.1. *Saturation magnetization*

The maximum attainable intensity of magnetization per unit volume is termed the saturation magnetization of a ferromagnetic material. This quantity is denoted by $M_s(T)$, signifying that it varies as a function of temperature. As $T \to 0°K$, $M_s(T) \to M_0$, which is called the absolute saturation magnetization. For each material, M_0 is the largest value of M_s, because rising temperature generally exerts a decreasing effect on M_s, as is indicated by Bloch's $T^{3/2}$ law expressed in (2.72). When we state the value of M_s, we should specify the temperature; otherwise the statement is incomplete. This is particularly true in the temperature range above $0.8\theta_C$, where the decreasing effect of temperature on M_s becomes severe.

The spontaneous magnetization attained in each domain is identical with M_s. This identity, however, offers no advantage in the determination of M_s because it is impossible to measure the magnetization in individual domains. The usual experimental procedure for determining M_s is to conduct a saturation measurement under a strong field. The absolute saturation magnetization is determined by conducting a series of saturation measurements at cryogenic temperatures centering around 4.2°K (the boiling point of liquid helium at one atmospheric pressure) and extrapolating the M_s values to 0°K. The experimental methods for the determination of saturation magnetization are reviewed by McGuire and Flanders [1969].

The intrinsic nature of saturation magnetization can be verified both theoretically and experimentally. For the theoretical verification, we refer to (2.19), which defines M_A as $M_A = N_0 Jg\mu_B$. This definition, though arrived at in the quantum-mechanical treatment for paramagnetism, should be valid for the absolute saturation magnetization by virtue of the absolute zero temperature and the existence of the intense molecular field in a ferromagnet. On the basis of unit volume, we may rewrite the definition of M_0 by switching N_0 to N,

$$M_0 = N\mu = NJg\mu_B. \tag{4.1}$$

None of the parameters, N, J and g, in this expression is structure sensitive. Thus, theoretically we expect saturation magnetization to be insensitive to the metallurgical conditions of the specimen.

This definition of M_0 works well for the heavy lanthanide metals. Table 4.1 lists pertinent data for all nine ferromagnetic elements. The observed absolute saturation magnetization values for the lanthanides are in good agreement (within 11%) with the values predicted from (4.1). For Fe, Co and Ni, however, the comparison between the theory and measurements indicates not only large differences (up to 38%*) in

*Iron, cobalt and nickel derive their magnetic moments only from the spins of electrons in the 3d subshell because of the quenching effect discussed in ch. 2, §1. Hence, these three metals have a g factor of 2.0, but their J values are not known theoretically. The comparison between the experimental results and

[continued on p. 101]

TABLE 4.1
Intrinsic magnetic properties of ferromagnetic metals*

Metal	n_{eff}, Bohr magneton number per atom at 0°K	Saturation magnetization, 10^5 A · m^{-1} or 10^2 G, at		Curie temperature	
		0°K	17°C	°K	°C
I. Transition elements:					
Iron	2.216	17.45	17.08	1043	770
Cobalt	1.716	14.30	13.98	1400	1127
Nickel	0.616	5.22	4.84	631	358
II. Lanthanides:					
Gadolinium	7.55 (7.0)†	21.16	10.90	293	20
Terbium	9.34 (9.0)†	27.14		220	−53
Dysprosium	10.20 (10.0)†	29.89		88	−185
Holmium	10.34 (10.0)†	30.79		19	−254.2
Erbium	8.0 (9.0)†	24.15		19.5	−253.7
Thulium	7.14 (7.0)†	22.01		32	−241.2

*References: Danan et al. [1968] and Elliott [1972].

†Theoretical values of Bohr magneton numbers per atom given by *Jg*.

the M_0 value, but also a basic discrepancy in that the observed Jg values in the three metals are all nonintegral, instead of integral numbers predicted by the Heisenberg theory of localized spins. As was explained in ch. 2, §2, this discrepancy is mainly responsible for the proposal of the band model for the iron-group transition metals. Also, for the latter metals, (4.1) is replaced by the empirical formula

$$M_0 = Nn_{\text{eff}}\mu_B, \tag{4.2}$$

where n_{eff} is the effective, or apparent mean, Bohr magneton number per atom, which is 2.216, 1.716 and 0.616 for Fe, Co and Ni, respectively.

Experimentally, the independence of saturation magnetization on metallurgical conditions of the specimen can be demonstrated easily. In conducting the saturation measurements, for instance, it is immaterial whether the specimen is a single crystal or polycrystalline, nor would the state of internal stresses affect the result so long as the applied field is sufficiently strong to saturate the material. The temperature dependence of saturation magnetization has been discussed in connection with spin waves in ch. 2, §2; it will be further discussed in the next section.

§1.2. *The Curie temperature*

All ferromagnetic materials exhibit a characteristic temperature known as the Curie temperature or point, θ_C. This is the critical temperature at which thermal energy is just enough to destroy the spontaneous magnetization. Thus, the Curie temperature marks the transition point at which a ferromagnet is converted into a paramagnet upon heating. Recent studies also have revealed that ferromagnetic and ferrimagnetic materials display singularities in their thermodynamic behavior in the vicinity of θ_C known as the critical phenomenon. Readers interested in this phenomenon are referred to the review article by Domb [1965] and a NBS symposium [1966].

---

theory in the $M_s(T)/M_0$ versus T/θ_C plot shown in fig. 4.2, coupled with the likely electronic configurations, would seem to suggest that $J = \frac{1}{2}$ for Ni and $J = 1$ for Co and Fe. On the basis of such J and g values, the atomic moments are calculated from $\mu = Jg\mu_B$ to be 1 μ_B for Ni and 2 μ_B for Co and Fe. Thus the calculated values differ from the experimentally deduced values by approximately -11% for Fe, $+14\%$ for Co and $+38\%$ for Ni.

The Curie temperatures of the nine ferromagnetic elements are listed in table 4.1. They cover a wide range from the lowest values of 19°K for Ho and 19.5°K for Er to the highest value of 1400°K (1127°C) for Co. Immediately we may ask, "What causes such a wide variation in θ_C?" To answer this question, we need to express θ_C in terms of basic physical parameters. For this purpose, we combine (2.25a) and (2.45) to obtain

$$\theta_C = zJ_{ex}J(J+1)/3k, \tag{4.3}$$

where J_{ex} is the exchange integral defined in (2.37), J is the total angular momentum quantum number of the atom and z is the coordination number. This expression should be applicable to the lanthanide metals, for which the Heisenberg model of localized electrons is effective, and the theoretical value of J is substantiated by experiments. Thus we may use (4.3) to evaluate J_{ex}. The deduced values of J_{ex}, along with other pertinent data, are given in table 4.2. Despite the $J(J+1)$ dependence of θ_C expressed in (4.3), the relationship between $J(J+1)$ and θ_C as seen in table 4.2. is erratic. On the other hand, there is a large fluctuation in the deduced values of J_{ex}; more significantly, the fluctuation is consistent, at least qualitatively, with the variation in θ_C. For example, the highest value of J_{ex} is deduced for Gd and the lowest values for Ho and Er; correspondingly Gd also has the highest and Ho and Er have the lowest Curie temperatures among the lanthanides. Also note that the values of both θ_C and J_{ex} decrease in the same sequence on going from Gd → Tb → Dy → Tm → Er → Ho. It is concluded that θ_C is closely related to J_{ex}, not to $J(J+1)$, for the lanthanides. No analysis of this sort can be made for the iron-group metals because of the inapplicability of (4.3). A more realistic interpretation of the exchange

TABLE 4.2

An analysis of the Curie temperatures of ferromagnetic lanthanides according to (4.3)

Lanthanide	Gd	Tb	Dy	Tm	Er	Ho
θ_C (°K)	292.7	220	88	32	19.5	19
J	3.5	6	7.5	6	7.5	8
$J(J+1)$	15.75	42	63.75	42	63.75	72
$J_{ex} \times 10^{23}$ (J)	6.42	1.81	0.477	0.262	0.106	0.091

integral in (4.3) will be given for the ferromagnetic lanthanides in terms of the indirect exchange interaction between localized magnetic electrons and itinerant conduction electrons in ch. 5, §1.

In the quantum-mechanical treatment of the Langevin-type paramagnetism, we have obtained an expression, (2.18), for the attainable magnetization in a paramagnet,

$$M(T) = NJg\mu_B B_J(a),$$

where $B_J(a)$ is the Brillouin function expressed in (2.17), $a = Jg\mu_B\mu_0 H/kT$ and N is the number of magnetic atoms per unit volume. For a specific system with fixed J and g, a series of M–H magnetization curves can be plotted for various temperatures ranging from 0°K to $T > \theta_C$, as shown in fig. 4.1a. A straight line 0a is drawn to represent the Weiss field H_W based on the relation $H_W = \lambda M$. Because the Weiss field only exists below θ_C, 0a will or will not intersect the magnetizations curve, depending on whether $T < \theta_C$ or $T > \theta_C$. At $T < \theta_C$, 0a will intersect the curve at a point P_i other than the origin. It will be tangent to the curve at the origin at $T = \theta_C$ and will not intersect all other curves at $T > \theta_C$. Figure 4.1b is an enlarged drawing of the line 0a and the magnetization curve at T_2. The latter figure explains why

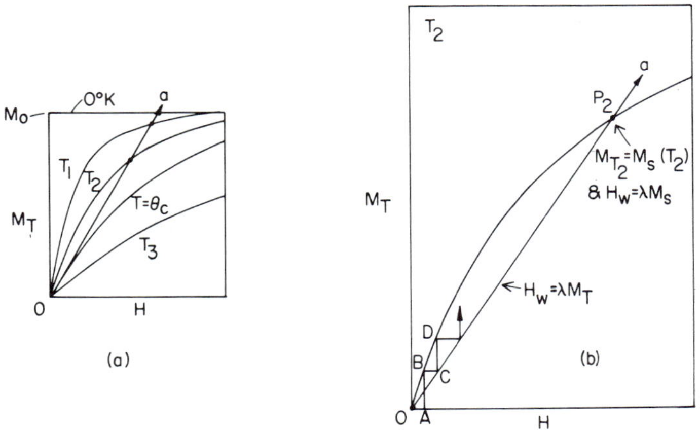

Fig. 4.1. (a) Superposition of the Weiss molecular field represented by line Oa to a series of M–H curves at various temperatures. (b) Illustrating the onset of spontaneous magnetization of value $M_s(T_2)$ in a ferromagnet at $T_2 < \theta_C$ (Brailsford [1966]).

only the spontaneous magnetization represents a stable situation for a ferromagnet at $T < \theta_C$. Suppose a small field A is accidentally applied, the material will be magnetized to value B according to the curve. The magnetization at level B, however, will spontaneously induce a Weiss field of strength C according to line 0a. In turn, the Weiss field will raise the magnetization to level D. Such a chain reaction will continue until magnetization finally reaches a value $M_s(T)$ determined by the point of intersection P_2 between the curve and the line. We thus have shown that the existence of the molecular field will induce a spontaneous magnetization to the saturation value $M_s(T)$ in a material that would otherwise be paramagnetic. Moreover, the value of $M_s(T)$ will increase steadily with decreasing temperature from θ_C.

For a quantitative examination of the temperature dependence of saturation magnetization in connection with the Curie temperature, let us first combine (2.45), (4.1) and (4.3) to obtain another expression for θ_C:

$$\theta_C = \mu_0(J+1)g\mu_B\lambda M_0/3k. \tag{4.4}$$

Referring to (2.20), first we replace $M_A(T)$ by $M_s(T)$, thus

$$M_s(T)/M_0 = B_J(a). \tag{2.20.1}$$

Then we apply (2.21) to set $H = H_W = \lambda M_s(T)$ in relation to (2.16). Now we have $a = gJ\mu_B\mu_0\lambda M_s(T)/kT$. Combining this expression of a and (4.4) into (2.20.1) yields

$$\begin{aligned}\frac{M_s(T)}{M_0} = \frac{2J+1}{2J}\coth\left\{\frac{3(2J+1)}{2(J+1)}\left[\frac{M_s(T)}{M_0}\right]\frac{\theta_C}{T}\right\} \\ -\frac{1}{2J}\coth\left\{\frac{3}{2(J+1)}\left[\frac{M_s(T)}{M_0}\right]\frac{\theta_C}{T}\right\}.\end{aligned} \tag{4.5}$$

This is the general expression for the temperature dependence of reduced saturation magnetization (i.e., $M_s(T)/M_0$). The expression is reduced to simpler forms for particular values of J. We cite three cases:

$$\begin{aligned}(1)\quad J=\frac{1}{2},\quad \frac{M_s(T)}{M_0} &= 2\coth\left[2\frac{M_s(T)}{M_0}\left(\frac{\theta_C}{T}\right)\right]-\coth\left[\frac{M_s(T)}{M_0}\left(\frac{\theta_C}{T}\right)\right] \\ &= \tanh\left[\frac{M_s(T)}{M_0}\left(\frac{\theta_C}{T}\right)\right],\end{aligned} \tag{4.6}$$

(2) $J = 1$, $$\frac{M_s(T)}{M_0} = \frac{3}{2}\coth\left[\frac{9M_s(T)}{4M_0}\left(\frac{\theta_C}{T}\right)\right] - \frac{1}{2}\coth\left[\frac{3M_s(T)}{4M_0}\left(\frac{\theta_C}{T}\right)\right], \tag{4.7}$$

(3) $J = \infty$, corresponding to the classical case in which all positions of the magnetic axes are equally allowed,

$$\frac{M_s(T)}{M_0} = \coth\left[3\frac{M_s(T)}{M_0}\left(\frac{\theta_C}{T}\right)\right] - \frac{1}{3}\frac{M_0}{M_s(T)}\left(\frac{T}{\theta_C}\right). \tag{4.8}$$

The value of $M_s(T)/M_0$ may be determined as a function of the reduced temperature T/θ_C by a graphical procedure or by algebraic substitution. The plot of $M_s(T)/M_0$ versus T/θ_C is shown in fig. 4.2 for the three J values. Also included in the figure are data points for Fe, Co and Ni. Experimental results are seen to fit best with the curve for $J = \frac{1}{2}$. Unfortunately, none of the three metals should have a J value of $\frac{1}{2}$, even for their free atoms with complete orbital quenching. This is an area where our present knowledge of ferromagnetism is still inadequate.

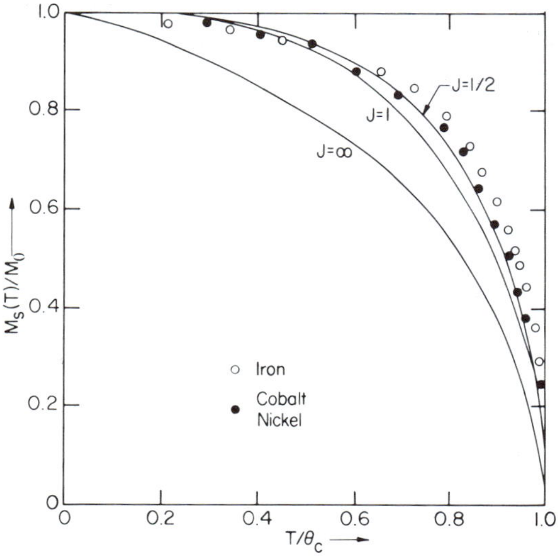

Fig. 4.2. Temperature dependence of saturation magnetization. The curves are obtained from the theory with $J = \frac{1}{2}$, 1 and ∞ and the points represent experimental data (Brailsford [1966]).

§2. STRUCTURE-SENSITIVE PROPERTIES

§2.1. *Static properties*

§2.1.1. *Induction and approach to saturation*

If the performance of a magnetic component or device depends on the magnetization of the ferromagnetic material involved, induction, rather than magnetization, is used to assess the performance. This is so because the definition of induction, $\boldsymbol{B} = \mu_0(\boldsymbol{H} + \boldsymbol{M})$ expressed in (1.6), has made it an inherently structure-sensitive static property through the $\mu_0\boldsymbol{H}$ term; and therefore induction is more useful than magnetization in practice. Of course, $\boldsymbol{B}$ and $\boldsymbol{M}$ are closely related as (1.6) indicates. In the B–H plot, the third part of the curve above the knee will eventually become a straight line with a small positive slope, implying that induction will never reach a limiting value because of the $\mu_0\boldsymbol{H}$ term involved in its definition. Figure 4.3, Curve I, illustrates this situation with a B–H curve. To overcome such an awkward situation, we rewrite the definition of B as

$$B - \mu_0 H = \mu_0 M \tag{4.9}$$

and plot the magnitude of $B - \mu_0 H$ against H as shown in fig. 4.3, curve II. The quantity $B - \mu_0 H$ is termed the intrinsic induction. Now we see that the third part of Curve II eventually reaches a ceiling value called

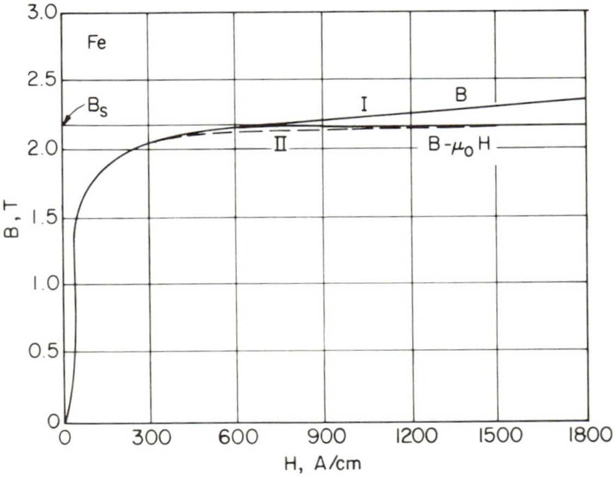

Fig. 4.3. Induction (curve I) and "reduced" induction, B–$\mu_0 H$, (curve II) plotted against the field showing the emergence of saturation induction, B_s.

the saturation induction B_s, i.e., $B - \mu_0 H \to B_s$. Because $B_s = \mu_0 M_s$, saturation induction is a structure-insensitive property, whereas induction is a structure-sensitive property.

Although saturation induction or magnetization per se is an intrinsic property, the magnetic field required to produce the saturation value varies according to the relative geometry of the field to the easy axes and other metallurgical conditions of the material. In other words, the approach to saturation is a structure-sensitive process. The problems associated with the approach to saturation were discussed as early as 1881, when Frölich established an empirical relation between M and H. Several other empirical equations have since been suggested, among which the ones proposed by Becker and Döring [1939] are considered most complete

$$M(T) = M_s(T)\left(1 - \frac{a}{H} - \frac{b}{H^2} - \cdots\right) + \chi_0 H, \tag{4.10a}$$

$$M(0°\mathrm{K}) = M_0\left(1 - \frac{a}{H} - \frac{b}{H^2} - \cdots\right), \tag{4.10b}$$

where $a, b, \ldots$ and χ_0 are constants. The term $\chi_0 H$ is inserted in (4.10a) to indicate an increase in spontaneous magnetization when a high field is applied to reduce the misalignment of spins caused by thermal agitation. Hence, this term is not included in (4.10b) for M at 0°K.

To demonstrate the validity of the empirical formulae, we first differentiate (4.10a) with respect to H,

$$\frac{\mathrm{d}M(T)}{\mathrm{d}H} = M_s(T)\left(\frac{a}{H^2} + \frac{2b}{H^3} + \cdots\right) + \chi_0. \tag{4.11}$$

Before (4.10a) and (4.10b) were introduced, Czerlinsky (1932, see Chikazumi [1964c]) had assumed that $M(T) \approx - M_s(T)/H^2$, which led to the relation: $\mathrm{d}M(T)/\mathrm{d}H \sim 2M_s(T)/H^3$. Thus he analyzed the data for Fe and Ni by plotting $\mathrm{d}M/\mathrm{d}H$ against H^{-3}. The results of his analysis are shown in fig. 4.4. His analysis showed that the approach to saturation is dominated by the $1/H^2$ term. The physical meaning of the constant b associated with this term is shown as follows.

The discussion on the processes of magnetization in ch. 3, §2 has attributed the third part of a magnetization curve to the rotation of the magnetization vector toward the field direction. Thus, at the final stage of magnetization, we have the relation

$$M(T) = M_s(T) \cos\theta = M_s(T)\left(1 - \frac{\theta^2}{2!} + \cdots\right), \tag{4.12}$$

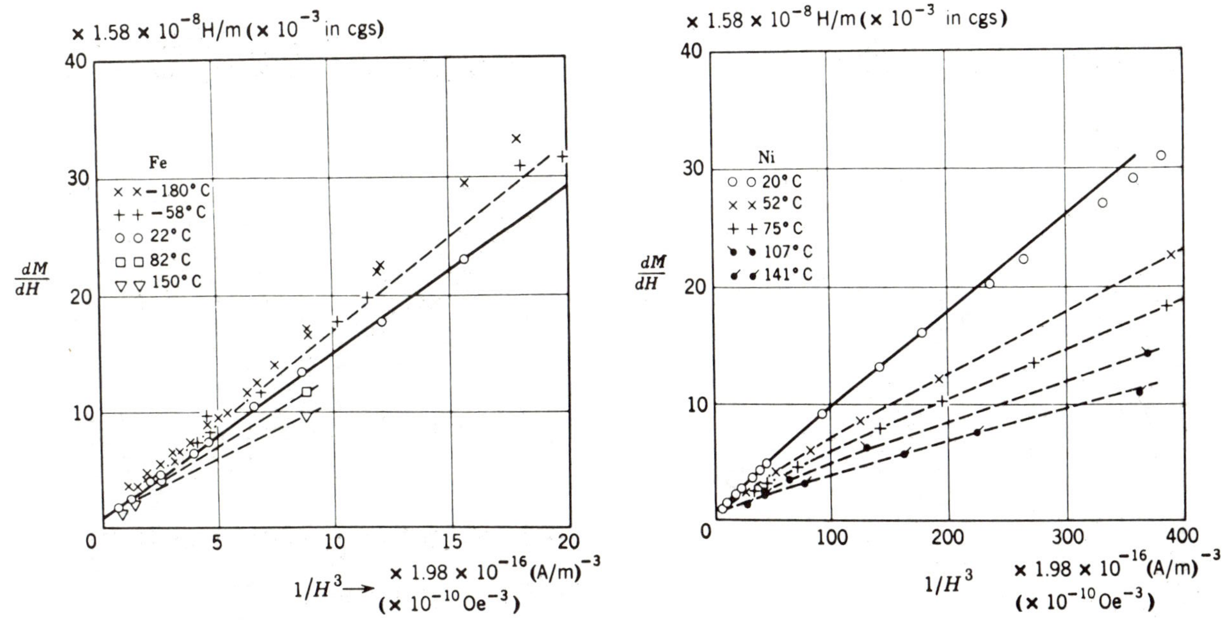

Fig. 4.4 Czerlinsky's analysis of the approach to saturation for (a) iron and (b) nickel. The lines are plotted by assuming $M(T) \sim M_s(T)/H^2$ and the points are experimental data (Chikazumi [1964]).

where θ is the angle between the magnetization vector and the applied field. As the magnetization vector is forced to turn away from the easy axis, the torque exerted by the field must be counterbalanced by the torque caused by the anisotropy. Hence

$$M_s H \mu_0 \sin\theta = -\,\partial E_a/\partial\theta, \tag{4.13}$$

where E_a is the anisotropy energy. If θ is small, $\sin\theta \approx \theta$; thus,

$$\theta = -\left(\frac{\partial E_a}{\partial\theta}\right)_{\theta\to 0}\bigg/ M_s H \mu_0. \tag{4.14}$$

Combining (4.12) and (4.14) gives

$$M(T) = M_s(T)\left(1 - \frac{b}{H^2} - \cdots\right), \tag{4.15}$$

with

$$b = \left(\frac{\partial E_a}{\partial\theta}\right)^2_{\theta\to 0}\bigg/ 2M_s^2(T)\mu_0^2. \tag{4.16}$$

For cubic systems it has been shown (see Bozorth [1951b] and Chikazumi [1964d]) that

$$(\partial E_a/\partial\theta)^2_{\theta\to 0} = 4K_1^2[(\alpha_1^6 + \alpha_2^6 + \alpha_3^6) - (\alpha_1^8 + \alpha_2^8 + \alpha_3^8) - 2(\alpha_1^4\alpha_2^4 + \alpha_2^4\alpha_3^4 + \alpha_3^4\alpha_1^4)], \tag{4.17}$$

where K_1 is the first anisotropy constant and $\alpha_i(\alpha_1, \alpha_2$ and $\alpha_3)$ are direction cosines of $\theta_i(\theta_1, \theta_2$ and θ_3 with respect to the three cubic axes). For polycrystalline specimens, $\overline{\alpha_i^6} = \frac{1}{7}$, $\overline{\alpha_i^8} = \frac{1}{9}$ and $\overline{\alpha_i^4\alpha_j^4} = \frac{1}{105}$, Therefore,

$$(\partial E_a/\partial\theta)^2_{\theta\to 0} = 4K_1^2(\tfrac{3}{7} - \tfrac{3}{9} - \tfrac{6}{105}) = \tfrac{16}{105}K_1^2, \tag{4.18}$$

and the constant b has the expression

$$b = \tfrac{8}{105}(K_1/M_s\mu_0)^2. \tag{4.19}$$

Attempts also have been made to explain the physical significance of other terms in (4.10a). The terms a/H and b/H^2 were attributed to the stress field around dislocations by Brown [1941, 1951] and the term a/H to voids or nonmagnetic inclusions by Néel [1948].

§2.1.2. *Permeabilities*

A basic definition of magnetic permeability was given in ch. 1, §3, which simply equate μ to B/H, the ratio of the magnetic induction to the amplitude or strength of the applied field. Specifically, this definition refers to the normal permeability when the material is in a "cyclic" magnetic state. A ferromagnet is said to be in such a state if it responds equally to a field applied to either of the two opposite directions. To attain a cyclic state, the specimen must be demagnetized completely by either of the following two methods:

(1) Heating to a temperature above the Curie point, followed by slow cooling in zero field.
(2) Magnetizing under an alternating field, first to a value close to saturation induction, then to zero induction by gradually decreasing the amplitude of the field.

The magnetization curve obtained under a d.c. field right after the specimen was treated by (1) is called the "virgin" curve. Meanwhile, a B (or $B-\mu_0 H$ or M) versus H curve obtained from the specimen treated by (1) after a virgin curve had been taken or by (2) is referred to as a "commutation" curve. The latter curve represents the locus of the tips of a series of hysteresis loops taken with increasing amplitudes of H, the field having been reversed several times at each amplitude. From a virgin or a commutation curve, various values of normal permeability are determined at different field strengths.

Since the relationship between B and H is far from linear in the first and second parts of a magnetization curve, the description of a permeability has little meaning unless the accompanying field is indicated. Unspecified values of permeability usually are assumed to be for the maximum permeability. There is no guarantee, however, that such practice is always in force. To circumvent this ambiguity, permeability has been expressed in different forms. The following four permeabilities are particularly important:

(A) Initial permeability, μ_i – the limiting value approached by normal permeability when H is reduced toward zero.
(B) Maximum permeability, μ_{max} – the largest value of normal permeability obtained from a virgin B–H curve.
(C) Incremental permeability, $\mu_\Delta = B_\Delta / H_\Delta$, where H_Δ is an alternating field applied in addition to a constant biasing field and B_Δ is the alternating induction caused by H_Δ. See figs. 3.2 and 4.5.
(D) Differential permeability, $\mu_d = dB/dH$ – simply the slope of the B–H curve at each point.

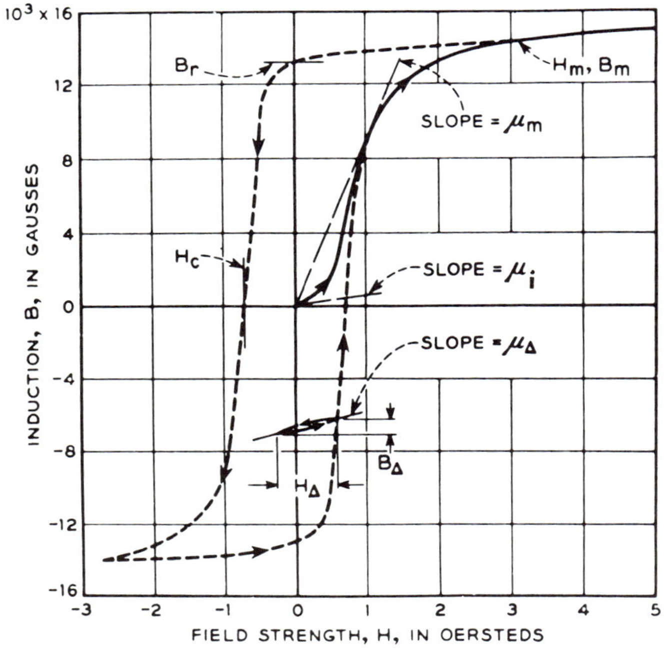

Fig. 4.5. A virgin magnetization curve (solid) and an associated hysteresis loop (dotted). Some structure-sensitive magnetic properties (H_c, B_r, μ_i, μ_{max} shown as μ_m and μ_Δ are illustrated (Bozorth [1951]).

Figure 4.5 shows that μ_{max}, μ_Δ and, of course, μ_d can be evaluated from appropriate B–H curves. The determination of μ_i, however, requires the use of the Rayleigh relation. According to (3.1), the relation can be expressed as

$$\mu = \mu_i + \nu H, \tag{4.20}$$

where μ is the normal permeability in the field H and $\nu = d\mu/dH$. This relation allows the initial permeability to be evaluated from the intercept on the ordinate axis in a μ–H plot. If a linear relationship exists between μ and H near zero field, extrapolation of the line drawn through data points will give a unique intercept on the ordinate axis, thus rendering a reliable determination of μ_i. Indeed, many soft magnetic materials behave in accordance with (4.20), as is shown in fig. 4.6.

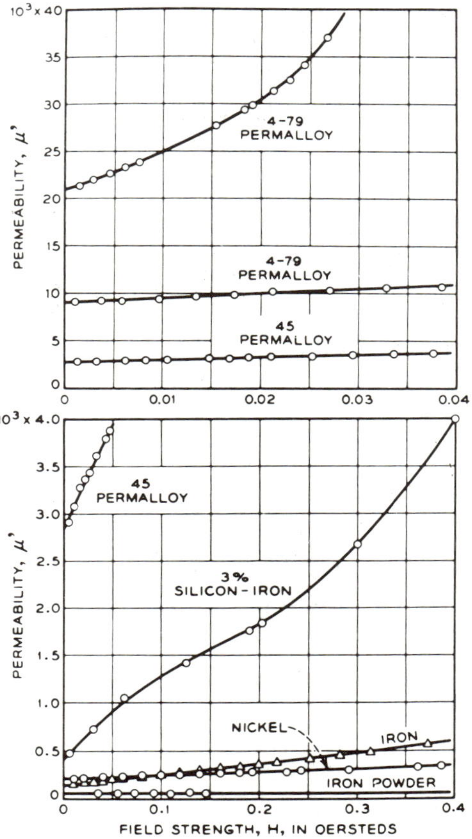

Fig. 4.6. Relative permeability curves of several prominent soft magnetic materials. The curves are plotted according to the Rayleigh relation on two sets of scales near zero field (Bozorth [1951]).

This group of "well-behaved" materials includes the 79% Ni–Fe alloy, whose unusually high value of initial permeability is the main attraction for applications. On the other hand, a few materials, including the important Fe–Si alloys, show nonlinear μ–H relationships near zero field. The evaluation of μ_i then becomes uncertain for the latter materials. Fortunately, the use of Si–Fe alloys generally is not motivated by their mediocre initial permeabilities; therefore, the uncertainty involved creates no major problem. A typical μ–H curve for the Si–Fe alloys also is shown in fig. 4.6.

To explore the physical significance of both the initial permeability and its structure dependence, we recall that early discussion in ch. 3, §1 has identified the process of magnetization in the initial part of the M–H or B–H curve to be the reversible movement of domain walls. In a perfect lattice, since the surface energy density, γ_w, of domain walls is independent of the position of the walls, the displacement of a domain wall will not cause any change in γ_w. This means that the displaced wall will not return to its original position when the field is withdrawn. In other words, if the crystal were completely free from defects and internal stresses, the movement could not be stopped, thereby leading to an infinite initial permeability. This suggests that the initial permeability is strongly dependent upon the structural conditions of the specimen. Also, from the finite, though sometimes very large, values of μ_i observed heretofore, we conclude that crystals can never be rid of defects or internal stresses completely.

A simple expression can be derived for μ_i as follows. For real crystals decorated randomly with defects, it is reasonable to assume that the surface energy density of domain walls varies with the displacement of the wall d, as shown in fig. 4.7a. The spatial distribution curve of γ_w features valleys and hills, representing equilibrium positions of domain walls and locations of defects, respectively. Now consider a plane 180° wall, and assume that the original position of the wall is at the minimum point, x_0, in a valley where $\partial\gamma_w/\partial d = 0$. Under these conditions, the increase in γ_w may be expressed approximately in the vicinity of x_0 by

$$\Delta\gamma_w = \tfrac{1}{2}\alpha d^2, \tag{4.21}$$

where α is a constant. When a magnetic field is applied in a direction slightly off the magnetization vector in the left domain by a small angle θ (fig. 4.7b), the wall will be displaced to the right by a distance d. This displacement produces a magnetization of $2M_s d$ per unit area of the wall and reduces the potential energy in $\mathrm{J \cdot m^{-2}}$ of the crystal by an amount per unit area of domain wall

$$E_H = -2M_s H\mu_0 \cos\theta \cdot d. \tag{4.22}$$

On the other hand, there will be an increase in γ_w given by (4.21). Hence, the total change in energy of the crystal is

$$\Delta E = \tfrac{1}{2}\alpha d^2 - 2M_s H\mu_0 \cos\theta \cdot d. \tag{4.23}$$

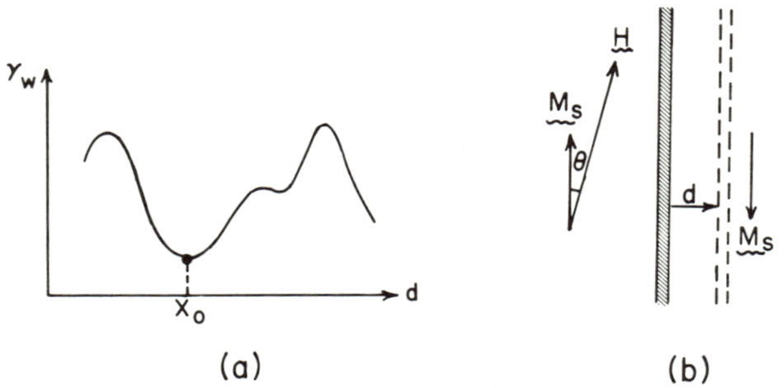

Fig. 4.7. Illustrating (a) nonuniform spatial distribution of the surface energy density of domain walls and (b) displacement of a 180° domain wall induced by a field.

Minimizing ΔE gives

$$d = (2M_s \cos \theta/\alpha)H\mu_0. \tag{4.24}$$

The total magnetization produced in the direction of H per unit volume is given by

$$M = 2M_s \cdot (2M_s \cos \theta/\alpha)H\mu_0 \cdot \cos \theta \cdot A, \tag{4.25}$$

where A is the total surface area of 180° walls in a unit volume and has the unit of m^2/m^3, thus m^{-1}. Thus the initial susceptibility is expressed as

$$\chi_i = (4M_s^2\mu_0 \cos^2 \theta/\alpha)A. \tag{4.26}$$

For polycrystals of cubic materials, $\cos^2 \theta$ has the average value of $\frac{1}{3}$. Hence,

$$\chi_i = 4M_s^2\mu_0 A/3\alpha \quad \text{and} \quad \mu_i = \mu_0(1 + 4M_s^2\mu_0 A/3\alpha). \tag{4.27}$$

These are general expressions of the initial suceptibility and initial permeability based on the movement of 180° walls. They clearly indicate that the structure dependence of χ_i and μ_i essentially is represented by the constant α because M_s is an intrinsic property and A only varies with the size and orientation of grains, not with other structural factors. If the initial magnetization involves the movement of 90° walls, the two equations in (4.27) remain in the same form except that the coefficient is changed from $\frac{4}{3}$ to $\frac{2}{3}$.

The physical meaning of α has been explored in two models. The inclusion model of Kersten will be discussed in conjunction with the coercive force in §2.1.5. The other model emphasizes the role played by internal stresses. For simplicity, let us assume the existence of an internal stress σ, whose magnitude varies sinusoidally with the distance d; that is,

$$\sigma = \Delta\sigma \cos 2\pi(d/l), \tag{4.28}$$

where $\Delta\sigma$ and l are the amplitude and wavelength of the varying stress, respectively. As has been discussed in ch. 3, §4, the presence of internal stresses in polycrystalline materials will alter the first anisotropy constant K_1 through the magnetoelastic effect by an amount $-\frac{3}{2}\lambda_0\sigma$, according to (3.23), where λ_0 is the isotropic longitudinal magnetostriction. In the internal stress model, therefore, we replace K_1 by an effective K in (3.36) for γ_w, where

$$K_{\text{eff}} = K_1 - \tfrac{3}{2}\lambda_0\Delta\sigma \cos 2\pi(d/l). \tag{4.29}$$

Now we have

$$\gamma_w = 2[(\pi^2 J_{ex} S^2/a)K_{\text{eff}}]^{\frac{1}{2}}. \tag{4.30}$$

If $K_1 \gg \lambda_0\sigma$, and if we use the power series $\cos\theta = 1 - \theta^2/2 + \theta^4/4 - \cdots$, (4.30) becomes

$$\begin{aligned}\gamma_w &= 2\pi S\left(\frac{J_{ex}K_1}{a}\right)^{\frac{1}{2}}\left[1 - \frac{3\lambda_0\Delta\sigma}{4K_1}\cos 2\pi\frac{d}{l} + \cdots\right]\\ &= 2\pi S\left(\frac{J_{ex}K_1}{a}\right)^{\frac{1}{2}}\left[1 - \frac{3\lambda_0\Delta\sigma}{4K_1}\left(1 - \frac{2\pi^2 d^2}{l^2} + \cdots\right) + \cdots\right].\end{aligned} \tag{4.31}$$

From the relation $\alpha = \partial^2\gamma_w/\partial d^2$, we obtain

$$\alpha = 6\pi^3 S(J_{ex}/K_1 a)(\lambda_0\Delta\sigma/l^2). \tag{4.32}$$

In terms of the thickness of the wall, d_w, expressed in (3.35),

$$\alpha = 6\pi^2(\lambda_0\Delta\sigma d_w/l^2). \tag{4.33}$$

Combining (4.27) and (4.33) yields

$$\chi_i = \frac{2M_s^2 l^2 \mu_0 A}{9\pi^2\lambda_0\Delta\sigma d_w} \quad \text{and} \quad \mu_i = \mu_0\left(1 + \frac{2M_s^2\mu_0 l^2 A}{9\pi^2\lambda_0\Delta\sigma d_w}\right) \tag{4.34}$$

which indicates that initial susceptibility and initial permeability would vary inversely with the amplitude (or the maximum magnitude) of the stress waves existing in the material.

§2.1.3. *Hysteresis loop and energy loss*

The word hysteresis has its origin in the Greek word hysterein which means "to lag behind." Physically, a hysteresis refers to a damping phenomenon in which two conjugate quantities such as stress and strain, magnetic induction and magnetizing field, etc. lag behind each other or become out of phase, thus damping some of the (mechanical, magnetic, . . .) energy and thus heating the lattice. Mechanical and magnetic hystereses are the most well-known damping phenomena.

The evidence for a magnetic hysteresis is the loop that appears in the four-quadrant plot of B versus H. A brief explanation of the hysteresis loop has already been given in ch. 3, §1. In the present section, we shall discuss the energy loss associated with hysteresis, followed by a general description of magnetization processes in various parts of a hysteresis loop.

In the B–H plot, the area enclosed by a hysteresis loop represents an energy given by

$$W_h = \oint H \, dB. \tag{4.35}$$

This magnetic energy is converted to heat which dissipates into the lattice immediately upon generation and is permanently lost. Thus W_h is called the hysteresis loss in units of joules per m^3 per cycle. With soft magnetic materials, hysteresis loss is very undesirable, not only because it wastes energy, but also because the heat involved could create further problems in the application of the material. Many industrial laboratories have devoted major research and development efforts to reducing hysteresis loss in such products as Fe–Si and Ni–Fe alloys.

In general, the most reliable method for the determination of the hysteresis loss is to obtain the pertinent loop experimentally and measure W_h from the loop area graphically. This method is applicable to any loop no matter how irregularly shaped. Indeed, the shape of hysteresis loops varies widely with (A) the magnetizing field (fig. 4.8), (B) the composition of the material (fig. 4.9), (C) the metallurgical conditions of the samples (fig. 4.10) and (D) the temperature (fig. 4.11). In these selected examples, we have shown six distinct shapes of hysteresis loops:

(1) nearly perfect ellipses at small magnetizations (fig. 4.8),
(2) distorted ellipses that are stretched and bent at the tips (figs. 4.10a and 3.2),
(3) squares (fig. 4.8c),

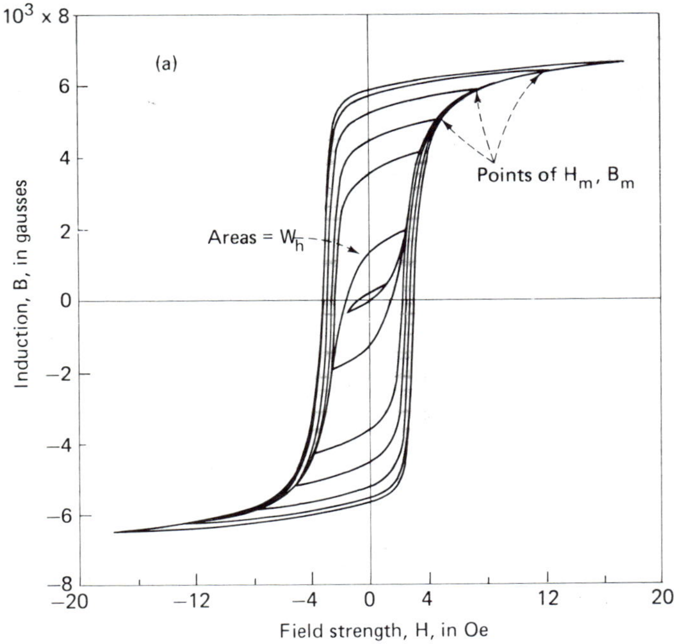

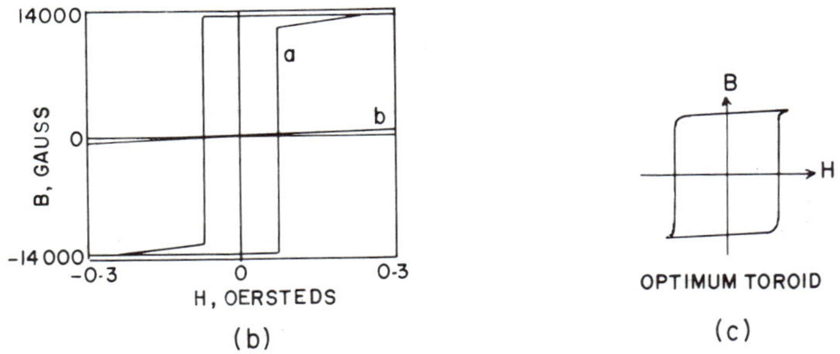

Fig. 4.8. Field dependence of hysteresis loops: (a) family of loops for 4-79 Permalloy (Bozorth [1951]), (b) loops for Perminvar (45% Ni, 30% Fe and 25% Co) measured with the field parallel (curve a) and perpendicular (curve b) to the direction of magnetic anneal-induced anisotropy and (c) square loop for a ferrite memory core with optimum toroid features.

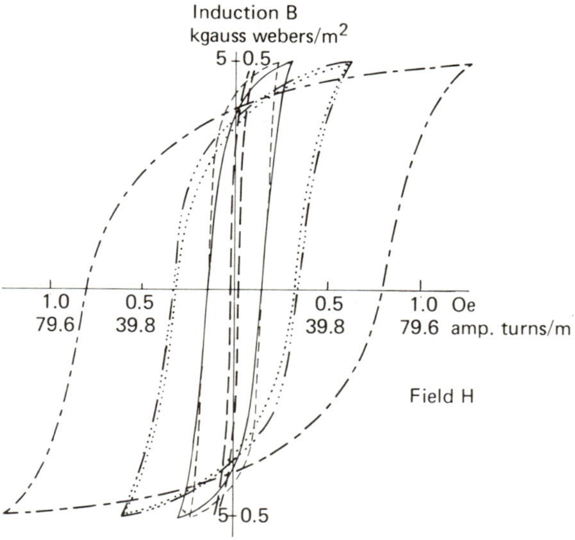

Fig. 4.9. Composition dependence of hysteresis loops (De Barr [1953]). 0.2% silicon–iron — — — — —; 4% silicon–iron · · · · · · · · · · · ·; 3% cold-reduced silicon–iron ————————; 78% nickel–iron* — — — — —; 50% nickel–iron* ---------; 36% nickel–iron* ——— · · · ———. (*With small amounts of other metals added).

(4) rectangles, fat or slender (fig. 4.10c),
(5) constricted loops that are unusually narrow near the origin (fig. 4.10b) and
(6) loops virtually converged to a straight line through the origin (figs. 4.8 and 4.11).

Obviously, the energy loss associated with square and rectangular loops is given by

$$W_h = 4B_rH_c, \tag{4.36}$$

where B_r and H_c are the characteristic remanence and coercive force, respectively. whenever a rectangular loop is required for applications of a soft magnetic material, efforts should be made to keep the coercive force, not the remanence, as small as possible to minimize the hysteresis loss.

The nearly perfectly elliptical loops produced at small magnetizations are known as Rayleigh loops. An enlarged Rayleigh loop is shown in fig.

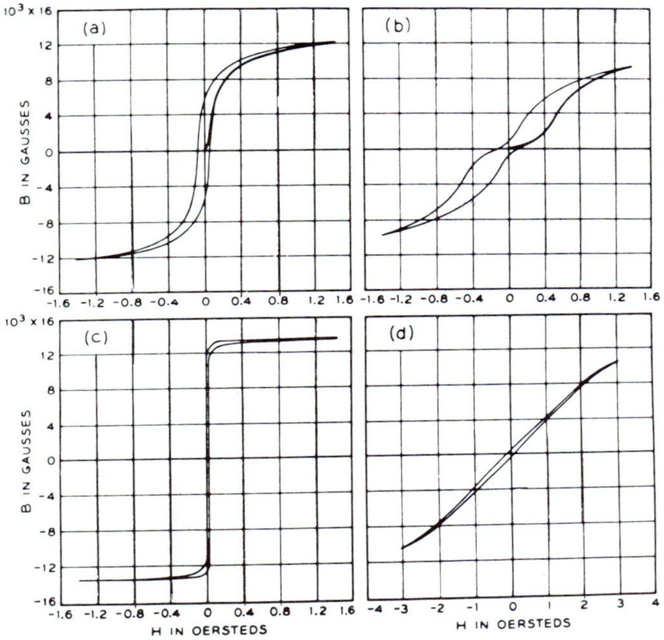

Fig. 4.10. Various shapes of hysteresis loops measured for 65 Permalloy (a) annealed at 1000°C, (b) baked at 425°C for 24 h, (c) annealed in a longitudinal field and (d) annealed in a transverse field.

4.12. According to Lord Rayleigh, the loop may be resolved into two branches C′B′C and CBC′, to each of which (3.1) applies. For a more convenient application of this equation, we set the origin of the lower branch C′B′C at C′. Consequently, this branch is represented by (3.1) in the form

$$B + B_m = \mu_i(H + H_m) + \nu(H + H_m)^2. \tag{4.37}$$

Since $B = B_m$ at $H = H_m$, we have

$$B_m = \mu_i H_m + 2\nu H_m^2. \tag{4.38}$$

Similarly, we set the origin of the upper branch CBC′ at C and rewrite (3.1) for this branch as

$$B - B_m = \mu_i(H - H_m) - \nu(H - H_m)^2. \tag{4.39}$$

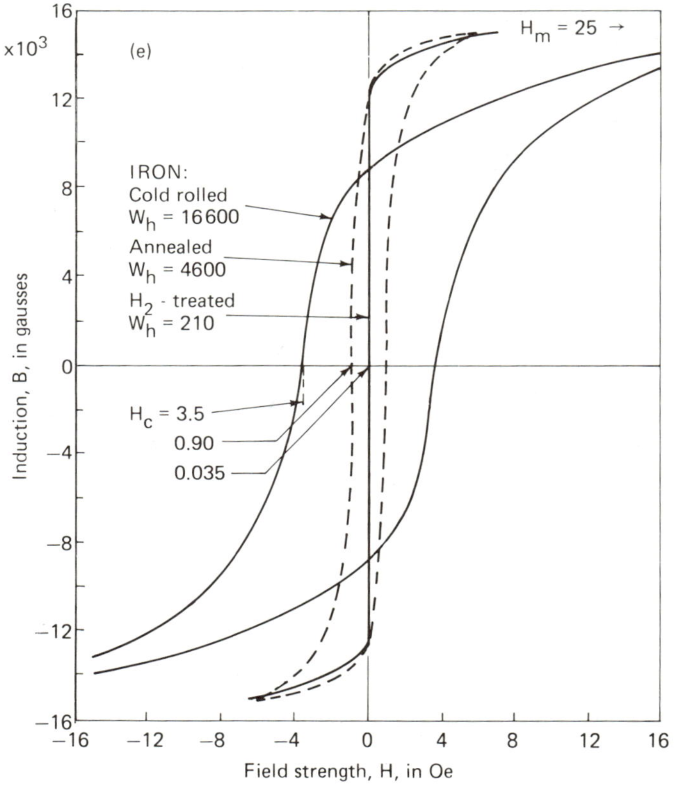

Fig. 4.10 (cont.). Various shapes of hysteresis loops measured for 65 Permalloy (e) for iron in three conditions (cold-rolled, annealed and hydrogen treated).

Note the opposite signs for the second term in the right-hand side of (4.37) and (4.39), which reflect the concave upward and downward shapes of the lower and upper branches, respectively. Substituting (4.38) in (4.37) and (4.39) gives

$$B = (\mu_i + 2H_m)H + \nu(H^2 - H_m^2), \tag{4.40}$$

$$B = (\mu_i + 2H_m)H - \nu(H^2 - H_m^2). \tag{4.41}$$

The hysteresis loss is divided into two parts: $\int_{-B_m}^{+B_m} H\,dB$ in association with the lower branch and $\int_{+B_m}^{-B_m} H\,dB$ with the upper branch. So

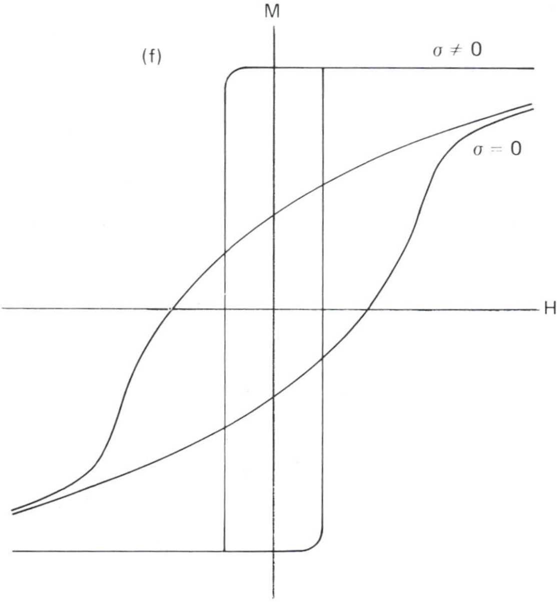

Fig. 4.10 (cont.). Various shapes of hysteresis loops measured for 65 Permalloy (f) for 68 Permalloy with and without applied tension (Bozorth [1951]).

$$W_h = \int_{-B_m}^{+B_m} H \, dB + \int_{+B_m}^{-B_m} H \, dB. \tag{4.42}$$

Substituting (4.40) in the first integral and (4.41) in the second integral of (4.42) gives

$$W_h = 4\nu \int_{-H_m}^{+H_m} H^2 \, dH = \tfrac{8}{3} \nu H_m^3. \tag{4.43}$$

This result of the Rayleigh analysis signifies that the energy loss associated with a Rayleigh loop is proportional to the third power of the amplitude of the applied field. Table 4.3 lists values for μ_i/μ_0, ν and W_h obtained at $B = 10^{-4}$ T from the Rayleigh analysis for a number of soft magnetic materials. The Rayleigh hysteresis loss fluctuates widely among various materials.

Under a very small magnetic field, the Rayleigh loop converges into a straight line through the origin. This is the case where the hysteresis loss is diminished to zero and the induction is given simply by $B = \mu_i H$. This case also prevails at temperatures near the Curie point, as is shown in fig. 4.11.

To understand the magnetization processes involved in various parts of a hysteresis loop, let us consider a stretched elliptical loop shown in fig. 4.13. Such a loop is representative of polycrystalline samples of Fe–Si and a host of other materials. At the tip C, the sample is in the state of a single domain with the magnetization vector $\boldsymbol{M}$ aligned close

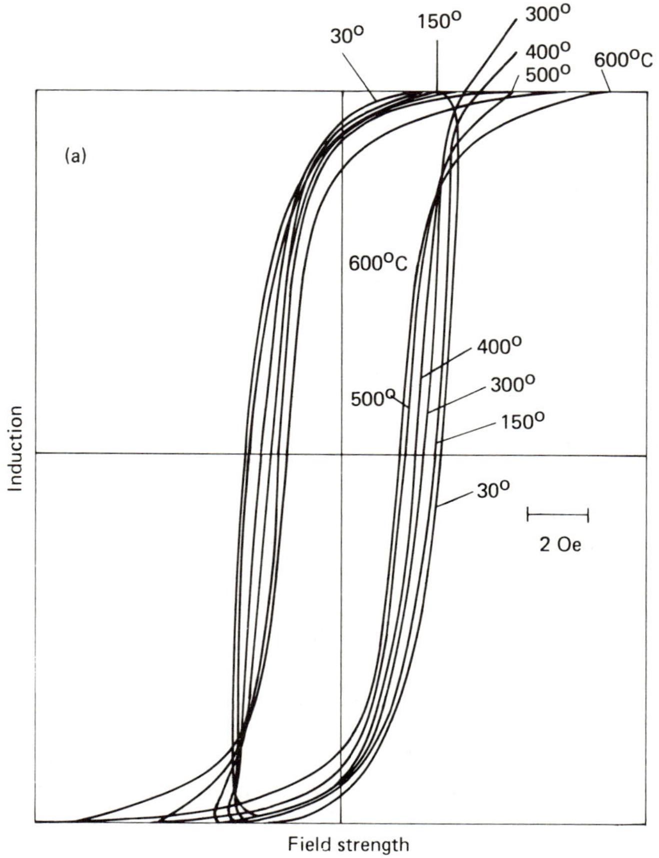

Fig. 4.11. Temperature dependence of hysteresis loops observed in grain-oriented 3% Si–Fe sheets at temperatures (a) between 30 and 600°C.

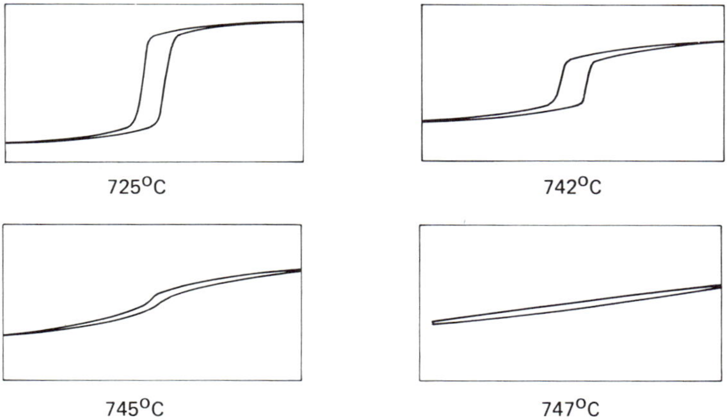

Fig. 4.11 (cont.). (b) close to the Curie temperature (Chen [1958]).

to the direction of the applied field $\boldsymbol{H}$. From C to B_r, as H is reduced from H_m to zero, the main event is the reversible rotation of $\boldsymbol{M}$ away from $\boldsymbol{H}$ so that $\boldsymbol{M}$ can be realigned into the nearest easy axis in individual grains. In the vicinity of B_r, the important event is the nucleation of new domains, some of which may be of the closure type. Along the knee of the upper branch of the loop below B_r, a small Barkhausen effect is detected, signifying the onset of irreversible jumps of newly-formed domain walls under the influence of a small reverse field. Between D and F, a pronounced Barkhausen effect is detected, indicating that some domain walls make large jumps. This segment is divided into two parts. Between D and E, the new domains grow

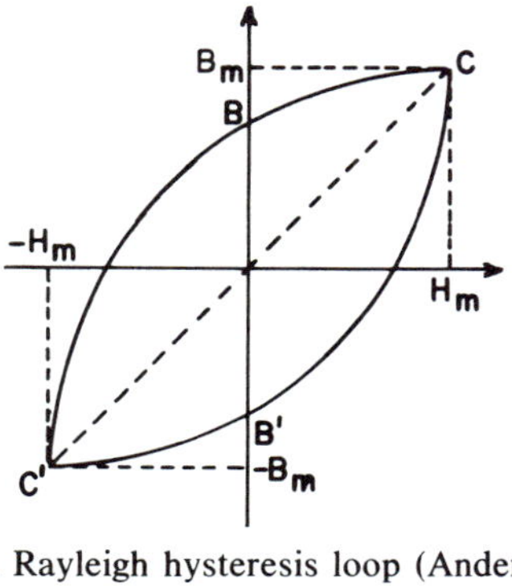

Fig. 4.12. A Rayleigh hysteresis loop (Anderson [1968]).

TABLE 4.3

Values for initial relative permeability (μ_i/μ_0), relative Rayleigh constant (ν/μ_0) and hysteresis energy loss calculated from (4.43) with $B = 10^{-4}$ T for select ferromagnet materials*

Material	μ_i/μ_0	ν/μ_0 (per $A \cdot m^{-1}$)	W_h (J/m^3/cycle)
Iron (sheets, annealed)	200	12.57	2.7×10^{-6}
Pressed iron powder	30	6.5×10^{-3}	0.39×10^{-6}
Cobalt	70	6.5×10^{-2}	0.03×10^{-6}
Nickel	220	1.55	0.25×10^{-6}
Supermalloy (≈80% Ni–Fe)	100,000	7.5×10^{4}	0.1×10^{-9}
45–25 Perminvar	400	6.5×10^{-4}	0.02×10^{-9}

*Reference: Anderson [1968].

through the irreversible movement of walls to the largest sizes at E, where the negative coercive force emerges. From E to F, the magnetization process remains the same as in D–E, but the growing domains begin to shrink until they are eliminated around F. The final

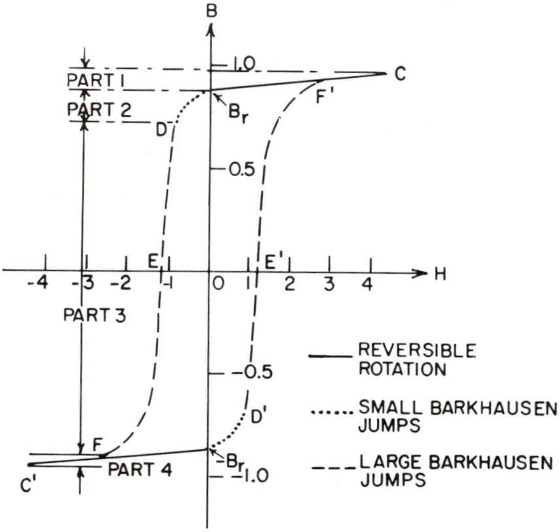

Fig. 4.13. The division of a half hysteresis loop into four parts.

segment FC′ involves the rotation process of $\boldsymbol{M}$ in reverse of that depicted for the segment C–B_r. To sum up, each half loop is divided roughly into four parts as shown in fig. 4.13. When the two branches are combined to form a complete loop, however, only two magnetization processes appear to dominate the loop – the irreversible movement of domain walls between $-B_r$ and B_r and the reversible rotation of magnetization beyond the remanence. This simplified approach to hysteresis loops is sketched in fig. 4.14. The sketch also demonstrates that hysteresis loss is approximately equal to $2B_r \cdot 2H_c$, as expressed in (4.36) for rectangular and square loops. Therefore, we conclude that the hysteresis energy loss is predominantly due to the irreversible movement of domain walls.

The foregoing discussion is an oversimplification of the actual situation. In reality, the loop division is seldom so clear-cut. Nevertheless, this simple approach serves to emphasize, at least qualitatively, that W_h is largely caused by the irreversible wall movements. More quantitative analyses of hysteresis loops have been made for special cases. For instance, Stoner and Wohlfarth [1948] followed closely the orientation of $\boldsymbol{M}$ in a single-domain ellipsoidal particle as a

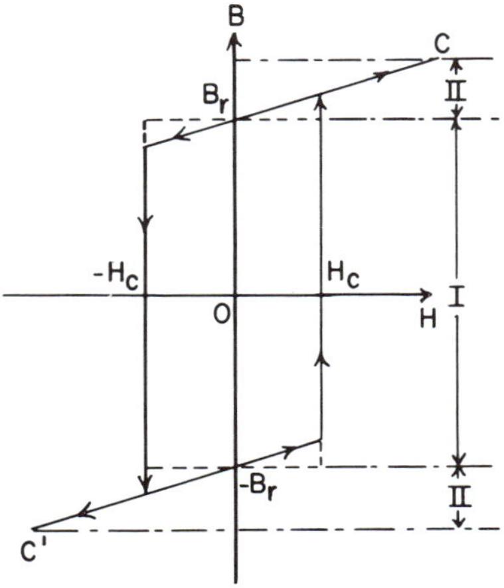

Fig. 4.14. In a simplified approach, the whole loop is shown to be dominated by (I) the irreversible movement of domain walls between remanences and (II) the rotation of magnetization vectors from an easy axis to the field direction between remanence and the tip.

function of the field applied perpendicular to the polar axis of the particle. Various shapes of hysteresis loop were predicted, and the associated energy losses were calculated. In their case, W_h is entirely due to the rotation of $\boldsymbol{M}$. Chikazumi [1964e] made a detailed analysis of hysteresis loops for aggregates of fine particles with uniaxial anisotropy. He showed that the dominant process of magnetization varies according to whether the particles have single or multiple domains.

§2.1.4. *Coercive force and coercivity*

Hysteresis loops were shown (ch. 3, §1 and the preceding section) to feature a characteristic field known as the coercive force, H_c, and a characteristic induction known as the remanence, B_r. If we start with a Rayleigh loop at a small magnetization and gradually strengthen the alternating field, the loop will grow in all directions, as shown in fig. 4.8a. Correspondingly, the values of H_c and B_r will rise steadily until they eventually reach their maximum, known as the coercivity, $H_{c,\infty}$, and the retentivity, $B_{r,\infty}$, respectively. The maxima emerge when the tips of the loop approach the value of saturation induction, B_s. The dependence on composition and structure and the physical significance of coercive force and coercivity will be discussed in this section, and those of remanence and retentivity will be dealt with in the next section.

Coercivity often is used as the most important single criterion for determining whether a ferromagnetic material is soft or hard. Although the classification is not sharp, materials with $H_{c,\infty}$ values less than $400\ \mathrm{A\cdot m^{-1}}$ (5 Oe) are definitely considered soft, and those with $H_{c,\infty}$ values greater than $8\,000\ \mathrm{A\cdot m^{-1}}$ (100 Oe) are labeled hard. For materials showing coercivity values between 400 and $8\,000\ \mathrm{A\cdot m^{-1}}$, other criteria, such as the energy product B_rH_c, must be considered to establish a clear classification. Keeping coercivity at a minimum is a primary goal in the preparation of soft magnetic materials. The lowest coercivity ever recorded in industrial products to date is approximately $0.4\ \mathrm{A\cdot m^{-1}}$ (0.005 Oe) in Supermumetal (73% Ni, 20% Fe, 5% Cu and 2% Cr and/or Mo).

Experimental data show that coercivity is closely related to other structure-sensitive magnetic properties. For instance, when $H_{c,\infty}$ is plotted against μ_i on logarithmic scales for a variety of materials, an inverse linear relationship emerges, as shown in fig. 4.15. Thus we have

$$\log H_{c,\infty} = A - \log \mu_i'; \quad \text{or} \quad H_{c,\infty} = A'/\mu_i', \tag{4.44}$$

where A and A' are constants independent of composition and metal-

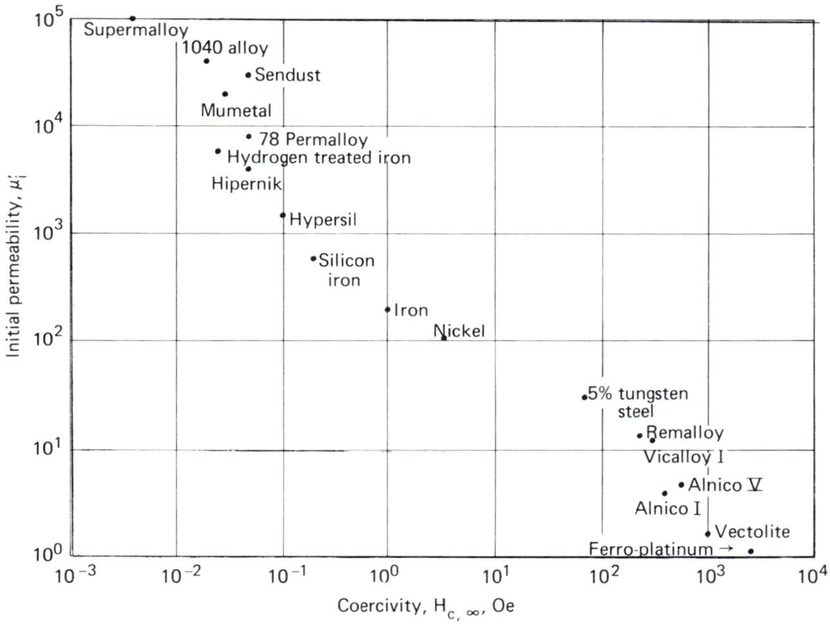

Fig. 4.15. Correlation between the coercivity and initial relative permeability in both soft and hard magnetic materials (Kittel [1949]).

lurgical conditions of the material. Since fig. 4.15 covers a wide range of coercivity, from 0.8 to 4×10^4 A · m^{-1} (0.01 to 5 000 Oe), these empirical formulae apply to all magnetic materials, both soft and hard.

The physical significance of the coercive force has been explained by three prominent theories:

(1) the Strain Theory developed by Bloch [1932], Kondorsky [1937] and Kersten [1938],
(2) the Inclusion Theory by Kersten [1943] and
(3) the Variable Internal Field Theory by Néel [1944, 1946].

All three theories were developed before the advent of the theory of dislocations. Some workers have thus suggested that a unified theory for the coercive force could be developed on the basis of the stress field associated with dislocations. One such theory was first proposed by Vicena [1955] and later by Seeger et al. [1964]. It will be discussed in ch. 5, §2.3.

The three theories share the basic principle that the irreversible movement of domain walls is responsible for the coercive force. This principle is consistent with the ideal model of the hysteresis loop shown in fig. 4.14. To better understand the theories, let us examine the physical meaning, in terms of energy changes, of the reversible or irreversible movement of domain walls.

For simplicity, we again consider a plane 180° wall initially located at x_0 in the yz plane shown in fig. 4.16a. Application of a field in the y-direction will drive the wall to the right by a distance, say x. Since the spatial distribution of the surface energy density of domain walls in real crystals is not uniform (§2.1.2), we may represent the variation in γ_w arbitrarily by a curve shown in fig. 4.16b. If we ignore the anisotropy energy and assume the absence of magnetostatic energy, the total change in energy of the crystal caused by the motion of the wall per unit area is

$$\Delta E = \gamma_w - 2M_s H \mu_0 x. \tag{4.45}$$

The value of x is determined by the equilibrium condition $\partial(\Delta E)/\partial x =$

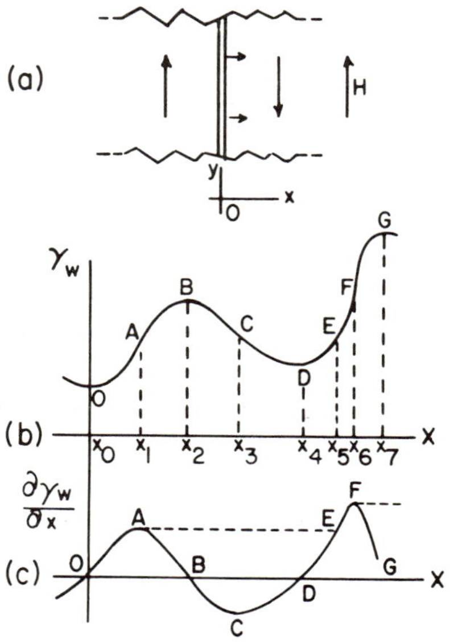

Fig. 4.16. (a) The movement of a 180° domain wall under the action of an applied field, (b) the spatial variation of the surface-energy density of domain walls in an imperfect crystal and (c) the plot of $\partial\gamma_w/\partial x$ versus x (Morrish [1965]).

0, which gives

$$2M_sH\mu_0 = \partial\gamma_w/\partial x. \tag{4.46}$$

The quantity $\partial\gamma_w/\partial x$ is the position gradient of the wall energy given by the slope of the γ_w versus x curve. It is also called the restoring force acting on the wall by the applied field. The latter term stresses the fact that, since a displaced wall tends to move back to its original position, the effect of an applied field is equivalent to exerting a force on the displaced wall sufficient to return the wall to its previous position. Equation (4.46) indicates that the stronger the applied field, the greater the tendency for a displaced wall to return to its initial position. On the other hand, this equation also signifies that the field will act to displace the wall to the right by a distance that depends on $\partial\gamma_w/\partial x$. The plot of $\partial\gamma_w/\partial x$ against x, as shown in fig. 4.16c, will then determine whether the mode of the wall movement is reversible or irreversible in the following manner. The $\partial\gamma_w/\partial x$ versus x curve displays valleys and hills. The hills represent barriers to the wall motion. Before climbing over a hill, the displacement of a wall remains reversible. Once a hill is overcome, the wall movement becomes irreversible up to a certain distance. For instance, in fig. 4.16c the wall movement is reversible between x_0 and x_1 because, when the field is withdrawn, the wall will return to x_0 without the hindrance of a barrier. As soon as the wall moves beyond x_1, where the first hill reaches its peak (A), it will move spontaneously to position x_5 corresponding to point E on the curve. The displacement from x_1 to x_5 is irreversible because, when the field is decreased to zero, the wall will only return to position x_4 corresponding to point D, where $\partial\gamma_w/\partial x = 0$. Applying a reverse field will drive the wall first toward position x_3, corresponding to a valley bottom (C). The movement between x_5 and x_6, corresponding to the segment E–F on the curve, is again reversible and beyond x_6 again irreversible.

Thus, we can consider the coercive force as a measure of the field required to move a wall past the hills in the $\partial\gamma_w/\partial x$ versus x curve. In other words, H_c corresponds to the maximum value of the restoring force; that is, $H_c \sim (\partial\gamma_w/\partial x)_{max}$. More precisely, according to (4.46),

$$H_c = \frac{1}{2M_s\mu_0}\left(\frac{\partial\gamma_w}{\partial x}\right)_{max}. \tag{4.47}$$

This is the common definition of the coercive force used by the three theories. We now proceed to show how different theories formulate $(\partial\gamma_w/\partial x)_{max}$ according to their adopted models.

(1) The strain theory

Historically, this is the first theory formally developed for the coercive force. In this theory, Bloch [1932] conceived the idea that the coercive force must be related to internal strains existing in the lattice as a result of faulty heat treatments, the presence of defects, etc. Later Kondorsky [1937] and Kersten [1938] developed Bloch's concept into a quantitative theory by invoking the associated magnetoelastic effect on the surface energy of domain walls previously discussed (ch. 3, §4). Following (4.29) and (4.30), we write

$$\gamma_w = 2[\pi^2 A_{ex}(K_1 + c\lambda_0\sigma)]^{\frac{1}{2}}, \tag{4.48}$$

where $A_{ex} = J_{ex}S^2/a$ and $c = -\frac{3}{2}$ (more generally, c is a numerical factor of order of unity and depends on the direction of σ with respect to the wall). In the treatment of Kondorsky, a sinusoidal spatial variation of the internal stress was assumed as we have done in the discussion of the initial susceptibility (§2.1.2), but the adopted expression,

$$\sigma = \sigma_0 + \Delta\sigma \sin(2\pi x/l), \tag{4.49}$$

is somewhat different from (4.28). Here σ_0 is the uniform component and $\Delta\sigma$ and l are the amplitude and wavelength of the varying component of the internal stress, respectively. From (4.48), we have

$$\frac{\partial\gamma_w}{\partial x} = c\lambda_0\left(\frac{\pi^2 A_{ex}}{K_1 + c\lambda_0\sigma}\right)^{\frac{1}{2}}\frac{d\sigma}{dx}. \tag{4.50}$$

If l is large compared with the wall thickness d_w expressed in (3.35), $\Delta\sigma$ may be considered a constant within the wall. We then have from (4.49)

$$\frac{d\sigma}{dx} = \frac{2\pi}{l}\Delta\sigma \cos\frac{2\pi x}{l}. \tag{4.51}$$

Substituting this in (4.50) gives

$$\frac{\partial\gamma_w}{\partial x} = \frac{2\pi c\lambda_0\Delta\sigma}{l}\left(\frac{\pi^2 A_{ex}}{K_1 + c\lambda_0\sigma}\right)\cos\frac{2\pi x}{l}, \tag{4.52}$$

which has the maximum value

$$\left(\frac{\partial\gamma_w}{\partial x}\right)_{max} = \frac{2\pi c\lambda_0\Delta\sigma}{l}\left(\frac{\pi^2 A_{ex}}{K_1 + c\lambda_0\sigma}\right)^{\frac{1}{2}}, \tag{4.53}$$

when $2\pi x/l = 2\pi, 4\pi, \ldots$ (or $x = l, 2l, \ldots$), and thus $\cos(2\pi x/l) = 1$. If $K_1 \gg c\lambda_0\sigma$, $[\pi^2 A_{ex}/(K_1 + c\lambda_0\sigma)]^{\frac{1}{2}} \sim (\pi^2 A_{ex}/K_1)^{\frac{1}{2}} = d_w$, according to (3.35).

Hence, combining (4.47) and (4.53) yields

$$H_c \simeq \pi c(\lambda_0 \Delta\sigma / M_s \mu_0)(d_w / l). \tag{4.54}$$

This is the expression derived for the coercive force in the strain theory involving the movement of one wall. For actual materials that have numerous walls, average values of $\Delta\sigma$, l and c must be used for a reliable estimate of H_c.

Equation (4.54) emphasizes the role played by the spatial variation of the internal stress in causing the coercive force as a result of the dependence of the wall energy on the stress condition of the lattice. We might expect the strain theory to be of particular importance for materials showing high magnetostrictions, such as nickel. Kersten made a comparison, shown in fig. 4.17, between theoretical and experimental values for H_c for two grades of nickel. An excellent agreement is seen for all stresses investigated in the hard drawn wires and for relatively low stresses in the recrystallized wires.

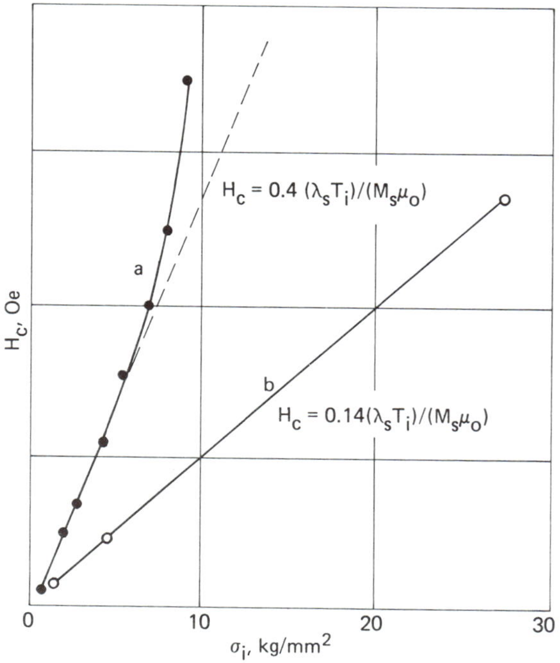

Fig. 4.17. Confirmation of the strain theory of the coercive force in (a) recrystallized and (b) hard drawn wires of nickel by Kersten (Kittel [1949]).

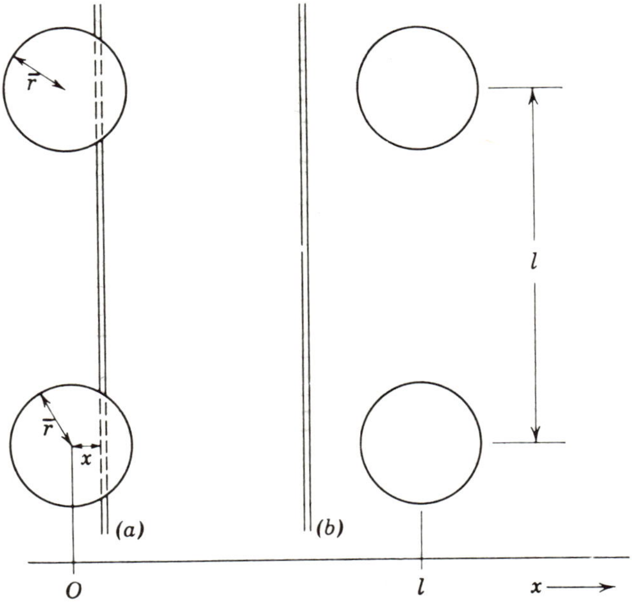

Fig. 4.18. Illustrating Kersten's theory of the coercive force. Inclusion particles form a simple cubic array and a 180° wall intersects some of the inclusions at (a), but stays clear of the particles at (b) (Morrish [1965]).

(2) The inclusion theory

This theory was developed by Kersten [1943], who had earlier helped establish the strain theory in connection with fig. 4.17. Apparently Kersten was not satisfied with the strain theory for the very reason that exceedingly high coercive force had been detected in materials containing appreciable amounts of inclusion particles. By emphasizing the inclusions, Kersten made the following analysis. Suppose that spherical nonmagnetic particles of average radius $\bar{r}$ are uniformly imbedded at the corners of unit cells of an adopted simple cubic lattice with lattice constant l (fig. 4.18). The intersection of a 180° domain wall with the inclusions at a distance x from the centers of particles will reduce the wall area by an amount $\pi(\bar{r}^2 - x^2)$ per particle. Since there are $1/l^2$ particles per unit area, the surface energy density of domain walls will be reduced from γ_w to

$$\gamma_{w'} = \gamma_w[1 - \pi(\bar{r}^2 - x^2)/l^2], \tag{4.55}$$

for $|x| \leq \bar{r}$. The position gradient of the reduced wall energy density, $d\gamma_{w'}/dx$, will have the maximum value at $x = \bar{r}$; that is,

$$(\partial\gamma_{w'}/\partial x)_{max} = 2\pi\bar{r}\gamma_w/l^2. \tag{4.56}$$

Now we introduce the inclusion packing factor f to represent the fraction of volume occupied by $1/l^3$ particles per unit volume; that is,

$$f = \left(\frac{4}{3}\pi\bar{r}^3\right) \cdot \left(\frac{1}{l^3}\right). \tag{4.57}$$

Combining (4.47), (4.56) and (4.57) gives

$$H_c = 1.2\frac{\gamma_w}{M_s\mu_0\bar{r}}f^{\frac{2}{3}}. \tag{4.58}$$

Using the relation $\gamma_w = 2d_wK_1$ expressed in (3.37), Kersten obtained an expression for the coercive force in the inclusion theory

$$H_c = 2.4\frac{d_wK_1}{M_s\mu_0\bar{r}}f^{\frac{2}{3}}. \tag{4.59}$$

To calculate the coercivity to the order of magnitude, we assume that H_c reaches a maximum value, $H_{c,\infty}$, when $\bar{r} = d_w$. Hence

$$H_{c,\infty} = 2.4(K_1/M_s\mu_0)f^{\frac{2}{3}}. \tag{4.60}$$

Kersten applied this expression to a series of dilute alloys of iron containing copper. In those alloys, nonmagnetic particles were formed of a Cu-rich second phase because the concentrations of Cu all exceeded the solubility limit of 0.5% at 600°C. The calculated values of $H_{c,\infty}$ are shown by the solid line in fig. 4.19. A comparison with the measured values revealed an overestimate of $H_{c,\infty}$ by a factor of 3.3. This discrepancy may be resolved, however, by setting $\bar{r} = 3.3d_w$ for the onset of coercivity, instead of $\bar{r} = d_w$.

In the same theory, Kersten also derived an expression for the initial susceptibility. Starting from the approximate relation

$$\chi_i \simeq (dM/dH)_{H\to 0} = [(dM/dx)/(dH/dx)]_{H\to 0}, \tag{4.61}$$

we need only to formulate dM/dx to define χ_i because dH/dx is readily derived from (4.46) and (4.56), which give

$$H = \frac{1}{2M_s\mu_0}\frac{d\gamma_{w'}}{dx} = \pi\left(\frac{\gamma_w}{M_s}\right)\left(\frac{x}{l^2}\right), \tag{4.62}$$

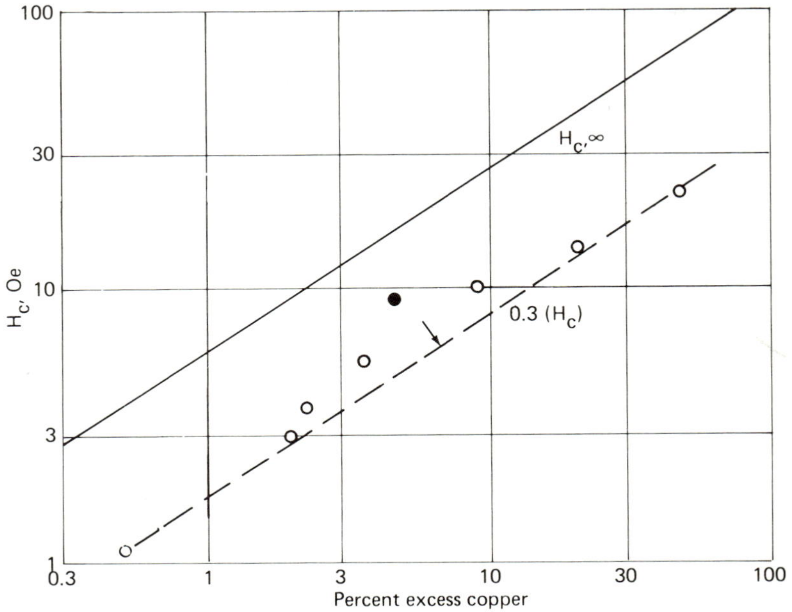

Fig. 4.19. Comparison between the inclusion theory and experimental values obtained for coercivity in dilute Fe–Cu alloys by Kersten (Kittel [1949]).

and

$$\frac{\mathrm{d}H}{\mathrm{d}x} = 2\pi\left(\frac{K_1}{M_s\mu_0}\right)\left(\frac{d_w}{l^2}\right), \tag{4.63}$$

using $\gamma_w = 2d_wK_1$ expressed in (3.37). For the quantity $\mathrm{d}M/\mathrm{d}x$, we note in fig. 4.16a that the magnetization change ΔM associated with the displacement Δx of a 180° wall per unit area is given by

$$\Delta M = 2M_s\Delta x/\bar{s}, \tag{4.64}$$

where $\bar{s}$ is the average width of domains. Combining the last four equations gives

$$\chi_i \approx (M_s^2 l^2 \mu_0/\pi K_1 d_w \bar{s})A, \tag{4.65}$$

where A is the total surface area of 180° walls in a unit volume. This expression of χ_i derived in the inclusion theory agrees with the version

(4.34) previously derived on the basis of internal stresses because both give $\chi_i \sim M_s^2 \mu_0 l^2 A/d_w$. On the other hand, the Kersten expression adopts the factor $1/\pi K_1 \bar{s}$ to replace $\frac{2}{9}\pi^2 \lambda_0 \Delta\sigma$ in the internal stress expression. Comparison between (4.27) and (4.65) reveals that

$$\alpha = 4\pi K_1 d_w \bar{s}/3l^2, \tag{4.66}$$

in Kersten's inclusion theory.

Néel [1946] criticized Kersten's treatment on three fronts. First, the assumption of a regular distribution of inclusion particles in a simple cubic array is obviously unrealistic. Néel argued that, had an irregular distribution of inclusions been considered, the value of H_c given by (4.59) would have been lowered appreciably because the size of the hills and valleys in the γ_w versus x curve shown in fig. 4.16b would have been reduced. Secondly, the validity of assuming domain walls to be rigid was questioned. Again this assumption would lead to an overestimate of H_c. Finally, the most serious criticism raised against Kersten's theory is the neglect of the magnetostatic energy associated with free poles at inclusions intersected by domain walls. Consequently, Néel [1946] proposed his own theory to stress the importance of the effect of the magnetostatic energy on the coercive force.

(3) The variable internal field theory

The main idea underlying Néel's theory is shown in fig. 4.20. Free poles are associated with a nonmagnetic inclusion as shown in fig. 4.20a. For a spherical particle of radius r, the magnetostatic energy is (see appendix 3, (A3.14) for derivation)

$$E_s = (\tfrac{1}{2}\mu_0 N_d M_s^2)V = \tfrac{1}{2}\mu_0 \tfrac{1}{3} M_s^2 \tfrac{4}{3}\pi r^3 = \tfrac{2}{9}\pi\mu_0 r^3 M_s^2. \tag{4.67}$$

This energy would be reduced by about 50% when the inclusion is intersected by a domain wall, shown in fig. 4.20b. The difference in E_s between positions a and b is a measure of the field that must be applied to make the wall move from a to b. Néel further envisaged that magnetization would not be uniform in regions around the inclusions and that uncompensated poles may also occur in the presence of internal stresses. Under these conditions, he carried out a rather complicated analysis on the variable internal field theory and arrived at the following expressions:

for Fe $\quad H_c = 168\,V_\sigma + 28\,800 f \quad$ in $\mathrm{A \cdot m^{-1}}$, (4.68)

for Ni $\quad H_c = 26\,400\,V_\sigma + 7\,760 f \quad$ in $\mathrm{A \cdot m^{-1}}$, (4.69)

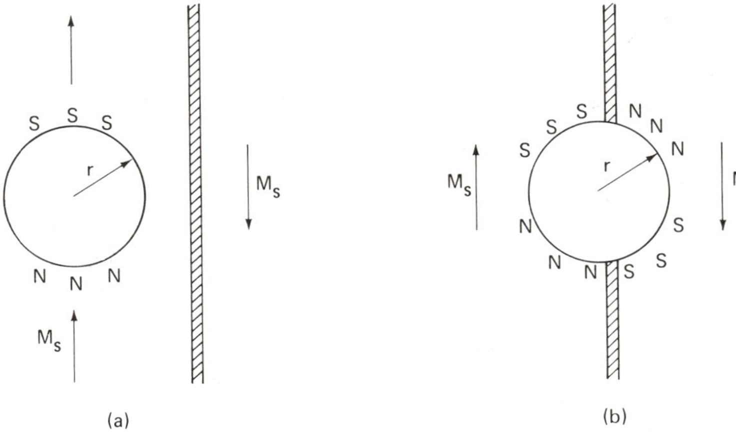

Fig. 4.20. The effect of an intersecting domain wall on the distribution of magnetic free poles at an inclusion particle (Kittel [1949]).

where V_σ is the fractional volume affected by irregular internal stress of magnitude $1.36 \times 10^{10}\,\mathrm{N \cdot m^{-2}}$ ($30\,\mathrm{kg \cdot mm^{-2}}$) and f is the fractional volume occupied by inclusions. The applicability of these equations depends on the detailed domain structure at inclusions and grain boundaries according to Kondorsky [1949] and Goodenough [1954]. Despite the complicated nature of the theory and the uncertainty concerning the domain-structure details, the variable field theory is able to account for high (≥ 800 A/m) coercive forces observed on semisoft and hard magnetic materials. Néel's theory was invoked in the development of a dislocation theory of coercive force by Vicena [1955].

§2.1.5. *Remanence and retentivity*

Previous discussion on hysteresis loops (§2.1.3) has revealed that the loop segment from the tip to the remanence is dominated by the rotation of magnetization vectors. If the tip has reached the saturation value, B_s, then the rotation refers exclusively to the turning of atomic spins from the direction of the applied field to the nearest easy axis in each individual grain. On the basis of such a process, the calculation of the retentivity, $B_{r,\infty}$, can be conducted easily in relation to B_s.

Let us consider the simple case wherein the magnetic anisotropy is uniaxial as in most hexagonal materials (Co, Gd, Co-rich Fe alloys, barium ferrites, etc.). If grains in the polycrystalline specimen are assumed to have completely random orientation and free from residual stresses, the demagnetized state then will have domains with magnetization vectors scattered uniformly in all directions. To represent the spatial distribution of domain magnetization, it is convenient to adopt a vector model in which unit vectors are drawn from the origin of a sphere of unit radius in directions corresponding to those of magnetization vectors in various domains. For the demagnetized state, the ends of the unit vectors will cover the entire surface of the sphere evenly. Hence, in a two-dimensional representation of the vector model, as shown at point O in fig. 4.21, the vector ends will cover uniformly the whole circumference of the great circle of the sphere.

When a magnetic field is applied in the horizontal direction OH, domains with magnetization vectors falling in the western hemisphere will be eliminated gradually. For instance, at point B in fig. 4.21, the left half of the circle is nearly empty. At the tip of the loop, where saturation induction is attained, all vectors will disappear except one, which will remain in the field direction. On removing the field, new vectors will emerge, first in the eastern hemisphere. At the point of retentivity, the eastern hemisphere will be more or less filled, whereas

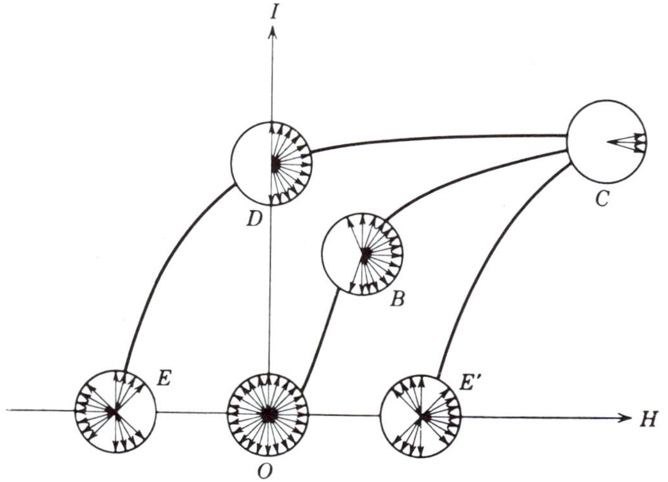

Fig. 4.21. Two-dimensional representations of the vector model showing the spatial distribution of domains at different stages of magnetization (Chikazumi [1964]).

the western hemisphere will still be empty. On this basis, since magnetization at an angle θ from the direction of the applied field contributes an amount $B_s \cos\theta \,d\Omega$ over a solid angle $d\Omega$, the retentivity is given by

$$B_{r,\infty} = \frac{1}{2\pi}\int_0^{\frac{1}{2}\pi} B_s \cos\theta \,d\Omega = \int_0^{\frac{1}{2}\pi} B_s \cos\theta \sin\theta \,d\theta = \tfrac{1}{2}B_s. \quad (4.70)$$

This result signifies that, for materials with uniaxial anisotropy, the maximum value of retentivity is just one half the saturation induction.

For cubic materials with multiple easy axes, the calculation of $B_{r,\infty}$ is somewhat complicated, but still it relies on the simple relation $(B_{r,\infty})_i = B_s \cos\theta_i$ in the ith domain, where θ_i is the angle between the applied field and the nearest easy axis and has the value between 0 and $\frac{1}{2}\pi$. From the expression $\overline{B_{r,\infty} = B_s \cos\theta_i}$, we obtain

$$B_{r,\infty} = B_s \overline{\cos\theta}. \quad (4.71)$$

Hence the calculation of retentivity is reduced to averaging $\cos\theta$ with respect to all possible easy axes in the eastern hemisphere. Analyses made for the ⟨100⟩ and ⟨111⟩ easy axes give $\overline{\cos\theta} = 0.832$ for Fe and other bcc materials and $\overline{\cos\theta} = 0.866$ for Ni and other fcc materials. (See Chikazumi [1966f] for detailed analysis.)

The foregoing discussion has fixed the maximum theoretical value of $B_{r,\infty}/B_s$ at 0.5, 0.832, and 0.866 for Co, Fe and Ni, respectively. Experimental results for these metals, however, are always less than the theoretical values. Table 4.4 compares the observed and theoretical values of the ratio for several prominent industrial materials prepared under different conditions. In the last column of the table, a theoretical value of 0.5 is listed in place of, or added to, the aforesaid value (0.832 or 0.866) for the cubic materials for the reason that the development of a preferred orientation of grains, known metallurgically as a texture, in the material would introduce an uniaxial anisotropy to replace the inherent multiaxial anisotropy. Indeed, if the material is well annealed after severe cold work, the observed value for the ratio is often close to 0.5 in Fe, 4% Si–Fe and Ni.

Several reasons have been given to account for the lower values observed for the $B_{r,\infty}/B_s$ ratio relative to the theoretical value. One main reason is that other processes than the rotation of spins might have occurred before the point of remanence. Processes suggested in this connection include the nucleation of new domains, both the reverse and

TABLE 4.4
Retentivity values observed in various materials

Material	Treatment	$B_{r,\infty}$(T)	B_s(T)	$B_{r,\infty}/B_s$	Theoretical $B_{r,\infty}/B_s$
Iron	Cold drawn	0.8–1.1	2.16	0.4–0.5	0.5
	Annealed	0.6–1.4	2.16	0.3–0.7	0.832, 0.5
Nickel	Cold drawn	0.29–0.39	0.61	0.5–0.65	0.5
	Annealed	0.2–0.4	0.61	0.3–0.65	0.866, 0.5
4% Si–Fe	Annealed	0.6–0.8	1.98	0.3–0.4	0.832, 0.5
	Cold rolled and annealed	1.4	2.02	0.7	0.832, 0.5
45 Permalloy	Annealed	0.75–0.95	1.6	0.45–0.6	0.5
	Cold rolled (95%) and annealed	0.7	1.6	0.45	0.5
65 Permalloy	Quenched	0.59	1.44	0.41	0.5
	Fast cooled	0.45	1.44	0.31	–
	Slowly cooled	0.16	1.44	0.11	–
	Baked	0.10	1.44	0.07	0
	Under tension	1.41	1.44	0.98	1.0
	Annealed in longitudinal field	1.30	1.44	0.90	1.0
	Annealed in transverse field	0.06	1.44	0.04	0

the closure types, and wall displacements of the newly-formed domains. The effect of these processes is equivalent to introducing vectors in the western hemisphere at point D in fig. 4.21, thereby lowering the $B_{r,\infty}/B_s$ ratio. Another reason is given at the end of this section. Readers interested in specific models proposed for the reduction of retentivity are referred to the book by Bozorth [1951c].

At this point, it is pertinent to ask, "How are domains nucleated?" Because domains are separated by Bloch walls of finite thickness, the formation of new domains necessitates the creation of domain walls, which, in turn, requires the rotation of atomic spins against the exchange and magnetocrystalline forces. If the creation of domain walls is accomplished without involving any structural singularities (grain and twin boundaries, dislocations, clusters of point defects, inclusion particles, cavities, etc.) we classify the domain nucleation as homogeneous, otherwise the nucleation process is termed heterogeneous. Another way to view the two modes of nucleation is that the probability of creating a domain wall is everywhere the same in the homogeneous mode, but fluctuates widely from region to region, depending on the availability of "preferred sites" for nucleation, in the heterogeneous mode. Hence, homogeneous and heterogeneous nucleations may be regarded as general and localized phenomena, respectively.

The internal field H_W in iron was shown (ch. 2, §2) to be 1.7×10^9 A · m^{-1} (2×10^7 Oe) at room temperature. This intense field virtually rules out the possibility of homogeneous nucleation because magnetic fields approaching the strength of H_W cannot be produced uniformly in the lattice. Even in the heterogeneous mode, which utilizes the localized fields induced through the demagnetizing effect at inclusions and cavities, the following estimate will show that these fields are still too weak, by about three orders of magnitude, to compensate the Weiss field.

The demagnetizing field induced by the magnetic free poles at the interface of an inclusion particle is shown (appendix 3) to be

$$H_d = -N_d M_s, \tag{4.72}$$

where N_d is the demagnetizing factor and M_s is the saturation magnetization of the matrix. For particles having the shape of a platelet or thin disc, $N_d \sim 1$. Based on this value, (4.72) yields $H_d = 1.7 \times 10^6$ A · m^{-1} (or 2.2×10^4 Oe) for Fe. Thus the magnetic field produced locally at the inclusions by the magnetostatic effect is lower than the Weiss field by roughly three orders of magnitude.

Despite the gap between the Weiss field and the demagnetizing field at a singularity site, experimental studies have established that the formation of domain walls often occurs preferentially at cavities, inclusions, and grain boundaries. This result implies that the localized fields induced at these sites must be more effective in promoting the formation of nuclei than (4.72) indicates for some unknown reason. Alternatively, domain nuclei could be generated at structural singularities at elevated temperatures below the Curie point during the final heat treatment. At high temperatures, the exchange and magnetocrystalline energies are considerably lower, thus allowing the Weiss field to be compensated more easily. This mechanism is strengthened by the argument that if domains cannot be nucleated by a magnetic field in the laboratory, neither will they be destroyed by the same field. Thus nuclei will persist and grow readily in certain stages of the cyclic magnetization once they are formed in the previous heat treatment. No experimental evidence has been reported to substantiate the mechanism, however.

In thin films, domains are seen to be nucleated frequently at the film edge and sometimes as a result of the development of a so-called "ripple structure" due to local deviation in anisotropy. A detailed discussion of the peculiar behavior of magnetic thin films will be given in ch. 6, §3.

The effects of stress, grain orientation and magnetic anneal on retentivity are reflected by the wide variation of the observed values of the $B_{r,\infty}/B_s$ ratio collected in table 4.4. Three factors are important in determining the value of the ratio; namely,

(1) The state of stress – whether the material remains strained after cold work, becomes stressed as a result of quenching, or is well-annealed.
(2) The orientation of grains – whether or not a texture is developed in the material; and
(3) Magnetic anneal – the application of a magnetic field during the final anneal and its direction relative to the outstanding geometric feature (e.g., longitudinal axes of wires and strips) of the material will affect the retentivity significantly.

The results obtained from 65 Permalloy (65% Ni–Fe alloy) dramatically demonstrate the pronounced influence of stress and magnetic anneal on retentivity. For instance, the value of the ratio (1) increases from 0.07 in "baked" specimens to 0.41 in quenched specimens; (2) jumps from 0.04 to 0.90 when the direction of the magnetic field applied during the final anneal is changed from transverse to longitudinal; and (3) approaches the perfect value of 1.0 when the alloy is under tension. To appreciate the sensitive change in the value of the ratio, six series of two-dimensional representations of the vector model are shown in fig. 4.22. Each series contains three stages with $B = 0$, B_s, and $B_{r,\infty}$. Series (a) predicts a value of 0.5 for the ratio in (1) well-annealed Co, (2) cold-worked and (3) properly textured Fe, 4% Si–Fe, and Ni. In all these materials, uniaxial anisotropy prevails. In series (b) and (c), the effect of stress on $B_{r,\infty}$ is exactly reversed because the ratio value

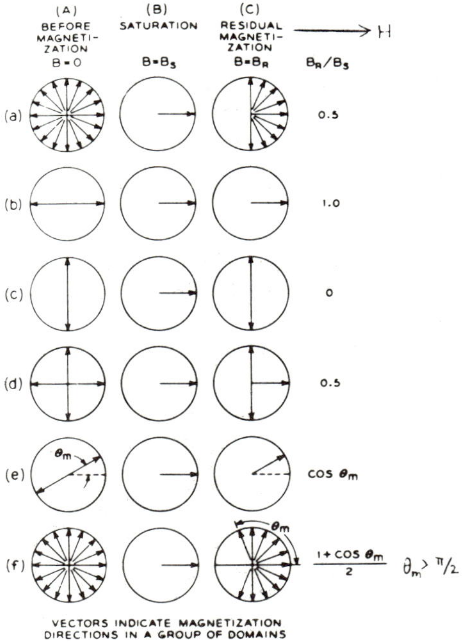

Fig. 4.22. Variation of domain distribution with induction, composition, stress, magnetic anneal and grain structure (Bozorth [1951]). (a): (1) Deformed and annealed Co, (2) Properly textured Fe, 3.5% Si–Fe and Ni, (3) Cold-worked Fe, 3.5% Si–Fe. (b): (1) 68 Permalloy under tension, (2) Heat treated in longitudinal field, (3) Ni under compression, (4) Co ∥ hexagonal axis. (c): (1) Ni under tension, (2) 68 Permalloy under compression, (3) Heat treated in transverse field, (4) Isoperm, (5) Co ⊥ hexagonal axis. (d): Ni or 68 Permalloy bent. (e): (1) Heat treated in oblique field, (2) Isoperm in oblique field, twisted. (f): See text. $B_{r,\infty}$ is represented by B_R in the figure.

changes from one extreme (1.0) to the other (zero). Note here that the stress effect is opposite for Ni and 65 Permalloy because of their opposite signs of magnetostriction. The magnetostrictive property of Isoperm will be described in ch. 5, §2.2. Series (d) deals with the combined effect of tension and compression in a bent specimen; thus the ratio attains the mean value 0.5 of the two extremes. Series (f) resembles series (a) except that the angle θ in (4.70) is allowed to extend to values beyond $\frac{1}{2}\pi$. For θ ranging from 0 to θ_m, (4.70) should be modified to the form

$$B_{r,\infty} = \frac{\int_0^{\theta_m} B_s \cos\theta \, 2\pi \sin\theta \, d\theta}{\int_0^{\theta_m} 2\pi \sin\theta \, d\theta} = \tfrac{1}{2}(1+\cos\theta_m)B_s. \tag{4.73}$$

Consequently, $B_{r,\infty}/B_s$ decreases progressively from 0.5 to 0 as θ_m extends from $\frac{1}{2}\pi$ to π. Values of θ_m greater than $\frac{1}{2}\pi$ are expected when the material is not free from residual stresses, of which the effect is equivalent to a demagnetizing field.

§2.2. *Dynamic properties*

In this section, we shall discuss the following magnetic properties of dynamic nature: (1) the eddy current and associated energy loss, the anomalous energy loss; (2) the motion and resonance of domain walls and (3) the resonance of spins. The first set of properties is of great importance in the application of soft magnetic metals and alloys. The resonances of Bloch walls and spins, on the other hand, are prominent phenomena detected in ferrites at high frequencies.

§2.2.1. *Eddy currents and anomalous energy losses*

When an alternating field is applied to magnetize a ferromagnetic body, an electromagnetic force (emf) is set up in the body. If the material is also a good conductor, as with metals and alloys, the induced emf will produce appreciable amounts of currents in different regions. These currents are known as the eddy currents, and their occurrence gives rise to an energy loss, W_e, through Joule (resistance) heating. Thus, in a.c. applications, metallic ferromagnets suffer the hysteresis as well as the eddy-current losses. Furthermore, as is shown in the following discussion, W_e increases with frequency according to the second power law. Hence, the eddy-current loss becomes even more burdensome than the hysteresis loss at intermediate and high frequencies.

In principle, the energy and power (P_e) losses caused by eddy currents can be calculated easily. The magnitude of the emf induced by a change in magnetic flux is given by Faraday's law; that is

$$V = -\mathrm{d}\Phi/\mathrm{d}t, \tag{4.74}$$

where V is the potential created by the emf and Φ is the instantaneous magnetic flux. The direction of the induced emf is governed by Lenz's law used earlier (ch. 2, §2.1). Here we may restate the law: "The direction of an induced emf is such as to oppose, through its magnetic effects, the change producing the emf." To link the induced electrical field E to the magnetic induction B, we utilize two electromagnetic relations:

$$(1)\ V = \oint E\,\mathrm{d}s \quad \text{and} \quad (2)\ \Phi = \int\!\!\int B\,\mathrm{d}S, \tag{4.75}$$

where the closed integral is taken over the path enclosing the electric field and the double integral is taken over the area in the closed path. Combining (4.74) and (4.75) gives

$$V = \oint E\,\mathrm{d}s = -\int\!\!\int \frac{\mathrm{d}B}{\mathrm{d}t}\,\mathrm{d}S. \tag{4.76}$$

This is the basic formula concerning the electric field responsible for the eddy currents. To compute P_e, we need to apply Ohm's law in order to determine the density of induced eddy currents, $J = E/\rho$ where ρ is the electrical resistivity in ohm-metre.

To illustrate how P_e is calculated, let us consider a metallic ferromagnet in cylindrical form. Figure 4.23a shows that the cylinder has been magnetized axially to the value M, and an induced eddy current is shown by a circular path flowing clockwise in accordance with Lenz's law. If we assume uniform magnetization, thus constant E and constant B, in the cylinder, (4.76) can be expressed as

$$V = E2\pi r = -(\mathrm{d}B/\mathrm{d}t)\pi r^2. \tag{4.77}$$

In the absence of a field, $\mathrm{d}B/\mathrm{d}t = \mu_0\mathrm{d}M/\mathrm{d}t$; so

$$E = -\tfrac{1}{2}r\mu_0\,\mathrm{d}M/\mathrm{d}t. \tag{4.78}$$

The power loss per unit volume is given by

$$P_e = \frac{1}{\pi r_0^2}\int_0^{r_0} EJ2\pi r\,\mathrm{d}r = \frac{\mu_0^2}{2\rho r_0^2}\left(\frac{\mathrm{d}M}{\mathrm{d}t}\right)^2\int_0^{r_0} r^3\,\mathrm{d}r$$
$$= \mu_0^2\frac{r_0^2}{8\rho}\left(\frac{\mathrm{d}M}{\mathrm{d}t}\right)^2, \tag{4.79}$$

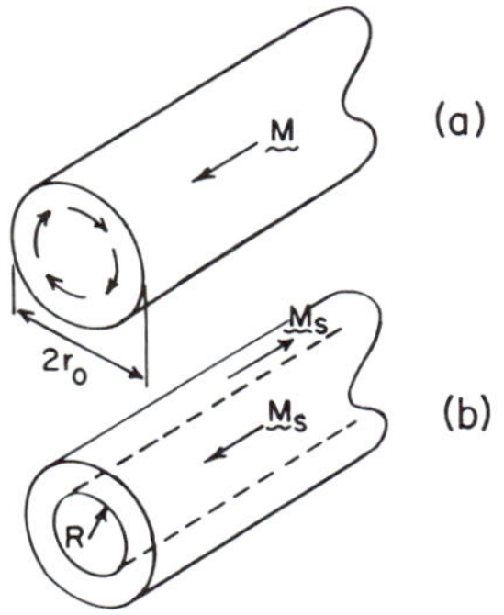

Fig. 4.23. Two models for the production of eddy currents in a cylinder: (a) featuring homogeneous magnetization and (b) involving the movement of a cylindrical domain wall (Chikazumi [1964]).

per unit length of the cylinder in watts per m^3. This expression is derived for a constant rate of change of magnetization in space. In the presence of an alternating driving field, $\mathrm{d}M/\mathrm{d}t$ may be assumed to vary linearly with frequency f up to a moderate value. Thus,

$$P_e = \mu_0^2 r_0^2 f^2 / 8\rho. \tag{4.80}$$

This formula, though only approximate, reveals the important relations that the eddy-current power loss is proportional directly to the square of the frequency of the driving field and inversely to the electric resistivity of the material.

In this analysis, however, the role played by the Bloch walls is ignored. Actually, eddy currents must be associated closely with domain walls because the change in magnetization is likely to be accomplished by the displacement of domain walls. A more realistic calculation of P_e therefore must be based on the pertinent domain structure. To demonstrate the importance of domain-wall motion in the eddy-current loss, suppose that the cylinder is divided into two antiparallel domains at radius R by a cylindrical 180° wall shown in fig. 4.23b. If a magnetic field is applied in the same direction as the magnetization in the inner domain, the wall will move outward radially. For the movement from R to $R + \mathrm{d}R$, the inner domain will gain in volume by the amount $2\pi R\,\mathrm{d}R$ per unit length. Correspondingly, there will be a change in magnetization by the amount $2\,2\pi R\,\mathrm{d}R\,M_s$. Consequently, we have

$$\frac{\mathrm{d}M}{\mathrm{d}t} = \frac{4\pi M_s R}{\pi r_0^2}\left(\frac{\mathrm{d}R}{\mathrm{d}t}\right) \text{ per unit volume,} \tag{4.81}$$

where $\mathrm{d}R/\mathrm{d}t$ is the velocity of the domain wall. Equation (4.76) then is expressed as

$$E 2\pi r = -\frac{4\pi M_s \mu_0 R}{\pi r_0^2}\frac{\mathrm{d}R}{\mathrm{d}t}\pi r_0^2. \tag{4.82}$$

Since the change in magnetization is confined to the outer domain, we can set the boundary conditions as follows:

$$E = 0 \qquad \text{for } r \leq R,$$

and

$$E(r) = -\frac{2M_s\mu_0 R}{r}\frac{\mathrm{d}R}{\mathrm{d}t} \qquad \text{for } r > R.$$

The power loss per unit volume is given by

$$P_e = \frac{1}{\pi r_0^2}\int_R^{r_0}\frac{E^2}{\rho}2\pi r\,\mathrm{d}r = \frac{8\mu_0^2 M_s^2 R^2}{\rho r_0^2}\left(\frac{\mathrm{d}R}{\mathrm{d}t}\right)^2 \log(r_0/R). \tag{4.84}$$

Combining with (4.81) yields

$$P_e = \frac{\mu_0^2 r_0^2}{2\rho}\left(\frac{\mathrm{d}M}{\mathrm{d}t}\right)^2 \log(r_0/R). \tag{4.85}$$

If we envisage a wall movement that starts at the center and continues until the wall reaches the surface, the movement will sweep the cylinder, thereby leading to a complete reversal of magnetization. In this way, the average value of the logarithmic factor is given by

$$\overline{\log(r_0/R)} = r_0^{-1}\int_R^{r_0}[\log(r_0/R)]\,\mathrm{d}R = 1. \tag{4.86}$$

Hence

$$P_e = (\mu_0^2 r_0^2/2\rho)(\mathrm{d}M/\mathrm{d}t)^2. \tag{4.87}$$

Comparison of (4.79) and (4.87) discloses that ignoring the effect of domain walls in the calculation would make the predicted value of P_e far too low (by a factor of four in this case). For the same reason, we may conclude that the value given by (4.87) is still too conservative because real domain structures are much more complex than that assumed in fig. 4.23b and the change in magnetization definitely involves more than just one wall. Unfortunately, it is impossible to make the theoretical calculation of the eddy-current loss much more accurate than we have done in relation to the real domain structure.

Other approaches have been tried to calculate P_e. For instance, the effect of domain walls may be considered in terms of the relationship between induction and the applied field or in terms of the wave form of the spatial distribution of flux density. For sheets of thickness t subjected to a sinusoidally varying field, analysis shows that the eddy-current power loss at low frequencies up to ~ 500 Hz is

$$P_e = \pi^2 t^2 f^2 B_{max}^2 / 6\rho. \tag{4.88}$$

Comparing this with (4.87), we note the same relations, $P_e \sim f^2/\rho$, in both expressions. Equation (4.88) is valid only if B is linearly related to H and if there exists no shielding effect to inhibit complete penetration of flux. Expressions for P_e under other conditions, such as incomplete flux penetration, high frequencies, etc., are found in Bozorth [1951d].

Experimental measurements on the hysteresis and total energy losses indicate a persistent discrepancy between the measured total energy loss W_{total} and the sum of the eddy-current loss, $W_e(= P_e/f)$ given by (4.87) or (4.88), plus the measured hysteresis loss. In almost all cases,

$$W_{total} > W_h + W_e, \tag{4.89}$$

well beyond the estimated experimental error. To reconcile this discrepancy, it is customary to resolve W_{total} into three components,

$$W_{total} = W_h + W_e + W_a, \tag{4.90}$$

where W_a is called the anomalous energy loss. Thus we simply introduce an unexplained loss component to account for the difference between W_{total} and $W_h + W_e$. The relative importance of the three components is shown for iron and two grades of Fe–Si in fig. 4.24. The anomalous loss becomes more prominent as the maximum induction is increased. At the highest value of B_{max}, the percentage anomalous loss jumps from 15% in iron to 47% in grain-oriented 3.2% Fe–Si. To further demonstrate the prominence of the anomalous loss, fig. 4.25 shows the variation of the three components when the frequencies are in the range 0–100 Hz for a grain-oriented Fe–Si used in two directions at $B_{max} =$ 1.3 or 1.6 T. Again, note the increasing dominance of W_a as the frequency goes up. Under the worst conditions ($f = 100$ Hz and the material used at 90° to the rolling direction), W_a amounts to a whopping 60% of the total loss.

The origin of the anomalous loss has been a subject for extensive investigation from different viewpoints. Before describing the investigation results, it is worth stressing that W_a was introduced largely

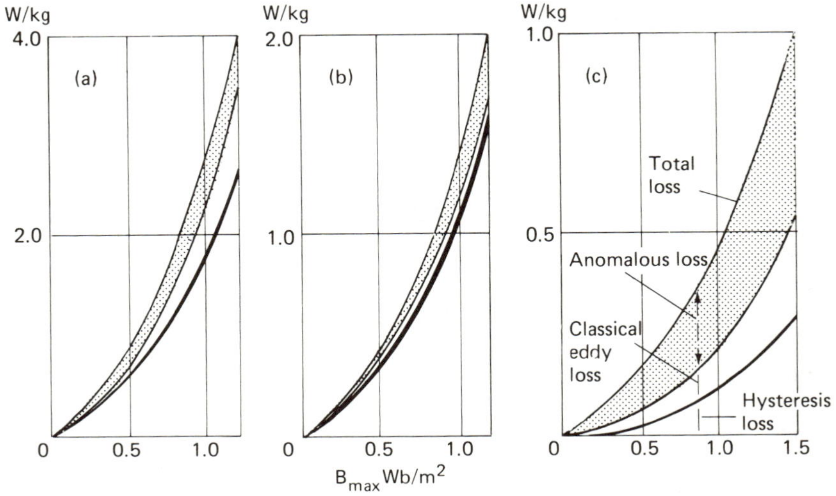

Fig. 4.24. Comparison of the three components (hysteresis, eddy-current and anomalous) of energy loss in three materials, 0.33 mm thick, at 50 Hz: (a) Fe, (b) hot-rolled 4% Si–Fe and (c) cold-rolled, singly-oriented 3.2% Si–Fe in the rolling direction (Brailsford [1966]).

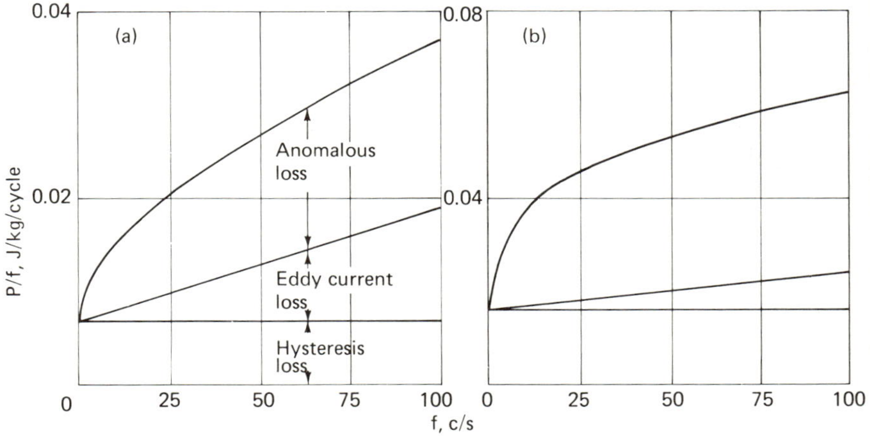

Fig. 4.25. Variation in distribution of the three components of energy loss with frequency of the applied alternating field in cold-rolled, singly-oriented Si–Fe (a) along the rolling direction, $B_{max} = 1.6T$ and (b) at 90° to the rolling direction, $B_{max} = 1.3T$ (Brailsford [1966]).

because all calculations of W_e tend to underestimate this loss. Many workers simply call W_a the eddy-current anomaly. In this premise, we may treat the value predicted for W_e from (4.87) or (4.88) as the classical component of the eddy-current loss and W_a as the anomalous component.

To evaluate the relative importance of the anomalous loss, it is useful to introduce an anomalous factor η defined by

$$\eta = (P_{total} - P_h)/P_e. \tag{4.91}$$

The three cases shown in fig. 4.24 have the approximate values of η at 1.5, 2.0 and 3.5. Assuming simple, but as realistic as possible, domain configurations in thin sheets, Lee [1958] found a sensitive dependence of η on the ratio of domain width d to sheet thickness t. Figure 4.26 shows that η begins to deviate from unity for $d/t > 0.16$, increases

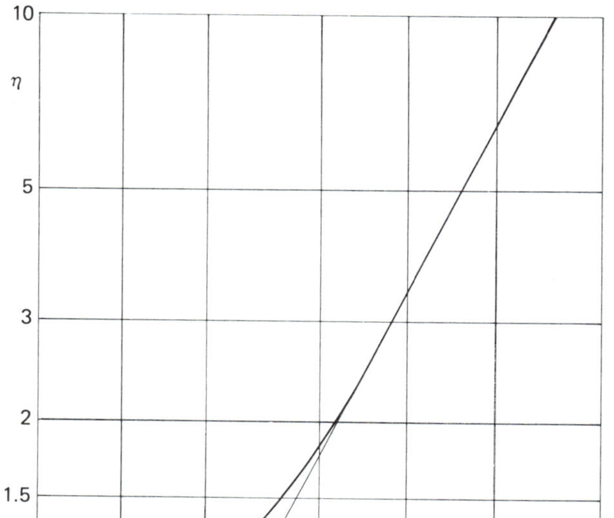

Fig. 4.26. Theoretical values for the anomaly factor η predicted as a function of the ratio of domain width d to sheet thickness t (Lee [1958]).

rapidly for $d/t > 0.4$ and reaches a value of 10 at $d/t \simeq 6$. Lee's results are consistent, at least qualitatively, with the belief that, as the domain width increases with respect to the sheet thickness, there will be fewer walls and, therefore, the distribution of eddy currents will be highly non-uniform, making the calculation of W_e extremely difficult.

Another interpretation of the three components of energy loss emerges when we assume two types of eddy currents: macrocurrents I_a resulting from homogeneous magnetization and highly localized microcurrents I_i resulting from domain wall motion. The total power loss then will be given by

$$P_{\text{total}} = \rho \int (I_i + I_a)^2 \, dV$$

$$= \rho \int I_i^2 \, dV + \rho \int I_a^2 \, dV + 2\rho \int I_i I_a \, dV, \qquad (4.92)$$

where the integration is taken over a unit volume. While the first and second terms are due to microcurrents and macrocurrents, respectively, the third term is a cross product of both currents. These three terms may be identified with the hysteresis loss, the classical eddy-current loss, and the anomalous loss. The relative importance of the various terms depends on the spatial distribution of the two types of currents. Three cases are worth citing:

(1) If the spatial distribution of the macrocurrents is fairly uniform compared with that of the microcurrents, the third term should be negligibly small because opposite components of I_i tend to cancel each other.
(2) If the domain structure is very fine, meaning the number of domain walls is enormously large, the velocity of an individual wall is expected to be small, as would be the induced microcurrents. The magnetization then should be quite homogeneous, and the classical eddy-current loss caused by the macrocurrents in the second term should dominate the power loss. This case corresponds to the situation with $d/t < 0.2$ in the analysis of Lee shown in fig. 4.26.
(3) If the domains are relatively large, corresponding to the case where $d/t > 1$ in Lee's analysis, or if the matrix of the material is decorated with nonmetallic inclusions, the distribution pattern of the micro- and macrocurrents will become highly complicated. In this case, the third

term plays a prominent role, leading to a large value of η in the steep portion of the curve shown in fig. 4.26.

It is shown in fig. 4.25 that the hysteresis component of P/f expressed in units of joules per kg (or per m^3) per cycle is independent of frequency, but the "classical" eddy-current component varies linearly with frequency and the anomalous component varies at higher order with frequency. Hence we conclude that the hysteresis, eddy-current and anomalous losses in units of watts per kg (or per m^3) are proportional to the first, second and higher power of frequency, respectively. According to Carr [1959], such sensitive loss dependence on frequency is discernible because at low frequencies the fraction of domain walls moving at a given instance is small; whereas at higher frequencies all the walls move. This also explains why the anomalous loss becomes more conspicuous at higher frequencies.

§2.2.2. *Motion and resonance of domain walls*

The motion of domain walls is not just a subject of scientific interest; it also plays an important role in the application of soft magnetic materials. For instance, we have seen from (4.84) that the eddy-current power loss is proportional to the square of the velocity of domain walls. Thus, to minimize the loss, we must try to keep the wall velocity at the lowest possible value. On the other hand, some applications of ferrites at microwave frequencies involve the resonance of domain walls. Again, the dynamic behavior of domain walls plays an essential role in the magnetic performance of ferrites in the latter case.

Since the present theory of domain wall motion was developed largely from experimental results, we will describe several crucial experiments first. Williams and Shockley [1949] developed an ingenious design of specimens with which the dynamics of domain walls could be studied quantitatively. Figure 4.27 shows a typical example of their specimens in the shape of a picture frame. For iron and Fe–Si alloys, each specimen was cut from a single crystal in such a way that the specimen surface is in the (100) plane and all four legs are in the ⟨010⟩ easy directions. A magnetizing coil is wound around each leg so that all 180° walls would move together. After attaining saturation magnetization, each leg would have only a single domain (fig. 4.27a).

Using such an experimental procedure, Williams et al. [1950] confirmed a linear relationship between the velocity of the wall and the driving field in Fe–Si alloys, that had been proposed much earlier by

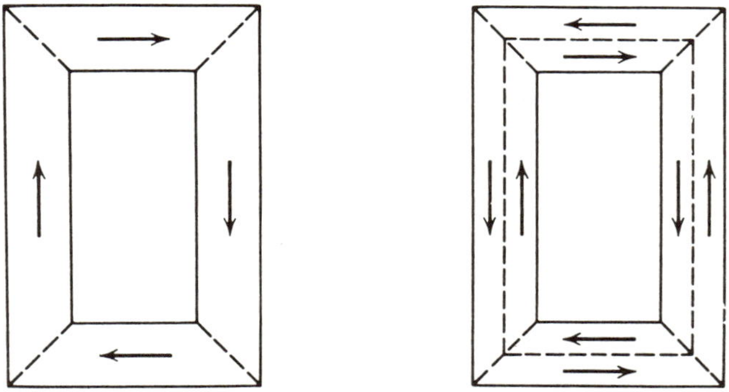

(a) *(b)*

Fig. 4.27. A typical "picture frame" specimen of Si–Fe prepared from a single crystal. With the top and bottom surfaces in the (100) plane, closure domains are formed in a loop having legs parallel to the [010] and [001] directions (Morrish [1966]).

Sixtus and Tonks ([1933], see Kittel [1949]),

$$v = \mathrm{d}x/\mathrm{d}t = \xi(H - H_t), \tag{4.93}$$

where x is the distance of travel, ξ is the constant representing the mobility of the wall, H is the applied field and H_t is the threshold or minimum field required to activate the wall.

Dillon and Earl [1959] applied the same procedure, coupled with some sophisticated electronic techniques, to record the motion of domain walls in a Mn-ferrite ($Mn_{1.4}Fe_{1.6}O_4$) at various cryogenic temperatures. Their study demonstrated vividly the validity of (4.93) in ferrites (see fig. 4.28) and revealed a sensitive temperature dependence of the constant ξ and the threshold field H_t (see table 4.5). From their data, we can calculate the apparent wall velocities at different temperatures for an arbitrarily chosen field. The results of such a calculation based on $H = 800\ \mathrm{A \cdot m^{-1}}$ (10 Oe) are plotted in fig. 4.29. The wall velocity is seen to increase monotonically with temperature by a factor of 870 from 4.2 to 195°K.

The theoretical justification of (4.93) and the physical interpretation of ξ and H_t are linked with the differential equation of the wall motion. For this purpose, let us first recall that, for a mechanical system wherein a body oscillates under an external force $F(t)$ in a viscous

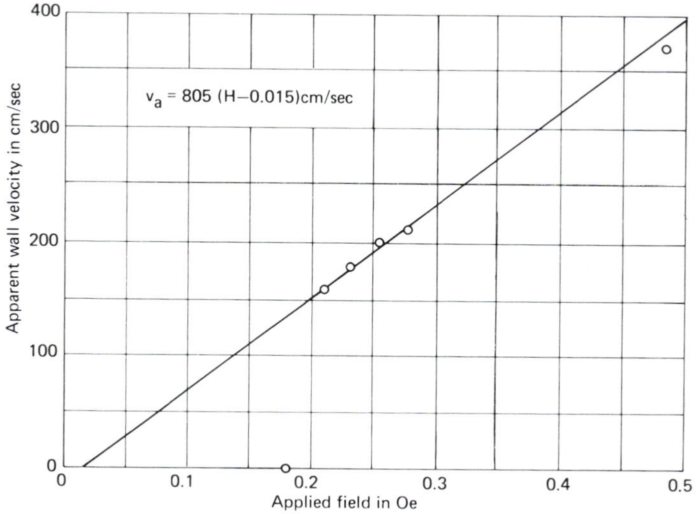

Fig. 4.28. Experimental data for the apparent wall velocity in $Mn_{1.4}Fe_{1.6}O_4$, demonstrating the validity of the relation $v = \xi(H - H_t)$ (Dillon and Earl [1959]).

medium, the oscillation is described completely by the following equation,

$$m\frac{d^2x}{dt^2} = -\beta\frac{dx}{dt} - \alpha x + F(t), \tag{4.94}$$

TABLE 4.5

Average values deduced for the mobility constant ξ and threshold field H_t in (4.93) of domain-wall motion in $Mn_{1.4}Fe_{1.6}O_4$

Temperature (°K)	ξ ($m \cdot s^{-1}/A \cdot m^{-1}$)	H_t ($A \cdot m^{-1}$)
4.2	0.12×10^{-2}	597
20	1.2×10^{-2}	111
65	7.9×10^{-2}	14
77	10.9×10^{-2}	4.8
87	13.1×10^{-2}	2.8
195	26.6×10^{-2}	2.4

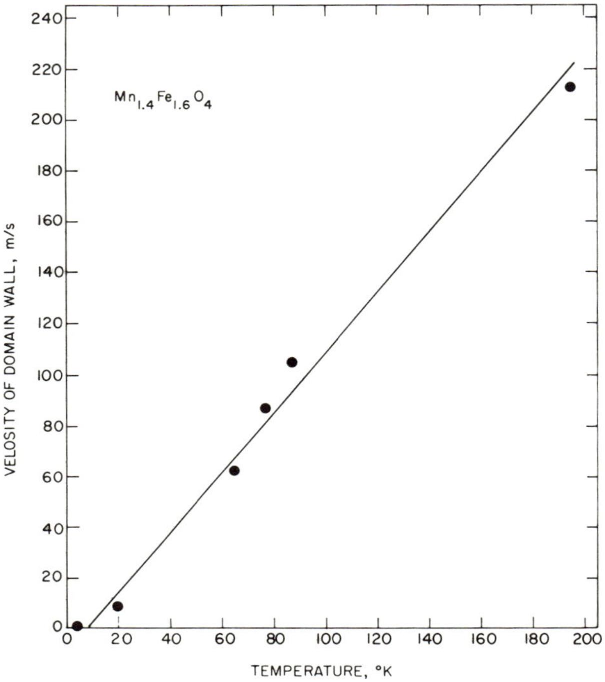

Fig. 4.29. The apparent wall velocities calculated from experimental data of Dillon and Earl [1959] for $Mn_{1.4}Fe_{1.6}O_4$ display a linear dependence on temperature in the cryogenic range.

where m is the mass of the oscillating body, β is the damping coefficient characterizing the viscous nature of the medium and α is the coefficient of stiffness. The motion of a domain wall in a defects-containing ferromagnetic matrix is strikingly similar to the oscillation in the mechanical system. Hence, by analogy, the equation of the domain-wall motion is

$$m_w \frac{d^2x}{dt^2} = -\beta \frac{dx}{dt} - \alpha x + bM_s H\mu_0, \tag{4.95}$$

where m_w is the apparent mass of the wall per unit area, α and β are the same coefficients, $bM_sH\mu_0$ represents the pressure, or force per unit area, acting on the wall by the applied field and the constant b has the value of 2 for the 180° walls and $\sqrt{2}$ for the 90° walls. For an a.c. field, it is convenient to express

$$H = H_0 \exp(\mathrm{i}\omega t), \tag{4.96}$$

where H_0 and ω are the amplitude and angular frequency of the field, respectively.

If the wall velocity is constant or only slowly varying, $\mathrm{d}^2x/\mathrm{d}t^2 \to 0$; so the left-side term in (4.95) is negligible. Moreover, we may replace the stiffness term αx by assuming that the effective driving field is less than the applied field by an amount H_t, where H_t is of the same order of magnitude as the coercivity. Under these conditions, (4.95) becomes

$$\beta \frac{\mathrm{d}x}{\mathrm{d}t} = 2M_\mathrm{s}\mu_0(H - H_\mathrm{t}), \tag{4.97}$$

which is identical with (4.93) with

$$\xi = 2M_\mathrm{s}\mu_0/\beta. \tag{4.98}$$

For the mobility constant ξ to exhibit a positive temperature dependence (table 4.5), as opposed to a negative temperature dependence of M_s, the damping coefficient β must be the dominating factor of ξ. Comparison between (4.95) and (4.97) gives

$$H_\mathrm{t} = \alpha x/2M_\mathrm{s}\mu_0. \tag{4.99}$$

Note the similarities between this equation for H_t and (4.47) for H_c. The identity of H_t with H_c will be utilized later for the interpretation of the stiffness coefficients α.

When the applied field included in (4.95) is alternating and at a high frequency, the activated domain walls will oscillate through small distances about their mean positions. To explain the resonance phenomenon of an oscillating wall, we again refer to (4.94). If the oscillation in the mechanical system occurs under zero external force in a frictionless medium ($\beta = 0$), (4.94) becomes

$$m(\mathrm{d}^2x/\mathrm{d}t^2) = -\alpha x, \tag{4.100}$$

which is known as the equation of simple harmonic motion proceeding at the natural frequency $\omega_0 = (\alpha/m)^{\frac{1}{2}}$. If the forced oscillation described by (4.94) occurs at the natural frequency, the body is said to be oscillating at resonance. By the same definition, when the oscillation of a domain wall proceeds at the frequency

$$\omega_\mathrm{R} = (\alpha/m_\mathrm{w})^{\frac{1}{2}}, \tag{4.101}$$

the wall is said to be oscillating at resonance. To calculate ω_R in specific cases, it is necessary to know the magnitudes of α and m_w, which are also important for calculating H_t and other parameters of domain-wall oscillation.

As the name implies, the coefficient of stiffness α measures the resistance force acting on the displaced wall by nearby defects. According to (4.95), this force is $-\alpha x$. Meanwhile we may associate all energy changes during the wall motion with changes in the wall energy γ_w. We have seen from (4.46) that the restoring force acting on the displaced wall per unit area is minus $d\gamma_w/dx$. Hence we obtain the relation

$$\alpha x = d\gamma_w/dx. \tag{4.102}$$

The same relation also is given by equating H_t in (4.99) to H_c in (4.47). Differentiating (4.102) with respect to x gives

$$\alpha = d^2\gamma_w/dx^2. \tag{4.103}$$

This is the basic interpretation of α. The same expression has been deduced from (4.21). Thus by applying (4.27), an explicit expression emerges for α in terms of the initial relative susceptibility χ_i';

$$\alpha = \Gamma(M_s^2 A/\chi_i'), \tag{4.104}$$

where A is the total surface area of domain-walls of a certain type (90° or 180°, for example) and Γ is a constant, being $\frac{2}{3}$ for the 90° walls and $\frac{4}{3}$ for the 180° walls. Based on the simple lamellar domain structure shown in fig. 3.13c, it is easy to see that the reciprocal of the numerical value of A gives the average width $\bar{d}$ of domains.

We now consider the physical nature of m_w. At first glance, the idea that a moving wall is associated with a mass, even in the virtual sense, seems strange because there is no transfer of mass during the motion. However, Döring [1948] pointed out that a moving wall should possess an inertia and, therefore, a mass. His reasoning was that the demagnetizing effect would make the energy of a moving wall differ from the wall at rest by a term proportional to v^2. If this term is identified as a kinetic energy, the mass of the moving wall is defined as twice the constant of proportionality. Alternatively, we may think of m_w in terms of the angular momenta of the spins forming the wall. For a calculation of m_w, we follow the simpler, though less rigorous, analysis of Becker [1951] reinterpreted by Brailsford [1966c].

Consider a 180° wall of unit area and thickness d_w parallel to the

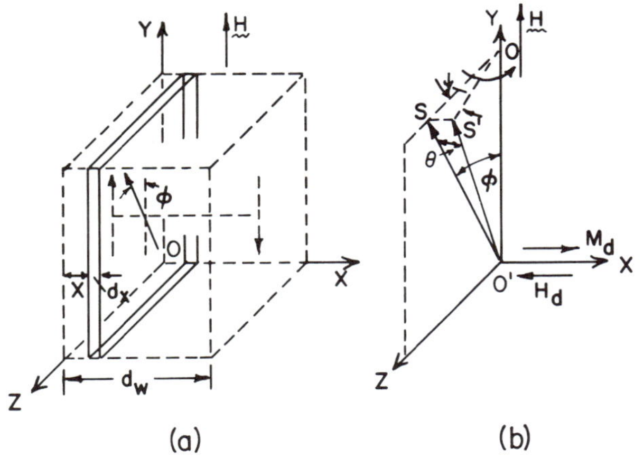

Fig. 4.30. A geometric analysis of the displacement of a domain wall (Brailsford [1966]).

YOZ plane (fig. 4.30a). This wall separates two domains with opposite spins in the $\pm Y$-axis. At rest, all spins in the wall are in planes parallel to YOZ, but they make an angle ϕ with the Y-axis from 0° at $x = 0$ to 180° at $x = d_w$. For a layer of thickness dx at a distance x from YOZ, the average spin direction is at an angle ϕ from OY, where

$$\phi = (x/d_w)\pi. \tag{4.105}$$

When a field $\boldsymbol{H}$ is applied along OY, the spin axis will precess about $\boldsymbol{H}$. At time t, suppose the average spin axis in the layer has precessed from O′S to O′S′, as shown in fig. 4.30b. O′S′ is out of the original plane by an angle θ, and the S′O′Y plane makes an angle ψ with the SO′Y plane, where

$$\theta = \psi \sin \phi. \tag{4.106}$$

The angular velocity of the precession is given by

$$\mathrm{d}\psi/\mathrm{d}t = (ge/2m_0)\mu_0 H = \gamma H \mu_0, \tag{4.107}$$

where $\gamma = ge/2m_0 = e/m_0$ for spins ($g = 2$). At time t, $\psi = \gamma H \mu_0 t$ and $\theta = \gamma H \mu_0 t \cdot \sin \phi$. Now the average spin axis in the layer has a component normal to the plane of the wall of the magnitude

$$M_s \sin \theta \simeq M_s \theta = M_s \gamma H \mu_0 t \sin \phi = M_s \gamma H \mu_0 t \sin (\pi x/t_w). \tag{4.108}$$

The total component in the OX direction is given by

$$\int_0^{d_w} M_s \gamma H \mu_0 t \sin(\pi x/d_w)\,dx = M_s \gamma H \mu_0 t(2d_w/\pi). \tag{4.109}$$

This component gives rise to a magnetization in the OX direction with its mean magnitude given by

$$\bar{M}_d = 2M_s \gamma H \mu_0 t/\pi. \tag{4.110}$$

This induced magnetization will produce free poles on the surface of the wall. Therefore, a demagnetizing field is induced in the XO direction, as shown in fig. 4.30b, of the intensity

$$H_d = N_d M_d = 2M_s \gamma H \mu_0 t N_d/\pi, \tag{4.111}$$

where N_d is the demagnetizing factor of the wall in the same direction, which has the value of unity in the present case.

The demagnetizing field again will induce a precession of the spin axis about H_d at the angular velocity

$$-\frac{d\phi}{dt} = \gamma H_d \mu_0. \tag{4.112}$$

While all these events take place inside the wall, the spinning electrons outside the wall on either side remain unaffected. Thus the latter precession of the spin axes about the demagnetizing field leads to a translation of the entire wall in the OX direction at a linear velocity which can be derived from (4.105),

$$\frac{dx}{dt} = (d_w/\pi)(d\phi/dt) = -(d_w/\pi)\gamma H_d \mu_0, \tag{4.113}$$

where the minus sign emerges to signify that the translation, represented by dx/dt, and the driving field, $\gamma H_d \mu_0$ are opposite to each other. Substituting (4.111) gives

$$dx/dt = (2M_s H d_w \gamma^2 \mu_0^2/\pi^2)t. \tag{4.114}$$

Differentiating and rearranging yields

$$(\pi^2/d_w \gamma^2 \mu_0)(d^2x/dt^2) = 2M_s H \mu_0. \tag{4.115}$$

Since $2M_s H \mu_0$ represents the total effective force per unit area acting on the wall (§2.1.4), the left-hand side term corresponds to $m_w \ddot{x}$, where m_w is the apparent mass of the wall per unit area in units of $kg \cdot m^{-2}$.

Hence we obtain

$$m_w = \pi^2/d_w\gamma^2\mu_0. \tag{4.116}$$

It is interesting to apply this expression for m_w to a typical material, say Fe. For a 180° wall lying on the (100) plane, table 3.2 gives $d_w = 1.41 \times 10^{-7}$ m. Substituting this and other appropriate values into (4.116) yields $m_w = 1.8 \times 10^{-9}$ kg/m^2, as compared with $m = 1.1 \times 10^{-3}$ kg/m^2 for the real material in the same wall.

By combining (4.101), (4.104) and (4.116), the resonance frequency is given by

$$f_R = \omega_R/2\pi = (M_s\gamma/2\pi^2)(\Gamma A\mu_0 d_w/\chi_i')^{\frac{1}{2}}, \tag{4.117}$$

where Ad_w gives the fraction of volume occupied by the 90° ($\Gamma = \frac{2}{3}$) or 180° ($\Gamma = \frac{4}{3}$) walls.

Before presenting experimental evidence for the resonances of domain wall motion and spins, it is useful to show how magnetic properties such as susceptibility and permeability can be resolved into a real and an imaginary component. Consider an a.c. field H expressed as $H = H_0\,e^{i\omega t}$ (4.96). If the magnetic induction B lags behind H by a phase angle δ, we may write

$$B = B_0\,e^{i(\omega t-\delta)}, \tag{4.118}$$

where B_0 is the amplitude of the induction. The corresponding absolute permeability is given by

$$\mu = B/H = B_0\,e^{i(\omega t-\delta)}/H_0\,e^{i\omega t} = (B_0/H_0)\,e^{-i\delta}. \tag{4.119}$$

Using the relation $e^{-ix} = \cos x - i\sin x$, we then have

$$\mu = (B_0/H_0)(\cos\delta - i\sin\delta) = \mu_R - i\mu_I, \tag{4.120}$$

where $\mu_R = (B_0/H_0)\cos\delta$ and $\mu_I = (B_0/H_0)\sin\delta$, representing the real and imaginary components of μ, respectively. Since $\tan\delta = \sin\delta/\cos\delta$,

$$\tan\delta = \mu_I/\mu_R, \tag{4.121}$$

which relates the phase angle to the two components of the involved property. If δ is sufficiently small, $\tan\delta \approx \delta$; thus $\delta \approx \mu_I/\mu_R$. Similar expressions apply to the magnetic susceptibility by simply replacing μ with χ in (4.118)–(4.121) and to the relative permeability and relative susceptibility by replacing μ with μ', etc.

Evidence for the resonance of domain-wall oscillation was found in a study on the Mg-ferrite by Rado et al. [1950]. Figure 4.31 shows the

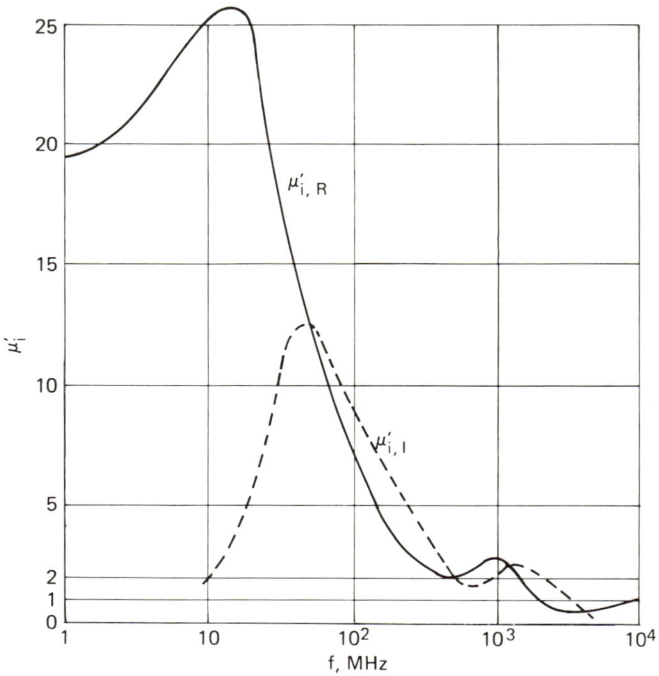

Fig. 4.31. Evidence for the resonance of domain wall oscillation observed in the plots of the real and imaginary components of initial complex relative permeability versus frequency. Data were obtained from a Mn-ferrite (Rado et al. [1950]).

initial observation on the frequency dependence of the real ($\mu'_{i,R}$) and imaginary ($\mu'_{i,I}$) parts of the initial relative permeability. Each curve displays two peaks at radio ($\approx$50 MHz) and microwave ($\approx$2 500 MHz) frequencies. Subsequently, the sample was crushed, and new samples were made by imbedding the fine powder particles in wax. Measurements on the new samples failed to detect the peaks at the radio frequencies. This result is difficult to explain except by assuming that domain walls were eliminated when the original sample was crushed into fine particles. Rado et al. thus considered the appearance of the large peaks at the radio frequencies as evidence for the resonance of wall motion and the appearance of the smaller peaks at higher frequencies as evidence for the resonance of spins.

To further substantiate their interpretation concerning the resonance of domain-walls, Rado et al. applied a formula for f_R similar to (4.117) to estimate the average domain size d. The following data were used: $M_s = 2.79 \times 10^5\ \mathrm{A \cdot m^{-1}}$, $d_w = 7 \times 10^{-8}$ m, $f_R = 45$ MHz taken at the large peak of the $\mu'_{i,I}$ curve in fig. 4.31 and $\chi'_i = 18.6$ (equivalent to $\chi_i = 1.48$ in CGS units). Their calculation estimated d at 5×10^{-5} m [$d \sim 3 \times 10^{-5}$ m deduced from (4.117)], which was considered in fair agreement with the wall resonance model because X-ray data indicated a grain-size range between 5×10^{-6} and 5×10^{-5} m in the original sample.

From a practical viewpoint, the damping coefficient β is probably the most important parameter in (4.95) because it deals with the energy loss caused by wall motion. Unfortunately, our present knowledge of the physical nature of β is still fragmentary. For metallic ferromagnets, it is reasonable to ascribe the viscous damping entirely to eddy currents. To derive an expression for β in this case, let us drop the $m_w(\mathrm{d}^2x/\mathrm{d}t^2)$ and αx terms in (4.95) and obtain

$$\mathrm{d}x/\mathrm{d}t = bM_s H\mu_0/\beta, \tag{4.122}$$

where $b = 2$ for the 180° walls. The power supplied by the field is given by $P_H = \mu_0 H(\mathrm{d}M/\mathrm{d}t)$, which, according to (4.81) in connection with fig. 4.23b, is

$$P_H = (4M_s R/r_0^2)/(\mathrm{d}R/\mathrm{d}t)H\mu_0. \tag{4.123}$$

In the absence of any change in the potential energy, P_H must be lost as heat through the eddy currents. Hence we can equate (4.123) to (4.84) and obtain

$$v = \mathrm{d}R/\mathrm{d}t = \rho H/2M_s R\mu_0 \log(r_0/R). \tag{4.124}$$

Comparing this with (4.122), we have

$$\beta = 2bM_s^2\mu_0^2 R \log(r_0/R)/\rho. \tag{4.125}$$

This is a simplified expression for β for metallic ferromagnetic materials. For Fe, we substitute $M_s = 1.71 \times 10^6\ \mathrm{A \cdot m^{-1}}$, $\rho = 1 \times 10^{-7}\ \Omega \cdot \mathrm{m}$, $R = 10^{-6}$ m and $\log(r_0/R) \approx 1$ and attain the value $\beta = 1.9 \times 10^2$.

For ferrites and other insulating materials, however, (4.125) does not apply because viscous damping due to eddy currents cannot be important. In these materials, the origin of domain-wall damping then must be related to the field-induced precessions of the spin axes shown in fig. 4.30. The problem was treated by Kittel [1950]. A general expression was derived for β, which has the simplified form in SI

$$\beta = 2\pi\lambda_r\gamma_w/\gamma^2\mu_0 A_{ex}, \tag{4.126}$$

where λ_r is a relaxation frequency, $\gamma = ge/2m_0$ as in (4.107) and $A_{ex} = 2J_{ex}S^2/a$ for bcc lattices and $4J_{ex}S^2/a$ for fcc lattices. For Ni-ferrite with $\lambda_r = 3.9 \times 10^7\ s^{-1}$, $\gamma_w = 4.75 \times 10^{-4}\ J \cdot m^{-2}$, $A_{ex} = 9.10 \times 10^{-12}\ J \cdot m^{-1}$ and $\gamma = 1.76 \times 10^{11}$ in SI units, $\beta = 0.33$ which is much smaller than the value given by the eddy-current model for Fe.

§2.2.3. *Resonance of spins*

The theory of the dynamic behavior of spins, including the resonance phenomenon, is developed using the Landau–Lifshitz equation of motion of magnetization, which is expressed by

$$\frac{\partial \boldsymbol{M}}{\partial t} = -\gamma\mu_0(\boldsymbol{M} \times \boldsymbol{H}) - \frac{4\pi\lambda_d}{M^2}(\boldsymbol{M} \times \boldsymbol{M} \times \boldsymbol{H}), \tag{4.127}$$

where $\boldsymbol{M}$ is the magnetization vector, $\boldsymbol{H}$ is the magnetic field vector, $\gamma = ge/2m_0$ is the magnetomechanical or gyromagnetic ratio, and λ_d is an adjustable parameter (in units of s^{-1}) that measures the degree of damping. Physically, this equation signifies the following features of the motion:

(1) The magnetization has a spatial dependence as well as a time dependence. Therefore, its time derivative should be properly written as a partial derivative.
(2) The first term on the right-hand side represents the precession of the magnetization about the applied field in the absence of damping. The precession causes a change in $\boldsymbol{M}$ in the direction normal to both $\boldsymbol{M}$ and $\boldsymbol{H}$, and the minus sign results because the magnetization vector always opposes the angular momentum vector.
(3) The second term on the right-hand side shows a second precessional motion of the magnetization caused by the damping force. This motion causes a further change in $\boldsymbol{M}$ in the direction perpendicular to both $\boldsymbol{M}$ and $\boldsymbol{M} \times \boldsymbol{H}$, which reflects the damping force.

Gilbert and Kelley [1955] proposed an alternative expression for the damping term

$$\frac{\sigma}{M_s}[\boldsymbol{M} \times (\partial \boldsymbol{M}/\partial t)],$$

where $\sigma = 4\pi\lambda_d/\gamma\mu_0 M_s$. Their expression emphasizes the important point that the damping acts on the motion of the resultant magnetization, $\partial \boldsymbol{M}/\partial t$, not just on $\boldsymbol{M}$. For this reason and other advantages, the latter expression has been incorporated in the Landau–Lifshitz equation if $\sigma^2 \ll 1$. Our discussion in the remainder of this section will be based on the so-called Gilbert equation

$$\partial \boldsymbol{M}/\partial t = -\gamma\mu_0(\boldsymbol{M} \times \boldsymbol{H}) + \frac{\sigma}{M_s}[\boldsymbol{M} \times (\partial \boldsymbol{M}/\partial t)]. \tag{4.128}$$

When the damping effect is absent, (4.128) is simplified to

$$\partial \boldsymbol{M}/\partial t = -\gamma\mu_0(\boldsymbol{M} \times \boldsymbol{H}). \tag{4.129}$$

This equation merely states that the rate of change of magnetization is proportional to the torque associated with the inherent precessional motion of magnetization. An unusual feature of ferromagnetic resonance is its sensitive dependence upon the shape of the specimen. This comes about because the magnetization and the corresponding demagnetizing field are very large in a ferromagnet. As shown in appendix 3, the demagnetizing factor N_d is a sensitive function of the specimen shape. Kittel [1976e] considered the ferromagnetic spin resonance in a damping-free single domain ellipsoidal specimen. For simplicity, he assumed that the principal axes coincided with the coordinate axes, as shown in fig. 4.32a. A strong static field H_z is applied along the z-axis, and an alternating field, $H_x = H_0\,\mathrm{e}^{\mathrm{i}\omega t}$, is applied along the $\pm x$-axis. The magnetization is assumed to be due only to those spins that are aligned in the direction at a small angle θ with $0z$ (fig. 4.32b). The presence of the static field will not pull the spin axis into the field axis, but causes the spins to precess about $0z$. The precession will result in components of alternating magnetization in the x and y axes. Let the demagnetizing factor be resolved into N_x, N_y and N_z. The effective magnetic fields in the axes then will be

$$\begin{aligned} H'_x &= H_0\,\mathrm{e}^{\mathrm{i}\omega t} - N_x M_x, \\ H'_y &= -N_y M_y, \\ H'_z &= H_z - N_z M_z, \end{aligned} \tag{4.130}$$

where M_x, M_y and M_z are the components of $\boldsymbol{M}_s$, as shown in fig. 4.32c. Recalling that, for a gyroscope, the rate of change of angular momentum about, say, the x-axis must be equal to the total torque acting on $\boldsymbol{M}_s$ about the same axis, we have

$$\mathrm{d}P_x/\mathrm{d}t = H'_z M_y - H'_y M_z. \tag{4.131.1}$$

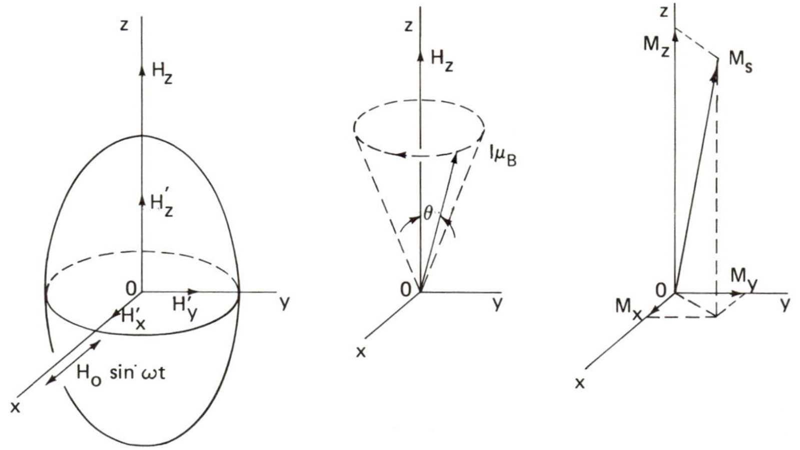

Fig. 4.32. Ferromagnetic spin resonance in a damping-free single-domain ellipsoidal specimen: (a) the coordinate system and the applied magnetic fields, (b) the precession of magnetization about the static field and (c) the three components of $\boldsymbol{M}_s$ (Brailsford [1966]).

Also

$$\mathrm{d}P_y/\mathrm{d}t = H'_x M_z - H'_z M_x. \tag{4.131.2}$$

Since $M_x/p_x = \gamma$, (4.131) lead to

$$\mathrm{d}M_x/\mathrm{d}t = \gamma[H_z + (N_y - N_z)M_z]M_y, \tag{4.132.1}$$

and

$$\mathrm{d}M_y/\mathrm{d}t = \gamma\{-[H_z + (N_x - N_z)M_z]M_x + H_0\,\mathrm{e}^{\mathrm{i}\omega t}M_z\}. \tag{4.132.2}$$

To first-order approximation, we may set $M_z = M_s$; so

$$\mathrm{d}M_z/\mathrm{d}t = 0. \tag{4.132.3}$$

Equations (4.132) are the explicit expressions of (4.129). To indicate the time dependence of the magnetization components, we express

$$\begin{aligned} M_x &= M'_x\,\mathrm{e}^{\mathrm{i}\omega t}, \\ M_y &= M'_y\,\mathrm{e}^{\mathrm{i}\omega t}, \\ M_z &= M'_z\,\mathrm{e}^{\mathrm{i}\omega t} + M_s \simeq M_s, \end{aligned} \tag{4.133}$$

where the primed symbols denote the components arising from the precession of the magnetization. By substitution (4.133) in (4.132), we obtain the following condition for (4.132) to have solutions:

$$\begin{vmatrix} \mathrm{i}\omega & \gamma[H_z+(N_y-N_z)M_z \\ -\gamma[H_z+(N_x-N_z)M_z & \mathrm{i}\omega \end{vmatrix} = 0. \qquad (4.134)$$

This condition specifies the frequency of ferromagnetic spin resonance in the absence of damping,

$$\omega_{\mathrm{R}}^2 = \gamma^2[H_z+(N_y-N_z)M_z][H_z+(N_x-N_z)M_z]. \qquad (4.135)$$

Solving (4.132) gives

$$M_x = \{M_z/[H_z+(N_x-N_z)M_z]\}\{1/[1-(\omega/\omega_{\mathrm{R}})^2]\}H_0\,\mathrm{e}^{\mathrm{i}\omega t}, \qquad (4.136)$$

which indicates that M_x is in phase with, or in phase opposition to the applied field, depending on whether ω is less or greater than ω_{R}. The corresponding absolute susceptibility is given by

$$\begin{aligned}\chi_x = M_x/H_x &= \{M_z/[H_z+(N_x-N_z)M_z]\}\{1/[1-(\omega/\omega_{\mathrm{R}})^2]\} \\ &= \chi_{x,0}/[1-(\omega/\omega_{\mathrm{R}})^2], \end{aligned} \qquad (4.137)$$

where

$$\chi_{x,0} = M_z/[H_z+(N_x-N_z)M_z]. \qquad (4.138)$$

Because the values of N_x, N_y and N_z vary with the shape of the specimen, the resonance frequency and the susceptibility also show sensitive shape dependence. Three prominent shapes are compared in table 4.6 in terms of N_{d}, ω_{R} and χ_x. These examples together with (4.135) clearly indicate the existence of a finite spin resonance frequency for ferromagnetic materials, even if $H_z = 0$, provided that the material is not in the shape of a sphere or a thin plate. Also note that for cylinders of infinite lengths, the value of f_{R} $(=\omega_{\mathrm{R}}/2\pi)$ varies slowly with H_z (e.g., from 30 000 MHz at zero field to 32 920 MHz at $H_z = 8\times10^4\ \mathrm{A\cdot m^{-1}}$).

If $\mu'_{x,\mathrm{R}}$ is the real component of the relative permeability in the x direction at an angular frequency ω and if μ_{x,R_0} gives the value of $\mu'_{x,\mathrm{R}}$ at zero frequency, we obtain from (4.137)

$$(\mu'_{x,\mathrm{R}}-1)/(\mu'_{x,\mathrm{R}_0}-1) = 1/[1-(\omega/\omega_{\mathrm{R}})^2], \qquad (4.139)$$

by setting $\chi_x = \mu'_{x,\mathrm{R}}-1$ and $\mu'_{x,\mathrm{I}} = 0$ on the basis that M_x is in phase with, or in phase opposition to H_x, as indicated by (4.136). This expression indicates two effects of the spin resonance on the real component of the permeability, as shown in fig. 4.33. First, at the

TABLE 4.6
Shape dependence of spin resonance and related properties

Specimen shape	(A) Spherical	(B) Plate in the x–z plane	(C) Cylinder of infinite length in the z-axis
1. Demagnetizing factor: N_x	$\frac{1}{3}$	0	$\frac{1}{2}$
N_y	$\frac{1}{3}$	1	$\frac{1}{2}$
N_z	$\frac{1}{3}$	0	0
2. Angular frequency of spin resonance, ω_R	γH_z	$\gamma[(H_z + M_z)H_z]^{\frac{1}{2}}$	$\gamma(H_z + \frac{1}{2}M_z)$
3. Absolute susceptibility χ_x in the direction of the applied alternating field	$\frac{M_z}{H_z}[1-(\omega/\omega_R)^2]^{-1}$	$\frac{M_z}{H_z}[1-(\omega/\omega_R)^2]^{-1}$	$\left[\frac{M_z}{H_z+\frac{1}{2}M_z}\right]\left[\frac{1}{1-(\omega/\omega_R)^2}\right]$
4. Example: Fe ($M_s = 1.7 \times 10^6$ A · m^{-1})			
Case (a). $H_z = 0$			
$f_R(=\omega_R/2\pi)$(MHz)	0	0	30 100
Case (b). $H_z = 8 \times 10^4$ A · m^{-1} (1000 Oe)			
f_R(MHz)	2820	13 320	32 920

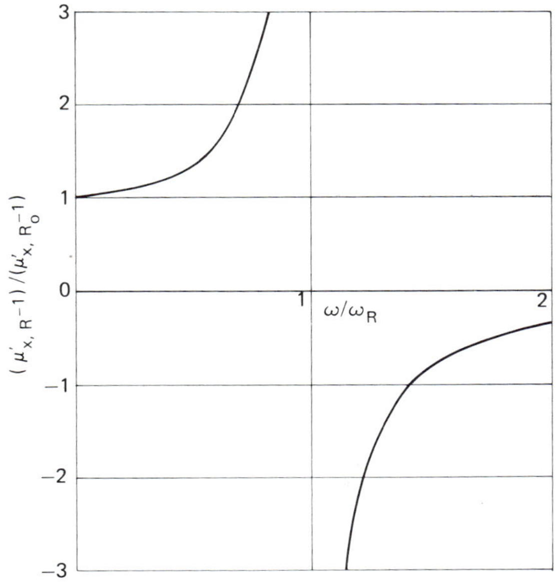

Fig. 4.33. The effect of spin resonance on the real component of complex relative permeability (Brailsford [1966]).

resonance with $\omega = \omega_R$, $\mu'_{x,R} \to \infty$. Secondly, when $\omega > \omega_R$, $\mu'_{x,R} - 1$, and similarly $\mu'_{x,R}$, becomes negative. A negative value of $\mu'_{x,R}$ means a phase opposition of M_x to the applied field. Indeed, Standley [1950] observed negative values of $\mu'_{x,R}$ in nickel film when the frequency of the alternating field was maintained at 24 400 MHz and the static field H_z was varied (fig. 4.34). In the same figure, the imaginary component of the relative permeability $\mu'_{x,I}$ reached a peak, signifying the onset of resonance at $H_z \sim 4400\ \mathrm{A \cdot cm^{-1}}$. To confirm Standley's result, we obtain the relation

$$\omega_R = 2\pi f_R = \gamma[(H_z + M_z)H_z]^{\frac{1}{2}}, \qquad (4.140)$$

from (4.135) for a thin plate specimen listed in table 4.6. For Ni, $M_z = 0.5 \times 10^6\ \mathrm{A \cdot m^{-1}}$, and substituting $f_R = 2.44 \times 10^{10}$ Hz in (4.140) gives $H_z = 4.9 \times 10^5\ \mathrm{A \cdot m^{-1}}$ (6150 Oe), which is in good agreement with the observed value of H_z at resonance in fig. 4.34.

Thus far, we have ignored the effect of damping. It is true that damping is practically absent either in insulating materials, such as

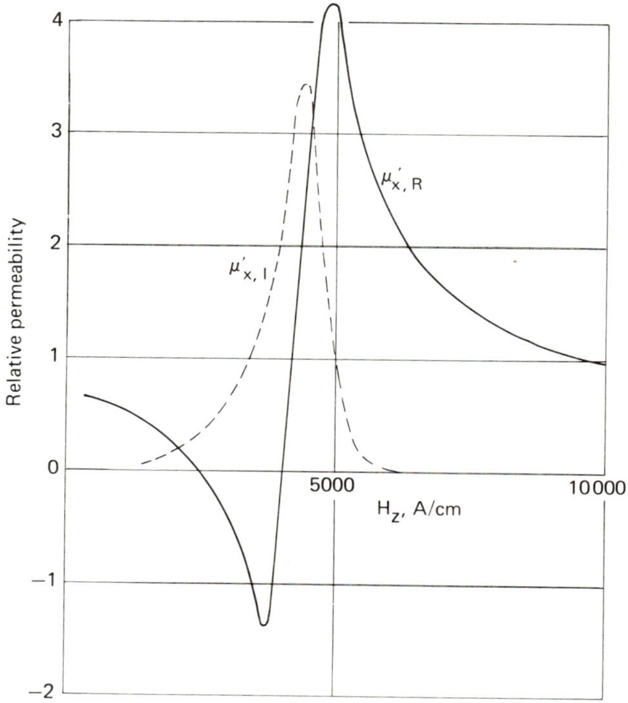

Fig. 4.34. Experimental evidence for the dependence of the real and imaginary components of complex relative permeability on the applied static field in Ni films at a constant frequency of 24 400 MHz (Standley [1950]).

ferrites, or in specimens that have attained saturation magnetization in a single-domain structure. In other cases, however, the existence of eddy currents in the material will cause appreciable damping. Consequently, there should be additional terms associated with the damping effect on the right-hand side of (4.132). For spherical specimens, for example, the more complete forms of (4.132) are

$$\partial M_x/\partial t = \omega_R M_y + (\alpha/M_s)[M_y(\partial M_z/\partial t) - M_z(\partial M_y/\partial t)],$$
$$\partial M_y/\partial t = -\omega_R M_x + (\alpha/M_s)[M_z(\partial M_x/\partial t) - M_x(\partial M_z/\partial t)], \quad (4.141)$$
$$\partial M_z/\partial t = (\alpha/M_s)[M_x(\partial M_y/\partial t) - M_y(\partial M_x/\partial t)],$$

where $\omega_R = \gamma H_z$. Solving them simultaneously gives

$$\frac{\partial M_x}{\partial t} = \frac{\omega_R}{1+\sigma^2} M_y + \frac{\omega_R \sigma}{1+\sigma^2} \frac{M_x M_z}{M_s},$$

$$\frac{\partial M_y}{\partial t} = -\frac{\omega_R}{1+\sigma^2} M_x + \frac{\omega_R \sigma}{1+\sigma^2} \frac{M_y M_z}{M_s}, \tag{4.142}$$

$$\frac{\partial M_z}{\partial t} = -\frac{\omega_R \sigma}{1+\sigma^2} M_s + \frac{\omega_R \sigma}{1+\sigma^2} \frac{M_z^2}{M_s}.$$

These equations have solutions of the form:

$$M_x = M_s \sin\theta \, e^{i\omega t},$$

$$My = M_s \sin\theta \, e^{i(\omega t + \pi/2)}, \tag{4.143}$$

$$M_z = M_s \cos\theta,$$

where θ is the instantaneous angle between the magnetization vector and the z-axis at time t. Figure 4.35 shows the motion of the resultant magnetization under three conditions: (a) zero damping, (b) small damping and (c) large damping. Damping is seen to cause the spin precession to decay eventually to a rest with $\theta = 180°$. The decay is

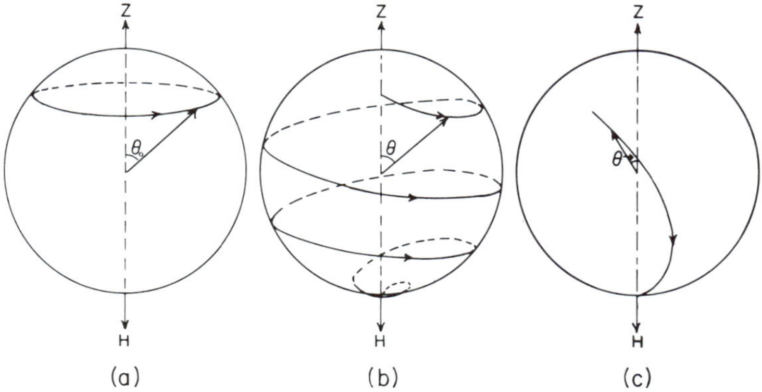

Fig. 4.35. The precession of spins in a lattice that is characterized by (a) the absence of damping, (b) small damping and (c) large damping (Chikazumi [1964]).

characterized by a relaxation time τ defined as

$$\tau = (1 + \sigma^2)/\omega_R\sigma. \tag{4.144}$$

Using this definition of τ and the solutions given in (4.143), it can be shown that

$$\tan \tfrac{1}{2}\theta = \tan \tfrac{1}{2}\theta_0 \, e^{t/\tau}, \tag{4.145}$$

where θ_0 is the initial inclination of the magnetization vector; and

$$\omega_d = 1/\sigma\tau = \omega_R/(1 + \sigma^2), \tag{4.146}$$

where ω_d is the angular frequency of the damped precession of the resultant magnetization. From (4.145), it is seen that for $\theta = 180°$ at the completion of the spin-precession decay, $\tan \frac{1}{2}\theta$ and hence t/τ tend to infinity. Thus, a short relaxation time implies an efficient damping, or equivalently a highly viscous lattice. On the other hand, a lattice free of damping features zero σ, and hence infinite τ. Between these two extremes lies a case in which the minimum relaxation time τ_{min} corresponds to critical damping. The value of τ_{min} is obtained by setting $\partial\tau/\partial\sigma = 0$ in (4.144), leading to $\sigma = 1$. Whence

$$\tau_{min} = 2/\omega_R = 2/\gamma H_z. \tag{4.147}$$

In a strong field, say, $H_z = 1.6 \times 10^6 \text{ A} \cdot \text{m}^{-1}$ (20 000 Oe),

$$\tau_{min} = \frac{2}{2.21 \times 10^5 \times 1.6 \times 10^6} = 5.7 \times 10^{-12} \text{ s}.$$

This value is much smaller than those observed, which fall in the range 10^{-6}–10^{-9} sec. Hence the condition for critical damping generally is not met. Means have been devised for promoting the damping effect, such as the enhancement of eddy currents. Critical damping is an important consideration for memory devices made of soft magnetic thin films because it controls the switching time of such devices.

Suggestions for further reading

Anderson, J. C., 1968, *Magnetism and Magnetic Materials* (Chapman and Hall, London). Chapters 6 and 7 on §2.

Bozorth, R. M., 1951, *Ferromagnetism* (van Nostrand, New York). Chapters 11, 13 and 17.

Chikazumi, S., 1964, *Physics of Magnetism* (Wiley, New York). Chapters 12 to 14 and 16.

CHAPTER 5

METALLURGY OF SOFT MAGNETIC MATERIALS

Metallurgy is the science of preparing and characterizing metals and alloys. Since 1955*, however, the scope of metallurgy has been extended to include nonmetallic materials, such as ceramics, polymers and composites. Hence in a broad sense, metallurgy and materials science are synonymous. In the same period, intensive research has considerably advanced the knowledge of solids, both crystalline and amorphous. Many obsolete theories have been replaced or corrected. The advance is especially conspicuous in the fields of solid-state theory and physical and mechanical metallurgy in connection with lattice imperfections.

Because the traditional soft magnetic materials are metallic** and their useful properties are structure sensitive, it is not surprising that metallurgists have played an important, and sometimes indispensable, role in the development of industrial magnetic products. The purpose of this chapter is to show how metallurgical approaches are fruitful or essential in advancing our understanding and in improving the performance of magnetic materials. The chapter is divided into two parts: the first part deals with soft magnetic materials from the standpoint of metallurgy or solid state theory, and the second part discusses various metallurgical factors and their effect on the magnetic properties of major industrial products.

*This was the year when the Department of Materials Science was installed at the Northwestern University, Evanston, Illinois, USA. For the first time, graduate work could be pursued for the MS or PhD degree with major in materials science in an independent department.

**Even in today's market, as ch. 6 will show, the total market value is split into $\approx$65% for metallic materials and $\approx$35% for nonmetallic materials.

§1. TECHNICAL MATERIALS

§1.1. *Pure metals*

It has been stated at the outset of the book that only nine elements out of a total of 106 thus far discovered exhibit ferromagnetism. All are metals, of which three (Fe, Co and Ni) are 3d transition metals forming the iron-group and the other six (Gd, Tb, Dy, Ho, Er and Tm) are lanthanides. To date, the iron-group metals have dominated the market of soft magnetic materials. This dominance will continue in the foreseeable future, although some applications have been found for lanthanide metals at cryogenic temperatures.

The scarcity of ferromagnetic elements naturally raises an intriguing question as to why ferromagnetism occurs so selectively. More specifically, the question may be resolved into two parts:

(1) Why can't other 3d transition metals such as V, Cr and Mn become ferromagnetic as Fe, Co and Ni do? All these metals have similar electronic configurations featuring an incompletely filled 3d subshell. Moreover, V, Cr and α-Fe have the same body-centered cubic crystal structure with comparable atomic sizes.
(2) Why is the ferromagnetism in lanthanides more or less confined to the elements (Gd to Tm) in the second half of the series?

The first question attracted the attention of many prominent physicists in the early 1930's. Bethe (1933, see Morrish [1965a]) considered the question as equivalent to asking why the exchange integral J_{ex} would not become positive in Mn, Cr, V, etc., but does so in Fe, Co and Ni below their Curie temperatures. For J_{ex} to be positive, let us examine (2.28). Clearly the two positive terms, $1/r_{12}$ and $1/r_{ab}$, must outweight the two negative terms, $-1/r_{a2}$ and $-1/r_{b1}$. Bethe made some rough calculations and concluded that the term $1/r_{12}$ is much more important than the other positive term $1/r_{ab}$ and that, for $1/r_{12}$ to attain large magnitudes, the wave functions ψ_a and ψ_b in (2.37) must satisfy three conditions: (1) small amplitude at their parent nuclei, (2) angular lobes pointing toward and overlapping one another, and (3) small radial extent compared with the internuclear spacing D.

All three conditions seem to be satisfied by Fe, Co and Ni. The 3d wave functions are small near the nucleus, so (1) is satisfied. Furthermore, the 3d wave functions in these three metals have strong angular lobes that can be made to overlap one another, so (2) is also satisfied. Earlier, Slater (1930, see Morrish [1965a]) made detailed estimates of the ratio of D to the diameter d of the 3d subshell for various metals.

Slater's results are listed in table 5.1, which Bethe used to plot J_{ex} against D/d in fig. 5.1. The three ferromagnetic metals indeed showed larger D/d ratios than all other transition metals, so (3) is thought to be satisfied. The value of 1.5 was adopted empirically in the Bethe–Slater criterion for ferromagnetic spin coupling as the demarcation line to separate the positive and negative J_{ex}.

The Bethe approach, however, encountered severe criticisms, of which the main ones are:

(1) The nonintegral values deduced for the mean Bohr magneton number per atom in Fe, Co and Ni (ch. 2, §2) practically rule out the applicability of the Heisenberg theory to the transition metals. Analyses based on that theory are therefore questionable.
(2) Even if the Heisenberg theory is assumed to work in these metals, Herring [1962] found great difficulties in accurately expressing the wave functions ψ_a and ψ_b consistent with the nonlocalizing characteristics of

TABLE 5.1
Interatomic distances (D) and diameters (d) of incompletely filled d-subshells of some transitional metals and alloys estimated by Slater* (see Bozorth [1951e])

Transitional metal	D(Å)	d(Å)	D/d ratio	Curie temperature θ_C(°K)
Mn	2.24	(3d) 1.71	1.31	
Fe	2.482	1.58	1.57	1 043
Co	2.50	1.38	1.81	1 400
Ni	2.49	1.27	1.96	631
Cu_2MnAl**	3.66	1.71	2.14	~720
Mo	2.73	(4d) 2.94	0.93	
Ru	2.65	2.33	1.14	
Rh	2.69	2.11	1.27	
Pd	2.75	1.93	1.42	
W	2.74	(5d) 3.44	0.80	
Os	2.68	2.72	0.99	
Ir	2.714	2.47	1.10	
Pt	2.78	2.25	1.24	

*Values for D are corrected with respect to more recent X-ray data given in *Structure of Metals*, by Barrett and Massalski [1966a].
**See discussion on p. 206.

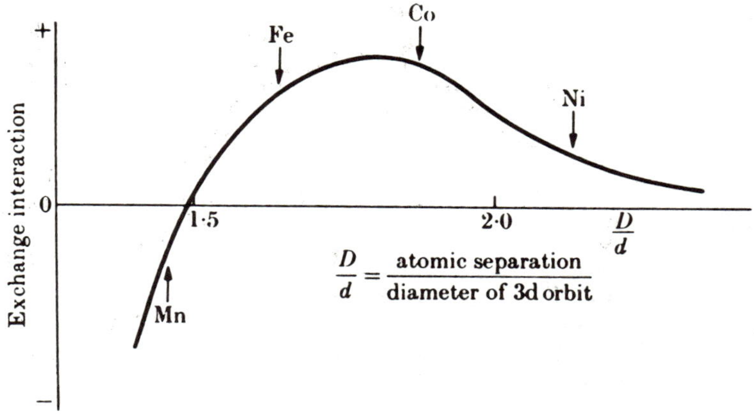

Fig. 5.1. The Slater–Bethe curve showing the magnitude and sign of the exchange integral as a function of D/d (Standley [1972]).

the d electrons. In an elegant way, he demonstrated that the original Heitler–London method of calculating the spin coupling is unreliable, even in the limit of large separations of atoms.
(3) Later attempts to determine the conditions under which J_{ex} would be positive were neither successful nor conclusive. Wohlfarth [1949] made the first explicit calculation using averaged spherical wave functions and concluded that J_{ex} would always be negative. However, the validity of his conclusion is in doubt because of the spherical wave functions used, which obviously violate Bethe's second condition. Stuart and Marshall [1960] made a computer calculation using more realistic wave functions and found that J_{ex} based on direct exchange would always be positive. Hence, they concluded that direct exchange is not responsible for ferromagnetism. Their conclusion was in accord with similar skepticisms expressed earlier by Slater [1936], Zener [1951a] and van Vleck [1953] (see Morrish [1965c]).

Several other theories were proposed to explain the selective appearance of ferromagnetism in the 3d transition metals. For instance, Vonsovsky [1953] and Zener [1951a] suggested an indirect exchange interaction between localized d electrons on neighboring atoms as a result of an induced polarization of the conduction electrons. Their theory, while avoiding some of the difficulties associated with the direct exchange mechanism, introduced new problems in its rigorous treatment. It eventually submerged into the Ruderman–Kittel–Kasuga–Yosida (RKKY) theory, which will be a major subject of discussion in the remainder of this section.

It is sad to say that despite numerous studies conducted interspersedly by many noted physicists in the past fifty years, the theory of ferromagnetism of the iron-group metals remains unsettled. Strictly speaking, neither the band theory nor the localized electron theory is satisfactory. Many workers now believe that an effective theory would have to embrace certain features of both theories. In other words, the new theory is probably a compromise or a combination of the band and the localized electron theories.

Ironically, the history of magnetic research on rare earths is relatively short. Research efforts had remained lukewarm until Spedding [1961], together with a group of metallurgists and chemists, succeeded in the preparation of high-purity ($\geqslant$99.9%) rare earths at Ames Laboratory of the Energy Research and Development Administration (formerly the Atomic Energy Commission), USA, in the late 1950's. In less than twenty years, however, the understanding of the magnetic properties of lanthanides has already surpassed that for the iron-group metals. The book edited by Elliott [1972] gives a long-needed, excellent and comprehensive review on this subject.

Formally, there are 14 elements in the series of lanthanides. (The 14 lanthanides plus Sc, Y and La are called rare earth elements or metals collectively.) The series includes Ce ($Z = 58$), the first member, Pr, Nd, Pm, Sm, Eu ¦ Gd, Tb, Dy, Ho, Er, Tm, Yb and Lu ($Z = 71$), the last member. As shown by the dashed lines between Eu and Gd, the series is customarily divided into two groups. The first group contains six elements from Ce to Eu, which are called the light lanthanides, and the second group contains eight elements from Gd to Lu, which are called the heavy lanthanides.

The most important characteristic of the lanthanides is, of course, the presence of 4f electrons in their normal electronic configurations. As is well known, these electrons are the sole source of magnetic moments. The number of the 4f electrons increases progressively from one in the Ce^{3+} ion to 14 in the Lu^{3+} ion, as was shown in table 2.2. The effective Bohr magneton numbers per ion deduced from paramagnetic susceptibility measurements are given in table 2.3. The magnetic data coupled with other pertinent physical measurements indicate that the tri-positive state represents all lanthanides except Ce, which is partly tetravalent (see Chen [1965]), and Eu and Yb, which are divalent.

TABLE 5.2
Crystal structures† and magnetic properties of lanthanide metals

Element	Z	Crystal structure and transition temperature (°C)	Free 3^+ ions*		Saturation moment (μ_B/atom)	
			g	$(g-1)^2J(J+1)$	gJ	Exp.
Ce	58	bcc $\xrightarrow[726^\circ\mathrm{C}]{}$ fcc $\xrightarrow[-23^\circ\mathrm{C}]{}$ dhcp $\xrightarrow[-157^\circ\mathrm{C}]{}$ fcc	$\frac{6}{7}$	0.18	2.14	–
Pr	59	bcc $\xrightarrow[795^\circ\mathrm{C}]{}$ dhcp	$\frac{4}{5}$	0.80	3.20	–
Nd	60	bcc $\xrightarrow[863^\circ\mathrm{C}]{}$ dhcp	$\frac{8}{11}$	1.84	3.27	–
Pm	61	bcc $\xrightarrow[890^\circ\mathrm{C}]{}$ dhcp	$\frac{3}{5}$	3.20	2.40	–
Sm	62	bcc $\xrightarrow[922^\circ\mathrm{C}]{}$ hcp $\xrightarrow[727^\circ\mathrm{C}]{}$ rhom	$\frac{2}{7}$	4.46	0.72	–
Eu	63	bcc	–	0	0	–
Gd	64	bcc $\xrightarrow[1235^\circ\mathrm{C}]{}$ hcp	2	15.75	7.0	7.55
Tb	65	bcc $\xrightarrow[1289^\circ\mathrm{C}]{}$ hcp	$\frac{3}{2}$	10.50	9.0	9.34
Dy	66	bcc $\xrightarrow[1381^\circ\mathrm{C}]{}$ hcp	$\frac{4}{3}$	7.08	10.0	10.33
Ho	67	hcp	$\frac{5}{4}$	4.50	10.0	10.34
Er	68	hcp	$\frac{6}{5}$	2.55	9.0	8.3
Tm	69	hcp	$\frac{7}{6}$	1.17	7.0	7.14
Yb	70	bcc $\xrightarrow[795^\circ\mathrm{C}]{}$ fcc $\xrightarrow[-3^\circ\mathrm{C}]{}$ hcp	$\frac{8}{7}$	0.32	4.0	–
Lu	71	hcp	0	0	0	–

†Reference: Beaudry and Gschneidner [1977].
*See table 2.2 for S, L and J values for free trivalent ions.

In the second question raised in this section, it was stated that ferromagnetism is more or less confined to the heavy lanthanides, specifically to Gd–Tm. Two further remarks are necessary on this point:

(1) Experimental studies of magnetic structures of light lanthanides are still incomplete. For instance, neutron diffraction experiments have not been conducted on single crystals of some of the light lanthanides (see Koehler [1972]). Nevertheless, from the available information, it is reasonable to conclude that none of the light lanthanides will show ferromagnetism in their stable structures; only antiferromagnetism has been detected in Ce, Pr, Nd and Sm. This conclusion applies to Pr and Nd whose stable structures below the phase transition temperatures (795°C for Pr and 863°C for Nd) are double hexagonal close-packed (dhcp) with a stacking sequence ABACABAC... in the *c*-axis, but the metastable face-centered cubic structures of Pr and Nd produced by quenching display ferromagnetism (Koehler [1965]).
(2) The emergence of ferromagnetism in the heavy lanthanides is complicated by (a) conical and pseudoferrimagnetic spin structures in Ho, Er and Tm, rather than a typical straight ferromagnetic spin coupling, and (b) transitions to more complex (helical in Tb, Dy and Ho; first quasiantiphase domain, then sinusoidal variation in Er; and antiferromagnetic sinusoidal variation in Tm) spin structures above θ_C. In other words, except Gd, all other "ferromagnetic" rare earths display a two- or three-step magnetic transition, as compared with the single-step, ferromagnetic→paramagnetic transition in the iron-group metals. The complicated magnetic transitions in lanthanide metals from Tb to Tm are shown in fig. 5.2.

A phenomenological theory has been developed that is capable of explaining many features of the magnetic behavior of lanthanides. Recall that the 4f subshell lies below the completed 5s and 5p subshells. Therefore, the 4f electrons are highly localized at individual atomic sites. Because of their localized nature, these electrons behave approximately as they would in a free ion. Thus the spin and the orbital motion contribute to the total magnetic moment per atom, as is reflected by the relation for the lanthanides

$$g\boldsymbol{J}\mu_B = (\boldsymbol{L} + 2\boldsymbol{S})\mu_B. \tag{5.1}$$

On the other hand, according to the third Hund's rule (ch. 2, §1), the

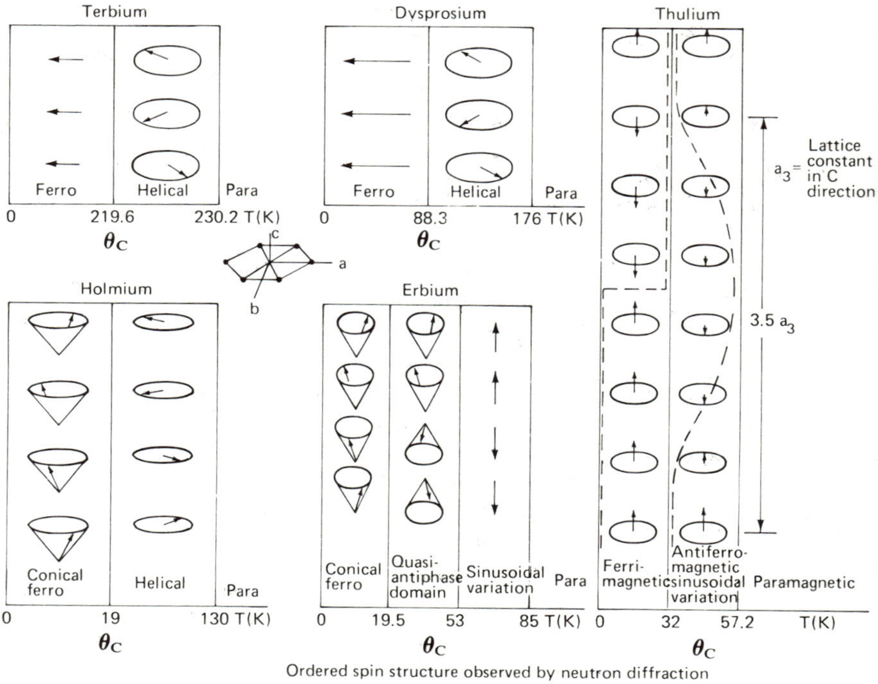

Fig. 5.2. Complex spin structures observed in Tb, Dy, Ho, Er and Tm by neutron diffraction (Legvold [1973]).

atomic total angular momentum quantum number $\boldsymbol{J}$ is given by

$$\boldsymbol{J} = \boldsymbol{L} + \boldsymbol{S} \quad \text{for Gd to Lu,} \tag{5.2.1}$$

$$\boldsymbol{J} = \boldsymbol{L} - \boldsymbol{S} \quad \text{for Ce to Eu.} \tag{5.2.2}$$

Combining (5.1) and (5.2) gives

$$(g-1)\boldsymbol{J} = \boldsymbol{S} \quad \text{for Gd to Lu,} \tag{5.3.1}$$

$$(g-1)\boldsymbol{J} = 3\boldsymbol{S} \quad \text{for Ce to Eu.} \tag{5.3.2}$$

The values of L, S, J, g and other basic constants for the lanthanides are given in table 5.2.

At first glance, the localization of the magnetic electrons seems to imply that the lanthanides would be ideal examples for the application of the Heisenberg theory. While this implication is substantiated by the fact that the phenomenological theory, too, is based on an exchange interaction of the type expressed by Heisenberg in (2.39), care must be exercised in accepting the $S_i \cdot S_j$ interaction. For convenience, let us rewrite (2.39) in terms of a non-relativistic Hamiltonian:

$$\mathcal{H}_{\mathrm{ff'}} = -\sum_{i,j} [J_{\mathrm{ex}}(\boldsymbol{R}_i - \boldsymbol{R}_j)]\boldsymbol{S}_i \cdot \boldsymbol{S}_j, \tag{5.4}$$

where $\boldsymbol{R}_i$ and $\boldsymbol{R}_j$ are the position vectors of two lattice sites and $J_{\mathrm{ex}}(\boldsymbol{R}_i - \boldsymbol{R}_j)$ represents the exchange interaction between the spin moments at $\boldsymbol{R}_i$ and $\boldsymbol{R}_j$. Two comments on the new expression are in order.

First, since the total magnetic moment per atom is due partly to the orbital motion, it is more complete to write $\boldsymbol{J}_i \cdot \boldsymbol{J}_j$ in place of $\boldsymbol{S}_i \cdot \boldsymbol{S}_j$, where the $\boldsymbol{J}$'s and the $\boldsymbol{S}$'s are related in (5.3). Thus, for the heavy lanthanides, we have

$$\begin{aligned}\mathcal{H}_{\mathrm{ff'}} &= -(g-1)^2 \sum [J_{\mathrm{ex}}(\boldsymbol{R}_i - \boldsymbol{R}_j)]\boldsymbol{J}_i \cdot \boldsymbol{J}_j \\ &= -\sum [\mathcal{J}_{\mathrm{ex}}(\boldsymbol{R}_i - \boldsymbol{R}_j)]\boldsymbol{J}_i \cdot \boldsymbol{J}_j,\end{aligned} \tag{5.5}$$

where

$$\mathcal{J}_{\mathrm{ex}}(\boldsymbol{R}_i - \boldsymbol{R}_j) = (g-1)^2[J_{\mathrm{ex}}(\boldsymbol{R}_i - \boldsymbol{R}_j)], \tag{5.6}$$

is the exchange integral based on the interaction between the total momenta $\boldsymbol{J}$. Equation (5.6) is known as the De Gennes [1966] relation.

The second remark is more serious in nature. It deals with the mechanism whereby $\boldsymbol{J}_i$ and $\boldsymbol{J}_j$ are coupled. Because the 4f electrons are localized and because the diameters of the ionic cores containing the magnetic electrons, d, are very small (approximately one-tenth of the interatomic spacing D*), direct exchange interaction between atomic

*Accordingly, the ratio D/d would have values around 10 for the lanthanides (Bozorth [1951e] used Slater's formula to calculate and obtained a value of $D/d = 3.1$ for Gd). More recent computation by Freeman [1972] gave $D/d = 3.583/0.636 = 5.7$. Applying the $D/d > 1.5$ criterion for the onset of ferromagnetism, it seems easy to understand why six lanthanides out of a possible total of 11 (three are excluded because Pm has no stable isotopes and Yb and Lu have completed 4f subshells) are ferromagnetic. This reasoning, however, must be rejected on the ground that the Bethe analysis is inapplicable to lanthanides (for instance, his first and second conditions are not satisfied).

moments is expected to be ineffective in lanthanides. To support (5.5), we must resort to another type of interaction.

It is now believed that lanthanide metals attain their various modes of magnetic ordering through an indirect interaction. This interaction is responsible for the $\boldsymbol{J}_i \cdot \boldsymbol{J}_j$ coupling and arises via an exchange interaction between the magnetic electrons in the 4f subshell and the conduction electrons. The latter s–f interaction comes about because of the polarization effect of the magnetic electrons on the conduction electrons. The indirect interaction, which is analogous to the d–d interaction via an s–d interaction advocated by Vonsovsky and Zener, is known as the RKKY interaction in tribute to Ruderman and Kittel [1954], who originally proposed a similar interaction to explain the long-range coupling between nuclear spins, and to Kasuya [1956] and Yosida [1957], who applied the interaction to lanthanides and transition metals, respectively. Quantitative treatment of the RKKY interaction requires the use of second-order perturbation theory and, therefore, is beyond the scope of this book. Here, we shall give a qualitative description of the interaction with emphasis on the applications.

The 4f electrons strongly overlap the conduction electrons. Thus these two groups of electrons interact, leading to the polarization of the conduction electrons. The polarization effect is a long-range phenomenon, which, in turn, gives rise to the f–f interaction. Moreover, the Fermi distribution of the conduction electrons causes the polarization to have an oscillatory component that restricts the wave vectors $\boldsymbol{K}$ of the s electrons. Consequently, the RKKY interaction depends on (1) the exchange integral between the magnetic cores and the conduction electron wave functions at the core sites; and (2) the energy bands and Fermi surface of the conduction electrons. Because of the second dependence, the spin arrangement in rare earths can be more complicated than simple ferromagnetic configuration. A convenient way to treat the problem of magnetic ordering in terms of the RKKY interaction is to introduce a wave vector $\boldsymbol{q}$ in $\boldsymbol{K}$-space for the conduction electrons. This wave vector is defined by

$$\boldsymbol{q} = \boldsymbol{K}' - \boldsymbol{K}, \tag{5.7}$$

where $\boldsymbol{K}$ and $\boldsymbol{K}'$ are the wave vectors of a conduction electron before and after the s–f interaction. If $\boldsymbol{q}$ is set in such a way that its corresponding direction in real space is parallel to the c-axis, then its magnitude will indicate the periodicity of the spin arrangement, and the mode of the magnetic order can be described by a set of three equations:

$$J_{n\zeta} = J \cos \theta, \quad J_{n\xi} = J \sin \theta \cos(\boldsymbol{q} \cdot \boldsymbol{R}_n), \quad J_{n\eta} = J \sin \theta \sin(\boldsymbol{q} \cdot \boldsymbol{R}_n) \tag{5.8}$$

where $J_{n\zeta}$, $J_{n\xi}$ and $J_{n\eta}$ are the components of J with ζ lying along the c-axis, ξ pointing toward a nearest neighbor in the basal plane and η in a direction orthogonal to ξ in the basal plane; and θ is the polar angle between J and the c-axis. The various values of q and θ lead to a variety of magnetic order observed in the lanthanides and are shown in the following table:

	Ferromagnetic				
	I	II	III	Conical	Planar spiral
q	0	0	0	$\neq 0$	$\neq 0$
θ	0	$\frac{1}{2}\pi$	Between 0 and π	$\neq 0$	$\frac{1}{2}\pi$
Example	Gd	Tb and Dy at low T		Er and Ho at low T	Tb, Dy and Ho at high T

Another important purpose of the wave vector $\boldsymbol{q}$ is to establish a $\boldsymbol{q}$-dependent generalized susceptibility function $\chi(\boldsymbol{q})$. This function represents the linear response of the conduction electrons to the effective fields associated with the magnetic cores. It plays a vital role in determining the magnetic structure. For example, when the exchange-energy term is much larger than the other (crystal field and magnetoelastic) terms, it will dominate the free energy of the magnetic system, as with the heavy lanthanides. In these cases, a maximum in $\chi(\boldsymbol{q})$ at a proper $\boldsymbol{q}$ value will signify the onset of magnetic ordering in conical, spiral or straight ferromagnetic form. This is so because a maximum $\chi(\boldsymbol{q})$ corresponds to a minimum in the exchange energy. More specifically, a peak in $\chi(\boldsymbol{q})$ at $\boldsymbol{q} = 0$ would indicate ferromagnetic ordering, and a peak at other $\boldsymbol{q}$ values would predict a conical or spiral spin structure.

The significance of the susceptibility function $\chi(\boldsymbol{q})$ is enhanced by the fact that it can be calculated. The calculation, however, involves tedious computer programming and necessitates precise knowledge of the Fermi surface and energy bands to achieve reliability. Liu et al.

[1971] made a series of approximate calculations for most lanthanides. Their results for Pr in the stable dhcp and metastable fcc structures explain the drastic difference between the magnetic ordering of the two phases of Pr and, possibly, of Nd.

The effect of the crystal field may also play a vital role in promoting the ferromagnetic ordering in the fcc Pr. Because of the relatively large radii of the 4f electrons in the light lanthanides, the crystal field splitting of magnetic energy levels is increased to be comparable with the exchange energies. For lanthanide ions having even numbers of 4f electrons (Pr, Tb, Ho and Tm), the crystal-field only ground state is a singlet. The first excited state is also a singlet for the hexagonal structure, but a triplet for the cubic structure. Thus, for the fcc Pr, the existence of a triplet lying above the singlet ground state makes it possible to induce magnetic moments with sufficiently strong coupling in the ferromagnetic mode.

To complete our discussion of the RKKY interaction, let us return to the typical Heisenberg theory expressed by (5.5). First, we want to write the Fourier transform of the exchange integral,

$$\mathcal{J}_{ex}(\boldsymbol{R}_n - \boldsymbol{R}_m) = \sum_{\boldsymbol{q}} \mathcal{J}_{ex}(\boldsymbol{q}) \exp[\mathrm{i}\boldsymbol{q} \cdot (\boldsymbol{R}_n - \boldsymbol{R}_m)]. \tag{5.9}$$

Moreover $\mathcal{J}_{ex}(\boldsymbol{q})$ is related to $\chi(\boldsymbol{q})$ by

$$\mathcal{J}_{ex}(\boldsymbol{q}) = (2/N)(g-1)^2 |\mathcal{J}_{sf}(\boldsymbol{q})|^2 \chi(\boldsymbol{q}), \tag{5.9.1}$$

where $\mathcal{J}_{sf}$ is the exchange integral of the s–f interaction. The last equation indicates that the RKKY interaction depends on (1) the driving exchange interaction represented by $\mathcal{J}_{sf}(\boldsymbol{q})$ and (2) the response function of the system represented by $\chi(\boldsymbol{q})$.

As a further application of the RKKY theory, we first correct (4.3) by replacing J_{ex} with $\mathcal{J}_{ex}$ to obtain

$$\theta_C = z\mathcal{J}_{ex}J(J+1)/3k. \tag{5.10}$$

Combining (5.6) with (5.10) yields

$$\theta_C = z(g-1)^2 J_{ex} J(J+1)/3k, \tag{5.10.1}$$

where $z \approx 12$ for the hcp lanthanides. Equations (5.10) and (5.10.1) permit a calculation of $\mathcal{J}_{ex}$ and J_{ex}, respectively, from known values of θ_C for various heavy lanthanides. The results of the calculation are shown in table 5.3.

TABLE 5.3
Analysis of the Curie temperatures of ferromagnetic lanthanides according to (5.10)

	Gd	Tb	Dy	Tm	Er	Ho
θ_c(°K)	292.7	220	85	32	19.5	19.0
$J_{ex}(10^{-23}$ J)[10^{-4} eV]	6.42 [4.00]	7.24 [4.52]	4.29 [2.68]	9.44 [5.89]	2.64 [1.65]	1.46 [0.91]
$\mathscr{J}_{ex}(10^{-23}$ J)[10^{-4} eV]	6.42 [4.00]	1.81 [1.13]	0.48 [0.30]	0.262 [0.16]	0.106 [0.07]	0.091 [0.06]

A comparison between tables 4.2 and 5.3 reveals that since the exchange integral in (4.3) should be the new integral based on the indirect interaction between the total momenta J, not just the spin momenta S, the old J_{ex} takes up new values given by the deduced values of $\mathscr{J}_{ex}$ using the de Gennes relation. The values listed for J_{ex} in table 4.2 are now for $\mathscr{J}_{ex}$ in table 5.3. Meanwhile, J_{ex} takes up new values given by the relation expressed in (5.6). The value of $\mathscr{J}_{ex}$ is seen to vary about ten times more with the species of the lanthanides than does J_{ex}. More significantly, as far as the Curie temperature is concerned, we see at least a qualitative parallelism between the variations with the species of lanthanides of θ_C and $\mathscr{J}_{ex}$, whereas the variation of J_{ex} is "erratic" and therefore unrelated to that of θ_C.

§1.2. *Alloys*

An alloy is a metallic solid or liquid formed from an intimate combination of two or more elements. The element that contributes the major portion of the composition is called the solvent, whereas the minor elements are called solutes. Although the solvents of alloys are always metals, the solutes can be metals, semimetals (Si, Ge, As, Se and Te) and nonmetals. In our discussion, the solvents are confined to ferromagnetic metals. Alloys may be classified according to the number of components present. A binary alloy is a two-component system, a ternary alloy contains three components, and so on. Alloys may also be classified as substitutional, when solute atoms replace solvent atoms at regular lattice sites, or interstitial, when solute atoms reside at interstices in the lattice. A dilute binary alloy contains the solute up to a few atomic percent, whereas an alloy with the solute content in the range roughly 20–50 at.% is considered concentrated.

A phase is a macroscopically homogeneous body of matter that is physically distinct. A metal exists as a pure phase in each state; but an alloy may exist in one phase or multiple phases in the same solid or liquid state. Thus, the word alloy only implies a metallic system whose composition is characterized by more than one component; it does not necessarily mean a single phase. This distinction is frequently ignored by physicists. For an accurate determination of the number of phases in the solid state of an alloy, one should examine the pertinent constitutional or phase diagram, which is a map of the alloy system showing the various phases present at different compositions or pressure and temperatures. In the phase diagram of a binary system, two single phases lie next to the pure metals at both terminals. They are called the primary or terminal solid solutions. Other single-phase solid solutions lying between the primary solid solutions are termed intermediate phases. An intermediate phase becomes an intermetallic compound if (1) the region representing the phase in the constitutional diagram is narrowed to essentially a vertical line; (2) its composition corresponds to a simple stoichiometric proportion indicated by A_xB_y, where x and y are integers, and (3) the atomic arrangement is orderly.

When using a phase diagram, it is important to know the Gibbs phase rule that deals with the degree of freedom, F, of a system in relation to the number of components, C, and the number of phases, P. The rule is expressed as

$$F = C - P + 2, \tag{5.11}$$

where the constant 2 comes from the temperature and pressure, which are considered as two separate variables in addition to the composition of each phase. For a system of a pure element, obviously $C = 1$; hence $F = 3 - P$. The phase rule then predicts that when the three phases (gas, liquid and solid) exist together in equilibrium, the system has zero degree of freedom; that is, both temperature and pressure are invariant. The phase diagram of a pure element is a plot of temperature versus pressure. In such a T–P phase diagram, the three-phase coexistence is shown by the triple point at which the temperature and the pressure are indeed fixed.

The phase diagrams of the binary systems are plots of temperature versus composition. In a T–C diagram, the pressure is assumed to be kept at the atmospheric value. Consequently, (5.11) is altered to $F = C - P + 1$. For binary systems with $C = 2$, the phase rule is also reduced to $F = 3 - P$. Therefore, there are regions in the phase diagrams where single phases exist wth two degrees of freedom (both temperature and composition can be changed simultaneously without disturbing the existence of the phase). Also, there are regions where two phases coexist in equilibrium with one degree of freedom (either temperature or composition can be changed). Whenever three phases coexist, since now $F = 0$, neither temperature nor composition can be changed. The latter situation is represented by a horizontal line in a binary phase diagram. In the Fe–Co phase diagram, for instance, a short horizontal line emerges between approximately 16.5 and 19.5 at.% Co at 1499°C to indicate the coexistence of γ, δ and the liquid phases at fixed compositions and at a fixed (1512°C) temperature.

The phase diagrams of the three major ferromagnetic binary systems are strikingly similar in the solid state. We note the following three common features in particular:

(1) All three systems display a continuous series of fcc solid solutions at intermediate temperatures. This situation arises because of the allotropies of Fe and Co shown below:

$$\text{Fe:}\quad \delta(\text{bcc}) \xrightarrow{1392°\text{C}} \gamma(\text{fcc}) \xrightarrow{911°\text{C}} \alpha(\text{bcc}); \tag{5.12.1}$$

$$\text{Co:}\quad \beta(\text{fcc}) \xrightarrow{450°\text{C}} \alpha(\text{hcp}). \tag{5.12.2}$$

Hence all these three elements exist in the fcc lattice, making it possible to form three series of isomorphous solid solutions at room temperature or above the pertinent phase-transition temperature. This feature also conforms with the Hume-Rothery 15% rule to be discussed shortly. It also demonstrates a well-known metallurgical phenomenon that when iron is alloyed with a fcc (or bcc) element such as Ni (or Cr), the γ- (or α-) phase field will be opened up by suppressing the other (α- or γ-phase) field.

(2) Because Fe, Co and Ni exist in the bcc, hcp and fcc crystal structures at room temperature, respectively, two primary solid solutions emerge with a two-phase region lying in the middle at temperature around 20°C in each system.

(3) The order $\rightleftarrows$ disorder transition occurs in all three systems at compositions centering around FeCo, Ni_3Fe and Ni_3Co. The reason why ordering occurs in these alloys, not in others such as FeNi and CoNi will be explained in §2.5.

In the development of ferromagnetic alloys, a primary concern is the solubility limit of the solute; that is, the maximum percentage of the solute that can be dissolved completely in the solvent. Once the solubility limit is known, a proper procedure can be devised to ensure the preparation of the alloy in the single-phase state. It is easy to appreciate that a real assessment of a new soft magnetic alloy system must rely on single-phase samples, rather than on two- or multiple-phase samples. Unfortunately, this principle has not always been observed. It is then not surprising to discover that some patent claims and experimental results reported in the literature turn out to be worthless or irreproducible.

The solubility limit of an element in a metal varies widely from practically nil to the maximum, in which case a continuous series of solid solutions are formed. Also, the limit (c_m) is usually a sensitive function of the temperature. (Roughly $c_m \sim \exp(-E_f/kT)$, where E_f is the appropriate energy for the formation of AA, BB and AB atom pairs in the nearest neighbors; see Cottrell [1967a]). If the phase diagram is available for the particular alloy system under study, the limit is readily determined from the solubility curve or solvus separating the appropriate single- and two-phase regions. If the alloy system is unknown, metallurgical rules must be applied to estimate the limit. One such rule states that, in the formation of binary substitutional solid solutions, if the crystal structures of the two elements are the same or have similar packing densities of atoms, there is a good chance for the formation of an extensive series of solid solutions. The systems of Ni–Cu, Pd–Pt, Gd–Sc and Gd–Y are cases in point. Hume-Rothery [1950a] proposed three semiempirical rules to further predict the solubility limits of substitutional primary solid solutions:

(1) The atomic size factor – If the difference between the atomic radii* (R_0) of the constituents does not exceed 15%, the solid solubility is likely to be extensive. Typical examples are given in table 5.4. Note the

*Several definitions have been suggested of the atomic radius. Hume-Rothery [1950b] defined R_0 as half the interatomic distance in the direction of the densest packing of atoms (e.g., [110] in fcc lattices and [111] in bcc lattices). His definition can be expressed by

$$R_0 = \tfrac{1}{2}Pa, \quad \text{for fcc and bcc lattices;} \tag{5.13.1}$$

or

$$\left.\begin{aligned} R_0 &= \tfrac{1}{2}a, \quad \text{if } c/a \geqslant \sqrt{(8/3)} \text{ or } 1.633 \\ R_0 &= \frac{1}{2}\left[\left(\frac{a}{\sqrt{3}}\right)^2 + \left(\frac{c}{2}\right)^2\right]^{\frac{1}{2}}, \quad \text{if } c/a < \sqrt{\frac{8}{3}} \end{aligned}\right\} \text{for hcp structures;} \tag{5.13.2}$$

where a and c are the lattice parameters and P ($=\sqrt{2}/2$ for fcc lattices and $\sqrt{3}/2$ for bcc lattices) is a geometric factor. This "closest approach" definition has since been adopted by metallurgists. The R_0 values given in table 5.4 are determined according to (5.13.1) or (5.13.2). Physicists, however, prefer the definition

$$R_0 = (3\Omega/4\pi)^{\frac{1}{3}}, \tag{5.13.3}$$

where Ω is the atomic volume. Chemists often use the ionic radius or Pauling's single-bond metallic radius for R_0. Further information on the atomic sizes in metallic solid solutions can be found in the book by Barrett and Massalski [1966b].

TABLE 5.4
Binary alloy systems of nickel showing the validity of the empirical "15% rule" of Hume-Rothery

	Solvent	Solute			
	Ni	Cu	Pd	Pt	Ag
R_0 at 20°C (Å)	1.246	1.278	1.366	1.388	1.445
$\Delta R_0/R_{0,\mathrm{solv.}}$ (%)	–	+2.6	+9.6	+11.4	+16.0
Obs. solubility limit at 20°C (at.%)	–	100	100	100	~0

complete solubility of Cu, Pd and Pt in Ni and the nearly complete insolubility of Ag in Ni, even though both metals have the same fcc crystal structure in the latter system.

(2) The electrochemical effect – The more electronegative the solute element and the more electropositive the solvent metal, the greater the tendency for the formation of compounds, rather than a series of solid solutions. Alternatively, this rule may be stated with reference to the Electromotive Series that the farther apart the relative positions are in the Series of the two components, the greater will be the tendency for the formation of compounds. The compounds formed under this effect are called the normal valence compounds. In general, all metals tend to form normal valence compounds with elements of Group IVB (C, Si, Ge, Sn and Pb), Group VB (N, P, As, Sb and Bi) and Group VIB (O, S, Se and Te).

(3) Relative valence effect – If two metals have the same valence, mutual dissolution is expected; otherwise, it is easier for a metal of higher valence to be dissolved in a metal of lower valence than vice versa. While examining this effect, Hume-Rothery discovered that a large number of intermediate phases were formed in several crystal structures, each of which is characterized by a fixed e/a ratio, where e and a are the numbers of electrons and atoms per formula, respectively. Bernal (see Barrett and Massalski [1966c]) suggested calling them "electron compounds". More properly, they should be called "electron phases". Features and examples of four prominent electron phases are given in table 5.5. In determining the e/a ratio, Hume-Rothery assigned zero valence to the transition metals, including Fe, Co and Ni. In his scheme, FeAl, NiAl and $CoZn_3$ then would all have the

TABLE 5.5
Characteristics and examples of three prominent electron phases.

Phase	β-A	β-B	γ	ε
Suggested chemical formula	CuZn	CuZn	Cu_5Zn_8	$CuZn_3$
Electron/atom ratio (e/a)	3:2	3:2	21:13	7:4
Crystal structure	bcc	Complex cubic (β-Mn type)	Complex cubic (γ-brass type)	cph
Examples	CuZn Cu_3Al (Fe, Co or Ni) Al	(Ag or Au)$_3$Al Cu_5Si $CoZn_3$	(Fe, Co, Ni, Rh, Pd or Pt)$_5$Zn$_{21}$ Cu_5(Zn or Cd)$_8$	Cu(Zn or Cd)$_3$ Cu_3(Ge, Si or Sn)

same e/a ratio of $\frac{3}{2}$. Hence, they appear as the β-electron-phase, crystallizing in either the bcc or β-Mn type complex cubic structure. There is little theoretical justification for the zero valence of all transition metals, however.

The motivation for the development of ferromagnetic alloys is to improve certain properties of the existing materials. Besides magnetic properties, emphasis could also be placed on electrical resistance to reduce the eddy-current loss, on mechanical properties to ensure the durability of the material under severe stressing conditions, or on corrosion resistance. In the remainder of this section, we shall focus our attention on the composition dependence of saturation magnetization of alloy systems. Furthermore, our discussion will be confined to alloys of the iron-group metals. This is not to imply that the lanthanide alloys are unimportant; only they are less known and their practical values have not yet been established.

When a ferromagnetic metal is alloyed, the effect of the solute on saturation magnetization can be resolved into two parts. First, the presence of foreign atoms is bound to change the atomic moment on an average basis. In other words, if we denote the mean atomic moment of the alloy by $\bar{\mu}$ and the atomic moment of the parent metal by μ_1, $\bar{\mu}$

always differs from μ_1, irrespective of the species of the solute. The practical goal in alloy work is, of course, to make $\bar{\mu} > \mu_1$. Now if the solute atoms also carry magnetic moments, there will be a second effect of the solute; namely, the mode of the coupling between moments of the solute and solvent atoms. Obviously, ferromagnetic coupling is desired if the alloy is to show an improved saturation. Both effects can be expressed by the empirical equation,

$$\bar{\mu} = \mu_1 + (\mathrm{d}\bar{\mu}/\mathrm{d}c)c, \tag{5.14}$$

where c is the concentration, in atomic percent (a/0), of the solute and $\mathrm{d}\bar{\mu}/\mathrm{d}c$ gives the rate at which the mean atomic moment changes with c. The magnitude and sign of $\mathrm{d}\bar{\mu}/\mathrm{d}c$ represent the first and second effects, respectively. A similar equation can be written

$$\theta_{\mathrm{C}} = \theta_{\mathrm{C},1} + (\mathrm{d}\theta_{\mathrm{C}}/\mathrm{d}c)c, \tag{5.15}$$

to indicate the effect of the solute on the Curie temperature, where $\theta_{\mathrm{C},1}$ is the Curie temperature of the parent metal.

Numerous binary systems of the iron-group metals have been studied for their magnetic properties. Data obtained at 20°C have been analyzed with respect to (5.14) and (5.15), and the results are listed for Fe, Co and Ni alloys, respectively, in tables 5.6–5.8. For each solvent, the alloy systems are divided into three groups according to the nature of the solute element; namely, (I) another ferromagnetic metal, (II) a nonferromagnetic transitional metal and (III) a nontransitional metal. Several systems (Fe–Pd, Fe–Cu, Co–La, Co–Ru, Ni–Al, Ni–Au, Ni–As and Ni–Sb) were discarded from the analysis on the ground that the solutes show practically no solubilities in the solvents, according to the established phase diagrams. Inclusion of these systems would not be sensible and it would also be misleading. The deduced $\mathrm{d}\bar{\mu}/\mathrm{d}c$ values shown in the tables can be summarized as follows:

(1) In five out of a total of six cases, alloying of two ferromagnetic metals together is beneficial because $\mathrm{d}\bar{\mu}/\mathrm{d}c$ is positive. The addition of Ni to Co, however, causes a negative $\mathrm{d}\bar{\mu}/\mathrm{d}c$ and, therefore, represents the only adverse case.

(2) In no case has the addition of a nontransitional element to Fe, Co and Ni been beneficial. The deduced values for $\mathrm{d}\bar{\mu}/\mathrm{d}c$ always display a negative sign.

(3) The effect of the nonferromagnetic transitional solutes is either beneficial (e.g., Fe–Os, –Rh, –Ir and –Pt and Ni–Mn and –Rh) or adverse for Fe and Ni; but the effect is always adverse for Co.

TABLE 5.6
Variation of mean magnetic moment per atom in dilute alloys of iron ($Z_1 = 8$, $n_s = 0.6$)

Alloying element	Z_2	ΔZ_n	$d\bar{\mu}/dc$ (μ_B/at.% solute) Observed	Calculated (1) $-(2n_d\downarrow + \Delta Z_n) = -1.6 + Z_2$	Calculated (2) $-\Delta Z_n + 10 - 2n_d\downarrow = -\Delta Z_n + 0.4$	Calculated (3) $-[2n_d\downarrow + 10 - (Z_1 + Z_2) + 2n_s] = -(12.8 - Z_2)$
(I) Magnetic						
Co	9	+1	+1.0		−0.6	
Ni	10	+2	+0.33		−1.6	
(II) Transitional, nonmagnetic						
Ti	4	−4	−2.9	−5.6		
V	5	−3	−2.73, −3.29	−6.6		
Cr	6	−2	−2.36			−6.8
Mn	7	−1	−2.0			−5.8
Ru	8	0	0		+0.4	
Os	8	0	+0.8		+0.4	
Rh	9	+1	+1.1		−0.6	
Ir	9	+1	+0.8		−0.6	
Pt	10	+2	+1.9		−1.6	
(III) Nontransitional						
Zn	2	−6	−2.2	Simple dilution effect: $\bar{\mu} = (1 - c)\mu_1$, or $d\bar{\mu}/dc = \mu_1$ (applies to Zn, Al, Si, Sn)		
Al	3	−5	−2.2			
Si	4	−4	−2.2			
Sn	4	−4	−2.2			

TABLE 5.7

Variation of mean magnetic moment per atom in dilute alloys of cobalt ($Z_1 = 9$, $n_s = 0.7$)

Alloying element	Z_2	ΔZ_n	μ_2^* (μ_B/solute atom)	$d\theta_C/dc$ (10^3 °K/at.% solute)	$d\bar{\mu}/dc$ (μ_B/at.% solute) Observ.	Calculated** (1) $-(10+\Delta Z_n)$	(2) $-[20-(Z_1+Z_2)+2n_s]$	(3) $-\Delta Z_n$
(I) Magnetic								
Fe	8	−1	+2.8	−0.77	+1.1			+1[a])
Ni	10	+1	+0.6	−0.4	−1.1			−1[a])
(II) Transitional, nonmagnetic								
Ti	4	−5		−1.9		−5		
V	5	−4		−3.7		−6		
Nb	5	−4		≈ −2.9		−6		
Cr	6	−3	−4.9	−3.2	−6.6		−6.4[a])	
Mo	6	−3		−5			−6.4	
W	6	−3		−6.3	−5.5		−6.4[b])	
Mn	7	−2	−2.9		−4.6		−5.4[b])	
Ru	8	−1	−3.15	−4.0	−4.9		−4.4[b])	
Os	8	−1	−7.7	−6.1	−9.5		−4.4[d])	
Rh	9	0	0	−1.8	−1.7		−3.4[c])	
Ir	9	0	−1.2	−2.8			−3.4	
Pd	10	+1	+0.8	−0.69	−0.95			−1[a])
Pt	10	+1	+0.5	−0.98	−1.25			−1[a])
(III) Nontransitional								
Cu	1	−8		−0.6	−2.3	−2[a])		
Be	2	−7		−1.9		−3		
Zn	2	−7		−1.7		−3		
Si	4	−5		≈ −3.0	−5.2	−5[a])		

*Bohr magneton numbers deduced from neutron diffraction data.

**The calculated values are in good[a]), fair[b]), somewhat questionable[c]) or no agreement[d]) with the observed values.

To understand the varying effects of solutes, the observed average values for $\bar{\mu}$ for many important binary alloys are plotted against the average number of the outer electrons $\bar{Z}_n$ in fig. 5.3, where $\bar{Z}_n = Z_1(1-c)+Z_2c$, with subscripts 1 and 2 referring to the solvent and solute, respectively. The value of Z_n ($n = 1$ or 2) is determined by the number of electrons outside the completed subshells. It changes systematically according to the position of the element in the Periodic Table with respect to the preceding inert gas atom up to the value $Z_n = 10$. Thereafter the Z_n values for the nontransitional elements are given by the numbers of the groups to which the elements belong. To facilitate quick computation, the Z_n values for various elements involved in this discussion are given in table 5.9. In tables 5.6–5.8, $\Delta Z_n = Z_2 - Z_1$, which may be positive integers, zero or negative integers. In the band model sketched in fig. 2.8, we may express the atomic moment $\mu_1 = 10-(Z_1-n_s)$ for Co and Ni and $\mu_1 = 10-(Z_1-n_s)-2\Delta n$ for Fe, where n_s is the average number of conduction electrons per atom and Δn is the number of holes in the lower half-d-band per atom of Fe.

Figure 5.3 includes the well-known Pauling–Slater (P–S) curve ABC. This curve commences at approximately $\bar{Z}_n = 6.65$, corresponding to the alloy $Fe_{0.325}Cr_{0.675}$, reaches a peak at $\bar{\mu} = 2.426\,\mu_B$ in the alloy $Fe_{0.65}Co_{0.35}$ ($\bar{Z}_n = 8.35$) and ends at $\bar{Z}_n = 10.6$ in the alloy $Ni_{0.4}Cu_{0.6}$. For a long time, the P–S curve had been considered as the ceiling of the $\bar{\mu}$ versus $\bar{Z}_n$ plot. It was thought impossible for any new alloys to penetrate through this curve. The present author [1963], however, found (fig. 5.4) that such a conviction is neither precise nor correct in two respects. First, the maximum value of the mean atomic moment attained in the alloy $Fe_{0.65}Co_{0.35}$ had been fixed at $2.426\,\mu_B$ in the P–S curve. This Bohr magneton number, however, was found to be sensitive to the degree of long-range order of atoms in the alloy. By subjecting the sample to prolonged annealing for more pronounced order, the moment would increase to $2.440\,\mu_B$ at 20°C. Most descriptions of the P–S curve were vague about this point. Secondly, the addition of Mn to FeCo and other Fe–Co alloys with the atomic ratio of Fe to Co less than unity would lead to an increase in $\bar{\mu}$ at the rate of $d\bar{\mu}/dc = +3\,\mu_B$ per atomic percent of Mn. For instance, in the state of nearly complete order, the value of $\bar{\mu}$ would rise from $2.40\,\mu_B$ in FeCo to $2.43\,\mu_B$ in $(FeCO)_{0.95}Mn_{0.05}$. Hence, a good portion of the P–S curve can be lifted to higher values of $\bar{\mu}$ by the moment-enhancing effect of Mn on Fe–Co alloys if the atomic ratio of Fe to Co is $\leqslant 1$. Chen's findings [1962a] will be discussed further at the end of this section.

TABLE 5.8

Variation of mean magnetic moment per atom in dilute alloys of nickel ($Z_1 = 10$, $n_s = 0.6$)

Alloying element	Z_2	ΔZ_n	μ_2^* (μ_B/solute atom)	$d\bar{\mu}/dc$ (μ_B/at.% solute) Observed	Calculated** (1) $-(10+\Delta Z_n)$	(2) $-[20-(Z_1+Z_2)+2n_s]$	(3) $-\Delta Z_n$
(I) Magnetic							
Fe	8	−2	+2.85	+2.25			+2[a])
Co	9	−1	+1.8	+1.18			+1[a])
(II) Transitional, nonmagnetic							
Ti	4	−6	−3.4	−4.0	−4[a])		
V	5	−5	−4.9	−5.5	−5[b])		
Ta	5	−5	−5.1	−5.7	−5[b])		
Cr	6	−4	−3.9	−4.5		−5.2[b])	
Mo	6	−4	−5.3	−5.9		−5.2[b])	
W	6	−4	−5.6	−6.2		−5.2[b])	
Mn	7	−3	+3.0	+2.4			+3[a])
Os	9	−1	−1.4	−2.0		−2.2[a])	
Rh	9	−1	+2.6	+2.0			+1[c])
Ir	9	−1	0	−0.61		−2.2[d])	
Pd	10	0	+0.61	0		−1.2[d])	
Pt	10	0	−0.61			−1.2	

(III) Nontransitional				
Cu	1	−9	−1.07	−1[a])
Au	1	−9	≈ −1.0	−1[a])
Be	2	−8	−2.1	−2[a])
Zn	2	−8	−2.1	−2[a])
Al	3	−7	≈ −3.0	−3[a])
Si	4	−6	≈ −3.7	−4[a])
Sn	4	−6	−4.2	−4[a])
Sb	5	−5	−3.6	−5[c])

*Bohr magneton numbers deduced from neutron diffraction data.

**The calculated values are in good[a]), fair[b]), somewhat questionable[c]) or no[d]) agreement with the observed values.

TABLE 5.9
Assigned values of the number of outer electrons $Z_n (n = 1 \text{ or } 2)$ for various elements

Z_n	Element	Z_n	Element
1	Li, Na, K and Rb	8	Fe, Ru and Os
2	Be, Mg, Sr and Ba	9	Co, Rh and Ir
3	Al, Sc, Y, La and the majority of lanthanides	10	Ni, Pd and Pt
		1	Cu, Ag and Au
4	Si, Ti, Zr and Hf	2	Zn and Cd
5	V, Nb and Ta	3	Ga, In and Tl
6	Cr, Mo and W	4	Ge, Sn and Pb
7	Mn and Re	5	As, Sb and Bi

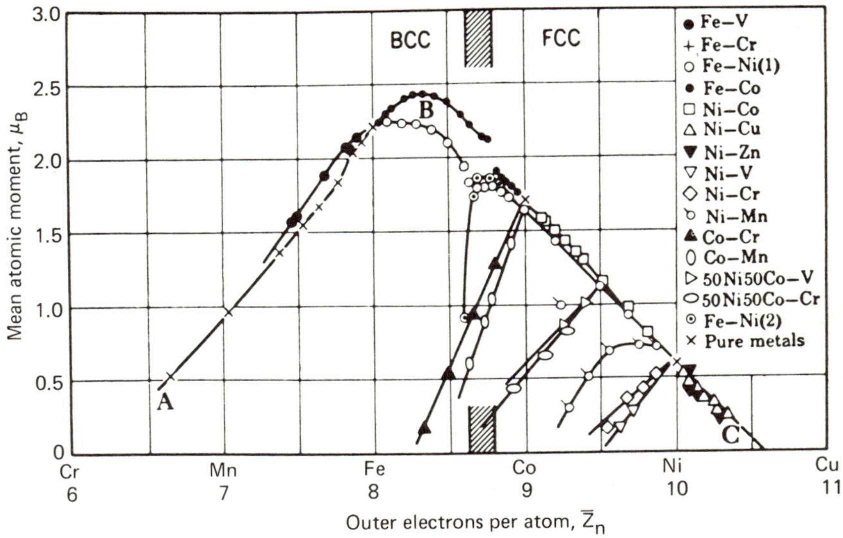

Fig. 5.3. The Slater–Pauling curve showing the variation of the mean atomic moment of 3d transitional metals and alloys with the average number of outer electrons per atom (Chikazumi [1964]).

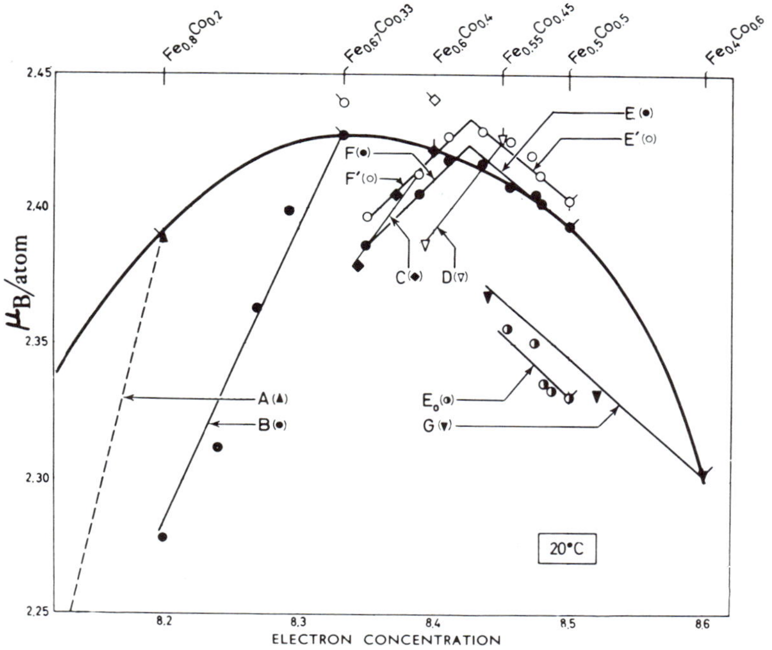

Fig. 5.4. Mean atomic moments of the Mn-substitutional Fe–Co alloys plotted against the average number of outer electrons per atom. Base alloys are: A – $Fe_{0.8}Co_{0.2}$, B – $Fe_{0.67}Co_{0.33}$, C – $Fe_{0.6}Co_{0.4}$, D – $Fe_{0.55}Co_{0.45}$, E and F – $Fe_{0.5}Co_{0.5}$ and G – $Fe_{0.4}Co_{0.6}$. Solid symbols are related to alloys in the partially ordered state (furnace cooled from 800°C); open symbols, to alloys in the pronouncedly ordered state (reheated to 800°C and cooled at a deliberately slow rate); half-open symbols, to alloys in the disordered state (reheated to 800°C and quenched). Data for the parent binary alloys are shown by symbols with short bars (Chen [1964]).

Perhaps the most significant implication of the P–S curve is that alloys of the iron-group metals behave magnetically as if they belong to two distinct groups. Alloys lying on the left side of the peak feature a positive slope ($d\bar{\mu}/dc > 0$), whereas those lying on the right side constitute another group displaying a negative slope ($d\bar{\mu}/dc < 0$). Moreover, the first group includes only alloys of iron, and the second group contains alloys of Co and Ni. Thus, we may conclude that the magnetic behavior of alloys of the iron-group metals varies largely according to the electronic structure characteristics of the parent metal.

Further substantiation of this conclusion can be seen in the fact that type III alloys of iron persistently display a common value of $-2.2\ \mu_B/(a/0)$ solute for $d\bar{\mu}/dc$ in the four systems examined. By contrast, type III alloys of Co and Ni show no sign of such harmonious behavior. The deduced values for $d\bar{\mu}/dc$ in the latter alloys vary widely. On the other hand, the following discussion will show that the magnetic behaviors of all types of alloys of Co and Ni can be accounted for satisfactorily in terms of a modified band theory, whereas those of types I and II alloys of Fe remain unexplained.

Although (5.14) is convenient for comparing the effects of solutes, it sheds no light on how the moments are distributed between solvent and solute atoms. To understand this distribution, it is more meaningful to assign separate moments μ_1 and μ_2 to the solvent and solute atoms, respectively, and to convert (5.14) to the form

$$\bar{\mu} = \mu_1(1-c) + \Gamma\mu_2 c = \mu_1 - \mu_1 c + \Gamma\mu_2 c, \tag{5.16}$$

where $\Gamma = \pm 1$, signifying the possibility of either parallel or antiparallel spin coupling between solute and solvent atoms. In the second form of (5.16), the effect of the solute, or the quantity $d\bar{\mu}/dc$ in (5.14), is resolved into two parts: the effect of simple dilution represented by $-\mu_1 c$ and the effect of a different moment carried by the solute atom represented by $\Gamma\mu_2 c$. When μ_2 is zero, (5.16) becomes

$$\bar{\mu} = \mu_1 - \mu_1 c, \tag{5.17}$$

which clearly indicates only the effect of simple dilution of the solute since $d\bar{\mu}/dc = -\mu_1$ in comparison with (5.15). In this case, the solute merely eliminates the magnetic moments on a one-to-one basis by substituting the "nonmagnetic" atoms for "magnetic" atoms. In this model, the value of ΔZ_n is significant because $d\bar{\mu}/dc$ is independent of Z_2. All type III alloys of iron listed in table 5.6 harmoniously display the effect of simple dilution.

For other alloys of iron and all alloys of Co and Ni, the effect of the solute is more than just simple dilution. Since the magnetic behaviors of pure Co and Ni can be explained satisfactorily to some extent by the band theory, it is reasonable to assume that the 3d band structure described in ch. 2, §2 still exists in the dilute Co and Ni alloys, only to be modified by the solute. We may further assume the existence of a common conduction band for both solvent and solute atoms. Beyond this point, the picture gets fuzzy because there are two ways in which the 3d band can be modified. As a general example, let us assume that each solute atom donates the same portion, say $n_s(<1)$ of the Z_2 outer electrons to the conduction band as does each solvent atom, leaving $Z_2 - n_s$ electrons available for the d-band. The latter electrons will either join the d-band by filling up the vacant states or else stay with the solute atom to form localized spins. The first possibility corresponds to a simple band model featuring a d-band common to all magnetic atoms. Here the effect of the solute is always adverse because the $Z_2 - n_s$ electrons donated by each solute atom can only go to the incomplete half d-band, thus always reducing the number of unpaired spins. The second possibility represents a mixed model in which the d-band applies only to the solvent atoms and there are localized moments at the solute atoms. In the mixed model, the effect of the solute is adverse under one of two conditions: (1) the spin coupling between solute and solvent atoms is antiparallel; and (2) the localized moment of μ_2 Bohr magnetons per atom is less than μ_1 of the solvent atom. If, however, the spin coupling is parallel and if $\mu_2 > \mu_1$, the solute effect is expected to be beneficial.

Neutron diffraction experiments conducted on types I and II alloys of Co and Ni furnished some evidence to support the mixed model. From the diffraction data, localized moments were deduced for various solute atoms, as shown in tables 5.7 and 5.8. Although this evidence is given by probably the most powerful technique, it still leaves some uncertainties. Unless the diffraction experiments are conducted on single crystals with polarized neutrons and unless certain precautions are exercised painstakingly in the experiment and in the analysis of the data so as to eliminate possible complications caused by other factors, the resolution of the observed magnetic moment into components of the solvent and solute atom is, at best, a deduction, not a direct observation.

In analyzing the deduced values for $d\bar{\mu}/dc$ listed in tables 5.7 and 5.8, the differentiation between the two models turns out to be immaterial because they give the same predictions for $d\bar{\mu}/dc$ when $Z_2 \leq 5$ and the spin coupling is antiparallel. In other cases, the mixed model is superior to the simple band model, at least analytically, because it predicts values for $d\bar{\mu}/dc$ unequivocally regardless of the value of Z_2. (Verification of the latter argument is left to the reader as an exercise.)

For reasons just cited, we shall use the mixed model for our analysis. This choice is more a matter of convenience than a physical preference.

In terms of the mixed model, we first consider the case of parallel spin coupling; that is, $\Gamma = +1$ with respect to (5.16). Since no alloy systems of Co and Ni have been found with $Z_2 \leq 5$ in this case, we need only to calculate $\bar{\mu}$ for systems with $Z_2 > 5$. Using $\mu_2 = 10 - (Z_2 - n_s)$, (5.16) becomes

$$\bar{\mu} = \mu_1 + (-\Delta Z_n)c. \tag{5.18}$$

In comparison with (5.14), we have $\mathrm{d}\bar{\mu}/\mathrm{d}c = -\Delta Z_n$. Theoretical values for $-\Delta Z_n$ are given for pertinent systems in the last column of tables 5.7 and 5.8. Excellent agreements are seen between the calculated values of $-\Delta Z_n$ and the deduced values for $\mathrm{d}\bar{\mu}/\mathrm{d}c$ in the systems of Co–Fe, –Ni, –Pd and –Pt and Ni–Fe and –Co, and a reasonable agreement is seen in the system of Ni–Mn.

The next case involves antiparallel spin coupling; that is, $\Gamma = -1$. When $Z_2 \leq 5$, we have the relation, $\mu_2 = Z_2 - n_s$, and from (5.16), we obtain

$$\bar{\mu} = \mu_1 - (10 + \Delta Z_n)c, \tag{5.19}$$

which gives $\mathrm{d}\bar{\mu}/\mathrm{d}c = -(10 + \Delta Z_n)$. Comparison of the calculated values for $-(10 + \Delta Z_n)$ with the deduced values for $\mathrm{d}\bar{\mu}/\mathrm{d}c$ indicates that this third mechanism explains very well the magnetic behavior of numerous Ni systems (with Ti, Cu, Au, Be, Zn, Al, Si and Sn as solutes) and the Co–Cu and –Si systems and explains reasonably well the behavior of the systems of Ni–V and –Ta.

The fourth mechanism involves $\Gamma = -1$ and $Z_2 > 5$. Now with $\mu_2 = 10 - (Z_2 - n_s)$, we obtain

$$\bar{\mu} = \mu_1 - [20 - (Z_1 + Z_2) + 2n_s]c, \tag{5.20}$$

and $\mathrm{d}\bar{\mu}/\mathrm{d}c = -[20 - (Z_1 + Z_2) + 2n_s]$. Again, comparison of the calculated values for $-[20 - (Z_1 + Z_2) + 2n_s]$ with the deduced values for $\mathrm{d}\bar{\mu}/\mathrm{d}c$ reveals satisfactory agreements for the systems of Co–Cr, –W, –Mn and –Ru and Ni–Os, –Cr, –Mo and –W.

Thus far we have explained the magnetic behaviors of the majority of alloy systems of Co and Ni listed in tables 5.4 and 5.5. No account can be given for the seemingly anomalous behaviors of a few systems. In view of the overwhelming success of the analysis, it is suggested that the composition dependence of saturation magnetization in the latter systems should be re-examined before an explanation is attempted.

Application of the last three mechanisms to alloys of iron is complicated by the presence of vacant states in both half d-bands. Applying the same diagram shown in fig. 2.8d for iron to the alloys of Fe, we obtain for $d\bar{\mu}/dc$ the theoretical values of $-(1.6 - Z_2)$, $-(\Delta Z_n - 0.4)$ and $-(12.8 - Z_2)$ for the three mechanisms. Table 5.3 shows that the theoretical values thus calculated for $d\bar{\mu}/dc$ in various Fe alloys reach no agreement at all with the deduced values for $d\bar{\mu}/dc$. In half of the systems, even the sign of $d\bar{\mu}/dc$ is in opposition. Such a total failure with type I and II alloys of Fe further justifies the separation of Fe alloys from those of Co and Ni and the conclusion that the magnetic behavior of dilute alloys depends largely upon the electronic-structure characteristics of the parent metal.

In the absence of a complete substantiation by neutron diffraction experiments, the foregoing analysis can only be considered as largely empirical. Rigorous theoretical analyses of the magnetic behavior of binary alloys of Fe, Co and Ni were attempted by several workers with some success. Worth mentioning among them is the coherent potential approximation (CPA) adapted for the ferromagnetic systems by Hasegawa and Kanamori [1971, 1972]. The CPA was originally introduced by Soven [1967] in the framework of the multiple scattering theory (see Lax [1951]), in which the propagation of an electron or a lattice wave in a binary alloy is regarded as a succession of elementary scatterings on the random atomic scatterers. By averaging over all configurations of atoms, an effective Hamiltonian is found, which represents the coherent potential. To apply the CPA to the ferromagnetic alloys, the exchange interaction must be included, which can be effected by the Hartree–Fock approximation. For concentrated ferromagnetic alloys A_cB_{1-c}, Hasegawa and Kanamori have shown that the CPA based on a non-degenerate tight-binding single band model is capable of calculating numerically the magnetic moments of A and B atoms and other pertinent electronic-structure properties, such as the density of states at the Fermi level and the electronic specific heat. The calculated results for the Ni–Fe and Ni–Co alloys are in reasonable agreement with the experimental data. The calculations performed on the Ni–Cr and Ni–Mn alloys were less successful, however, because the latter alloys do not follow the Pauling–Slater curve (fig. 5.3), according to these authors. Also, the applicability of the CPA is limited to disordered alloys because the coherent potential does not distinguish between the nearest neighbors and far neighbors of a given atom.

Ternary and multicomponent alloys are more often used in practice than pure metals and binary alloys. However, little basic knowledge had been available before the present author [1962a,b, 1964] undertook a series of studies on ternary Fe–Co alloys. These studies were motivated partly to gain information to fill the knowledge gap and partly to search for new alloys whose saturation magnetizations might exceed the peak value in the P–S curve. Chen's studies centered around two variables: the effect of the solute on the mean atomic moment per se and its dependence on the solvent. The first variable was evaluated by alloying the equiatomic alloy FeCo with Cu or other 3d transition metals such as Ti, V, Cr, Mn and Ni. Results of the first study are summarized in table 5.10, Part I. Two solutes, Mn and Ni, displayed ferromagnetic spin coupling. Moreover, each Mn atom carries 3 μ_B; thus the net effect of Mn is to increase $\bar{\mu}_1$ of FeCo at the rate

$$d\bar{\mu}/dc = -\bar{\mu}_1 c + 3c. \tag{5.21}$$

At 20°C $\bar{\mu}_1$ is approximately 2.4 μ_B, so $d\bar{\mu}/dc = +0.6c$. The first study thus established an increasing effect of Mn on $\bar{\mu}_1$, as opposed to the adverse effect of other solutes. Subsequently, Mn was retained as the sole solute in the study of the second variable, wherein the solvents were varied to cover a range of the Co/Fe ratio from 0:1 to 3:2. Results of the latter study are summarized in table 5.10, Part II, and in fig. 5.4. The effect of Mn is seen to change drastically according to the position of the base alloy in the P–S curve. For Fe–Co alloys lying on the left side of the peak, the Mn effect is harmful, whereas for those lying on the right side, a beneficial effect prevails. More specifically, Mn exerts on $\bar{\mu}$, (1) merely the effect of simple dilution in pure iron and $Fe_{0.8}Co_{0.2}$; (2) a progressively weakened decreasing effect in alloys having the Co/Fe ratio <1:1; and (3) a constant enhancing effect in other Fe–Co alloys including FeCo within the bcc region. The increasing effect of Mn in (3) is independent of the degree of atomic order. At the rate expressed by (5.21), the concentration of Mn required to increase $\bar{\mu}$ in ternary FeCo alloys to surpass the peak value of 2.44 μ_B recorded in $Fe_{0.67}Co_{0.33}$ is ~7 a/0. Unfortunately, the solid solubility of Mn in FeCo is less than 7 a/0. The maximum value of the mean atomic moment moment for the ternary system is 2.43μ_B observed in $(FeCo)_{0.95}Mn_{0.05}$.

Finally, a study was conducted to see whether any of the 4d and 5d transition elements would exhibit a behavior similar to that of Mn. Of particular interest are such elements as Re, which belongs to the same column as Mn in the Periodic Table, and Rh, Ir and Pt, the substitution of which increases the mean atomic moment of the dilute alloys of Fe. None showed an increasing effect similar to that of Mn. It is concluded that the chance of raising the saturation magnetization in Fe–Co alloys by further alloying is very slim, if not entirely impossible.

TABLE 5.10

Data pertaining to the magnetic and electronic structures of dilute ternary alloys of iron and cobalt

Part I – The effect of various solutes on the equiatomic alloy FeCo ($\bar{Z}_1 = 8.5$, $\bar{\mu}_1 = 2.40\ \mu_B$/atom)

Solute	$d\bar{\mu}/dc$ (μ_B/at.% solute)	$\Gamma^*_{\mu 2}$ (μ_B/solute atom)	Z_2	Possible distribution of outer electrons per solute atom	
				n_s	n_d
Ti	−4.3	−2	4	2	2
V	−5.5	−3.2	5	2	3
Cr	≈ −5.5		6		
Mn	+0.7	+3	7	0	7
Ni	−1.7	+0.6	10	0.6	9.4
Cu	−2.3	0	1	1	10

*A small correction to 0.1 is applied in order for $\Gamma\mu_2 = 0$ required for Cu.

Part II – The varying effects of Mn on different Fe–Co alloys

Solvent	$d\bar{\mu}/dc$ (μ_B/at.% Mn)	$\Gamma\mu_2$ (μ_B/Mn atom)	Possible distribution of outer electrons per Mn atom		
			n_s	$n_d\uparrow$	$n_d\downarrow$
Pure iron	$-\bar{\mu}_{Fe}$	0	0	3.5	3.5
$Fe_{0.8}Co_{0.2}$	$-\bar{\mu}_1$	0	0	3.5	3.5
				(simple dilution effect)	
$Fe_{0.67}Co_{0.33}$	−1.44	$<\bar{\mu}_1$	0	$n_1(<4)$	$7-n_1$
$Fe_{0.6}Co_{0.4}$	−1.0	$<\bar{\mu}_1$	0	n_2	$7-n_2$
$Fe_{0.55}Co_{0.45}$	−0.95	$<\bar{\mu}_1$	0	n_3	$7-n_3$
				$(n_1 < n_2 < n_3 < 5)$	
FeCo	+(0.6–0.7)	3	0	5	2
$Fe_{0.4}Co_{0.6}$	+0.7	3	0	5	2

Another group of ternary alloys that exhibit interesting ferromagnetic properties is known as the Heusler alloys. The history of these alloys itself is an amazing story in two respects. First, the ferromagnetism in certain Cu–Mn–Sn alloys was discovered by F. Heusler in 1898. Yet, as late as 1973, according to the Proceedings of that year's Conference on Magnetism and Magnetic Materials, work (Campbell and Leiper [1974]) was still continuing on a host of typical (Cu_2MnM' with M′ being Al, In, Sn or Bi) and new (MMnM′ with M being Co, Ni or Pd) Heusler alloys. Thus the Heusler alloys have been studied for more than 70 years, and an enormous number of papers have been published on them. Secondly, the alloys and their magnetic properties were so fascinating to members of the next two generations of F. Heusler (son, O. Heusler, and grandson, J. Heusler [1966]) that they also devoted some of their research efforts to the alloys, thus establishing a sort of family tradition.

Besides their fascinating magnetic properties, the Heusler alloys are also well-known to metallurgists for their wide variation in composition and to crystallographers for their superlattices. As a combination of their magnetic, metallurgical and crystallographic features, we characterize the Heusler alloys as follows:

(1) A Heusler alloy is a solid solution of the β phase of a ternary system of manganese with compositions centering around M_2MnM', where M may be Cu, Pd, Ni or Co and M′ may be Al, Ga, As, In, Sn, Sb or Bi. It may appear in more complex forms through partial substitution of the two elements M and M′, or in entirely different forms such as PtMnM′ and Ag_5MnAl.
(2) The alloy tends to order chemically into the $L2_1$-type superlattice and to order magnetically. The chemically ordered alloys usually display ferromagnetism, but antiferromagnetism is possible and has been found in Pd_2MnIn and Pd_2MnAl.
(3) For each ternary system, the highest saturation magnetization usually occurs in the alloys very near the composition M_2MnM', such as Cu_2MnAl and Cu_2MnSn.

Partly because of (3) and partly because of their early discovery by F. Heusler, Cu_2MnAl and Cu_2MnSn are the two most extensively studied Heusler alloys. Their saturation values are comparable with that of nickel, which is $\approx 5 \times 10^5\ A \cdot m^{-1}$ (≈ 500 G) at 17°C. At the time of F. Heusler's discovery, the detection of ferromagnetism in Cu_2MnSn was considered rather astonishing because none of the constituent elements was ferromagnetic, or more explicitly, because the alloys contained none of the iron-group ferromagnetic metals (Fe, Co and Ni). Moreover, the Heusler alloys were the first group of ferromagnets discovered to display sensitive dependence of saturation magnetization and Curie temperature upon the degree of long-range chemical order.

Despite the first characteristic cited, the number of Heusler alloys is enormous. This is because for each system of M_2MnM', the Heusler alloys could cover a wide range of composition. For instance, in the ternary system of Mn, Cu and Al, the composition range of the Heusler alloys is shown in fig. 5.5. This figure represents a partial isothermal section of the phase diagram of the ternary system at room temperature. Ferromagnetism exists in solid solutions of the β phase outlined by the solid curves. The dashed curves are isosaturation-induction contours. The highest B_s value was found at 0.55 T (5500 G) in Cu_2MnAl. As the composition deviates from Cu_2MnAl, the B_s value diminishes rapidly.

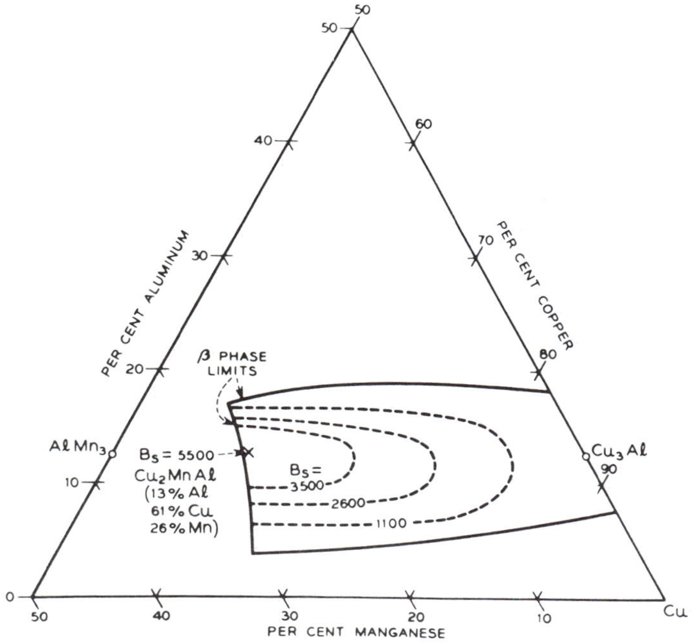

Fig. 5.5. Saturation induction (broken lines) and phase boundaries (solid lines) of Heusler alloys of the Mn–Cu–Al ternary system (Bozorth [1951]).

In most of the Heusler alloys, only Mn is a transitional element; the other constituents (Cu, Al, Ga, In, Sn, As, Sb, Bi and Ag) are nontransitional. Hence it had been reasoned, and later confirmed by neutron diffraction experiments (e.g., Webster and Tebble [1968]), that the Heusler alloys owe their magnetizations to the Mn atoms. On the basis of the $3d^6 4s^1$ electronic configuration and the first Hund's rule (p. 24), each Mn atom is expected to carry a moment of $4\,\mu_B$. The value deduced for each formula unit from neutron diffraction data is $4.3 \pm 0.2\,\mu_B$, in good agreement with the calculation. The fact that pure manganese and Mn-rich binary alloys are not ferromagnetic, but the presence of Mn in the Heusler alloys imparts ferromagnetism, must be considered as strong evidence for the Bethe–Slater criterion for ferromagnetic spin coupling (§1.1). Figure 5.6 shows a unit cell of Cu_2MnAl in the superlattice. On the average, each unit cell contains 16 atoms corresponding to four formula units. The simplest way to view such a large unit cell is to divide it into eight body-centered cubic cells with Cu atoms occupying all the corners and Mn and Al atoms occupying the alternate centers. In this scheme, the four Mn atoms form two pairs oriented in the [110] and [$\bar{1}$10] directions at a distance D of 3.66 Å. This distance is larger than the interatomic distance of 2.24 Å

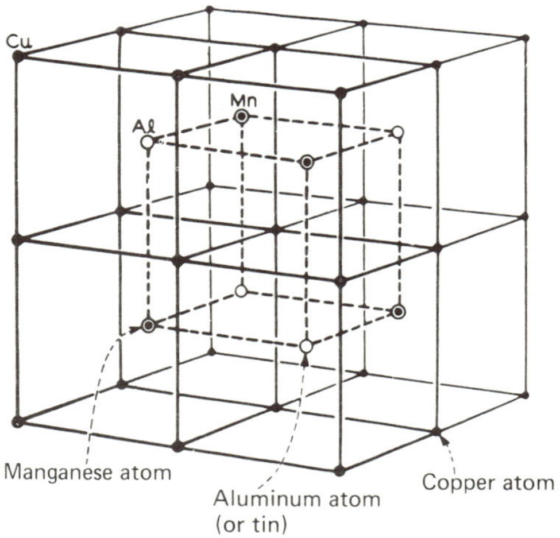

Fig. 5.6. A unit cell of the Heusler alloy Cu_2MnAl showing the ordered arrangement of the three species of atoms (Bozorth [1951]).

in pure α-Mn by $\approx$63%. Using the same value of 1.71 Å given in table 5.1 for the diameter d of the 3d subshell of Mn, we obtain $D/d=2.14$ for the Mn atoms in the ordered Cu_2MnAl, as compared with the ratio of 1.31 for pure Mn. Hence, in conformity with the Bethe–Slater criterion, spins of the Mn atoms in the Heusler alloys are coupled ferromagnetically. If the alloy Cu_2MnAl is disordered, most of the Mn–Mn atom pairs are at distances of ($\sqrt{3}a/2=$) 2.24 and ($a=$) 2.59 Å, giving D/d values of 1.31 and 1.51, respectively. This explains why the occurrence of ferromagnetism in the Heusler alloys depends so sensitively on the degree of chemical order.

The work of Webster and Tebble [1968] on the Pd_2MnM' Heusler alloys ($M'=$ Al, In, Sn or Sb) revealed some complexity in the magnetic behavior of the Heusler alloys. From the magnetic and neutron diffraction data obtained from ordered specimens, these authors established ferromagnetism in Pd_2MnSn ($\theta_C=189$°K) and Pd_2MnSb ($\theta_C=247$°K), but antiferromagnetism in Pd_2MnIn ($\theta_N=142$°K) and Pd_2MnAl ($\theta_N=240$°K). Although there is a progressive decrease in lattice parameter on going from the Sb alloy ($a=6.424$ Å in an $L2_1$ structure), through the Sn-alloy ($a=6.380$ Å in $L2_1$) to the In-alloy ($a=6.373$ Å in $L2_1$) and the Al-alloy ($2a=6.165$ Å in a B2, NaCl structure), the decreases were too small and inconsistent with the basic change in the mode of spin coupling. They suggested that, while small changes in lattice parameter may have some effect on the magnitude of the exchange interaction, it is the electron concentration which determines the type of magnetic order in the Pd–Mn alloys. Assuming that the contribution of conduction electrons per atom is 1 for Mn, 3 for Al and In, 4 for Sn, 5 for Sb and n (<1) for Pd, the spin coupling of the Mn atoms would seem to change from ferromagnetic to antiferromagnetic upon the increase in electron concentration somewhere between $1+0.5n$ and $1.5+0.5n$ electrons per atom in the Pd–Mn Heusler alloys.

§1.3. *Ferrites and other magnetic compounds*

A substance is called a compound if it contains two or more elements in a composition range that can be expressed in terms of a simple stoichiometric formula and if all its constituent ions are perfectly ordered in the lattice. For example, MnAs, Co_5Nd and Fe_3O_4 are compounds, whereas FeCo and Ni_3Fe are only solid solutions, not compounds, because the latter may change their compositions considerably without destroying the crystal structure or the phase.

A compound may be classified as intermetallic or nonintermetallic, depending on whether or not its constituent elements are all metallic. Some intermetallic compounds, such as Co_5Nd and MnBi, possess attractive magnetic properties, but their high magnetic hardness, as measured by coercivity and the $(HB)_{max}$ product, only qualifies them as permanent magnets. As far as soft magnetic materials are concerned, only the nonintermetallic compounds are worth considering.

Among the large number of nonintermetallic compounds, the ferrites and the chalcogenides of lanthanides are important. These compounds show useful magnetic properties and magnetic softness, and, unlike metallic ferromagnets, also display high electrical resistivities in the insulating or semiconducting range. Hence, they are ideal soft magnetic materials for use at high frequencies in electronic devices. Indeed, many applications have been found for magnetic compounds, especially the ferrites. The bulk of this section will thus deal with ferrites.

§1.3.1. *Ferrites*

Ferrites are complex magnetic oxides that contain the ferric oxide (Fe_2O_3) as their basic magnetic component. The crystallographies of ferrites fall in a natural manner into three types: (I) the cubic ferrites of the spinel type, (II) the cubic ferrites of the garnet type and (III) the hexagonal ferrites. The chemistry and magnetic characteristics of each type will be described separately first, followed by a discussion of the methods developed for the preparation of polycrystalline and single-crystal ferrites.

(1) The cubic ferrites of the spinel type. These ferrites are also called ferrospinels because they crystallize in the same crystal structure as the mineral spinel and they derive their general formula MFe_2O_4 from that ($MgAl_2O_4$) of spinel. In this formula, M represents a divalent ion of metal. Besides the divalency, another condition to qualify a metal in ferrospinels is its ionic radius which should fall between 0.6 and 1.0 Å. Mg, Fe, Co, Ni, Cu, Zn and Cd all satisfy these two conditions and thus form various single cubic ferrites. Magnetite, which contains one ferrous ion and two ferric ions in each formula unit is a typical ferrite. A special ferrite is $Li_{0.5}Fe_{2.5}O_4$ in which the monovalent Li cation replaces a divalent cation. To maintain the charge neutrality in the Li-ferrite, the concentration of the ferric ions is increased proportionally.

The crystal structure of ferrites is based on a face-centered cubic lattice of the oxygen ions. Each unit cell contains eight formula units.

Therefore, there are 32 O^{2-} anions, 16 Fe^{3+} cations and 8 M^{2+} cations in the unit cell, and the lattice constant is rather large, of the order of 8.5 Å. In each unit cell, there are 64 tetrahedral, or (A), sites and 32 octahedral, or [B] sites. These sites are so named because they are surrounded by four and six oxygen ions at equal distances, respectively. In the mineral spinel, the 16 trivalent cations occupy one-half of the [B] sites, and the eight divalent cations occupy one-eight of the (A) sites. If ferrospinels display the same occupany of (A) and [B] sites by the M^{II} and Fe^{III} ions, respectively, they are described as having the normal spinel structure. Both $ZnFe_2O_4$ and $CdFe_2O_4$ have the normal spinel structure. However, more ferrites, including magnetite and Li-ferrite, crystallize in an inverse spinel structure, in which the eight divalent cations occupy [B] sites and the 16 ferric ions are evenly divided to occupy (A) and [B] sites. The occupancy of the two kinds of sites by different cations is compared between the normal and inverse spinel structures in table 5.11. According to the coupling scheme of spins in the normal spinel structure, the antiparallel spin alignment of Fe^{III} ions and the zero moment of Zn^{II} ions should make Zn-ferrite an

TABLE 5.11
Comparison between normal and inverse spinel structures

	Normal spinel structure	Inverse spinel structure
1. Formula units/unit cell	$8(M^{II})[Fe_2^{III}]O_4$	$8(Fe^{III})[M^{II}Fe^{III}]O_4$
8 M^{II} ions occupy	(A) sites	[B] sites
16 Fe^{III} ions occupy	[B] sites	8 (A) sites 8 [B] sites
2. Typical example	$8\{ZnFe_2O_4\}$	$8\{NiFe_2O_4\}$
Spin coupling:		
Spins on 8 (A) sites	Zn^{II} – 00000000	Fe^{III} – ↓↓↓↓↓↓↓↓
Spins on 8 [B] sites	Fe^{III} – ↑↑↑↑↑↑↑↑	Fe^{III} – ↑↑↑↑↑↑↑↑
Spins on 8 [B] sites	Fe^{III} – ↓↓↓↓↓↓↓↓	Ni^{II} – ↑↑↑↑↑↑↑↑
Resultant moment	0	$2\times8 = 16\ \mu_B$/unit cell
3. Other examples	$CdFe_2O_4$	$MnFe_2O_4$, Fe_3O_4, $CoFe_2O_4$, $CuFe_2O_4$, $MgFe_2O_4$, and $Li_{0.5}Fe_{2.5}O_4$

antiferromagnet. No evidence has been found for such magnetism in $ZnFe_2O_4$, however.

X-ray and neutron diffraction experiments and magnetization measurements show that there is a whole range of cation distribution between the normal and inverse structures. This situation is represented by the formula $(M^{II}_{\delta}Fe^{III}_{1-\delta})[M^{II}_{1-\delta}Fe^{III}_{1+\delta}]O_4$, where cations in the parentheses are at (A) sites and those in the brackets are at [B] sites. The extreme case $\delta = 1$ corresponds to the normal spinel structure, and the other extreme case $\delta = 0$ corresponds to the inverse structure. On a statistical basis, $\delta = \frac{1}{3}$ corresponds to a random distribution. The factors that determine the distribution of cations are ionic radius, electronic configuration, electrostatic energies and polarization effects (Smit [1968]). A sensitive temperature dependence of the distribution of the M^{II} cations was observed in $MgFe_2O_4$ and $CuFe_2O_4$ in conformity with the Boltzmann distribution expressed by

$$\delta(1+\delta)/(1-\delta)^2 = e^{-E_{\delta}/kT}, \tag{5.22}$$

where E_{δ} is the activation energy for the movement of an M^{II} ion from a [B] site to an (A) site and for returning an Fe^{III} ion from the (A) site to the [B] site and has the value of 0.11–0.14 eV (Bertaut [1951]).

The ferrimagnetic mode of spin coupling in Fe_3O_4 has been discussed in ch. 2, §4. A comparison is given between spin couplings in the normal and inverse spinel structures in table 5.11. These two known modes of spin coupling, together with the known magnetic moments of various transitional metal ions, allow the magnetic moment per formula unit (μ_B/f.u.) to be calculated for all single ferrites. Table 5.12 lists the calculated and observed values of μ_B/f.u. along with other magnetic properties of important single ferrites. A few comments on these data are in order. First, a reasonably good agreement is reached between experimental and theoretical values for the moment per formula unit for all ferrites except $MgFe_2O_4$, which is expected to have zero moment, but measurements show a value of 1.1 μ_B per formula unit.

Secondly, all ferrites save $CoFe_2O_4$ are like nickel in having ⟨111⟩ easy axes of magnetization and a negative first constant of magnetocrystalline anisotropy. The Co-ferrite, however, has ⟨100⟩ easy axes and an enormously large, positive K_1.

Thirdly, all ferrites except Fe_3O_4 exhibit a negative average coefficient of saturation longitudinal magnetostriction at room temperature. The $\bar{\lambda}_s$ value for the Co-ferrite is particularly large. Use has

TABLE 5.12

Magnetic properties of single cubic ferrites of the spinel type (Brailsford [1966])

Ferrite	μ_B/f.u. Calc.	μ_B/f.u. Obs.	M_s at 20°C, (10^5 A · m^{-1} or 10^2 G)	θ_C (°C)	K_1 at 20°C, (10^3 J · m^{-3})	Easy axes	λ_{100} ($\times 10^6$)	λ_{111} ($\times 10^6$)	$\bar{\lambda}_s$ ($\times 10^6$)
$MnFe_2O_4$	5	4.85–5.0	4.0	300	−4	⟨111⟩	−35*	−1*	−14.6
$FeFe_2O_4(Fe_3O_4)$	4	4.2	4.8	585	−11	⟨111⟩	−17	+91	+47.8
$CoFe_2O_4$	3	3.3	4.2	520	~220	⟨100⟩	−600	+120	−168
$NiFe_2O_4$	2	2.3	2.7	585	−6.7	⟨111⟩	−36†	−4†	
$CuFe_2O_4$	1	1.3	1.4	455	−6	⟨111⟩			
$MgFe_2O_4$	0	1.1	1.2	440	−3.9	⟨111⟩			
$Li_{0.5}Fe_{2.5}O_4$	2	2.5	3.1	670					

*Values for $Mn_{0.98}Fe_{1.86}O_4$. For $Mn_{0.95}Fe_{2.05}O_4$, $\lambda_{100} = -22 \times 10^{-6}$ and $\lambda_{111} = +5 \times 10^{-6}$.

†Values for $Ni_{0.8}Fe_{2.2}O_4$.

thus been made of $CoFe_2O_4$ in performing magnetomechanical functions in transducers and similar devices.

Thus far, we have described only the single ferrites, each of which contains a single species of M^{II} ions. Nearly all single ferrites display versatile mixing capability in the sense that two single ferrites are capable of dissolving each other to form a partial or complete series of solid solutions known as mixed ferrites. An interesting and technically important phenomenon of the mixed ferrites is the enhancement of saturation magnetization by Zn-ferrite when it is added to other ferrites of the inverse structure. Mixed ferrites with the Zn–Mn, Zn–Fe, Zn–Co and Zn–Ni combination all show such an enhancement. At first glance this enhancement seems rather odd because Zn-ferrite itself has no net moment per formula unit. The effect, however, can be explained in terms of the formula previously given for the mixed ferrites. Based on the formula $(Zn^{II}_{\delta}Fe^{III}_{1-\delta})[M^{II}_{1-\delta}Fe^{III}_{1+\delta}]O_4$ and the known moments of 0, 5.0 and m μ_B per ion of Zn^{II}, Fe^{III} and M^{II}, respectively, the total moment on the (A) and [B] sites per formula unit would be $5(1-\delta)$ and $m(1-\delta)+5(1+\delta)$ in units of μ_B, respectively. Because these moments are antiparallel, the resultant moment is given by taking their difference. Hence, the net magnetic moment of such a mixed ferrite is expected to be

$$\mu_{m.f.} = 10\delta + m(1-\delta) \tag{5.23}$$

in units of μ_B per formula unit. If we use $m = 5$, 4, 3 and 2 for Mn^{II}, Fe^{II}, Co^{II} and Ni^{II}, respectively, the calculated values for $\mu_{m.f.}$ are shown as a function of the concentration of the Zn-ferrite, δ, by the dashed lines in fig. 5.7. The observed values for $\mu_{m.f.}$, shown by the solid curves, are in fair agreement with the calculated values in the range of δ from 0 to 0.5. The experimental results thus indicate a maximum benefit of the presence of Zn-ferrite in the mixed ferrites at about 1:1 ratio with the other component ferrite.

(2) The cubic ferrites of the garnet type. The mineral garnet refers to a group of mixed oxides, of which the widely known one has the chemical formula $Mn_3Al_2Si_3O_{12}$, or equivalently, $3MnO \cdot Al_2O_3 \cdot 3SiO_2$. Single magnetic garnets have the general formula $3M_2O_3 \cdot 5Fe_2O_3 = 2M^{III}_3Fe^{III}_5O_{12}$. Note that, in magnetic garnets, the 24 positive charge units per formula unit are divided unequally between the ferric ions (15 units) and another species of trivalent ions (9 units). Technically useful garnets are those with M = Sm, Eu, Gd, Tb, Dy, Ho, Er, Tm, Yb, or Y.

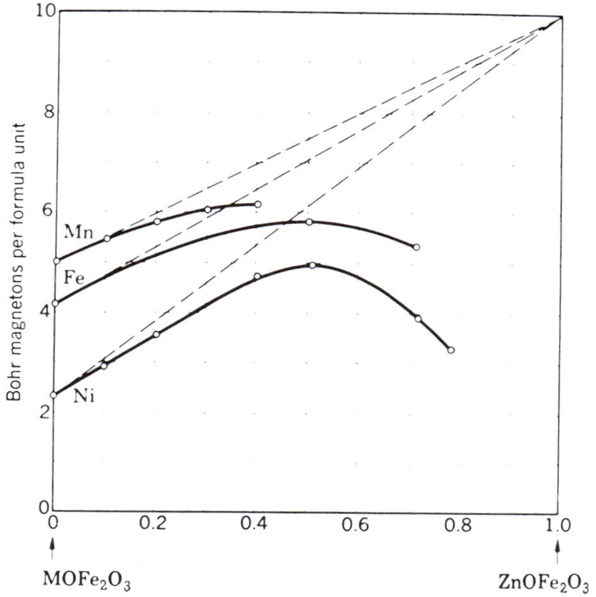

Fig. 5.7. Saturation moments of mixed Mn–Zn, Fe–Zn and Ni–Zn ferrites as a function of the zinc concentration (Went and Gorter [1952]).

They are known as the rare-earth garnets. A code system has been adopted to name them: REG stands for the rare-earth garnets, GdIG for the gadolinium-iron garnet ($Gd_3Fe_5O_{12}$), YIG for the yttrium-iron garnet ($Y_3Fe_5O_{12}$), etc.

Garnets crystallize in the cubic system with two-fifths of the ferric ions forming a body-centered cubic lattice. Like the ferrospinels, the garnets, too, pack a large number (160) of ions in eight formula units into a unit cell. The lattice constant is ≈12.5 Å, about 50% larger than those of ferrospinels. Also, the crystal structure of garnets is more complicated than the spinel structure because of the size (0.85–1.10 Å) of the M^{III} ions. They are too large to be accommodated at the interstitial sites between the oxygen ions. Hence, the oxygen ions are prohibited from forming a close-packed structure as in the spinel. The arrangement of the 24 M^{III} ions and 40 Fe^{III} ions in a unit cell of the garnet structure is shown in fig. 5.8a. In each unit cell, which contains

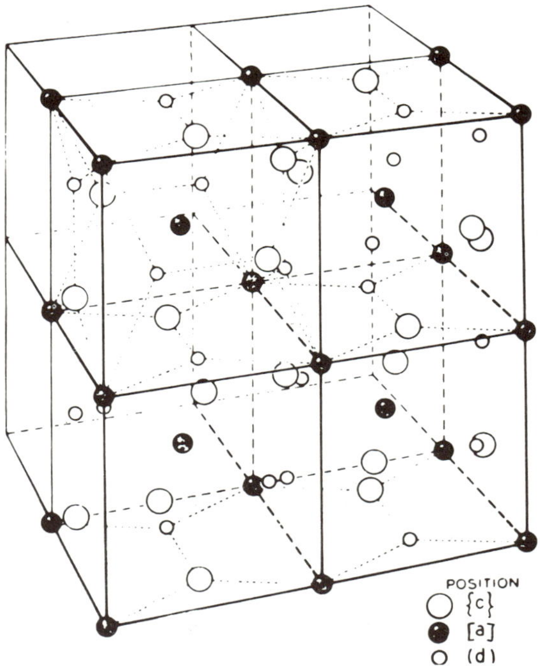

Fig. 5.8. (a) Arrangement of cations in c, a and d sites in four octants of the garnet unit cell (Geller and Gilleo [1957]).

eight formula units, there are three kinds of cation sites, of which

16 octahedral [a] sites are occupied by Fe^{III} ions,
24 tetrahedral (d) sites are also occupied by Fe^{III} ions and
24 dodecahedral {c} sites are occupied by M^{III} ions.

Also note that each unit cell consists of 8 body-centered cubic unit cells formed by the Fe^{III} ions occupying the [a] sites. From the above list, the formula of the magnetic garnets can be written more descriptively as $\{M_3^{III}\}[Fe_2^{III}](Fe_3^{III})O_{12}$, where the distribution of the cations to the three kinds of lattice sites is indicated both in the c, a, d order and by different enclosures, { }, [ ] and ( ), respectively. The surroundings of the three cation sites by oxygen ions are shown in fig. 5.8b.

Because the three cation sites and their occupancies are well defined, it is reasonable to envisage the division of the crystal structure into

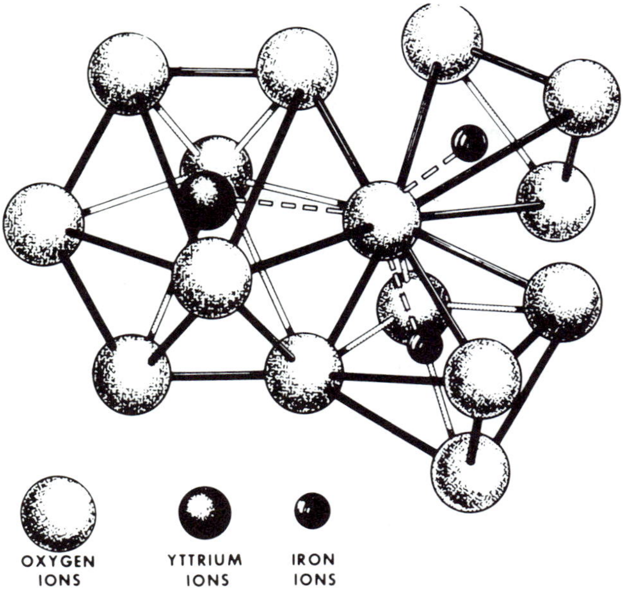

Fig. 5.8 (cont.). (b) The surroundings of a dodecahedral (left) site by eight oxygen ions, an octahedral site (lower right) by six oxygen ions and a tetrahedral site (upper right) by four oxygen ions (Bobeck and Torre [1975]).

three separate magnetic lattices. The two lattices formed by the ferric ions at [a] and (d) sites are ferromagnetic within each lattice because the a–a and d–d interactions are strong enough to align the moment in one direction within each lattice. Since the 4f electrons are so deeply seated in the ionic cores, there is hardly any c–c interaction. Hence, the lattice formed by the rare-earth ions would be either paramagnetic, for the Sm^{III}–Yb^{III} ions, or diamagnetic, for the Y^{III} and Lu^{III} ions. But there are weak c–a and c–d interactions that, at sufficiently low temperatures, could align the magnetic moments of the lanthanide ions ferromagnetically. Neutron diffraction experiments also reveal that, while the moments on the [a] and (d) sites are aligned antiparallel to each other, those on the {c} and [a] sites are aligned parallel. (This is unfortunate from the standpoint of saturation magnetization but fortunate from the standpoint of bubble devices discussed in ch. 7, §2 because, had the ferromagnetic alignment been between {c} and (d), the REG would have shown much larger spontaneous magnetizations.) A simple way to represent the overall spin coupling in a REG is

$$\{\uparrow \uparrow \uparrow\} \quad [\uparrow \uparrow] \quad (\downarrow \downarrow \downarrow),$$

where the dashed arrows signify the varying nature of the magnetic moments of the lanthanide ions. On this basis, it is easy to calculate the net moment per formula unit. For example, in GdIG, because each Gd^{III} ion has a moment of 7 μ_B, the net moment per formula unit is $3\times 7+2\times 5+3\times 5=16\ \mu_B$/f.u. at 0°K.

When the temperature rises, the {c} lattice of the lanthanide ions loses its magnetization M_c rapidly because the weak c–a and c–d interactions responsible for M_c are seriously affected by thermal agitation. In the meantime, the magnetizations M_a and M_d in the two Fe^{III} ion lattices remain relatively unaffected up to a moderate temperature. The temperature variation in magnetization of the three lattices in GdIG is shown in fig. 5.9. There exists a temperature at which spontaneous magnetization of the REG vanishes because $M_{REG}=M_a+M_c-M_d=0$. This temperature is known as the compensation point, T_p, which is well below the Curie point. As the temperature continues to rise from T_p, the garnet first regains its magnetization slowly until a small peak is reached, then loses it again permanently at the Curie temperature. The temperature dependencies of spontaneous magnetization of various REG are shown in fig. 5.10. Pertinent magnetic and physical data for rare-earth garnets are listed in table 5.13.

As might be expected, YIG and LuIG show no compensation point because Y^{III} and Lu^{III} ions have zero moments. These two garnets differ from the rest of REG in still another aspect. Their spontaneous magnetizations rank at the lower end at 0°K, but reach the top at room temperature. This reversal in the rank of magnetizations of REG is caused by the onset of T_p, which appears as a result of the rapid decrease in magnetization of the lanthanide-ion lattice.

For YIG the most attractive property is the extremely narrow ferromagnetic resonance linewidth, of the order of $\sim 50\ A\cdot m^{-1}$ (0.6 Oe). YIG attains this outstanding property because it derives all its magnetic properties from the ferric ions, for which $L=0$ and $S=\frac{5}{2}$. The absence of orbital contribution to magnetization for the ferric ions, coupled with the spherical charge distribution, produced a weak interaction of the spins with lattice deformations and phonons, thus the narrow linewidth in the resonance experiments. Advantages have been taken of this attractive property of YIG in microwave devices to achieve low energy losses by virtue of the limited damping of spin resonance.

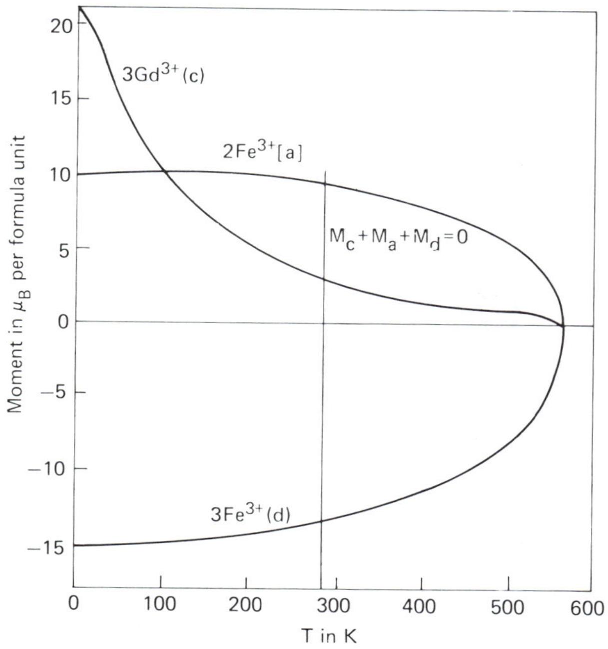

Fig. 5.9. Magnetization of the Gd–Fe garnet sublattices per formula unit as a function of temperature (Pauthenet [1958]).

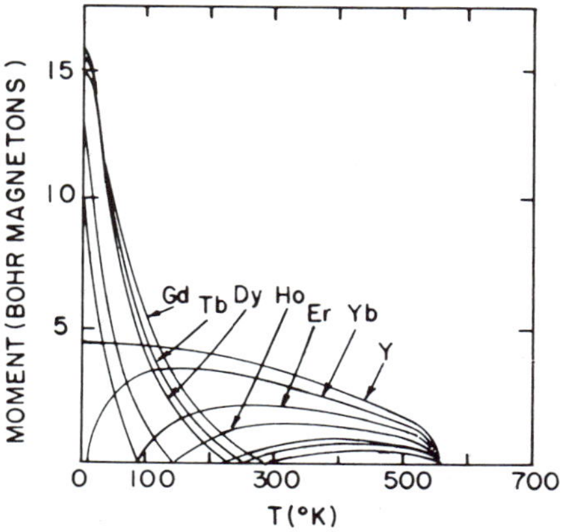

Fig. 5.10. Experimental data showing the temperature dependence of saturation moments of various iron garnets. The unit formula is $M_3Fe_5O_{12}$, where M is a trivalent rare-earth ion indicated (Pauthenet [1956]).

TABLE
Magnetic and physical

Material formula	Curie temp. θ_C (°K)	Compensation temp. T_p (°K)	Magnetic moment per formula unit at 0°K (μ_B)	Saturation magnetization at 293°K (10^5 A · m^{-1})
$Y_3Fe_5O_{12}$	553		5.01	1.39
$Sm_3Fe_5O_{12}$	578		5.43	1.35
$Eu_3Fe_5O_{12}$	566		2.78	0.93
$Gd_3Fe_5O_{12}$	564	286	16.00	0.14
$Tb_3Fe_5O_{12}$	568	246	18.20	0.04
$Dy_3Fe_5O_{12}$	563	226	16.90	0.43
$Ho_3Fe_5O_{12}$	567	137	15.20	0.78
$Er_3Fe_5O_{12}$	556	83	10.20	1.03
$Tm_3Fe_5O_{12}$	549		1.20	1.10
$Yb_3Fe_5O_{12}$	548	6(?)	0.00	1.30
$Lu_3Fe_5O_{12}$	549		5.07	1.40

(3) The hexagonal ferrites. The third type of ferrites is often called the barium ferrites because these compounds usually contain BaO, in addition to Fe_2O_3, as the basic component oxide. They are also known as the magnetoplumbites. The common chemical formula of barium ferrites is $l(BaO) \cdot m(MO) \cdot n(Fe_2O_3)$, or $Ba_l^{II}M_m^{II}Fe_{2n}^{III}O_{l+m+3n}$, where l, m and n are integers. Despite the simple formula, this group of ferrites is much more complex than the previous two in both composition and crystallography. Partly for this reason and partly because they are used largely as ceramic-type permanent magnets, our discussion on barium ferrites will be brief.

There are four ways in which the composition of barium ferrites may be changed. One is to vary the M^{II} ions. Mg, Mn, Fe, Co, Ni, Cu and Zn are found suitable for the formation of hexagonal ferrites. Another way is to alter the values for l, m and n. Basic combinations are found at 1–0–6 (the ferrite is designated M), 1–2–8(M_2W), 2–2–6(M_2Y) and 3–2–12(M_2Z). Compounds formed under these combinations are termed classical hexagonal ferrites. Still another way is to substitute for Ba with Pb or Sr and (or) substitute for Fe with Al, Ga, Cr or Mn. A fourth way to vary the composition of the barium ferrites is to mix two or more of the classical hexagonal ferrites in different proportions.

5.13
properties of rare earth garnets

Wavelength-independent magnetic rotation at 300°K (deg · cm^{-1})	Linear magnetic birefringence at 300°K		a (Å)	Density (g · cm^{-3})
	$R_{[111]}$	$R_{[100]}$		
	0.450	0.45	12.376	5.17
			12.529	6.23
	1.1	1.1	12.498	6.31
3	0.50	0.45	12.471	6.46
	0.40	0.02	12.436	6.55
40	0.2	0.05	12.405	6.61
42	0.52	0.40	12.375	6.77
47	0.80	0.40	12.347	6.94
56	0.475	0.55	12.323	7.06
	0.50	0.32	12.302	7.14
			12.283	

The crystallography of the barium ferrites is also very complex. The hexagonal structure is based on layers of oxygen ions. The combination in a basic unit of (1) the number of the oxygen layers, (2) the relative orientations of these layers and (3) the manner in which the cations are packed in or between the oxygen layers results in three basic structure units denoted by $S(2M^{II}Fe_2O_4)$, $M(BaFe_{12}O_{19})$ and $Y(2BaM^{II}Fe_6O_{11})$. The exact crystal structure of each hexagonal ferrite then can be described in terms of the stacking sequence of S, M and Y. The notable examples are M_2Y_2 in $Ba_6M_4^{II}Fe_{48}O_{82}$ and several stacking isomers of M_2Y_8(MY–MY_7, MY_2–MY_6, etc.). For detailed information on the crystallography of hexagonal ferrites, the reader is referred to the paper by Braun [1957].

Techniques for the preparation of ferrites

Ferrites are prepared for use in magnetic devices other than bubble devices discussed in ch. 7, §2 in either polycrystalline or single-crystal form. Different techniques have been developed for the preparation of practical ferrites in these two forms. Accordingly, our discussion on the preparation techniques should also be divided into two parts. For background information, however, we must first review some pertinent physical chemistry of iron oxides. It will become clear that the techniques suitable for the preparation of ferrites of controlled composition and microstructure depend upon the stability and reactions of Fe_2O_3 with other oxides at elevated temperatures.

Iron oxides exist in three stable phases; namely, ϵ-FeO_x (wüstite)*, ζ-Fe_3O_4 (magnetite) and η-Fe_2O_3 (hematite). Each of these oxides has a characteristic dissociation oxygen pressure, which varies with temperature. Their equilibrium states are shown in a plot of $\log(p_{CO_2}/p_{CO})$ versus $1/T$ in fig. 5.11. This plot was based on data obtained from

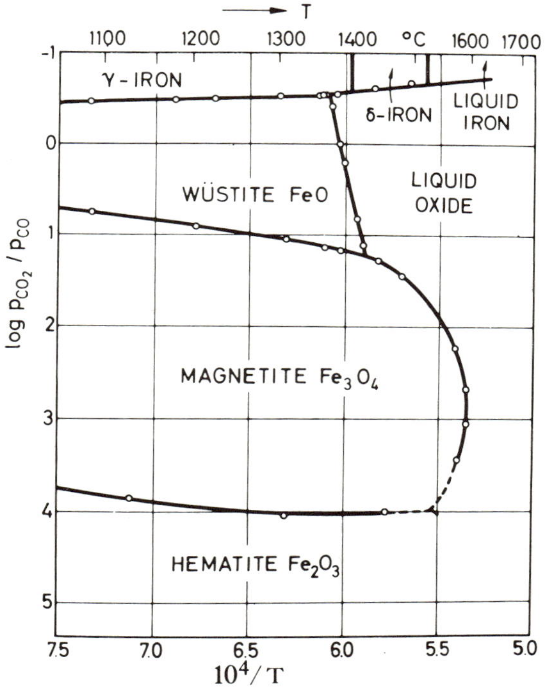

Fig. 5.11. Partial pressure of oxygen in equilibrium with different iron oxides plotted in terms of p_{CO_2}/p_{CO} versus temperature (Darken and Gurry [1946]).

*Wüstite is always deficient in iron with respect to the stoichiometric composition FeO. Therefore the correct formula for wüstite is FeO_x, where x is slightly greater than unity.

studies in which the atmosphere over the oxides was varied in terms of the partial pressure of oxygen, p_{O_2}, through the dissociation of CO_2,

$$2CO_2 \rightleftarrows 2CO + O_2 \tag{5.24}$$

This reaction defines p_{O_2} by $p_{O_2} = K_p(p_{CO_2}/p_{CO})^2$, where K_p is the equilibrium constant expressed in partial pressures. Applying this definition, we see from fig. 5.11 that, at a fixed oxygen partial pressure, magnetite tends to dissociate to wüstite at elevated temperatures via the reaction approximately expressed by

$$2Fe_3O_4 \rightarrow 6FeO + O_2\uparrow . \tag{5.25}$$

This dissociation is equivalent to reducing Fe^{III} to Fe^{II} in the magnetite. At the oxygen partial pressure corresponding to $\log(p_{CO_2}/p_{CO}) = 1$ in fig. 5.11, for instance, Fe_3O_4 becomes unstable and transforms to FeO at ~1270°C. Obviously, the dissociation will proceed at much faster rates when melting of Fe_3O_4 or Fe_2O_3 is attempted at higher temperatures.

It can be safely predicted that melting of a mixture of oxides, including Fe_2O_3, to form a ferrite under normal conditions (in air or at one atmospheric oxygen pressure) will result in an appreciable conversion of Fe^{III} ions to Fe^{II} ions. For example, melting of a ferrospinel mixture will involve a reaction similar to that expressed in (5.25),

$$2MO + 2Fe_2O_3 \rightarrow 2MFe_2O_3(2MO \cdot 4FeO) + O_2\uparrow , \tag{5.26}$$

which yields $MFe_2^{II}O_3$ instead of $MFe_2^{III}O_4$. Moreover, according to Latsh [1961], the dissociation temperature of Fe_2O_3 is considerably lowered when the ferric oxide is dissolved in the spinel. Latsh's results are shown in fig. 5.12, which indicates a decrease of the dissociation temperature for Fe_2O_3 from ≈1480°C in the absence of spinel to ≈1050°C when Fe_2O_3 is dissolved in $(Ni, Zn)Fe_2O_4$ at the ratio of 1 to 4.

(I) Polycrystalline ferrites. Because ferrites tend to be dissociated at elevated temperatures, the conventional technique of melting and casting so widely used for metals and alloys is impractical for ferrites. Polycrystalline ferrites are prepared normally by powder metallurgy or ceramic technology. This means that ferrites attain their homogeneous compositions by reactions in the solid state, and the shapes of the ferrite products are produced by pressing and subsequent sintering. The preparation procedure generally consists of four major steps:

1. Preparing a mixture of materials with the cations more or less in the ratio corresponding to that in the final product.

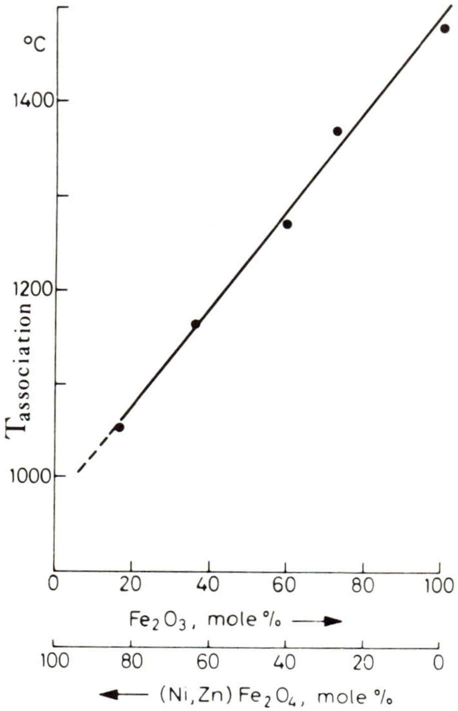

Fig. 5.12. Temperature of complete dissociation in air of excessive Fe_2O_3 in the system (Ni, Zn)Fe_2O_4–Fe_2O_3 to a stoichiometric spinel with a cation-to-anion ratio of 3:4 (Latsch [1961]).

2. Prefiring the mixture to form the ferrite.
3. Converting the "raw" ferrite into powder and pressing the powder into the required shape.
4. Sintering to produce a highly densified product.

The first two steps are special for ferrites, whereas ball-milling and pressing in step (3) and sintering in step (4) are standard powder metallurgy processes. Of course, not all these steps are necessary, nor should they be taken separately in every case. For instance, hot pressing may be chosen to replace step (3) and step (4). The following discussion will cover the important features of each step.

1. Preparing a mixture of materials having the right ratio of cations.

The extent of work in this step varies greatly, depending on the starting materials. When component oxides are used, this step involves a mere mixing of the oxides by wet milling. To avoid iron contamination, mixing is done in a rubber-lined mill with stainless-steel balls and a fluid such as water, alcohol or kerosene to prepare the mixture in a slurry.

Often, however, high cost, insufficient purity or unavailability of the component oxides necessitates the use of starting materials other than oxides. Choice can then be made between the methods of thermal decomposition and precipitation. The decomposition method involves the heating of appropriate carbonates, nitrates, sulfates, or oxalates in air. Some workers prefer to prepare oxides from these salts in situ because the mixtures are claimed to be more ready for solid-state reactions in the next step. The precipitation method produces a mixture of two or more hydroxides or oxalates that precipitate simultaneously from a solution. The mixture of precipitates should contain the required cations in the correct proportions with respect to the final composition. This method is further explained in a comparison between the preparations of precipitations of hydroxides and oxalates for the formation of the Mg-ferrite (see next page) (Economos [1955]). In the hydroxide precipitation, $Fe(OH)_3$ is completely precipitated in an acid solution ($pH = 3.8$), but $Mg(OH)_2$ precipitates only in a strongly basic solution of $pH \approx 10.2$. To promote the precipitation of $Mg(OH)_2$, NaOH is added to the hydroxide solution so that the pH value in the solution is adjusted properly. The addition must be made simultaneous when the solution enters a container. The use of NaOH, however, has the inherent drawback that the sodium ions are difficult to remove completely, thus remaining as impurities.

The oxalate precipitation is preferred for two other reasons. First, precipitation can be made easily with ammonium oxalate, which leaves no residue after ignition. Secondly, most metal oxalates are similar in crystal structure; therefore the precipitates can be mixed not only intimately, but also on a molecular scale.

2. Prefiring the mixture to form the ferrite. The slurry prepared in step (1) is filtered or dried and transferred to a porcelain crucible for prefiring at temperatures between 900 and 1200°C. For garnets, which require high prefiring temperatures up to 1450°C, platinum crucibles are used.

As far as the final composition of the ferrite is concerned, step (2) is most crucial because subsequent steps will not change the composition substantially. For this reason, it is important to understand how a

Hydroxide precipitation	Oxalate precipitation
$3FeSO_4 + 9(OH)^- \rightarrow 3Fe(OH)_3 + 3(SO_4)^{--} + 3e^-$ (solubility product = 1.1×10^{-36})	$FeSO_4 + (C_2O_4)^{--} + 2H_2O \rightarrow FeC_2O_4 \cdot 2H_2O \downarrow + (SO_4)^{--}$ (solubility product = 2.1×10^{-7}) (solubility of $FeC_2O_4 \cdot 2H_2O = 4.6 \times 10^{-4}$ mol/liter)
$MgSO_4 + 2(OH)^- \rightarrow Mg(OH)_2 + (SO_4)^{--}$ (solubility product = 1.2×10^{-11})	$MgSO_4 + (C_2O_4)^{--} + 2H_2O \rightarrow MgC_2O_4 \cdot 2H_2O \downarrow + (SO_4)^{--}$ (solubility product = 8.6×10^{-5}) (solubility of $MgC_2O_4 \cdot 2H_2O = 9.3 \times 10^{-3}$ mol/liter)

ferrite is formed from its component oxides. The solid-state reactions leading to the formation of a ferrite actually are achieved by counterdiffusion. This means that the diffusion involves two or more species of ions which move in opposite directions, initially across the interface of two contacting particles of different component oxides. Carter [1961], Kooy [1965] and Reijnen [1965] made careful studies of the position of inert markers in diffusion couples and concluded that the counterdiffusion essentially involves the movement of cations through a more or less rigid lattice of oxygen anions.

To understand the mechanism more specifically, the reactions involved in the formation of $MgAl_2O_4$ and $MgFe_2O_4$ are summarized and compared in fig. 5.13. For $MgAl_2O_4$, the counterdiffusion involves the Mg^{2+} and Al^{3+} ions, which migrate to the right and left sides, respectively, with respect to the original interface. To preserve charge neutrality, every three Mg^{2+} ions are coupled with two Al^{3+} ions; therefore, three spinel molecules are formed at the Al_2O_3–$MgAl_2O_4$ interface for every spinel molecule formed at the MgO–$MgAl_2O_4$ interface. The Kirkendall effect is absent in fig. 5.13a because the interface markers remain stationary.

The formation of $MgFe_2O_4$ is complicated by the fact that some of the iron ions diffuse in the divalent state. Figure 5.13b shows that the reduction of Fe^{3+} to Fe^{2+} occurs at the Fe_2O_3–$MgFe_2O_4$ interface and is accompanied by a loss of oxygen through the reaction, $2O^{2-} \rightarrow O_2\uparrow + 4e^-$. At the MgO–$MgFe_2O_4$ interface, Fe^{2+} ions are oxidized to restore the trivalency, and there is a gain of oxygen from the reaction $O_2 + 4e^- \rightarrow 2O^{2-}$. Hence the formation of $MgFe_2O_4$ also involves a mass transfer of oxygen from Fe_2O_3 to MgO through two processes equivalent to evaporation and condensation, in addition to the counterdiffusion of cations. Consequently, the Kirkendall effect was observed during the formation of $MgFe_2O_4$ with the interface markers moving into the Fe_2O_3 side. On the basis of maintaining a charge balance during the diffusion process, it is easy to show that, if all Fe^{3+} ions are reduced to Fe^{2+} before migration, (1) the ratio of the counterdiffusing ions of Mg^{2+} and Fe^{2+} will be 3:3; (2) the ratio of the relative amounts of $MgFe_2O_4$ formed on the MgO and Fe_2O_3 sides will be 1:2 and (3) the markers will move toward the Fe_2O_3 side. The observed ratio of the relative amounts of $MgFe_2O_4$ formed on the two sides is 1:2.7, indicating that the reduction of Fe^{3+} ions occurs on a limited scale. Despite the incomplete reduction, it is important to realize that the formation of ferrites depends on the oxygen partial pressure.

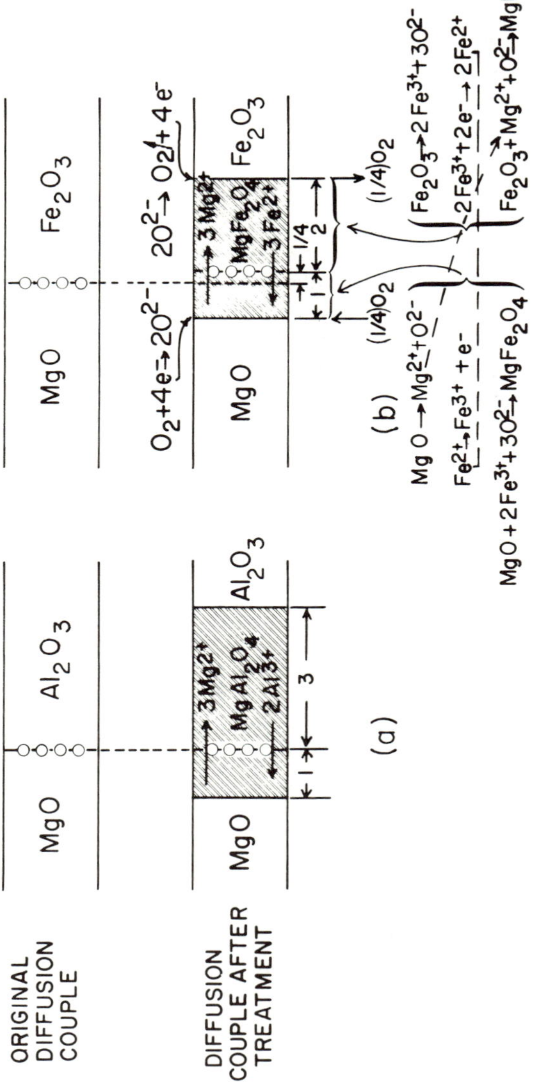

Fig. 5.13. Counter-diffusion mechanisms involved in the formation of (a) $MgAl_2O_4$ and (b) $MgFe_2O_4$ (Winkler [1971]).

Reducing conditions will promote the rate of ferrite formation because larger concentration gradients will be built between the two interfaces in fig. 5.13b and, also, more Fe_2O_3 can be dissolved into the ferrite (Reijnen [1967]).

The mechanism just described for the formation of ferrites justifies the separate prefiring and sintering steps taken in the preparation procedures. The ferrite is formed essentially in step (2), but the "raw" ferrite thus formed has poor qualities in two respects. First, its composition is not expected to be homogeneous because the recovery of the Fe^{3+} ions at the MgO–$MgFe_2O_4$ interface is not likely to be complete. Secondly, the ferrite formed on the Fe_2O_3 side could contain pores as a result of the evolution of oxygen. Indeed, experimental results obtained from the formation of $MgFe_2O_4$ and $NiFe_2O_4$ indicate an appreciable weight loss of samples. This loss is identified with the loss of oxygen and implies an incomplete recovery of the Fe^{3+} ions. These undesirable qualities of the ferrite must be eliminated in the remaining steps of the procedure.

Another important observation in the prefiring step deals with the critical temperature below which the formation of ferrites would not be completed regardless of the length of prefiring time. Figure 5.14 shows the maximum fraction of $NiFe_2O_4$ formed from NiO and Fe_2O_3 at 800–1200°C. Clearly, a complete formation requires a prefiring temperature at and above 1200°C in this case. For ferrites that contain cations of such metals as Zn and Li, care must be taken in the selection of temperature and atmosphere for prefiring and sintering so as to avoid the evaporation of these volatile elements.

3. Converting the raw ferrite into powder and pressing the powder. The ferrite produced in the previous step is usually in lump form.

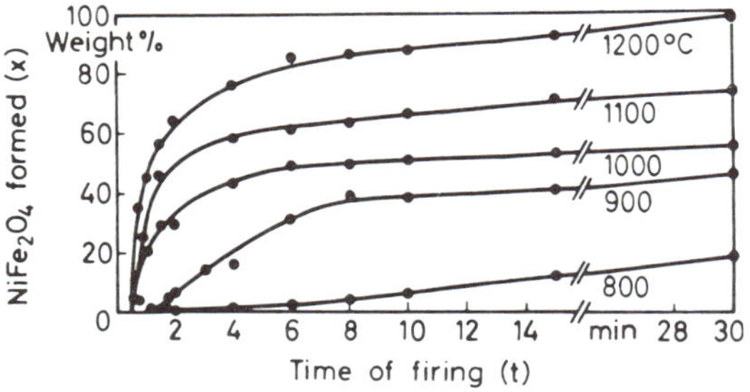

Fig. 5.14. Percent $NiFe_2O_4$ formed versus time of firing at temperatures between 800 and 1200°C (Blum and Li [1961]).

It must be ground into powder in a steel ball mill. Besides reducing the particle size to $\leq 1\ \mu m$, grinding also eliminates intraparticle pores and homogenizes the ferrite by mixing. To promote successful sintering in the next step, the powder must be well characterized after grinding with respect to such factors as particle size and distribution, particle shape, homogeneity, adsorbed gases, impurities and intraparticle porosity. Iron contamination due to continuous wear of the mill wall and steel balls should be closely watched and minimized.

Pressing the powder into compacts of desired shapes is done either by the conventional method in a die-punch assembly or by hydrostatic or isostatic compaction. The latter method uses a sealed rubber or plastic mold to hold the powder and applies a hydrostatic pressure of the order of 7–21 kg/mm^2 to the powder assembly in a vessel filled with a suitable fluid such as glycerin, water or hydraulic oil. This method offers several advantages over the conventional method, notably higher and more uniform densities attained and no need of granulating the powder, which produces a spherical particle shape necessary to ensure a smooth flow during die punch pressing. The conventional method, on the other hand, is faster and more economical; hence, it is more widely used in the industry, particularly for mass productions. Hydrostatic or isostatic compaction is preferred for small-scale productions and laboratory projects.

4. Sintering. This is a heat treatment by which a mass of compacted powder is transformed into a dense object. As the final major step in the preparation of ferrite products, sintering must fulfill three requirements: (1) to bond the particles together so as to impart sufficient strength to the products; (2) to densify the green compacts by eliminating the pores and (3) to homogenize the materials by completing the reactions left unfinished in the prefiring step. The first two requirements are closely related as far as their mechanisms are concerned. For instance, as shown in fig. 5.15, a simple way to bond two spherical particles is to form a neck at the contact by volume diffusion. The formation of necks is also a key process leading to the shrinkage of pores. Precisely how pores disappear during sintering has long been a subject of intensive investigation. The theory has been changed from the traditional, but now obsolete, theory based on the reduction of surface energy as the driving force for eliminating the pores, through the microcreep diffusion theory of Naborro and Herring to the current theory of Kuczynski [1961]. The current theory emphasizes the formation of necks as a key step in bonding the particles and in eliminating

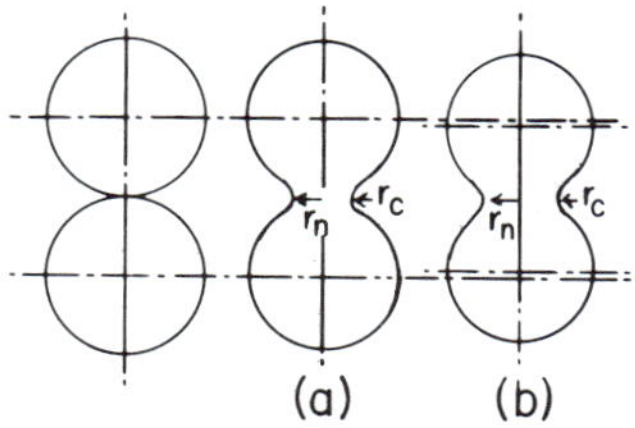

Fig. 5.15. Neck formation during sintering of spherical particles: (a) when there is no overall shrinkage and (b) when shrinkage occurs (Sands and Shakespeare [1966]).

the pores and considers volume diffusion under surface tension as the main mechanism of neck formation. Other processes, such as plastic flow, surface or grain-boundary diffusion and evaporation-condensation, could play important supporting roles. On the basis of volume diffusion induced under surface tension, an important equation for the initial stage of the sintering process is given by

$$\Delta L/L = [(AD^*\gamma\Omega/\bar{r}^3kT)t]^{\frac{1}{2}}, \tag{5.27}$$

where $\Delta L/L$ is the shrinkage of the compact, Ω the volume of a single vacancy, D^* the coefficient of self-diffusion for the slowest moving species, γ the surface energy, $\bar{r}$ the average radius of the particles, t the sintering time and A a constant, of the approximate value of unity. Equation (5.27) indicates that sintering fulfills requirements (1) and (2) more efficiently when the compact features high surface energy and self-diffusivity and the particles are fine. Detailed discussion of the sintering theory is beyond the scope of this book. For further information on the sintering behavior of powder compacts of alloy systems of iron, cobalt and other elements, the article by Thümmler and Thomma [1967] is recommended.

For densification, Burke [1959] made two important observations. One is that the formation of necks only marks the initial stage of sintering; an intermediate stage begins when grain growth occurs with the compact density at around 60% of the theoretical value and ends when the pore phase becomes discontinuous and the density reaches a value of ≈ 95%. Another observation of Burke deals with the final stage of sintering. If discontinuous grain growth occurs after the intermediate stage, migration of grain boundaries will leave the remaining isolated pores trapped inside the grains, thus preventing further shrinkage of these intraparticle pores and practically stopping the sintering process. On the other hand, if discontinuous grain growth is prohibited, the last few percentages of porosity are eliminated by emitting vacancies at grain boundaries. Hence, to obtain high densities close to the theoretical value, measures to hinder discontinuous grain growth, such as using pure raw materials, avoiding granulation of the powder before pressing and applying a grain-growth inhibiter (e.g., V_2O_5 or ThO_2 in the preparation of ferrospinels), should be adopted.

Although only limited chemical reactions are expected during sintering in connection with requirement (3), proper atmosphere should be provided to keep the correct stoichiometry of the ferrite produced in the prefiring step intact. Winkler [1971] gave the following general rules for the selection of sintering atmosphere:

(1) All high-frequency and microwave ferrites should be sintered in pure oxygen during the entire period, including cooling, to avoid the loss of oxygen and the reduction of the Fe^{3+} ions. Otherwise, reoxidation of densely sintered ferrites during cooling often causes cracking.
(2) All divalent Mn-ferrites should be treated in nitrogen during the cooling period from 900–1100°C to room temperature to avoid oxidation of Mn^{2+} to Mn^{3+} or Mn^{4+} and precipitation of Mn_2O_3, which often causes microcracking.
(3) All ferrites containing Fe^{2+} ions in their final stoichiometries, such as $MnZnFe^{II}$-spinel, require carefully controlled inert atmospheres with varying oxygen content to establish the correct concentration of Fe^{2+} ions according to the equilibrium value of the oxygen partial pressure at the chosen temperature (Blank [1961] and Macklen [1965]).

(II) Single crystals of ferrites. The growth of single crystals of ferrites is also a complicated phenomenon, and in some cases it is seriously hampered by the great tendency of Fe_2O_3 to dissociate at the temperature of direct melting. By direct melting is meant that the component oxides are heated to form a liquid solution without the use of fluxes. Despite the problem of dissociation, two methods have been developed to grow single crystals of certain ferrites from the melt. One is known as the Verneuil or flame-fusion method. This method was introduced by Verneuil for the preparation of ruby and has been used on a commercial scale for the production of artificial sapphires. It uses an oxy-hydrogen flame to melt fine powder particles, 1 or 2 microns in diameter, which are dropped into the flame continually. After passing through the flame, the material is in droplet form and falls upon a seed

crystal, which grows into a column or boule form. During growth, the support of the crystal descends at a suitable speed in relation to the rate at which droplets are accumulated. A burner of the tri-cone type is convenient for maintaining the required atmosphere for melting. The nozzle of the burner that supplies the hydrogen to the flame has an oxygen supply both inside (carrying the feed powder) and outside the nozzle. Usually the crystal support is made of a ceramic rod and heated in a tube furnace to prevent too rapid solidification of the droplets.

The Verneuil method has been applied successfully for growing large crystals of manganese ferrite. In one instance, a crystal of Mn-ferrite, 4 cm long and 1.5 cm in mean diameter, was grown in 3 hours. Ferrite crystals grown by this method, however, often contain many cracks and other macroscopic-size imperfections. The method has another drawback in that the possibility of the dissociation of Fe_2O_3 always exists, which will result in a deficiency of the Fe^{3+} ions as well as nonstoichiometric compositions of the ferrites.

The second method that invokes direct melting was first demonstrated by Smiltens [1952] for the growth of crystals of synthetic magnetite. In this method, the starting material was heated inside a covered platinum crucible with a conical bottom in an atmosphere of CO and CO_2. After the charge had been melted and homogenized, the crucible was lowered at a controlled rate into a steep temperature gradient. Obviously this is a Bridgman method in which the pointed tip of the conical crucible would allow only one nucleus to be produced and the motion of the crucible in a temperature gradient would permit the nucleus to grow into a single crystal. The key to success in the growth of a single crystal is, of course, the inhibition of the formation of other nuclei during crystallization of the melt. Using a temperature gradient of 20°C per cm and a lowering rate of about 5 mm per hr, Smilten was able to grow crystals 4 cm long and 1 cm in diameter, and the crystals were of high purity and superior quality.

To circumvent the problem of the dissociation of Fe_2O_3; or if the ferrite composition must meet stringent specifications, the flux method is favored over the direct melting methods. A substance is qualified to be a suitable flux if it satisfies three conditions: (1) It melts at a temperature lower than the melting points of the component oxides. (2) It is capable of dissolving the ferrite material in the liquid state. And (3) it can be separated from the ferrite during crystal growth by either the Bridgman technique or the crystal-pulling (Czochralski) technique. To meet the third condition, the radii of the flux ions must be such that these ions will not be dissolved in the crystal lattice of the ferrite. In executing the flux method, the flux will be melted first, which, in turn, will dissolve the ferrite material at much lower temperatures than those required to melt the ferrite directly (e.g. 1100°C versus 1575°C for magnetite). At such lower temperatures, the dissociation of Fe_2O_3 is largely avoided.

Many fluxes have been tried for various ferrites with success. Information on the specific amount of flux and other conditions required for best results has been reported for commercially important ferrites. For instance, Borax ($Na_2B_4O_7$) was used for Ni- [and (Zn, Mn)] ferrites at the proportions of 16–17 gm of Borax, 6.4 gm of Fe_2O_3 and 3.0 gm of NiO. The mixture was heated to 1330°C, held there for 10 hr, and cooled, first slowly at 2°C per hr for 25–40 hr, then rapidly by turning off the furnace completely (Galt et al. [1950]). Lead oxide (PbO) was initially used as a flux for YIG. Later an improved flux was found by Nielsen [1960] by adding lead fluoride (PbF_2) to PbO at the ratio: 30 mole % PbO, 40 mole % PbF_2, 22 mole % Fe_2O_3, and 8 mole % Y_2O_3 for YIG; and 39.4 mole % PbO, 40.8 mole % PbF_2, 14 mole % Ga_2O_3 and 5.8 mole % Y_2O_3 for YGaG. There are two advantages of adding PbF_2 to PbO: to further lower the temperature required to melt the flux through the formation of a low-melting eutectic between the two lead compounds and to reduce the viscosity of the melt. According to Linares [1964], however, neither PbO nor PbO–PbF_2 can be used in a completely open system owing to the high vapor pressure of these materials at the temperature ($\approx$ 1250°C) of crystallization. Instead, a BaO–B_2O_3 flux was found to give satisfactory results for YIG.

Before closing this section, it should be mentioned that some spinel ferrites are naturally occurring, and therefore there is always the possibility that large crystals of such minerals might be available. Minerals in this category include magnetite, magnesis ferrite (Mg-ferrite), jacobsite (Mn-ferrite), trevorite (Ni-ferrite) and franklinite (Zn-ferrite).

§1.3.2. *Europium chalcogenides*

In 1960, Matthias et al. [1961] discovered ferromagnetism in the divalent europium oxide, EuO. Their discovery has aroused considerable interest in this and other chalcogenides of europium. Three reasons can be cited to explain the interest. First, pure EuO is also an insulator and has the highest magnetic moment per unit volume ($232\,G \cdot cm^3/g$ at 0°K) of any nonconducting materials. The combination of intense magnetization and high electrical resistivity in EuO would make the compound particularly attractive for magnetic applications at high frequencies. Secondly, subsequent studies detected excellent magneto-optical properties in some of the Eu^{2+} compounds. For instance, EuO exhibits high Faraday and Kerr rotation; thus the compound is considered to be a promising material for performing magnetic functions in memory and magneto-optical devices at liquid nitrogen temperature. Another similar europium compound, namely, the orthosilicate (Eu_2SiO_4), is transparent in the form of thin single crystals and also ferromagnetic. It shows a Verdet constant of -2.5 min/Oe · cm at 20°C, which is almost an order of magnitude greater than those previously reported for other transparent materials (McGuire et al. [1964]). Thirdly, the Eu^{2+} chalcogenides are unusual ferromagnets from the scientific viewpoint because their ferromagnetism exists in a simple, NaCl-type crystal structure with the magnetic moments unquestionably localized in the lattice of pure, ionic crystals. Thus, for the first time, we have a group of nearly ideal Heisenberg ferromagnets with which to conduct critical studies, both experimental and theoretical, to test the Heisenberg theory.

Among the four Eu^{2+} chalcogenides, EuO, EuS and EuSe become ferromagnetic at 69, 16.5 and 7°K, respectively, and EuTe becomes antiferromagnetic at 7.8°K. Their crystal structural and magnetic data are listed in table 5.14. The Eu^{2+} ion has a $4f^75S^25P^6$–$^8S_{\frac{7}{2}}$ configuration like Gd^{3+}. Its magnetic moment is expected to be $7\,\mu_B$/ion. Experimental values obtained for EuO, EuS and EuSe are about 6.7–6.8 μ_B/ion. The small discrepancy is attributed to a small fraction (1–2%) of the Eu ions in the trivalent state. For the Eu^{3+} ion, the electronic configuration is characterized by the $4f^65S^25P^6$ distribution and a $^7F_0$ ground-state term with zero total angular momentum. Hence, the presence of trivalent Eu ions will dilute the magnetization of the Eu^{2+} chalcogenides.

The most interesting feature of the data listed in table 5.14 is the sharp drop of the Curie or paramagnetic transition temperatures of the chalcogenides. To understand the physical significance of such magnetic behavior, first let us see how the molecular-field theory is applied here. Besides the reasons already cited, another reason that the Eu^{2+} chalcogenides are considered nearly ideal Heisenberg ferromagnets is the $^8S_{\frac{7}{2}}$ ground state of the cations, which is relatively insensitive to the crystalline field. In a NaCl structure, the magnetic lattice of the chalcogenides is face-centered cubic as shown in fig. 5.16. Each Eu^{2+} ion has 12 nearest neighbors (nn) of Eu^{2+} ions at a distance of $a/\sqrt{2}$ and 6

TABLE 5.14

Crystal structure and magnetic data for divalent europium chalcogenides (Shafer and McGuire [1968]).

Compound	Crystal structure	a (Å)	$\theta_C(\theta_N)$ (°K)	Θ (°K)	Molar Curie constant	Saturation moment observed per Eu^{2+} ion at 0°K (μ_B)
EuO	fcc (NaCl-type)	5.14	69	76	7.60	6.80
EuS	fcc (NaCl-type)	5.96	16.5	19	7.80	6.87
EuSe	fcc (NaCl-type)	6.19	7	9	7.33	6.70
EuTe	fcc (NaCl-type)	6.60	(7.8)	−6	7.04	

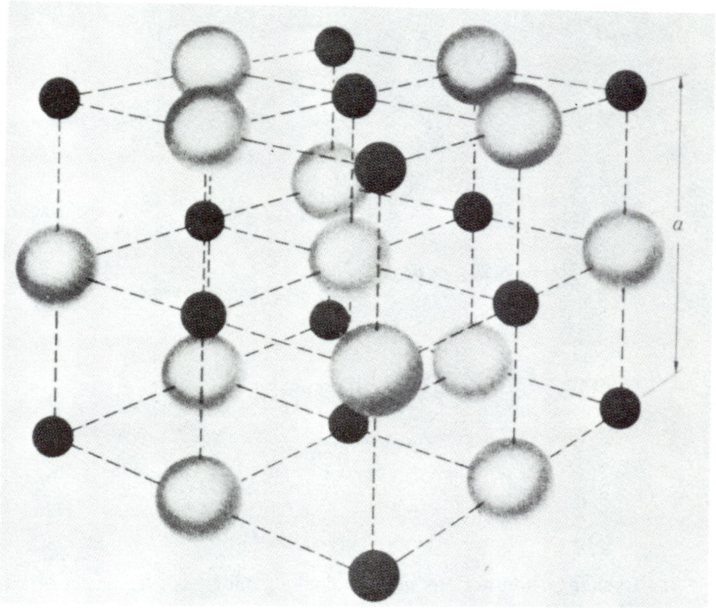

Fig. 5.16. A unit cell of the sodium chloride crystal structure of europium chalcogenides. The space lattice is fcc, and the basis has one Eu^{2+} ion at (000) and one anion at $(\frac{1}{2}\frac{1}{2}\frac{1}{2})$ (Kittel [1976]).

next nearest neighbors (nnn) of Eu^{2+} ions at a distance of a. The nnn Eu^{2+}–Eu^{2+} pairing, however, is intervened by an O^{2-} anion. On the basis of the Heisenberg theory and the molecular-field approximation, McGuire et al. [1963, 1964] obtained the relations

$$\Theta = \tfrac{2}{3}S(S+1)(12J_1 + 6J_2) = 126J_1 + 63J_2, \tag{5.28}$$

$$\theta_N = \tfrac{2}{3}S(S+1)(-6J_2) = -63J_2, \tag{5.29}$$

where Θ is the paramagnetic Curie or Néel temperature and θ_N is the Néel temperature. In these relations, J_1 and J_2 are the exchange interactions between nn and nnn Eu^{2+} ions, respectively. The dependence of the observed values for Θ upon a is shown by the solid curve in fig. 5.17. This experimental curve is discernible with respect to (5.28) if J_1 and J_2 are assumed to be positive and negative, respectively. Two hypothetical curves can then be drawn to represent $126J_1$ and $-63J_2$, which together would give the solid curve. This interpretation

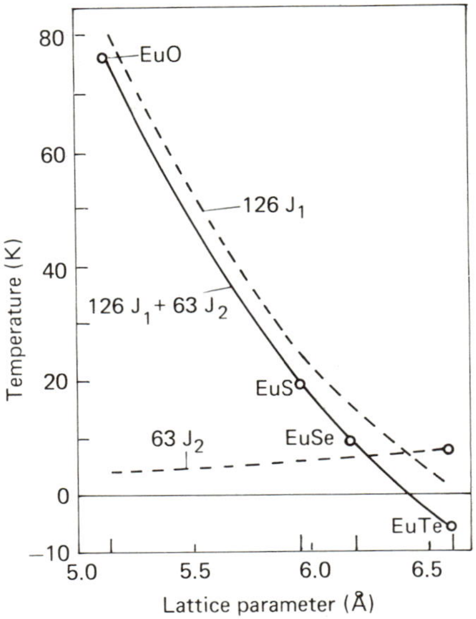

Fig. 5.17. Exchange interactions in Eu^{2+} chalcogenides as a function of lattice parameter. The solid curve is deduced from experimental points; the broken curves are based on assumptions derived from molecular field theory (McGuire and Shafer [1964]).

further implies that (1) $|J_2| \ll J_1$ and (2) J_1 is strongly dependent upon the nn separation, but J_2 is insensitive to the nnn separation. Experimental studies using different techniques, such as measurements of magnetization, specific heat, and nuclear magnetic and electron paramagnetic resonances gave strong support to the interpretation. For EuO, the best deduced values for J_1/k and J_2/k are $+0.20 \pm 0.10$°K and -0.08 ± 0.02°K, respectively, where k is Boltzmann's constant.

It is generally agreed that the negative J_2 in the Eu^{2+}–O^{2}–Eu^{2+} interaction along the cube edge is due to the super-exchange mechanism previously discussed (ch. 2, §4) for the case of antiferromagnetic MnO. The mechanism leading to $J_1 > 0$ has not been clearly understood, however. The direct exchange mechanism proposed in the original Heisenberg theory is unlikely to be operative in the chalcogenides for the same argument used for the lanthanide metals (p. 179); namely, the cation core size is too small in comparison with the nn distance. According to Smit [1966], J_1 could be positive if there is an excitation of the core electrons to outer (5d) orbits as a result of zero-point lattice vibration. Such an electronic excitation could lead to direct orbital overlap, thus making the direct exchange possible. Smit justified his proposal by citing the large decrease in the value of J_1 with increasing nn separation.

Another interesting, as well as practical, aspect of the ferromagnetic chalcogenides is the ready doping of these compounds with certain elements or compounds. The purpose of doping is to introduce new cations of different (M^{1+} or M^{3+}) valencies so that the doped chalcogenides will have either a deficiency or a surplus of electrons. For example, doping EuO or EuS with another rare earth will introduce trivalent ions, each of which will donate one electron in excess of the two electrons donated by a Eu^{2+} ion. Hence, the rare-earth-doped EuO and EuS will have free electrons wandering in the otherwise ionic lattices. Indeed, Heikes and Chen [1964] observed a huge (up to seven orders of magnitude) increase in electrical conductivity and a small rise ($\sim 10°$) of the paramagnetic Curie temperature in the slightly doped EuS by La ($La_{0.05}Eu_{0.95}S$). These changes are taken as evidence for the formation of a magnetization-dependent impurity band by the excess electrons.

The effect of doping with other rare earth compounds on the magnetic behavior of the chalcogenides was investigated even more actively because of the interest of the computer industry to increase the Curie temperatures of the compounds for possible applications. Dopants tried include gadolinium sulfide and selenide and several rare earth sesquioxides (RE_2O_3). According to Shafer and McGuire [1968], the RE_2O_3 doping yielded the most promising results. For instance, melting 5 mole % Gd_2O_3, 10 mole % Eu and 85 mole % EuO under europium pressures up to ≈ 20 atm in a sealed molybdenum crucible at $\approx 2000°C$ produced small crystals of the composition $Eu_{0.966}Gd_{0.034}O$. This compound showed θ_C at 135°K and $M_s \sim 170\ (A \cdot m^{-1}) \cdot (m^3 \cdot kg^{-1})$ at 77°K. Similar results were obtained from doping with La_2O_3 and Ho_2O_3, as shown in table 5.15. Eu is used in the doping process to prevent the dissociation of EuO and to enable the trivalent rare-earth ions to be accepted as a substitute for Eu^{2+} in the EuO lattice via the reaction $Eu + RE_2O_3 \rightarrow 3(EuRE)O$. The large increase in θ_C is attributed to an effective enhancement of the direct exchange interaction by the excess electrons. Meanwhile the presence of these electrons sharply lowers the resistivity of the doped EuO from $\sim 10^5$ to $\sim 1\ \Omega \cdot m^{-1}$ at room temperature.

TABLE 5.15

Results of doping EuO with RE_2O_3 and Eu (Shafer and McGuire [1968])

Doping cation	Starting composition			Final dopant concentration (mole %)	θ_C (°K)	Θ (°K)	Electrical resistivity at 20°C ($\Omega \cdot m^{-1}$)
	RE_2O_3	Eu	EuO				
La^{3+}	5	10	85	0.4 ± 0.2	132	133	6
Gd^{3+}	5	10	85	3.4 ± 0.5	135	137	2
	10	5	85	7.8 ± 0.3		136	
Ho^{3+}	5	10	85	5.0 ± 0.2	134	136	4
	8	10	82			135	2

§2. THE EFFECT OF METALLURGICAL FACTORS ON PROPERTIES OF SOFT MAGNETIC MATERIALS

We have dealt with the major magnetic properties in chapter 4 and with the classification of soft magnetic materials in the preceding section. Now we proceed to show how various metallurgical factors affect the magnetic and other application-related properties of soft magnetic materials. To demonstrate the importance of this discussion, let us recall that the magnetic materials are seldom used in the elemental form and that the magnetic properties that dominate the applications are most likely structure sensitive. Therefore, great opportunities exist for metallurgists to apply their knowledge and experience to the design, preparation and treatments of the materials to best meet the requirements for specific magnetic performance.

From the metallurgical standpoint, the quality or characterization of a material is not just composition and purity. It also means the state of crystal defects, the grain structure, the state of stress and strain and, with concentrated solid solutions, the degree of chemical order. Accordingly, this section is divided into six parts: (1) purity, (2) alloying, (3) crystal defects, (4) recovery and recrystallization; texture, (5) chemical order and (6) magnetic anneal. In each subsection the main metallurgical factor will be described, followed by a discussion of its effects on the magnetic properties.

§2.1. *Purity*

Traditionally, the purity required for the starting materials of an industrial product is at the level of 99.5–99.9% (see Bozorth [1951f]). The advancement of technology in the past two decades, however, has changed this tradition in two respects. First, the development of modern techniques for melting and purification has made it economically feasible to raise the standard purity to better than 99.9% even for materials consumed in tonnage quantities. Secondly, for electronic and computer components requiring thin films and single crystals, for which the quantity of the magnetic material is small but the quality of the material must be superior, it is possible to maintain a purity of 99.99% or greater.

For magnetic materials used on a massive scale, it may be argued that the gain by raising the purity from, say, 99.5% to 99.95% may not justify the increase in cost. As far as the intrinsic magnetic properties are concerned, the argument has some merit. Indeed the magnetic gain is small, but other considerations must not be ignored. The nature and seriousness of any adverse effect that impurities might exert on the magnetic material depend on the identity and concentration of the impurities. In general, substitutional impurities such as Si, V, Cu and Sn in Fe up to a few percent will not reduce the saturation magnetization more than the amount corresponding to simple dilution. Hence, the presence of 1% substitutional impurities is acceptable in most cases. However, if the impurities are interstitial, such as H, C, N and O, then one must be careful in accepting impurities of even 0.1%. In the first place, the percentages quoted by engineers and analytical chemists are usually based on weight. Because of the large differences between the atomic weights of the interstitial elements and the iron-group magnetic elements, one ppm by weight is equivalent to 3.5–58.5 ppm by atom, as shown in table 5.16. Suppose that the impurity of a 99.7% Fe is essentially oxygen. This iron actually contains 1.04 atomic % O, which is substantial. Moreover, as shown in table 5.17, the solid solubilities of the interstitial elements in Fe, Co and Ni are often highly limited, particularly at room and moderate temperatures. When the contents of the interstitial impurities exceed the solubility limits, the impurities tend to precipitate out in particle form of hydrides, carbides, nitrides and oxides at the expense of the host metal. Under these circumstances, the presence of interstitial impurities in excessive quantities means a double loss – one represented by the percentage of the impurity itself and the other associated with the host metal lost in the precipitates. Let us again refer to the 99.7% Fe with ~0.3% O. According to Chaudron [1962], oxygen is almost insoluble (<10 atomic ppm) in zone-refined iron in the α-phase (up to 911°C) and in less pure grades of electrolytic iron below 500°C. Hence, it is reasonable to conclude that the 0.3% O

TABLE 5.16
Conversion of one part per million by weight to atomic ppm

Host metal (atomic weight)	Atomic ppm in equivalence to 1 wt. ppm of			
	H(1.008)	C(12.011)	H(14.008)	O(16.00)
Iron(55.85)	55.4	4.6	4.0	3.5
Cobalt(58.94)	58.5	4.9	4.2	3.7
Nickel(58.71)	58.2	4.9	4.2	3.7

will appear as precipitates of magnetite in the equilibrium state below 560°C (see Hansen [1958]), equivalent to 1.08 wt % in the form of Fe_3O_4 in this iron. As far as magnetization is concerned, the 99.7% Fe is effectively 98.9% pure. The presence of Fe_3O_4 particles on the iron also will have a detrimental effect on structure-sensitive properties such as coercivity and permeabilities.

Within the solid solubility limits, interstitial impurities generally increase the electrical resistivity of metals markedly. A trace amount (10–100 atomic ppm) of carbon, oxygen and nitrogen in an otherwise ultrapure metal will increase the resistivity at 4.2°K dramatically. This effect is reflected in the research practice that the resistance ratio between values at 273°K or room temperature (~300°K) and 4.2°K serves as a measure of the purity of the sample. Although the correspondence of the resistance ratio to actual concentrations of impurities must be critically evaluated for individual cases, it is generally recognized that a ratio of 10 to 50 for an iron sample indicates the presence of a relatively large total concentration of interstitial impurities, perhaps in excess of 1000 atomic ppm, whereas a ratio of ≈2000 signifies the meagre presence of interstitial impurities, say, less than 20 atomic ppm. The increase in resistivity due to impurities, however, is somewhat overshadowed by the resistivity caused by phonon-induced scattering of electrons at and above room temperature. Nevertheless, the impurity-induced resistivity is considered beneficial to a magnetic material with respect to the eddy-current loss.

Interstitial impurities could exert similarly pronounced effects on the mechanical behavior of metals and alloys. The mechanical effects are more variant in nature and desirability, however. Strengthening by the interstitial atoms in metals is beneficial, of course. A case in point is plain-carbon steel for which carbon is the principal strengthening agent. Adding 0.40 wt.% C to iron, for instance, will increase the yield strength at room temperature to $4.83 \times 10^8\ N \cdot m^{-2}$ (70 000 psi) from $2.41 \times 10^8\ N \cdot m^{-2}$ (35 000 psi) measured for electrolytic iron. Another example to illustrate the strengthening power of carbon is found in the high-strength structural steels having the bainitic structure. An empirical formula was suggested (Pickering [1964]) to predict the tensile strength, in units of tons per in^2, in terms of weight percentage of various alloying elements: T.S. = 16 + 125(% C) + 15(% Mn + Cr) + 12(% Mo) + 8(% Ni) + 6(% W) + 4(% Cu) + 25(% V + Ti). The carbon concentration in these steels is kept below 0.2 wt %. Clearly carbon is the most effective element to strengthen the bainitic steels.

TABLE

Solid solubilities of hydrogen, carbon, nitrogen

Metal	Solubility (wt.%) at temperatures: Hydrogen		Carbon
Fe	(500°C) 0.0001 (900°C) 0.00025 (1 000°C) 0.00045 (1 200°C) 0.00075 (1 545°C) 0.0012	α-Fe	(444°C) 0.0022 (500°C) 0.0036 (568°C) 0.0068 (660°C) 0.0130 (723°C) 0.020
		γ-Fe	max. solubility ≈ 2.03% at 1147°C, progressively decreasing to 0.15% at 1493°C and to 0.80% at 723°C
Co	(800°C) 0.00017* (1 200°C) 0.00048*		(220°C) 0 (1 000°C) ≈ 0.33 (1 309°C) ≈ 1
Ni	(27°C) 0.000014 (500°C) 0.00015 (900°C) 0.00029 (1 000°C) 0.0012		(700°C) 0.08 (745°C) 0.1 (925°C) 0.2 (1 055°C) 0.3 (1 177°C) 0.4 (1 265°C) 0.5

*Reference: Smith [1948].

Interstitial impurities may cause serious problems of embrittlement in metals and alloys. This adverse effect is particularly prevalent with iron. Three notorious cases may be cited in which iron and steels show sensitive susceptibility to brittle fracture: (1) hydrogen embrittlement, (2) the promotion of cleavage cracking due to dislocation pinning by carbon, oxygen and nitrogen atoms and (3) temper and blue brittleness

5.17
and oxygen in iron, cobalt and nickel

indicated in parentheses, of

Nitrogen	Oxygen
(590°C) 0.1*	γ-Fe (1 345°C) ≤ 0.001
(*Maximum solubility in α-Fe, progressively decreasing to zero solubility at 910°C and at ≈ 400°C)	γ-Fe (950°C) < 0.007
	α-Fe (700°C) 0.008
Nitrogen is practically insoluble in Co up to 1200°C	(600°C) 0.006
	(700°C) 0.009
	(810°C) 0.016
	(875°C) 0.010
	(945°C) 0.007
	(1 000°C) 0.008
	(1 200°C) 0.013
Nitrogen is insoluble in Ni up to 1 400°C	(600°C) 0.020
	(800°C) 0.019
	(1 000°C) 0.014
	(1 200°C) 0.012

caused by carbides. Also worth mentioning in this connection is the hot shortness caused by sulfur.

In view of the pronounced and varying effects of impurities on electrical and mechanical properties of metals just discussed, the best policy when preparing soft magnetic materials is to start with the purest raw materials available and permissible from the economical standpoint. In this way, the choice and concentrations of impurities can be controlled according to the requirements of the application. For example, carbon is preferred and hydrogen should be avoided for the strengthening of iron. To increase the resistivity of iron, nitrogen is superior to oxygen by virtue of the extensive solid solubility of nitrogen in iron. Listed in table 5.18 are the impurities of several grades of iron available as starting materials.

TABLE 5.18
Impurities of several grades of iron available as starting materials for the preparation of magnetic alloys*

Impurity element	Impurity content (wt. ppm)			
	Ingot iron	Electrolytic iron Grade B	Electrolytic iron Grade A	Spectroscopic grade iron
O	900	100	100	100
N	10	100	100	100
H	–	100	100	
C	200	50	40	300
S	150	50	30	
P	100	30	20	ND**
Si	400	50	30	7
Mn	500	20	15	3
Mg	–		<5	2
Ni	100	300	10	2
Cu	700	50	40	1
Cr	100	20	10	ND
Co		70	10	ND
Mo	<100	20	<40	ND
Al		20	<10	ND
V	<100	30	<10	ND
Pb			<10	ND
Ta			<10	ND
Sn		50	<30	ND
W			<10	ND
Zn			<10	ND
Zr			<10	ND

*Reference: Moore [1968].
**ND: not detected.

Many techniques have been developed for the preparation of high-purity metals. Among them, the vacuum-melting and zone-refining techniques are particularly effective and suitable for Fe, Co and Ni. The principles and main features of these two techniques are described below. For information on other methods, such as vacuum arc melting and electroflux remelting processes, the book edited by Bunshah [1968] should be consulted.

Vacuum melting. Melting of a metal in vacuum serves two special purposes: to remove the normal atmosphere surrounding the metal, so that there will be no possibility for the molten metal to be contaminated by oxygen and nitrogen, and to remove certain impurities from the metal through one of the following processes:

(1) Dissociation – Some of the compounds of nitrogen (e.g., Fe_4N and Fe_2N), carbon (e.g., Fe_3C, Co_3C and Ni_3C) and oxygen (e.g., Fe_2O_3 and Fe_3O_4) are unstable and tend to be dissociated at elevated temperatures. Once the compounds are dissociated, the gaseous elements will be pumped out of the melt.
(2) Deoxidation – Both carbon and hydrogen are ideal deoxidizers because they form gaseous deoxidation products through the reactions:

$$2C + O_2 \rightarrow 2CO\uparrow, \tag{5.30}$$

$$2H_2 + O_2 \rightarrow 2H_2O\uparrow. \tag{5.31}$$

The CO and water vapor are continuously pumped away. Because carbon is a more potent deoxidizer than hydrogen, it is more commonly used. For small-scale melting, a combination of carbon and hydrogen is used in the following manner: Carbon is added to the initial charge. Melting proceeds under a pressure of about 5–10 torr (1 torr = 133.322 $N \cdot m^{-2}$ or 1 mm Hg or 1/760 normal atmospheric pressure) of hydrogen, which is continuously pumped through the chamber. When melting is completed, hydrogen is slowly pumped out of the chamber, and eventually the chamber is under vacuum. The carbon–hydrogen deoxidation process offers the added advantage of subduing carbon-boiling and -splattering, which could result in a substantial loss of the metal. Solid deoxidizers are not used in vacuum melting because the slag formed cannot be easily removed.
(3) Degassing – The principle underlying the degassing process is Sievert's law, which states that the amount of a gas dissolved in a melt is proportional to the square root of the partial pressure of the gas

above the metal. Thus the use of vacuum in the chamber will result in lower equilibrium concentrations of various gaseous impurities in the molten metal. Kinetically, however, the rate at which the gaseous impurity is removed depends upon the diffusivity of the impurity atoms in the particular liquid metal. Thus, hydrogen will be removed much faster than nitrogen. To aid in the removal of nitrogen, argon is bubbled through the molten bath via a perforated ceramic tube.

(4) Vaporization – Metallic impurities with high vapor pressures, such as zinc, magnesium, calcium and lithium, can be removed by the pumping action. The extent to which a metal will be evaporated from the melt depends upon pressure, temperature and time. For a metallic impurity in solution, the vapor pressure is given by multiplying the vapor pressure of the pure metal by its activity or, as an approximation, by its mole fraction. According to Nesmeyanov [1953], vapor pressures of elements are given approximately by the semi-empirical formula $P = AT^{\alpha} e^{B/T}$, where P is in torr, T is in degrees K and A, B and α are constants. For solid Fe and Zn, $A = 7.76 \times 10^{16}$ and 1.74×10^{11}, $B = -48\,547$ and $-15\,776$ and $\alpha = -2.14$ and -0.755, respectively. If we assume that the melting of iron occurs at its triple point of 1810°K and that the iron contains 0.05 wt.% or 0.00043 mole fraction of Zn, the vapor pressures of Fe and Zn are 0.0187 and 42.6 torr, respectively, or at the ratio of 1:2278.

The effectiveness of the vacuum-melting technique in removing the gaseous elements (O, N and H) and sulfur is shown in table 5.19. Typical contents of these impurities before and after a vacuum-induction melting in iron, cobalt and nickel are compared. Also the use of vacuum melting in the preparation of 2%V–FeCo is partly responsible for the vast improvement of permeabilities and other structure-sensitive magnetic properties of Supermendur over those of Permendur.

The equipment required for vacuum melting consists of a vacuum-tight chamber for housing the crucible and the pouring mechanism, a pumping system and necessary melting facilities. The pumping system should be capable of pumping down the chamber to 10^{-3} torr or less during the melting. For small jobs, the system contains a mechanical pump and an oil diffusion pump. Larger pumping systems may contain a bank of mechanical pumps connected in parallel, an oil ejection pump or a mechanical booster pump inserted between the roughing pump and the oil diffusion pump. The crucibles used for melting Fe, Co, Ni and their alloys are made of magnesia, alumina and (or) zirconia. In some cases, it is advantageous to remelt the metal in the same crucible

TABLE 5.19
Typical impurity contents of iron, cobalt and nickel before and after vacuum-induction melting*

Metal	Impurity content (wt. ppm)				
	Oxygen	Nitrogen	Hydrogen	Carbon	Sulfur
Before vacuum-induction melting					
Iron(ingot)	900	10	$\leqslant 10$	200	150
Iron(electrolytic)	100	100	100	50	50
Cobalt(electrolytic)	70	10	8		
Nickel(carbonyl)	30	10	$\leqslant 100$		
Nickel(270)	$\leqslant 100$	$\leqslant 100$	$\leqslant 100$		
After vacuum-induction melting					
Iron	20	2	–	< 10	30
Cobalt	50	–	–	50	30
Nickel	10	1	–	< 10	30

*Reference: Moore [1968].

several times. Repeated melting of iron in a new magnesia crucible in vacuum, for instance, results in progressive decreases of the content of the gaseous impurities as seen from the chemical analysis data shown in table 5.20. Because no slag is formed in vacuum melting and because there is no really leakproof or trouble-free stopper for bottom pouring, it is more convenient to use a tilt-pouring arrangement. Other descrip-

TABLE 5.20
Chemical analysis data for gaseous impurities in iron showing the benefit of repeated melting in the same crucible

Heat No.	Crucible condition	Analysis (wt. ppm)			
		Oxygen	Nitrogen	Hydrogen	Carbon
1	New	66	20	1	< 10
2	2nd heat	30	3	1	< 10
3	3rd heat	21	1	0	< 10
4	4th heat	10	1	0	< 10

tions of the vacuum-melting equipment may be found in a paper by Moore [1968].

Melting in vacuum of the iron-group metals and alloys is best effected by induction heating or an electron beam. Although other methods, such as arc melting, should not be ruled out, their applications are of secondary importance as far as purification is concerned.

Induction heating is ideally suited for vacuum melting because the electromagnetic force set up in the molten metal produces a stirring action. This stirring action will continuously expose new surfaces of the molten metal to the action of pumping, thus promoting the degassing process. Also, it will help homogenize the liquid mixture when an alloy is being prepared. The frequency of the alternating current flowing through the coil will determine the depth of penetration of the induced current and, thus, the part of the charge to be heated. Lower frequencies provide not only deeper penetration but also more vigorous stirring. Frequencies up to 10 000 Hz are useful in vacuum induction melting and are obtained conveniently from motor-generator sets. Typically, for 45–225 kg (100–500 lb) capacity furnaces, a motor-generator set of 150–200 kW at 400 V and 1920 Hz is used, whereas for a 910–1360 kg (2000–3000 lb) furnace, the power supply would be about 500–700 kW at 800 V and 960 Hz.

The use of energetic electrons in a focused beam to melt metals was first tried on tantalum by von Pirani at Siemens in 1905. The method remained obscure, however, until the late 1950's, when the strong demand to melt and purify reactive (Be, Ti, Zr and Hf) and refractory (W, Ta, Mo, Nb, Cr and V) metals pushed the development of the techniqũe to near completion. Today, electron beam melting is a proven technique for melting and purifying many of the above mentioned metals on a commercial scale. The versatility of the technique is reflected in (1) the availability of a choice between two types of electron gun, the thermionic emission type and the plasma-beam type, and (2) the wide range of applications, including melting, refining, annealing, welding, distillation and machining. The principles and practice of electron-beam melting are described in the book edited by Bakish [1962] and hence will not be discussed here. Some of the advantages cited by Bunshah and Cocca [1968] are:

(1) The beam intensity, focus, shape and position are controllable precisely. Thus the heat can be applied to any defined area on the charge.

(2) Unlike gas flames and electric arcs, no medium is needed to transfer the energy from the source to the charge. In fact, to prevent scattering of electrons and to prolong the life of the filament, a medium-quality ($<10^{-4}$ torr) vacuum is required for the operation of the thermionic emission-type gun and for heating, thus almost automatically providing a reasonably clean environment.
(3) If any crucibles are used, they are almost always made from water-cooled copper, thus avoiding any contamination from the container.
(4) The conversion of electron energy to heat is efficient because the losses due to scattering, generation of secondary electrons and production of X-rays are very small at the usual operating voltages, which range from a few kV up to 150 kV.
(5) Power can be applied and controlled independently of melting rate, thus allowing melting and superheating of the material at a rate consistent with its impurity content to achieve optimum purification.

The effectiveness in removing the interstitial impurities in Fe, Co and Ni by electron-beam melting is demonstrated by the analytic data compiled in table 5.21.

TABLE 5.21
Removal of interstitial impurities in iron, cobalt and nickel by electron-beam (eb) melting*

Metal (condition)	Impurity content (wt. ppm)			
	Oxygen	Nitrogen	Hydrogen	Carbon
Electrolytic iron				
(Before eb melting)	110	10	5	10
(After eb melting)	10	9	0.8	10
Nickel-base alloy				
(Vacuum arc remelting)	180–220	80–110	1–10	
(eb melting)	10–20	15–25	1–2	
Cobalt				
(arc melting)	220	100	6	140
(eb melting-fast)	240	30	5	50
(eb melting-medium)	120	20	4	60
(eb melting-slow)	170		4	60

*Reference: Bunshah and Cocca [1968].

Zone refining. This technique was invented by Pfann [1952] for the purification of elemental semiconductors, silicon and germanium. Soon after his invention, both theoretical and experimental work demonstrated the applicability of the technique to numerous systems of metals and compounds under certain conditions. The principle of the zone-refining technique is originally based on a simple metallurgical phenomenon. Figure 5.18a and b show portions of the equilibrium diagrams of two binary systems near the component A. A liquid alloy of the average composition $A_{1-c_l}B_{c_l}$ will initially freeze out at a different composition $A_{1-c_s}B_{c_s}$ at the liquidus point. In fig. 5.18a, $c_s < c_l$ and in fig. 5.18b $c_s > c_l$; otherwise, c_l and c_s are $\ll 1$ in both cases, corresponding to dilute alloys or to the presence of an impurity element in a metal. We define the distribution or partition coefficient k as the ratio of the solute concentration in the solid to that in the bulk liquid; that is, $k = c_s/c_l$. The coefficient has its equilibrium value k_0 given by values for c_s and c_l taken from the equilibrium diagram. Figure 5.18a corresponds to the case $k_0 < 1$ and fig. 5.18b to $k_0 > 1$. Under normal rates of freezing, equilibrium conditions are not attained, and the effective value must be used for k. From the solute concentration versus distance plot shown in fig. 5.19, Burton et al. [1953] arrived at the following relation between k and k_0:

$$k = k_0/[k_0 + (1 - k_0)\,e^{-f\delta/D}], \qquad (5.32)$$

where f is the freezing velocity, D is the diffusion coefficient of the

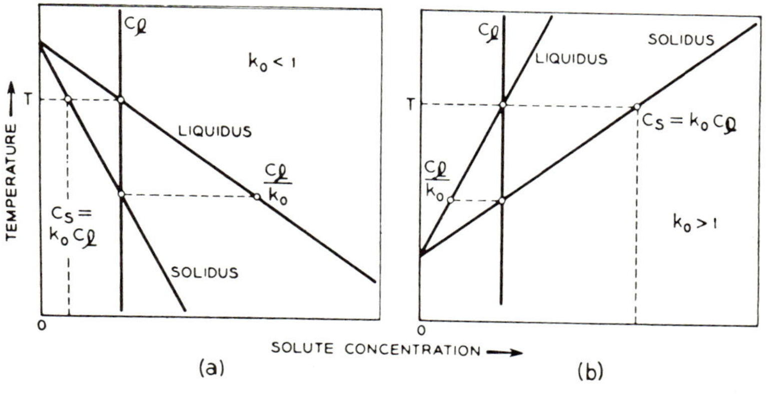

Fig. 5.18. Idealized parts of equilibrium diagrams in which the solute (a) lowers and (b) raises the melting point (Pfann [1966]).

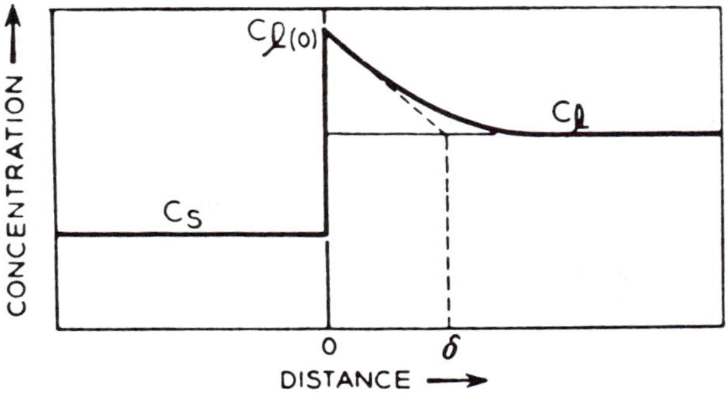

Fig. 5.19. Solute concentration near an advancing solid–liquid interface for a finite growth rate and for $k < 1$.

solute in the liquid and δ is the thickness of the enriched or depleted solute layer at the solid–liquid interface.

In the normal mode of freezing, the liquid is allowed to freeze from one end with only one solid–liquid interface (fig. 5.20a). Even in this

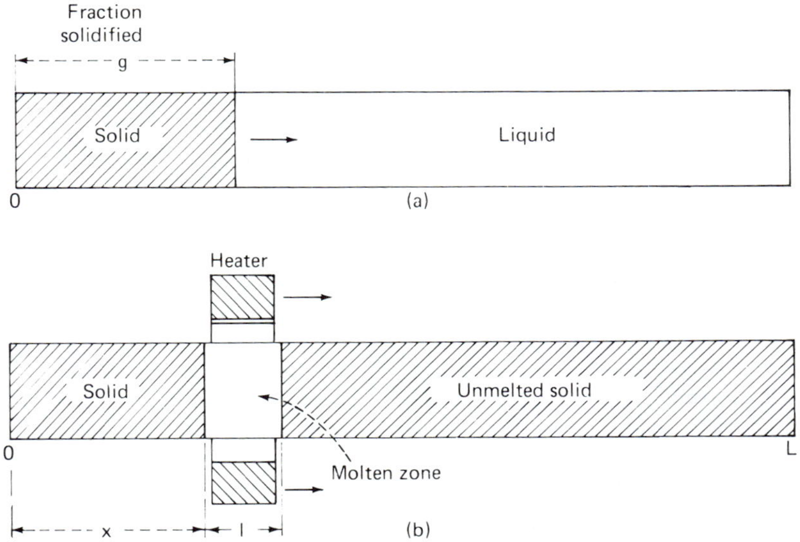

Fig. 5.20. (a) Solidification by normal freezing and (b) melting and freezing conditions in zone-refining (Pfann [1966]).

mode, as long as $k \neq 1$, there will be a segregation of the solute given by

$$c_s/c_0 = k(1-g)^{k-1}, \tag{5.33}$$

where c_0 is the initial average solute concentration in the system and g is the fraction of the bar that has solidified. The invention of Pfann was to melt a bar in a small section and move the molten zone slowly from one end to the other (see fig. 5.20b). Such a zone melting technique can enrich or deplete the solute more efficiently than can normal freezing. Pfann showed that a single pass of zone melting will produce solute segregation of the value

$$c_s/c_0 = 1-(1-k)\,e^{-kx/l}, \tag{5.34}$$

where x is the distance of the first solid–liquid interface and l is the zone length. Theoretical curves of the solute distribution as a function of x in a single pass are plotted for various k values in fig. 5.21. The impurity travels either with the zone or opposite to the zone, depending on whether $k<1$ or >1. The larger the deviation of the value for k from unity, the more efficient the zone refining. Other factors that improve the efficiency are (1) the number of passes n, (2) a small diffusion coefficient D of the solute, and (3) an optimum value of f.

Zone refining can be conducted in a number of ways with respect to the principal parameters:

(1) Position of the charge – either horizontal or vertical, depending upon whether or not (A) the liquid metal is allowed to be in contact with a container in connection with the possibility of contamination, and (B) the surface tension and the density of the liquid are such as to allow a floating zone to be formed without dripping or collapsing in the vertical position.
(2) Melting method – induction, resistance, or electron-beam heating.
(3) Atmosphere – vacuum, inert or reducing atmosphere.
(4) Moving part – the motion of the molten zone can be accomplished by moving either the heating element (fig. 5.20b) or the charge.
(5) Stirring – by keeping the thickness of the diffusion layer (δ in (5.32) and fig. 5.19) at a minimum; stirring of the liquid by induction heating or by mechanical or magnetic means was found helpful in increasing the efficiency of purification.

Zone-refining techniques have been applied to Fe, Co and Ni under different conditions with varying degrees of success. Data on selected experiments were compiled by Lawley [1968] and those for Fe, Co and

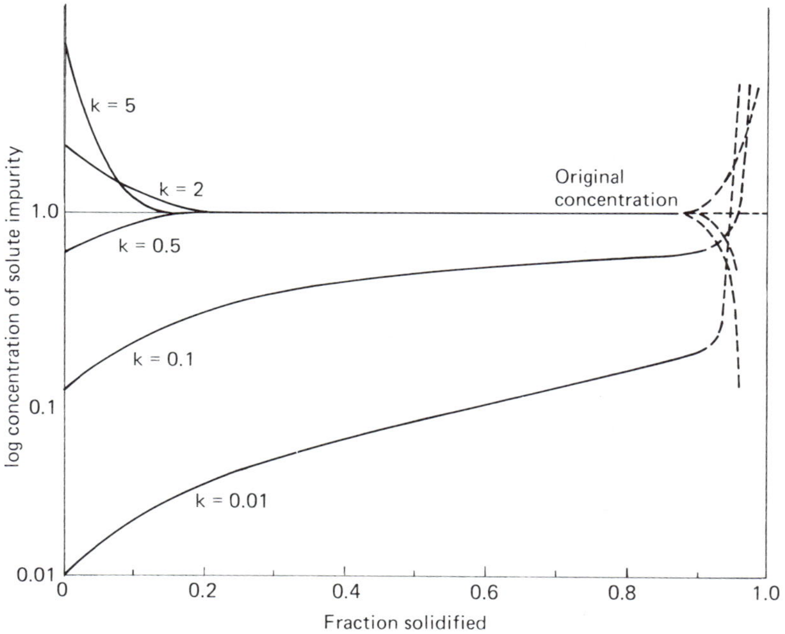

Fig. 5.21. Variation in solute distribution after one zone pass with the value of k indicated (Parr [1960]).

Ni are reproduced in table 5.22. The remarkably high efficiency of zone refining of Puron iron (>99.999% grade used as a standard in spectrographic analysis) is shown in table 5.23.

Finally, for iron and its dilute alloys used as soft magnetic materials, it has long been a standard practice to subject the final products to a hydrogen anneal. The treatment involves thorough heating in an atmosphere of wet, pure hydrogen above 1350°C, followed by slow cooling through the $\gamma \rightarrow \alpha$ phase transition. This treatment is effective in removing oxygen, carbon and nitrogen and, to some extent, sulfur. Drastic improvements were achieved in the hysteresis loss and permeabilities (Bozorth [1951h]). If hydrogen entrapped in the products is undesirable, vacuum should be used to replace the hydrogen atmosphere at the end of the anneal before cooling.

TABLE 5.22
Application of zone-refining techniques to ferromagnetic transitional metals (Lawley [1968])

Metal	Experimental details	Remarks	Ref.
Fe	Water-cooled hearth, induction heating, 4–8 in./h, 6 and 12 passes, gas atmosphere	Effective segregation of phosphorous ($P \approx 0.016$ ppm), requires ~ 2 times more zone passes than floating zone for same degree of purification	[1, 2]
	Ceramic boat, induction heating, argon atmosphere, up to ~ 12 in./h	Distribution of Cr, Si, P, Mn studied in iron and steel	[3]
	Floating zone, induction heating, gas atmosphere	Quantitative determination of segregation of several impurities with $k <$ and $k > 1$	[4–7]
	Levitation zone melting, wet hydrogen atmosphere	Studies of immobilization of C in Fe, zone leveling of B in Fe	[8–10]
	Electron beam floating zone, 0.64 cm diam bar, 0.2 in. zone length, 138 W for zoning	–	[11]
Ni	Water-cooled copper hearth, induction heating	No contamination from copper crucible	[1]
	Floating zone, induction heating, 2.25 in./h $\frac{3}{16}$ in. diam rod, vacuum $\sim 10^{-5}$ mm Hg, 4 passes	Purification by volatilization, $R_{4.2°K}/R_{273°K} \simeq 0.0014$	[12]
	Floating zone, induction heating	~ 20 ppm carbon, $R_{4.2°K}/R_{273°K} \simeq 0.0005$	[6, 13]
	Electron beam floating zone, vacuum $\sim 10^{-5}$ mm Hg	–	[8, 14–16]
Co	Vertical water-cooled copper crucible, induction heating	Sound, high-purity ingots	[1, 17]

[1] A. Berghezan and E. Bull-Simonsen, Trans. AIME, **221**, 1029 (1961).
[2] E. Bull-Simonsen, J. Iron Steel Inst., **200**, 193 (1962).
[3] W. A. Fischer and R. Uberoi, Arch. Eisenhuettenw., **33**, 661 (1962).
[4] T. Ooka, H. Mimura, S. Yano, and S. Soeda, Proc. Japan. Acad., **39**, 294 (1963).
[5] E. J. Koepel and B. Park, A New Method of Using the Zone Refining Technique, ASM Tech. Rept., p. 17-3-64 (1964).
[6] R. L. Smith and J. L. Rutherford, J. Metals, **9**, 478 (1957).
[7] W. M. Williams, G. B. Craig, and W. C. Winegard, Can. Mining Met. Bull., **55**, 35 (1962).
[8] B. F. Oliver and A. I. Shaler, Trans. AIME, **218**, 194 (1960).
[9] B. F. Oliver, Trans. AIME, **227**, 960 (1963).
[10] B. F. Oliver and F. Garofalo, Trans. AIME, **233**, 1318 (1965).
[11] D. K. Donald, Rev. Sci. Instr., **32**, 811 (1961).
[12] J. H. Wernick, D. Dorsi, and J. J. Byrnes, J. Electrochem. Soc., **106**, 245 (1959).
[13] V. J. Albano and R. R. Soden, J. Electrochem. Soc., **113**, 511 (1966).
[14] A. Calverley, M. Davis, and R. F. Lever, J. Sci. Instr., **34**, 142 (1957).
[15] H. G. Sell and W. H. Grimes, Rev. Sci. Instr., **35**, 64 (1964).
[16] R. Brownsword and J. P. G. Farr, J. Sci. Instr., **41**, 350 (1964).
[17] V. G. Epifanov and A. G. Lesnik, Akad. Nauk Ukr. SSR, **20**, 185 (1964).

TABLE 5.23
Purification of Puron iron by zone refining*

Element	Weight ppm Before zone refining	After 10 zone-refining passes
C	50	7
N	40	< 10
O	< 0.9	2
Ca	1	–
Co	10	–
Cr	50	–
Cu	800	1.5
Mn	10	0.5
Mo	100	–
Ni	1 500	0.5
P	1 000	< 0.1
S	30	0.2
Si	10	–
V	50	2
W	50	< 10

*Reference: Smith and Hendrickson [1962].

§2.2. *Alloying*

The important role played by alloying in the development of soft magnetic materials is reflected in the fact that rarely does a pure element perform satisfactorily in all respects in a magnetic device. The simple reason is that pure elements do not possess all the desired properties, both magnetic and nonmagnetic, to meet the manifold requirements for industrial applications. Take iron as an example. This element probably comes closest to having the desired properties for applications. Certainly it has excellent saturation magnetization and Curie temperature, and its maximum relative permeability and coercive force could be made spectacularly attractive (250 000 and 3.2 $A \cdot m^{-1}$ (0.04 Oe), respectively) by the famous hydrogen treatment developed by Cioffi et al. (Bozorth [1951g]). But its electrical resistivity at room temperature is not sufficiently high to keep the eddy-current loss low, and its corrosion resistance is notoriously poor. Thus even for iron, there is ample room for property improvement by alloying and other metallurgical means.

Because high values of saturation magnetization are almost always appreciated by the applications engineers* and also satisfy the curiosity of scientists, it is not surprising that considerable efforts were made in the early developmental work on magnetic alloys to seek a new record value for saturation magnetization at room temperature. Intensive searches led to the discovery of the highest mean saturation moment of 2.43 μ_B per atom in the 35% Co–Fe alloy at 20°C. The results of this and other searches are discussed in §1.2 with a summary in the form of Pauling–Slater curve shown in figs. 5.3 and 5.4.

Attempts to improve the structure-sensitive magnetic properties by alloying had been started earlier and with greater intensity. Significant achievements and "breakthrough" types of discoveries were made by many workers, notably Yensen on a variety of Fe–Co and Fe–Ni alloys at Westinghouse, Elmen on Permalloy and Permendur at Bell Laboratories, and Snoek and Gorter on ferrites at Philips Laboratories (Bechtold and Wiener [1965]).

Since 1934 (see discussion in ch. 6, §1), iron alloys with 2–4.5% Si have been used most extensively for electrical power generation, transmission and utilization, and their annual consumption is the largest among all soft magnetic materials. Undoubtedly the development of iron–silicon alloys has been the most significant achievement in this field. Ironically, Hadfield discovered the Fe–Si alloys by accident, and the first attractive property recorded was mechanical, not magnetic. After Hadfield's discovery, other outstanding properties of the alloys were detected in polycrystalline samples with random orientation, followed by the development of textures, first the (110)[001], then the (100)[001] preferred orientation, in Fe–Si sheets (Walter [1965]). The addition of Si to Fe improves not just the magnetic properties, but other properties as well. The prominent advantages of the Fe–Si alloys over pure Fe are the following:

(1) An improvement in magnetic softness. For instance, the 2–2.5% Si–Fe alloys show coercive forces that are about one-half that of iron.

*In a rare case connected with the magnetic bubble-domain technology (ch. 7, § 2), the value of M_s is required to be at a minimum in order to keep the so-called quality factor $Q = H_k/M_s$ (H_k being the anisotropy field) at values greater than unity for efficient operations of the bubble devices.

Other magnetic properties improved include the initial and maximum relative permeabilities and the hysteresis loss.
(2) Large increase in electrical resistivity. This effect is as important, if not more so, as the improvement in magnetic properties. At room temperature, the resistivity is increased from $1\times10^{-7}\ \Omega$-m ($10\ \mu\Omega$-cm) in pure iron to $\sim8\times10^{-7}\ \Omega$-m in the 6% Si–Fe at approximately a linear rate. Consequently, the eddy-current loss is proportionally reduced according to the relation, $P_e\sim1/\rho$, expressed in (4.88).
(3) The elimination of the face-centered cubic γ phase by the addition of Si. The partial Fe–Si phase diagram shown in fig. 5.22 indicates that Fe alloys containing 2.5% or more Si will not undergo the $\gamma\rightarrow\alpha$ phase transition. The absence of the transition would demand less critical heat treatments for the latter alloys.
(4) The strengthening of Fe by Si. The yield strength is reported to be increased to a maximum value of $5.1\times10^{8}\ \mathrm{N\cdot m^{-2}}$ (74 000 psi) in the

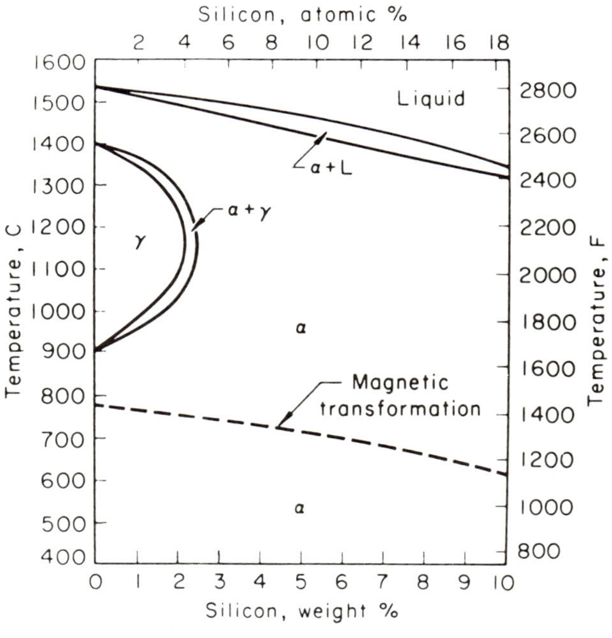

Fig. 5.22. Constitutional diagram of the iron–silicon system near the iron terminal (Stanley [1963]).

4.5% Si–Fe from 1.8×10^8 N · m^{-2} (26 000 psi) in pure iron. This amounts to $\sim 0.73 \times 10^8$ N · m^{-2} (10 700 psi) per weight percent of Si.
(5) A decrease in density at the rate of 0.813% per weight percent of Si. Hence, 5% Si–Fe has the density of 7.55×10^3 kg · m^{-3} (7.55 g · cm^{-3}) as compared with 7.87×10^3 kg · m^{-3} of pure iron at 20°C. This density change results in a reduction in weight of the machinery.
(6) An improvement of the corrosion resistance.

The addition of Si, however, has two adverse effects:

(1) It decreases the saturation magnetization at a rate corresponding to simple dilution. Also, correspondingly, it lowers the Curie temperature at an initial rate of 7.5° per weight percent of Si up to 2%, and at accelerating rates thereafter.
(2) The strengthening of Si is accompanied by an embrittlement when the Si content exceeds 2%. This effect becomes appreciable beyond 2.5%. For instance, 2.7% Si–Fe exhibits only 20% reduction in area in tensile tests at room temperature as compared with more than 80% recorded in pure iron. The industrial products of Fe–Si alloys containing more than 5% Si are described as totally brittle in impact or bending tests. This is an unfortunate situation because the 6.5% Si–Fe alloy displays zero longitudinal magnetostriction in the [100] direction, as shown in fig. 5.23. Had the brittleness of this alloy been less severe, its sheet products would have been an ideal material for the construction of low noise transformer cores.

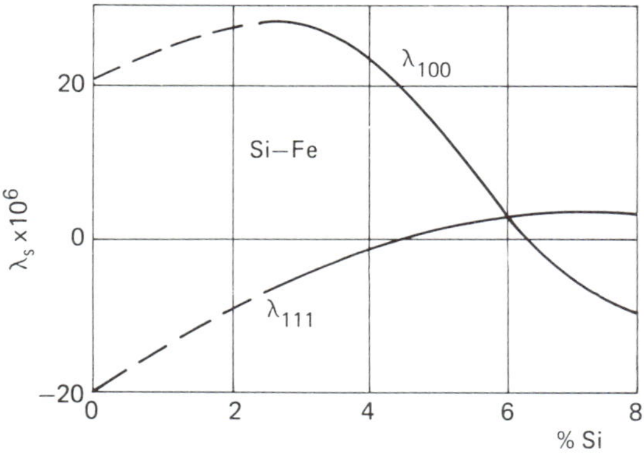

Fig. 5.23. Composition dependence of magnetostrictive coefficients at saturation in the [100] and [111] directions of iron–silicon single crystals (Brailsford [1966]).

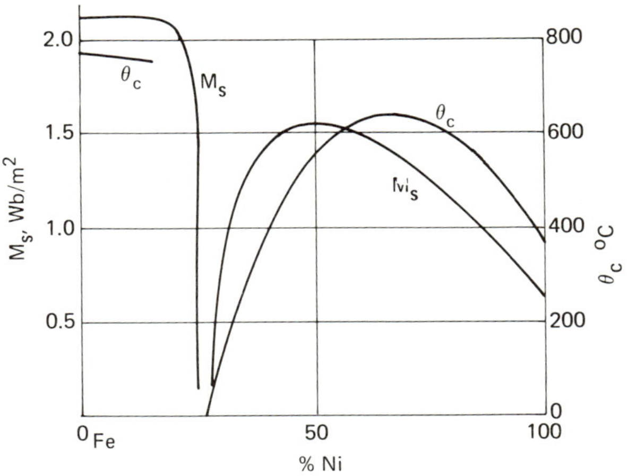

Fig. 5.24. Variation in saturation magnetization and the Curie temperature of alloys in the Fe–Ni system (Brailsford [1966]).

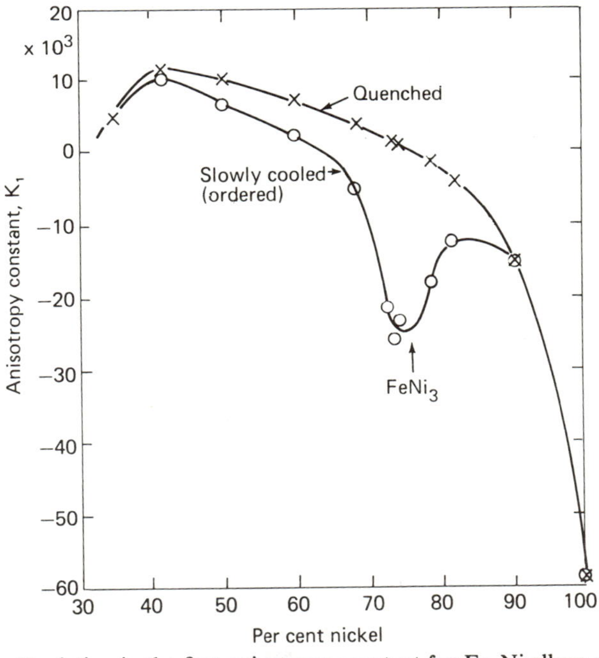

Fig. 5.25. Deviation in the first anisotropy constant for Fe–Ni alloys caused by heat treatment (Bozorth [1951]).

The advantages of alloying in the Fe–Ni alloys are somewhat different from those realized in the Fe–Si alloys. For the Fe–Ni system, alloying exerts pronounced effects on nearly all basic magnetic properties over the entire composition range. The dependence of various magnetic properties upon composition is shown for saturation magnetization and Curie temperature in fig. 5.24; for the first and second magnetocrystalline anisotropy constants (K_1 and K_2) in figs. 5.25 and 5.26; for the saturation longitudinal magnetostriction in three major directions ($\lambda_{s,[100]}$, $\lambda_{s,[110]}$ and $\lambda_{s,[111]}$) in fig. 5.27 and for the electrical resistivity in fig. 5.28. The property variations seen in these figures explain why the Fe–Ni system offers three attractive magnetic alloys

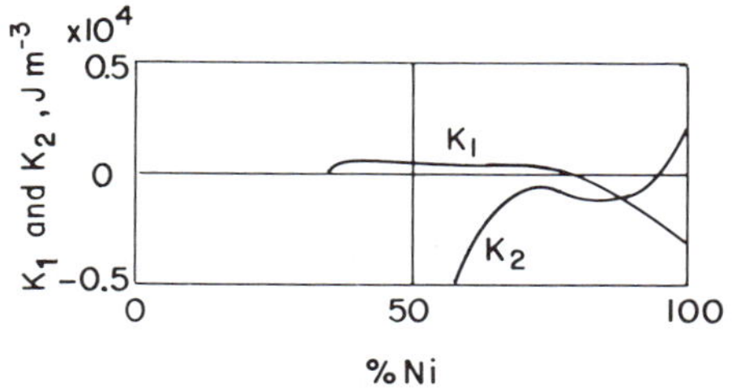

Fig. 5.26. Magnetocrystalline anisotropy constants for the Fe–Ni alloys (Brailsford [1966]).

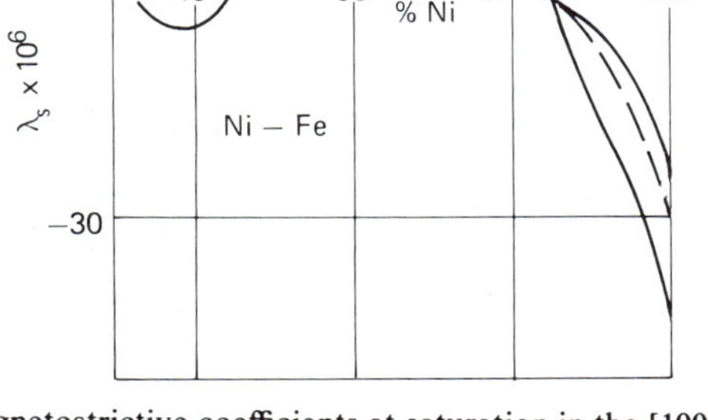

Fig. 5.27. Magnetostrictive coefficients at saturation in the [100], [110] and [111] directions of single crystals of Fe–Ni alloys (Brailsford [1966]).

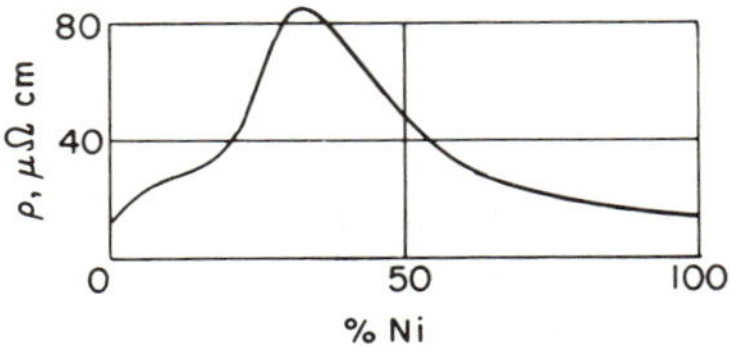

Fig. 5.28. Electrical resistivities of Fe–Ni alloys at room temperature (Brailsford [1966]).

known as Hipernik, Rhometal and Permalloy in the industry. The saturation magnetization of the face-centered cubic alloys between pure Ni and ~30% Ni–Fe reaches a maximum value at the equiatomic composition (fig. 5.24). This was the basis for the development of Hipernik by Yensen. A similar situation arose with the resistivity, which, too, displays a maximum at 36% Ni (fig. 5.28), thus leading to the development of Rhometal. The Fe–Ni system at this composition, however, shows another striking phenomenon: the mean coefficient of linear thermal expansion reverses its sign from a small negative value (-0.5×10^{-6} per °C) to a large positive value (20×10^{-6} per °C). The alloy containing 36% Ni with other minor solutes (Mn, Si and C) amounting to less than 1% has a coefficient of expansion so low that its length was almost invariant with temperature. Hence the alloy was named Invar, a famous Fe–Ni alloy of numerous applications (as absolute standards of length; in rod and tape form for geodetic work; in clocks, watches and automobile pistons and as thermostatic strips). The value of K_1 is shown to vanish at ~76% Ni (fig. 5.25), whereas all three λ_s vanish between 80 and 84% Ni (fig. 5.27). The zero value of K_1 and and the small values of λ_s are held responsible for the sudden jump of the permeability to a maximum at ≈78% Ni in the plots of μ_i and μ_{max} versus composition shown in figs. 5.29 and 5.30. The enormous values of μ_i and μ_{max} for the 78% Ni–Fe alloy motivated Elmen and Arnold to develop the 78 Permalloy, which is doubtless the most famous magnetic alloy of Ni.

After his initial work on the 78 Permalloy, Elmen further discovered that an addition of molybdenum of about 4% would increase the resistivity almost threefold (from $\sim2\times10^{-7}\ \Omega\cdot$m in 78.5% Ni–Fe to $5.7\times10^{-7}\ \Omega\cdot$m in 4% Mo–78.5% Ni–Fe) at the expense (~20% decrease) of saturation magnetization. The presence of Mo also simplifies the heat treatment necessary to obtain high initial permeability. Thus,

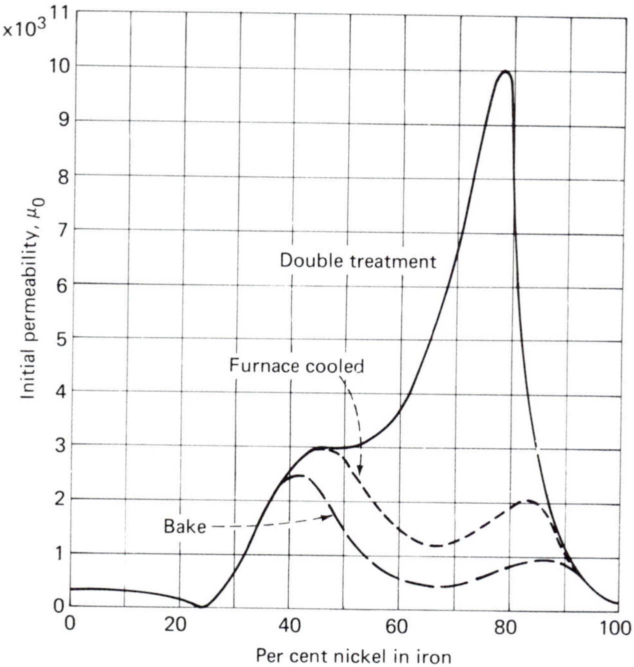

Fig. 5.29. Composition dependence of and effect of heat treatment on initial relative permeability of the Fe–Ni alloys (Bozorth [1951]).

Elmen introduced the 4-79 Molybdenum Permalloy to replace the 78 Permalloy. High permeability nickel-base alloys used today are based on the composition of the 4-79 Mo-Permalloy at approximately 4% Mo, 79% Ni and 15% Fe. We cite the series of developments of the high-permeability Ni-base alloys here to illustrate the important point that industrial products are seldom simple alloys, let alone pure metals. The alloys often contain more than three elements. The 4-79 Molybdenum Permalloy is a typical example. In addition to Mo, Fe and Ni, it may also contain 0.5% Mn for the improvement of ductility by desulfurization and possibly 0.2% Al for the purpose of deoxidation. If constancy of permeability with varying field strengths is desired, a small amount of Co is added to Permalloy.

The development of the Fe–Co alloys resembles that of the Fe–Ni alloys in many respects. For instance, the binary system of Fe and Co

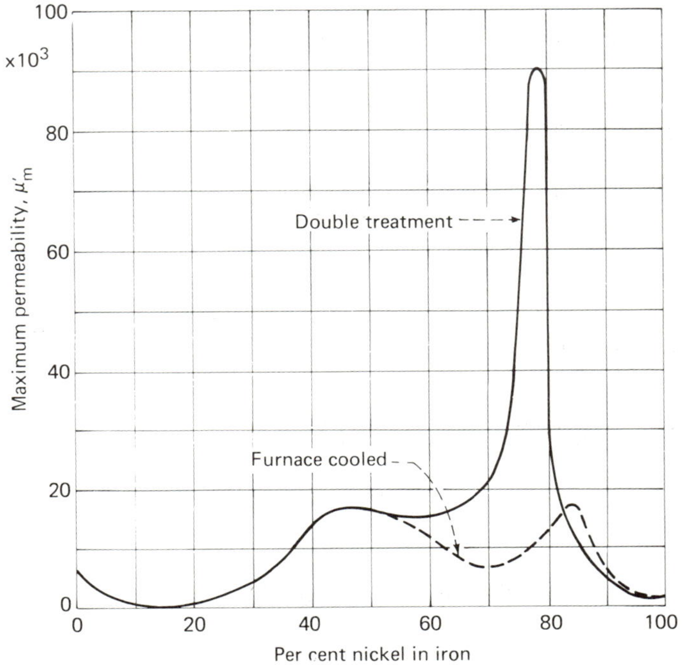

Fig. 5.30. Composition dependence of and effect of heat treatment on maximum relative permeability of the Fe–Ni alloys (Bozorth [1951]).

reaches a maximum value for M_s at 35% Co, thus leading to the development of Hiperco by Yensen (who also added 0.5% Cr for higher resistivity). Meanwhile, the system showed peak values for the initial and maximum permeabilities at the equiatomic composition (fig. 5.31). Hence, the alloy FeCo is in a similar situation to the 78 Permalloy in the Fe–Ni system – the large values of permeabilities, coupled with high saturation magnetization, present a great potential for applications. Unfortunately, FeCo undergoes the order–disorder transition, which cannot be suppressed by rapid cooling no matter how severely the alloy is quenched. The ordering of FeCo causes two serious problems in the attempt to convert the alloy into a useful magnetic material. One is the considerable reduction in resistivity and the other is the extreme brittleness. To solve the first problem, Elmen prepared a series of ternary alloys by doping FeCo with various elements (Ti, V, Cr, Mn,

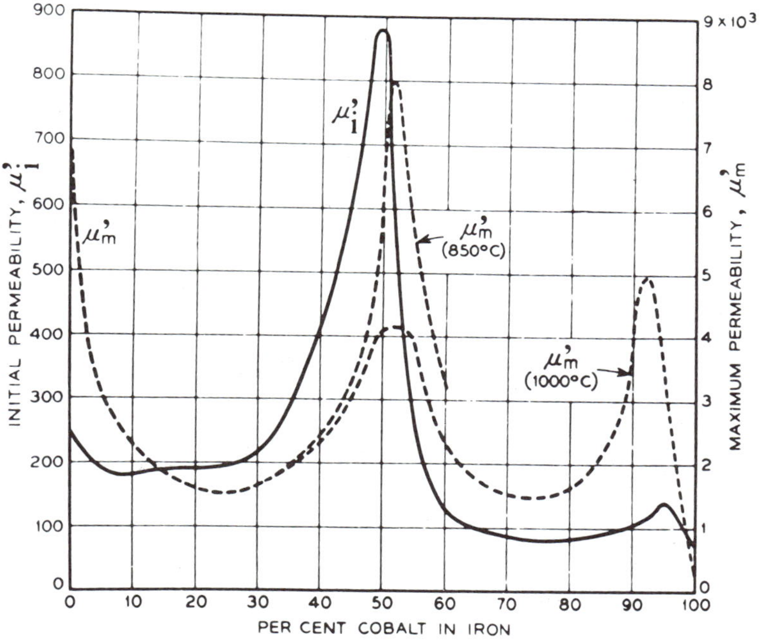

Fig. 5.31. Initial and maximum relative permeabilities of the Fe–Co alloys (Bozorth [1951]).

Ni, Cu, Cd and W) and compared their resistivities. His results are shown in fig. 5.32. All solutes tended to increase the resistivity of FeCo, but V showed the greatest efficiency beyond 0.5%. Moreover, the V-doped FeCo exhibited much improved fabricability for rolling in the quenched state. On the basis of these results, Elmen proposed the ternary alloy, 2% V–FeCo, to replace FeCo and called it Permendur to indicate the combination of good values of permeability and endurability, or, more appropriately, ductility. The reason for the improvement of mechanical properties will be discussed later.

The foregoing discussion of the Fe–Si, Fe–Ni and Fe–Co systems has demonstrated that alloying could indeed produce pronounced effects on various magnetic properties, both intrinsic and structure-sensitive, as well as on other application-related properties such as electrical resistivity and mechanical behavior. The discussion also implies that alloying effects are complex and difficult to predict. This is certainly true in

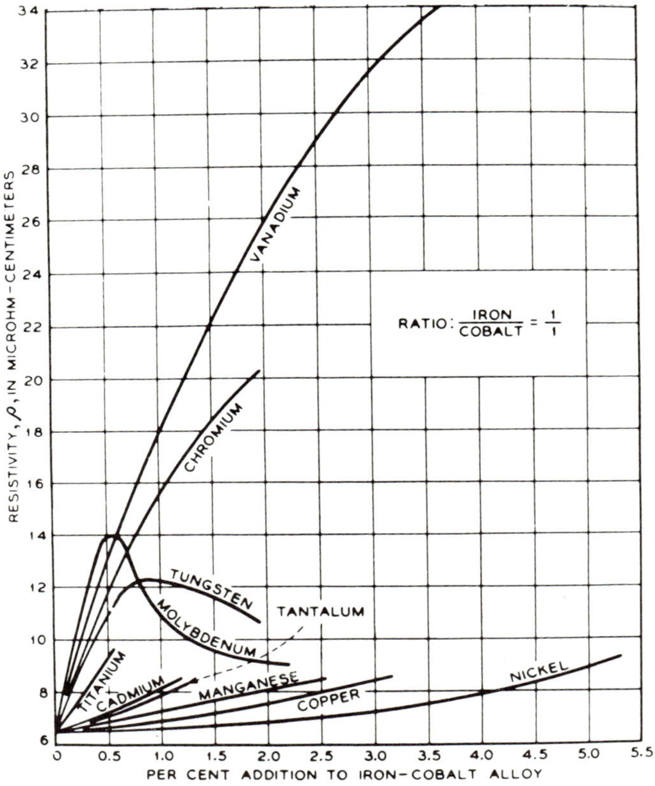

Fig. 5.32. Effect of various solutes on the room-temperature resistivity of the equiatomic Fe–Co alloy (Bozorth [1951]).

reference to figs. 5.23 to 5.31. Because of the difficulty in predicting the alloying effects, the traditional industrial approach to the development of a new alloy intended for the improvement of a particular property is to prepare many series of alloys, each of which features a solute element, and to evaluate the property of these alloys. After the evaluation, the sample that shows the best value of the property is chosen as the new alloy. This approach has produced some remarkable results, as in the case of Elmen's work on the Permendur. However, it is inevitably tedious and often fruitless. Although no reliable statistics are available, this kind of developmental work unquestionably earns a low score of success. It is perhaps not too harsh to state that thousands of dollars have been wasted in the field of magnetic materials over the years because of the failure of the trial-and-error approach to the development of new products.

In lieu of better methods and because our understanding of alloys is still very incomplete, the traditional approach to alloy development will continue to be practised by industry. Nevertheless, some good measures are now available for adoption to improve the chance of success and to avoid unnecessary wastes and mistakes. Three pertinent suggestions are offered here:

1. A huge body of information now exists in the literature concerning the alloying effects on properties. The data could be sorted out for their reliability and reproducibility, classified, and stored in an appropriate computer program. Then such information would be ready for retrieval in assisting the selection of solutes with respect to the objective of the development. The computerization of information on magnetic materials has not been done to date in the fashion just cited.* Hopefully, large laboratories, either industry- or government-supported, will undertake such a task separately or jointly.
2. As a first rule for soft magnetic materials, whether they are alloys or mixed ferrites, the final products should be single-phase solid solutions in the equilibrium state. There are several reasons for observing this rule. First, the equilibrium structure of a single-phase solid solution will not undergo any phase change, thereby ensuring performance stability of the magnetic component. Secondly, the absence of second-phase precipitates in an alloy or in a mixed ferrite will provide the ideal base for good structure-sensitive properties, especially the coercive force and permeabilities. Thirdly, the beneficial effects of solutes on the mechanical and electrical properties discussed in the remainder of this section are at the maximum when the solute atoms are dissolved in the matrix. To comply with this rule, it is imperative to know the pertinent

*Since 1972, a computerized literature search program has been in use at the library of the author's University. Though not prepared merely for magnetic materials, the program can be used to retrieve journal articles and books on a magnetic phenonemon or a specific magnetic material. Similar programs are available at several other (e.g., Georgia Institute of Technology, Lehigh and Ohio State) universities and two commercial "teletype terminal" programs can be purchased for similar purposes.

portion of the phase diagram of the alloy system under study. Such knowledge will then guide the selection of alloys, of which all but one or two should be within the limit of the solid solubility.

3. Although alloy theory is still far from complete, our understanding of most important properties of alloys has reached the point at which certain simple principles are now available. These principles should be followed whenever possible. The principles underlying the variation of saturation magnetization with composition in dilute alloys have been discussed in §1.2. In what follows, we shall briefly describe the principles governing two important nonmagnetic properties; namely, the mechanical behavior and the electrical resistivity of dilute alloys.

For a metallic substance, the term "mechanical behavior" means two things of equal importance; strength and ductility. The job that a mechanical metallurgist strives to do on a new alloy thus is either to raise the strength at little or no expense of the ductility or to improve the ductility. To begin the discussion of the strengthening effect of solute atoms, the term strength should be defined in light of its different meanings. For materials used at room and subzero temperatures, strength is most likely referred to as the yield strength at which plastic (permanent) deformation begins. However, if dynamic loading occurs, which subjects the material to cyclic stressing, endurance limit (minimum stress under which the material can endure an infinite number of stress cycles without failure) or fatigue strength (stress causing fracture of the material at the end of a specific number of stress cycles) are more important than yield strength. Finally for magnetic materials used at elevated temperatures, creep resistance should dominate the strength consideration; this refers to the stress required to produce a specific value of the steady-state creep rate (e.g., 1% per 10^3 or 10^4 hr). Unless further specified, the following discussion is focused on the yield strength.

The increase in strength caused by dissolving a small amount of a solute element in a metal is termed solid-solution strengthening. The strengthening mechanism is termed either interstitial or substitutional according to the nature of the solute. Interstitial solutes such as carbon, nitrogen and oxygen are capable of strengthening the matrix because these atoms tend to be closely associated with dislocations to form the so-called Cottrell atmospheres. Once in the atmosphere, activation of the dislocation requires higher shearing stress because the dislocation has to first break away from the impurity atoms. The strengthening effect of carbon in plain-carbon steels is a vivid example of this mechanism.

The strengthening mechanism for dilute substitutional solid solutions involves one or more of the following three effects:

(1) An elastic effect of the strain field associated with the solute atom due either to a difference between the atomic sizes of the solute and solvent atoms or to a change in the shear modulus (Dieter [1976]). In a metallurgical consideration, the atomic size is customarily based on the closest distance of approach (d) in the hard-sphere model discussed on p. 187.

(2) An electronic effect caused by an increase in the average number of valence electrons per atom (e/a). Experimental values for the critical resolved shear stress for several series of nonmagnetic alloys such as Cu (e/a = 1) with Zn (e/a = 2), Al or Ga (e/a = 3) and Ge (e/a = 4) confirmed such a strengthening effect (Hibbard [1958]).

(3) A chemical effect caused by the segregation of certain substitutional solute atoms in the faulted regions between the partial components of extended dislocations (see discussion in §2.3). The close association of the solute atoms with extended dislocations is said to form a Suzuki atmosphere, which tends to strengthen the crystal in the same way as does the Cottrell atmosphere. The Suzuki mechanism is an interesting and theoretically valid concept but it has not been substantiated by convincing experimental evidence. Also, its operation is somewhat in doubt for the body-centered cubic alloys because the formation of extended dislocations in the bcc lattice is shown to be highly restricted (Chang and Chen [1975]).

The effectiveness of the strengthening mechanisms varies widely with both solvent and solute. Interstitial strengthening based on the Cottrell atmosphere seems to be best known for the bcc alloys of iron and Group VB metals (V, Nb and Ta), but there is no real reason to expect the mechanism to be inapplicable to fcc and hcp alloys except that the latter metals generally show limited solubilities for the interstitial elements. Hydrogen should not be adopted as a strengthening agent for iron because hydrogen in Fe can cause catastrophic brittle fracture.

For iron and its bcc alloys, perhaps the most important consideration of their mechanical behavior is whether or not alloying would enhance the susceptibility to brittle transcrystalline cleavage fracture. To assess this susceptibility, Cottrell [1958] arrived at the following criterion for the ductile-to-brittle transition of the fracture behavior of bcc crystals:

$$\left.\begin{array}{ll} \sigma_y K_y \bar{d}^{-\frac{1}{2}} & \\ \text{or equivalently} \quad (\sigma_i \bar{d}^{-\frac{1}{2}} + K_y) K_y & \end{array}\right\} \gtrless \beta G \gamma \quad \begin{array}{l}\text{(brittle)}\\ \text{(ductile)}\end{array} \tag{5.35}$$

where σ_y is the yield stress, σ_i is the inherent "friction" stress that resist the motion of dislocations, K_y is a constant relating to the localized stress required to unlock dislocations held up at a grain boundary, $\bar{d}$ is the average grain diameter, β is a constant of the stress system induced by the test ($\beta = 1$ for tensile tests, 2 for torsion tests, $\sim\frac{2}{3}$ for tests on notched bars), G is the shear modulus and γ is the surface energy. The derivation of this criterion involved the semi-empirical Petch–Hall relation,

$$\sigma_y = \sigma_i + K_y \bar{d}^{-\frac{1}{2}}, \tag{5.36}$$

which expresses the strengthening effect in polycrystalline materials through the piling up of dislocations by grain boundaries. From this criterion, we can categorically set the conditions under which an alloying element will promote the cleavage-type brittle fracture:

(1) An enormous increase in σ_y;
(2) A larger increase in K_y than that in G; i.e., $\Delta K_y/\Delta G > 1$;
(3) Grain coarsening promoted by the solute; and
(4) An appreciable decrease in the surface energy.

The extreme brittleness of Fe–Si alloys with more than 5% Si is an excellent example to demonstrate the embrittling effect of certain solutes. The addition of Si to Fe is believed to cause (A) a substantial increase in σ_y, (B) grain coarsening and (C) a large decrease in γ.

The principles governing the creep and fatigue strengths are not expected to be the same. Because dislocation climb is a prominent process of high-temperature creep, an effective way to improve the creep strength by alloying is to reduce the diffusion rate of vacancies so as to slow down the climb process. It can be shown (see Reed-Hill [1973]) that the rate of steady-state creep is expressed by

$$\dot{\epsilon}_c = AF(\sigma)\,e^{-Q_c/kT} \tag{5.37}$$

where Q_c is the activation energy, in units of eV per atom, for the process leading to creep, A is a constant characteristic of the crystal structure of the material and $F(\sigma)$ is an appropriate function of the applied stress. If creep is controlled by the climb process, which is performed largely by the motion of vacancies, then $Q_c \sim Q_{D^*}$ is the appropriate activation energy for self-diffusion. Thus, when the solute atoms exert an effect on the motion of vacancies equivalent to an increase in Q_{D^*}, $\dot{\epsilon}_c$ will be reduced as if the creep resistance of the alloy is improved over the pure metal. Turning to the practical side, the alloys of Fe, Co and Ni fare particularly well as creep-resisting materials for high temperature applications. The so-called superalloys developed for the construction of high temperature machines and devices are essentially complex allloys of Fe (13 to 35%), Co ($\approx$20%), Ni (20 to 40%) and Cr ($\sim$20%).

The main process of fatigue failure is the formation of the so-called intrusions and extrusions on the surface. These are shallow grooves and thin narrow ridges, respectively, produced by persistent localized back-and-forth movements of dislocations along definite crystallographic directions. Because the mechanism by which these surface defects are produced is not precisely known, it is only possible to suggest that solutes that tend to impede such persistent, reversible movements of dislocations should enhance the resistance to fatigue. In practice, alloying that tends to raise the endurance limit should also help delay the fatigue failure.

At the outset of the discussion on electrical resistivity, ρ, a word of caution about the definition of ρ is in order. In the free-electron theory, it is customary to define ρ as $\rho = m_0/ne^2\tau$, where τ is the relaxation time of the conduction electrons and n is the number of conduction electrons per unit volume. This definition is misleading because it seems to imply that ρ would be inversely proportional to n. We know that electrical conductivity of a metal depends only on those electrons at the Fermi level, not on the total number of conduction electrons. In other words, metals are good conductors only because of the high velocity of those electrons at the top of the Fermi distribution. A more precise formula of ρ is (Ziman [1964c])

$$\rho = 12\pi^3\hbar/e^2 \int \Lambda \, dS_F, \tag{5.38}$$

where Λ is the mean free path of the conduction electron at the Fermi surface and the integration is over the Fermi surface. For the calculation of the extra residual resistivity, $\Delta\rho_0$, induced by impurity or solute atoms, it is convenient and more accurate to use the partial wave method (Schiff [1955]). We first define

$$\Delta\rho_0 = (k_F c/n_s e^2)\langle\sigma\rangle, \tag{5.39}$$

where c is the concentration of the solute, k_F is the wave number of electrons at the Fermi surface, n_s is the average number of conduction electrons per atom of the host metal and $\langle\sigma\rangle$ is the effective average scattering cross-section of the impurity atom. In terms of phase shifts in the partial wave method (Kittel [1963a]),

$$\langle\sigma\rangle = (4\pi/k_F^2)\sum_{l=0}^{\infty}(2l+1)\sin^2(\eta_l - \eta_{l-1}), \tag{5.40}$$

where l is the orbital angular-momentum quantum number and η_l is the phase shift of the lth partial wave suffered by the incident electrons at the solute ion. To calculate the phase shift in substitutional solid solutions, Friedel [1954, 1958] derived a useful relation between η_l and the nuclear charge of the solute ion in excess of that of the host ion; i.e., $\Delta Z = Z_{\text{solute}} - Z_{\text{solvent}}$. This relation is known as the sum rule,

$$(2/\pi)\sum(2l+1)\eta_l(k_F) = \Delta Z. \tag{5.41}$$

If only one phase shift η_l is large, then the sum rule is reduced to

$$\eta_l = \pi\Delta Z/(4l+2). \tag{5.42}$$

Combining (5.39), (5.40) and (5.42) gives

$$\Delta\rho_0 = \frac{4\pi c}{n_s e^2 k_F}(2l+1)\sin^2\left(\frac{\pi\Delta Z}{4l+2}\right). \tag{5.43}$$

The calculated values given by (5.43) for Cu alloys containing polyvalent impurities such as Zn($\Delta Z = 1$), Ga($\Delta Z = 2$), Ge($\Delta Z = 3$) and As($\Delta Z = 4$) are in excellent agreement with the experimental results (Kittel [1963b]). For the Cu–Zn alloys with $l = 0$, we have $\langle\sigma\rangle = 4\pi/k_F^2 \simeq 10^{-15}\ \text{cm}^2$; thus the extra resistivity induced by Zn is of the order of $4\times 10^{-8}\ \Omega\cdot\text{m}$ (or $4\ \mu\Omega\cdot\text{cm}$) per atomic percent solute.

In connection with (5.43), it is worth noting that, if the argument of the sine function in (5.43) is sufficiently small, say, $\pi\Delta Z/(4l+2) \leqslant 0.35$, then $\sin^2[\pi\Delta Z/(4l+2)] \approx [\pi\Delta Z/(4l+2)]^2$ and

$$\Delta\rho_0 \sim (\Delta Z)^2. \tag{5.44}$$

The relation $\Delta\rho_0 \sim (\Delta Z)^2$ is known as the Linde rule.

The applicability of (5.43) to the dilute alloys of Fe, Co and Ni, however, is hampered by the unsettled values for the valence of the transition metals. For example, Hume-Rothery (1926, see Barrett and Massalski [1966b]) arbitrarily assigned zero valence to all transition metals in his proposal of various series (β, γ and G) of the so-called electron phases. Later Raynor [1949] adopted different valences for Cr ($Z = 4.66$), Mn ($Z = 3.16$), Fe ($Z = 2.66$), Co ($Z = 1.71$) and Ni ($Z = 0.61$) dissolved in Al ($Z = 3$) in an effort to establish the M_2Al_{11} phase. The problem of the valence of transition elements is twofold – their valences are unknown and do not stay at the same value in different alloy systems. Multiple valences are generally the rule, rather than the exception, for most transition metals.

The increasing effect of various solutes on resistivity of iron is shown in fig. 5.33. These experimental results show an inconsistent trend

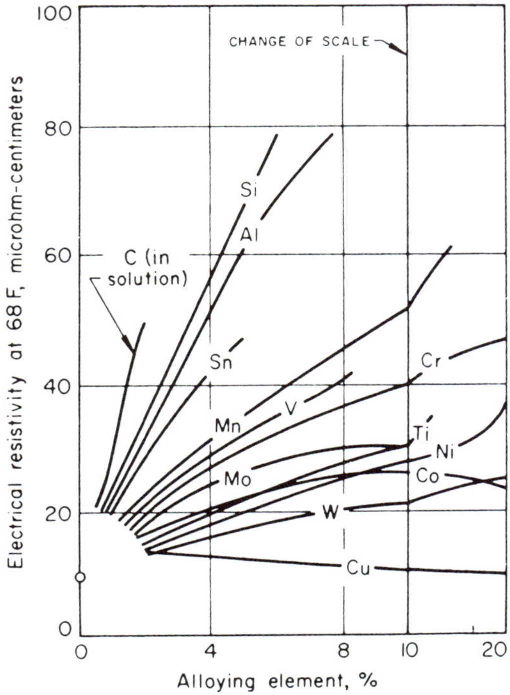

Fig. 5.33. Effect of solutes on the room-temperature resistivity of iron (Metals Handbook [1961]).

toward the establishment of $\Delta\rho_0 \approx (\Delta Z)^2$, irrespective of the value chosen for the valence of the ferromagnetic host metal. Meanwhile C, Si and Al are seen to increase ρ of Fe most effectively at room temperature.

Attempts to analyze the resistivities of ferromagnetic alloys when both solvent and solute are transition metals encounter still other difficulties. Not only is ΔZ totally unknown, but also the electronic characteristics of the magnetic matrix are so complex that even the principles underlying the elegant treatments of localized magnetic states in transition metals by Wolff [1961], Anderson [1961] and Clogston [1962] were incapable of dealing with the situation. Chen [1963] conducted a systematic study of the effect of solutes of the first row of transition elements (Ti to Ni) and Cu on the residual resistivity of Ni and four prominent ferromagnetic alloys, Ni_3Fe, FeCo, CoNi and $Ni_{0.6}Cu_{0.4}$. The choice of these solvents was partly restricted by a minimum average value of 8.4 outer electrons per atom so that all host materials would lie to the right side of the peak of the Pauling–Slater curve shown in fig. 5.3. The results are shown in fig. 5.34. The increases in resistivity caused by one atomic percent of Mn, Fe, Co, Ni and Cu are uniformly small, whereas the increases caused by Ti, V and Cr are much larger, especially those caused by Cr. The large increases are ascribed to a phenomenon known as the resonance scattering of electrons, according to a theory proposed by Friedel [1958]. In terms of the phase shifts expressed in (5.41), scattering of electrons by the impurity atoms of Mn, Fe, Co, Ni and Cu is considered normal in the sense that $\eta_0 > \eta_1 > \eta_2 \ldots$, and even η_0 is small. Therefore, the average effective scattering cross sections $\langle\sigma\rangle$ associated with these impurity atoms are uniformly small, and so are their induced resistivities. For Cr and, to a lesser degree, for Ti and V, the perturbing potentials of these impurity atoms create a "virtual" energy level above the Fermi level. Consequently, η_2 tends to $\frac{1}{2}\pi$, whereas other η_l are negligibly small. The virtual state leads to a maximum value of $\langle\sigma\rangle$ and thus a large impurity-induced resistivity. In the resonance scattering, $\Delta\rho$ expressed in (5.43) reaches a maximum value,

$$(\Delta\rho_0)_{\max} = 20\pi c/n_s e^2 k_F. \tag{5.45}$$

Comparison between the predicted and experimental values for $(\Delta\rho_0)_{\max}$ in Chen's analysis gives reasonable support to the Friedel theory. Neutron diffraction experiments conducted by Low and Collins [1963] also shed light on the physical significance of the drastic drop in the

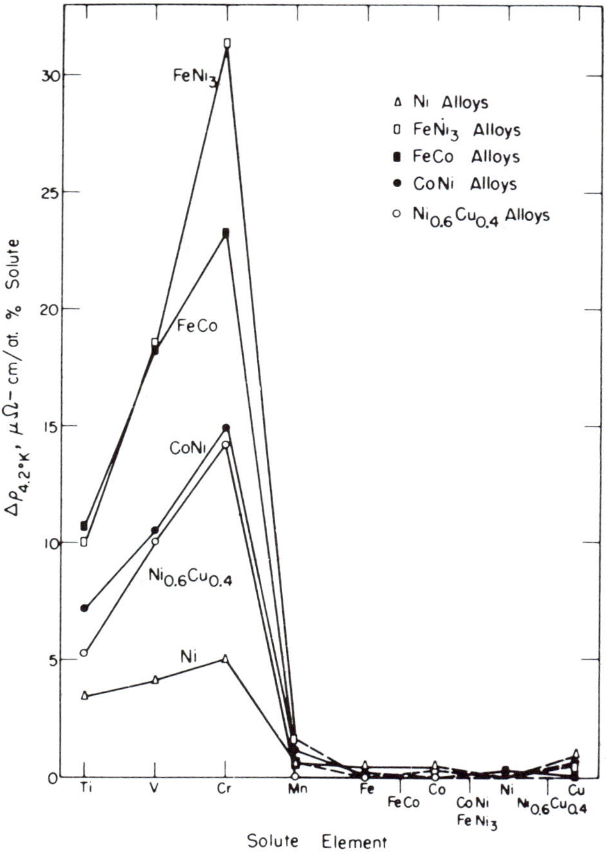

Fig. 5.34. Effect of various solutes on the resistivities at 4.2°K of Ni and four ferromagnetic alloys (Chen [1963]).

specific residual resistivity increase on changing the solute from Cr to Mn in fig. 5.34. From the neutron data these workers concluded that the presence of Cr atoms in Ni causes a large disturbance of the magnetic moment over several interatomic distances, whereas the disturbance caused by a Mn atom in Ni is both small and spatially confined to the impurity site.

A final point on the impurity-induced resistivity of alloys deals with the relation $\Delta\rho_0 \approx c$ implied in (5.43). This relation applies only to dilute

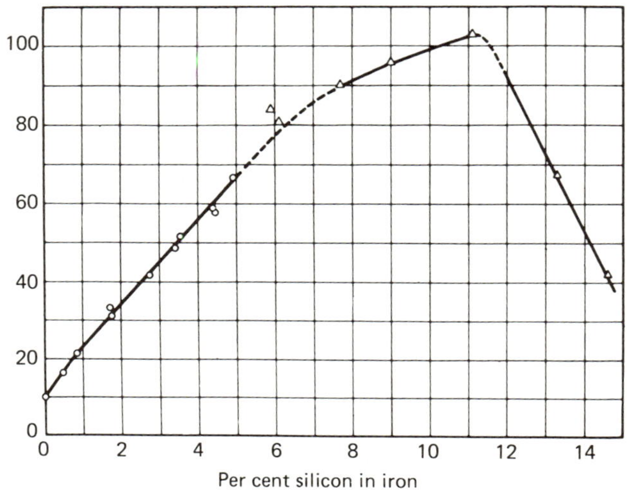

Fig. 5.35. Electrical resistivities of iron-rich Fe–Si alloys at room temperature (Bozorth [1951]).

alloys. As the solute concentration exceeds a few percent, the relation should be modified to

$$\Delta\rho_0 \approx c(1-c), \tag{5.46}$$

in conformity with the Nordheim rule. The experimental results for the Fe–Si system are shown in fig. 5.35. Indeed, the resistivity increases induced by Si first show a linear dependence on the solute concentration up to ≈5%, in agreement with $\Delta\rho_0 \approx c$. At high concentrations between 5 and 11% Si, the curve bends over in agreement with the $\Delta\rho_0 \approx c(1-c)$ relation. At 11% Si, the curve reverses the sign of its slope promptly, and $\Delta\rho_0$ diminishes rapidly. The cause of the latter drastic changes is attributed to the ordering of the Si atoms. A minimum of $\Delta\rho_0$ was found (not shown) at 15% Si, which coincides with the limit of solid solubility of Si in Fe.

§2.3. *Crystal defects*

The discussions we have had thus far deal with perfect crystals, or else we assume that the effect of lattice defects on magnetic properties is unimportant. In reality, crystals are not perfect, nor can the effect of crystal defects be ignored completely. There has been ample evidence to indicate that no matter how carefully the crystals are prepared, imperfections always exist. Also, as later discussion will show, vacancies are an integral part of the crystal in thermal equilibrium at temperatures above absolute zero. The very presence of defects in the lattice thus makes it necessary to consider the structure-sensitive magnetic properties in real, imperfect crystals.

Many different kinds of defects have been found in crystalline materials. A general classification of crystal defects is based on their dimensions. Hence they may be (A) point defects, each of which occupies a single, regular lattice site or an interstice; (B) line defects, each of which extends over a row (not necessarily straight) of atoms, (C) planar defects, which have two-dimensional configurations and (D) volume defects, each of which occupies a volume many times greater than the atomic volume. Important species of the four classes of defects are listed as follows:

(A) Point defects – For metallic crystals, point defects are further classified:

- (a) Intrinsic point defects:
  - (i) vacancies.
  - (ii) self-interstitial atoms, or simply interstitials.
- (b) Extrinsic point defects – impurity atoms:
  - (i) substitutional impurity atoms.
  - (ii) interstitial impurity atoms.

(B) Line defects – dislocations.
(C) Planar defects – grain boundaries, subgrain boundaries, twin boundaries, stacking faults, magnetic domain boundaries, antiphase domain boundaries and planar precipitates.
(D) Volume defects – voids, gas bubbles, three-dimensional precipitates, inclusion particles and cracks.

The listing covers all possible defects in metallic crystals but leaves out many varieties of point defects in ionic crystals. In the latter materials, because of the polarity of the ionic charges, an unoccupied cation site is treated differently from a vacant anion site. Furthermore, the principle of charge neutrality demands a delicate balance of charges associated with the point defects, thus creating many additional species of point defects not encountered in metals. For further information on point defects in ionic crystals, the book edited by Crawford and Slifkin [1972] should be consulted.

To explain the various defects, let us consider a crystal of α-Fe. An iron crystal of macroscopic size consists of millions of cubic-shaped unit cells stacked together to completely fill the space in the crystal. If this body-centered cubic crystal is perfect, every regular lattice site, either at the corners or in the centers of the unit cells, will be occupied by iron atoms as shown in fig. 5.36a. However, if the atom at the corner site A is displaced to an octahedral interstitial site O_1, this process of atomic displacement simultaneously creates two point defects – a vacancy at A and a self-interstitial atom at O_1, as shown in fig. 5.36b. An impurity atom of such elements as H, B, C, N and O is shown at another octahedral interstitial site O_1' in fig. 5.36a. Note that O_1 and O_1' are equivalent interstitial sites of the octahedral type just as A and A′ are equivalent substitutional sites.

A dislocation is a line of discontinuity in the crystal. Figure 5.37a is a three-dimensional drawing of a portion of a cubic crystal. A dislocation is shown to exist at the edge of the middle plane, which is discontinued at A. Because of the geometry, this dislocation is classified as the edge type. Dislocations can be of the screw type shown in fig. 5.37b. To create a screw dislocation, it may be imagined that the crystal is cut in the ABC plane halfway through, followed by shearing the front left half of the crystal upward relative to the front right half by one atomic

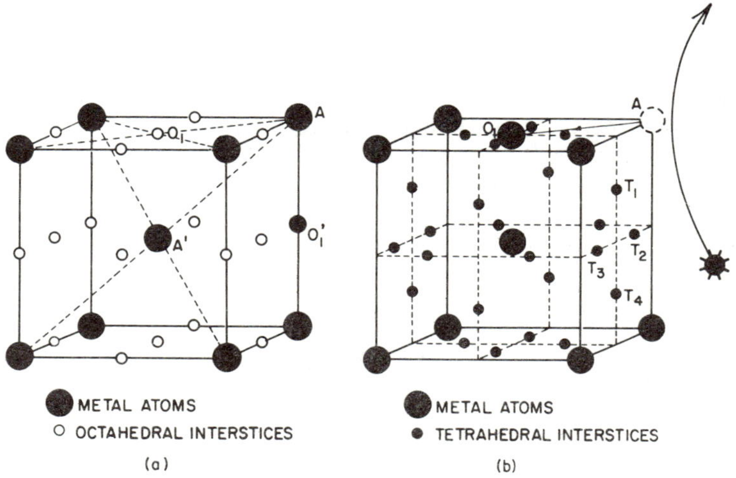

Fig. 5.36. (a) Octahedral and (b) tetrahedral interstitial sites in a unit cell of the bcc lattice.

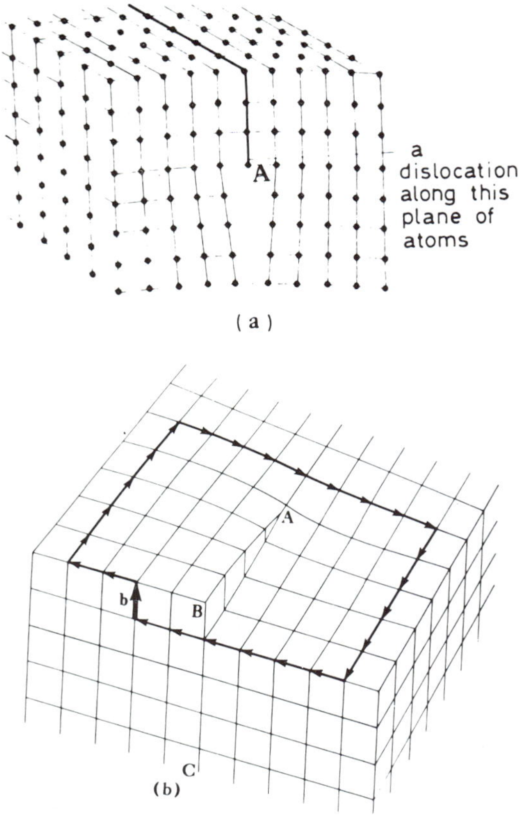

Fig. 5.37. (a) An edge and (b) a screw dislocation in crystals. The series of arrows forms a Burgers circuit, which determines the Burgers vector ***b*** of the screw dislocation in (b) ((a) Smallman [1970]; (b) Read [1953]).

spacing. The dislocation is a screw because it converts the crystal from a deck of parallel atomic planes into a single atomic plane in a helical or spiral form like the threads of a screw. Dislocations are characterized by a vector known as the Burgers vector ***b***, which is defined by constructing a Burgers circuit as shown in fig. 5.37b. A basic difference between an edge and a screw dislocation is that a pure edge has a Burgers vector normal to the dislocation line, whereas a pure screw has ***b*** parallel to the dislocation line. If a dislocation is in the loop form, it still can be a pure edge if its Burgers vector is normal to the plane of the dislocation loop; otherwise, it is a mixed dislocation when the Burgers vector lies in the plane of the loop or in a direction inclined to the loop plane. Another important application of the Burgers vector is to classify the dislocations as either unit or partial. If $\boldsymbol{b}$ is a vector of the crystal lattice connecting two regular sites, the dislocation is unit; but if $\boldsymbol{b}$ is not connecting two regular sites, the dislocation is partial. Thus in fcc lattices, dislocations of $\boldsymbol{b} = a[100]$, $\frac{1}{2}a[110]$ or $\frac{1}{2}a[111]$ are unit dislocations, whereas those having $\boldsymbol{b} = \frac{1}{3}a[111]$ or $\frac{1}{6}a[111]$ are partial dislocations.

Although grain boundaries exist in all polycrystalline materials, other boundaries, such as subgrain, twin, antiphase and magnetic domain, may exist in single crystals as well as polycrystals. The difference between grain and subgrain boundaries lies in the degree of misorientation between the adjacent regions. Usually boundaries of misorientations less than 1° are called subgrain boundaries. A stacking fault is a region in which the sequence of stacking the atomic planes is erratic. For instance, the normal stacking sequence for a face-centered cubic crystal is ABCABCA . . . in any of the ⟨111⟩ directions as shown in fig. 5.38a. Now, if a portion of the second "B" plane is removed, as shown in fig. 5.38b, a stacking fault is created there together with an edge dislocation in loop form because the stacking sequence becomes ABCA/CA . . . . In fact, this is called an intrinsic stacking fault because the removal of the partial "B" plane is equivalent to condensing a group of vacancies there. In contrast, an extrinsic stacking fault is created by inserting a group of interstitial atoms to produce a partial plane between two regular planes ("B" and "C" in fig. 5.38c). Again, the stacking becomes faulty with the sequence ABCAB/A/CAB . . . . Note that an

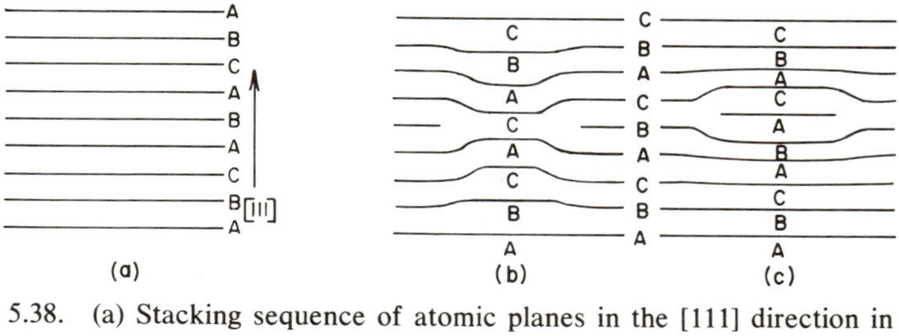

Fig. 5.38. (a) Stacking sequence of atomic planes in the [111] direction in fcc crystals. (b) and (c) Single or intrinsic and double or extrinsic stacking faults in fcc crystals, respectively.

extrinsic stacking fault consists of a double fault as indicated by . . ./A/ . . . , whereas an intrinsic stacking fault is a single fault as indicated by . . . A/C . . . in the above example. In a close-packed hexagonal metal like α-Co, the normal stacking sequence along the c-axis is ABABA . . . , and a stacking fault is created in the intrinsic type with the faulty sequence ABA/A . . . and of the extrinsic type with the same faulty sequence. Here both types of stacking faults each has a single fault.

Voids are cavities with some gaseous atoms filling the space. Voids are produced by aggregation of vacancies in certain diffusion specimens because of the well-known Kirkendall effect, in crystalline materials irradiated to relatively high fluence ($>10^{18}$ per cm^2) of fast neutrons (with $E > 0.1$ MeV) and in sintered compacts. Gas bubbles, on the other hand, are clusters of atoms of hydrogen, helium, argon and other inert gases. Hydrogen bubbles are produced during the process of converting hydrogen atoms into hydrogen molecules. Alpha-iron and copper are two metals known for their high susceptibilities to the formation of hydrogen bubbles. Gas bubbles of helium are produced in materials irradiated by thermal ($E = 0.025$ eV) or fast neutrons if the (n, α) reaction occurs during the irradiation to the extent that sufficient helium atoms ($\geqslant 1$ atomic ppm) are yielded. Because helium (and other inert gases) is extremely insoluble in metals, the α-particles thus produced by the transmutation process rapidly supersaturate the lattice and precipitate to form gas bubbles on any suitable nucleation sites, such as vacancy clusters, dislocation loops and grain boundaries. In general, the absorption cross section for the (n, α) reaction is negligibly small (<0.1 barn, that is, $<10^{-25}$ cm^2) for nuclei of most elements. This is especially true with fast neutrons. With thermal neutrons, however, there are two well-known cases in which unusually large absorption cross sections are found:

Reaction	σ(b)
$^{10}B(n, \alpha)^7Li$	3 990
$^6Li(n, \alpha)^3H$	950

These two reactions are of some practical importance in our consideration because the naturally occurring ^{10}B and ^6Li isotopes with 19.6% and 7.4% relative abundance are possible impurities in iron. In fact, boron is frequently used to improve the hardenability of steels in industry. The boron-containing iron and steels, therefore, will be highly susceptible to the formation of helium bubbles in thermal neutron environments.

Lattice defects are interrelated in many ways. In the first place, intrinsic point defects can conglomerate to form dislocations, which, in turn, can account for the formation of many of the planar defects such as subgrain, twin and low-angle ($<15°$) grain boundaries. Conversely, the motion of dislocations can produce intrinsic point defects under certain geometric conditions when they cut through each other. Also, as is shown in fig. 5.38b and c, the periphery of the faulted region is a dislocation. It is a well-tested rule that a partial dislocation is always associated with a strip of stacking fault.

Another important consideration of lattice defects is whether or not the existence of each species of defects is in thermodynamic equilibrium. Intrinsic point defects are inherently present in the crystal at finite temperatures, but other defects are not inherent. Among the latter defects, however, several species, notably dislocations, are practically impossible to eliminate completely. Even in the extreme cases where extraordinary care is exercised to grow crystals from the vapor or liquid phase, each crystal is found to contain at least a screw dislocation, which acts as an effective mechanism for crystal growth. Iron whiskers are good examples. Also, inert gas bubbles are difficult to remove by annealing once they are introduced into the lattice.

The reason that vacancies and self interstitials exist as integral parts of the crystal can be explained by a simple thermodynamical analysis. At a finite temperature T, a crystal of a chemically pure metal may be considered to be a binary system of the atoms and point defects of a prominent species, say single vacancies. The Gibbs free energy of the system is expressed by

$$G = H_0 + \int_0^T C_p \,\mathrm{d}T - T\left(\int_0^T (C_p/T)\,\mathrm{d}T + S_{\mathrm{mixing}}\right), \tag{5.47}$$

where H_0 is the enthalpy at 0°K, C_p is the heat capacity at constant pressure and S_{mixing} is the entropy associated with mixing the two components. Because S_{mixing} is always a positive quantity, the presence of vacancies tends to lower the free energy of the system by virtue of the $-TS_{\mathrm{mixing}}$ term. This explains why vacancies are thermally introduced into the crystal so as to stabilize the system when the temperature is raised.

To calculate the concentration of vacancies in thermal equilibrium at a fixed T, let us assume that there are N atoms and n vacancies in the system, the concentration of vacancies is defined by $c_v = n/N$. The entropy of mixing the atoms and the vacancies may be calculated by applying the Boltzmann relation, $S = k \ln p$, where p is the probability of the state of the system, or, more explicitly, the number of possible independent ways to arrange the n vacancies over the total lattice sites $N + n$. Because vacancies in the metal are indistinguishable, $p = (N+n)(N+n-1)\ldots(N+1)/n!$, which is equivalent to

$$p = (N+n)!/N!n!. \tag{5.48}$$

By using the Stirling approximation ($\ln N! = N \ln N - N$) (5.47) becomes

$$G = H_0 + \int_0^T C_p \, dT - T \int_0^T (C_p/T) \, dT + TNk[(1-c_v)\ln(1-c_v) + c_v \ln c_v]. \tag{5.49}$$

To obtain the equilibrium value for c_v, we set $\partial G/\partial n = 0$, which yields

$$c_v \simeq \exp(-\Delta H_v/kT), \tag{5.50}$$

where ΔH_v is the enthalpy for the formation of a single vacancy. For Fe, ΔH_v is about 2.1 eV; the concentration of single vacancies is 1×10^{-20}, or equivalently 1×10^{17} vacancies per m^3, at 300°K and 2×10^{-4} (0.02%), or 4×10^{22} vacancies per m^3, at 1800°K (8° below the melting point).

The above analysis assumes that all vacancies are isolated. Experimental results indicate that vacancies are also clustered in double (divacancy), triple (trivacancy), or even more complex forms. From a simple statistical argument invoking the Boltzmann relation, the concentration of divacancies is shown to be given by

$$c_{v_2} \simeq \tfrac{1}{2} z c_v^2 \exp(B/kT), \tag{5.51}$$

where z is the coordination number (6, 8 and 12 for nonideal hcp, bcc and fcc structures) and B is the energy of binding two single vacancies into a divacancy. The value for B must be positive for the divacancy to be stable, but in general is small, of the order of 0.1–0.2 eV. Adopting $B = 0.15$ eV for Fe, (5.51) predicts that, at 1800°C, $c_{v_2}/c_v \simeq 5 \times 10^{-3}$, which indicates the insignificant fraction of vacancies in the double form.

Similar analysis could be made for self-interstitials. The formation of enthalpy of interstitials, ΔH_i, is much larger, however, and c_i is negligibly small even at temperatures close to T_m. Hence, the inherent existence of self-interstitials is generally ignored.

The advent of the theory of dislocations marked the beginning of a new era of physical and mechanical metallurgy. Although dislocations were originally proposed to account for the huge (four orders of magnitude) discrepancy between the theoretical and observed values of the critical resolved shear stress for slip, the very existence of dislocations in crystalline materials and their versatility explain almost all major mechanical properties of metals and many metallurgical phenomena such as precipitation, martensitic transformation, polygonization, etc. Each dislocation is associated with a strain field, thereby possessing a self-energy, E_d, which represents the amount of work required to create the dislocation. Typically E_d has the value of 3–5 eV per atomic length of dislocation. In comparison with ΔH_v, obviously dislocations are associated with self-energies too high to exist in thermal equilibrium. Indeed, a prolonged anneal of a ultra-pure metal at temperatures close to T_m tends to reduce the density of dislocations to as low as 10–1000 lines per cm^2 (Young et al. [1968]). Normally, otherwise, metal crystals contain dislocations of the density of 10^6–10^7 lines per cm^2 in the annealed state and 10^8–10^{10} lines per cm^2 in the plastically deformed state.

Because of the large variety of crystal defects, the effect of lattice imperfections on magnetic properties is expected to vary widely. Actually the dependence of the magnetization process upon the condition of the crystal lattice had been recognized long before the advent of the dislocation theory. The inclusion theory of coercivity and initial permeability proposed by Kersten (p. 132) is clearly a case in point. The more general strain theory of coercivity conceived by Bloch and Kondorsky (p. 130) is also an example. However, the correct physical model of the internal strain had not emerged in any of the theories until the dislocation theory was developed.

Point defects, both intrinsic and extrinsic, do not exert detectable effect on even the two most structure-sensitive properties of magnetization; namely, the initial permeability and the coercivity. Nevertheless, they may affect μ_1, H_c and other magnetic properties appreciably if they are clustered in complex forms. When point-defect clusters are converted to dislocation loops, voids, gas bubbles, or inclusion particles, they are effective deteriorators to the properties of soft magnetic materials. In the extreme situation, a hard magnet may eventually be developed by introducing precipitates of suitable size and shape dispersed uniformly in the lattice. Dykstra [1969] compiled experimental data of several studies showing the dependence of the maximum increase in coercive force upon the volume fraction of various precipitates. His observations are shown in fig. 5.39.

Interstitial impurities may exert a dynamic effect upon magnetic properties. The phenomenon is known as the magnetic aftereffect. It refers to a steady increase in magnetization of a material after a magnetic field is applied. The rate of change is too slow to be explained in terms of an eddy-current effect. This time-dependent magnetization change is attributed to a redistribution of interstitial impurity atoms induced by the field, as viewed from other experimental results which indicate the onset of short-range diffusion of these atoms. A relaxation

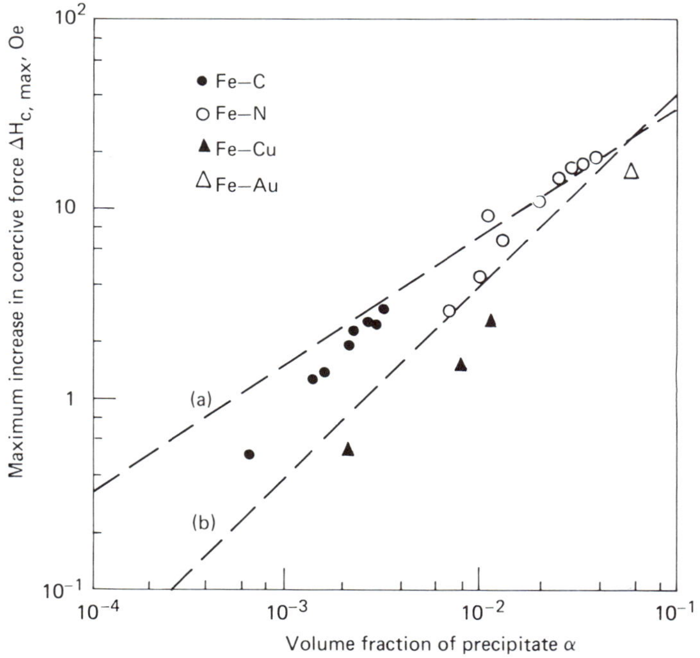

Fig. 5.39. Maximum increases in coercivity of iron caused by precipitates of either interstitial or substitutional solutes. Data points are shown by different symbols and curves (a) and (b) represent theoretical estimates according to Kersten and to Néel, respectively (Dykstra [1969]).

time ξ characterizes the magnetic aftereffect, where ξ is defined by

$$\xi(T) = \xi_0 \exp(Q_i/kT), \tag{5.52}$$

with Q_i being the activation energy for the diffusion process and ξ_0 a constant equal to ξ at the infinite temperature. Like other relaxation phenomena, the magnetic aftereffect will steadily change the magnetization at a rate that depends on the relaxation time, as indicated by the following approximate expression:

$$M(t) = M(\infty)[1 - \exp(-t/\xi)]. \tag{5.53}$$

Two theories have been proposed to account for the migration of interstitial impurity atoms leading to a magnetic after-effect. The first theory was proposed by Snoek [1939], who noted that, when a metal exhibited a magnetic aftereffect, it would also exhibit a similar time-dependent mechanical phenomenon known as the internal friction. The latter phenomenon can be explained as follows: Figure 5.40 shows four octahedral interstitial sites, W, X, Y and Z, in a bcc lattice. Each site is surrounded by six host atoms, four at a distance of $a/\sqrt{2} = 0.707a$ and

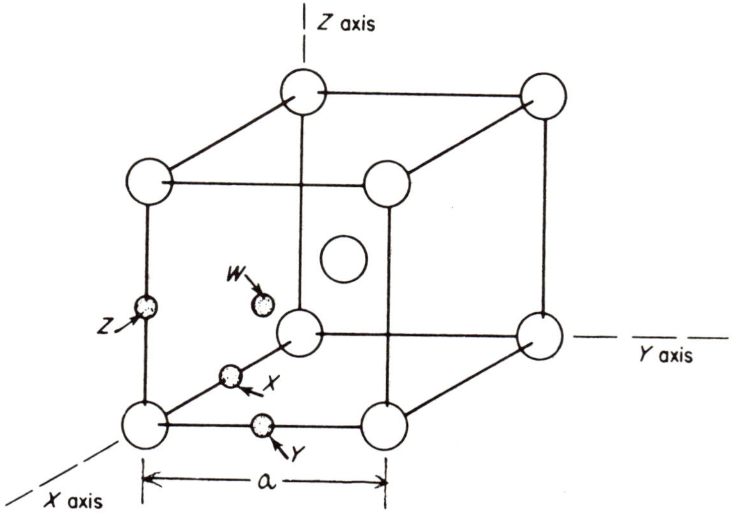

Fig. 5.40. Four orientations of crystallographically equivalent interstices, marked by X, Y, Z and W, in the bcc lattice to explain the Snoek effect (Reed-Hill [1973]).

two at a shorter distance of $a/2 = 0.5a$. When an external force is applied to the crystal in such a manner as to cause an internal tensile stress to oscillate between the [001] and [010] directions the Z and Y sites would be opened up alternately, whereas the other sites would be squeezed. Hence, interstitial impurity atoms of H, C, O and N would be moving alternately into the Z and Y sites in response to the cyclical distribution of internal stresses.

Snoek thus proposed a similar movement of interstitial impurity atoms into the preferential sites as a result of a magnetoelastic coupling between the applied field and magnetostriction. This movement produces a directional alignment of interstitial impurity atoms which is responsible for the anisotropic behavior similar to the magneto-crystalline anisotropy discussed in ch. 3, §3, and thus the magnetic aftereffect. The Snoek theory was substantiated by some specially designed experiments, but obviously its applicability is limited largely to bcc systems. (Rarely has the Snoek-type internal friction been observed in fcc and hcp crystals.)

In light of the inadequacy of the Snoek theory, Néel [1952] proposed a more basic theory of the magnetic aftereffect. Néel conceived his theory on a modified spin–orbit coupling between two magnetic atoms near an interstitial impurity atom. This coupling is assumed to occur whenever the presence of an impurity atom or a defect causes asymmetrical strain in the lattice. Experimental results reported by De Vries et al. [1959] seem to support the Néel theory.

As far as magnetic properties are concerned, dislocations are doubtless the most influential class of lattice defects. This is partly because dislocations are always present in relatively high density and partly because the stress fields associated with dislocations interact importantly with the domain walls, thereby causing changes in the magnetization process performed by the wall movement.

The effect of dislocations on the magnetization process has its origin in the magnetoelastic coupling between the stress field and magnetostriction. Brown ([1941], see Bozorth [1951*l*]) was the first to examine the effect of dislocations on magnetization near saturation. As was cited in ch. 4, §2.1.1, his theoretical analysis led to the conclusion that the magnetostrictive forces induced by the stress field around a dislocation indeed give rise to the terms a/H and b/H^2 in (4.10). The latter equation is a fair representation of the law of approach to saturation when $H_a \gg M/\mu_0$, but is a poor approximation if $H_a < M/\mu_0$. Vicena [1955] made the first attempt to define the coercivity and permeability in terms of dislocations by formulating the potential of interactions between domain walls and dislocations. Later, Seeger et al. [1964] found it simpler to calculate the force $\boldsymbol{p}^{\mathrm{w}}$ exerted by a dislocation on the domain wall by the formula:

$$\boldsymbol{p}^{\mathrm{w}} = \int_S (\mathrm{d}\boldsymbol{S} \times \nabla) \times \boldsymbol{\sigma}^{\mathrm{M}} \cdot \boldsymbol{b}, \tag{5.54}$$

where $\boldsymbol{S}$ is the vector of the surface enclosed by the dislocation line in loop form, ∇ is the del operator and $\boldsymbol{\sigma}^{\mathrm{M}}$ is the tensor of the extra stress associated with magnetostrictive strains due to the variation of the direction of magnetization within the domain wall. By using Stokes' theorem, (5.54) is reduced to

$$\boldsymbol{p}^{\mathrm{w}} = \int_l \mathrm{d}\boldsymbol{l} \times \boldsymbol{\sigma}^{\mathrm{M}} \cdot \boldsymbol{b}, \tag{5.55}$$

where $\mathrm{d}\boldsymbol{l}$ is a line element of the dislocation. For an isolated plane wall parallel to the x–y plane (fig. 5.41a) (i.e., the wall is assumed to be unaffected by the dislocations), $\boldsymbol{\sigma}^{\mathrm{M}}$ will have the form

$$\boldsymbol{\sigma}^{\mathrm{M}} = \begin{bmatrix} \sigma_{11}^{\mathrm{M}} & \tau_{12}^{\mathrm{M}} & 0 \\ \tau_{12}^{\mathrm{M}} & \sigma_{22}^{\mathrm{M}} & 0 \\ 0 & 0 & 0 \end{bmatrix}, \tag{5.56}$$

where σ_{ii} and τ_{ij} denote the normal and shear components, respectively. Rieder [1959] has calculated $\boldsymbol{\sigma}^{\mathrm{M}}$ for the prominent domain walls in

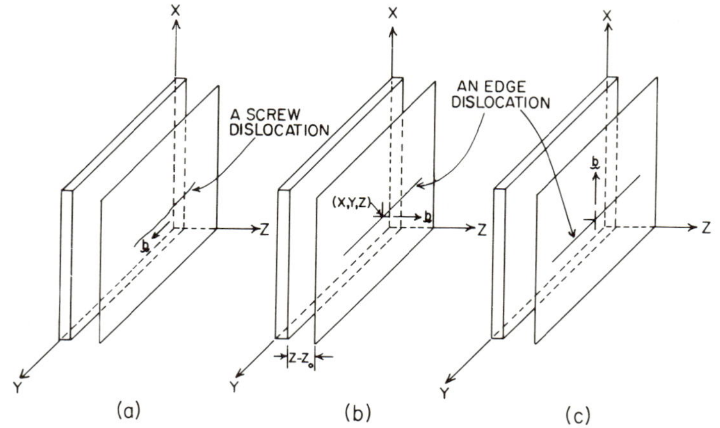

Fig. 5.41. Three dislocations of different type or orientation assumed in the analysis on coercivity by Seeger et al. [1964].

elastically anisotropic, ferromagnetic metals. To simplify the analysis, however, Seeger et al. considered the metals to be isotropic, thus characterizing each metal by two elastic constants, or equivalently, by the shear modulus G and Poisson's ratio ν.

For the 180° wall shown in fig. 5.41a and assuming a straight dislocation lying parallel to the wall, the force p_z exerted on the wall by the dislocation in the z-direction is shown to have the magnitude

$$p_z = \sigma_{11}^{M} l_y b_x - \sigma_{22}^{M} l_x b_y + \tau_{12}^{M}(l_y b_y - l_x b_x), \tag{5.57}$$

where $\boldsymbol{l} = l_x \boldsymbol{i} + l_y \boldsymbol{j} + l_z \boldsymbol{k}$ and $\boldsymbol{b} = b_x \boldsymbol{i} + b_y \boldsymbol{j} + b_z \boldsymbol{k}$. The values for p_z are compared in table 5.24 for three simple cases of dislocations. Calculations of p_z for other single and coupled dislocations were described by Traüble [1969].

In the analysis of Seeger et al, the coercive force is expressed for the case of single crystals by

$$H_c = (1/2M_s A \cos\phi)\left\langle\left|\sum (z)_{\max}\right|\right\rangle_{av}, \tag{5.58}$$

and the initial susceptibility by

$$\chi_i = 4\cos^2\phi M_s^2 (A/L\langle\alpha\rangle_{av}) \tag{5.59}$$

where A is the area of a domain wall, L is the separation between neighboring walls, ϕ is the angle between the applied field and the magnetization vector, $\langle|\Sigma (z)_{\max}|\rangle$ is the expectation value for the maximum resultant force acting on the wall at some position z and α is the zero-point slope of the $\Sigma(z)$ versus z curve. The average is taken over all domain walls in the crystal. $\Sigma(z)$ is related to p_z by the relation

$$\sum (z) = \sum_j m_j p_j (z_0 - z), \tag{5.60}$$

where m_j is the number of dislocations of j-type and p_j is the force acting on a domain wall by the dislocation at the distance $z - z_0$. Equations (5.58)–(5.60) show how the stresses exerted on the domain walls by dislocations determine the coercive force and the initial susceptibility. The application of these equations for the theoretical calculation of H_c and χ_i in actual materials, however, is complicated by the fact that dislocations are seldom in simple form, nor are they uniformly distributed. Therefore, the calculation of $\langle|\Sigma(z)|\rangle_{av}$ and $\langle\alpha\rangle_{av}$ is difficult even approximately. Thus the theory has little practical value except that

TABLE 5.24

Forces exerted on a 180° domain wall by dislocations in three orientations shown in fig. 5.41

Type of dislocation	A screw	An edge	An edge
(1) As shown in	Fig. 5.41a	Fig. 5.41b	Fig. 5.41c
(2) $\boldsymbol{b}$ lying in the direction of	y-axis	z-axis	x-axis
(3) Component of $\boldsymbol{b}$	$\boldsymbol{b} = b_y\boldsymbol{j}$	$\boldsymbol{b} = \boldsymbol{b}_z\boldsymbol{k}$	$\boldsymbol{b} = b_x\boldsymbol{i}$
(4) p_z	$p_z = -\sigma_{22}^{M}l_xb_y + \tau_{12}^{M}l_yb_y$ $= (-\sigma_{22}^{M}l_x + \tau_{12}^{M}l_y)b$	$p_z = 0$	$p_z = \sigma_{11}^{M}l_yb_x - \tau_{12}^{M}l_xb_x$ $= (\sigma_{11}^{M}l_y - \tau_{12}l_x)b$

it reveals the following important relations:

$$H_c \approx \rho_d^{\frac{1}{2}}; \qquad \chi_i \approx \rho_d^{-\frac{1}{2}}, \tag{5.61}$$

where ρ_d is the dislocation density expressed in number of dislocation lines per cm^2, or in total length of dislocations per cm^3.

Voluminous experimental results on work-hardening in single crystals of appropriate orientations establish that plastic deformation of metals by slip through the movement of dislocations causes an increase in the resolved shear stress from the initial value τ_0 to τ as a function of shear strain in three distinct stages. A typical shear stress–strain curve is shown for Fe in fig. 5.42b. In stage I, where a single slip system operates, and in stage II, where multiple slip systems operate, examination of the deformed single-crystal specimens by transmission electron microscopy has revealed that

$$\text{stage I:} \qquad (\tau - \tau_0) \approx \rho_d, \tag{5.62a}$$

$$\text{stage II:} \qquad (\tau - \tau_0)^2 \approx \rho_d. \tag{5.62b}$$

Combining (5.61) and (5.62), we obtain the following relationships:

	Coercive force, H_c	Initial susceptibility, χ_i	
stage I	$H_c \approx (\tau - \tau_0)^{\frac{1}{2}}$,	$\chi_i \approx (\tau - \tau_0)^{-\frac{1}{2}}$,	(5.63a)
stage II	$H_c \approx (\tau - \tau_0)$,	$\chi_i \approx (\tau - \tau_0)^{-1}$.	(5.63b)

Good agreement has been obtained between experimental and predicted values for Fe, Co and Ni to support these relationships, as shown in figs. (5.42a), (5.43) and (5.44).

The effect of internal boundaries on the magnetization process may be examined in two respects. First, the existence of the boundary means a change in the direction of the magnetization vector in regions adjacent to the boundary. The degree of change depends on the misorientation of the boundary, of course. This effect could become important when the boundary is of a high angle, say 20°. The second effect of boundaries is on the distribution and motion of domain walls. If the boundary is relatively clean (i.e., free from impurities), a boundary may not be capable of disrupting a domain wall. As an example, fig. 5.45 shows the continuation of domain walls across a grain boundary (but there is a change in the direction of magnetization). Segregation of impurities at boundaries, however, will enhance the effect, sometimes drastically. For instance, the presence of inclusion particles

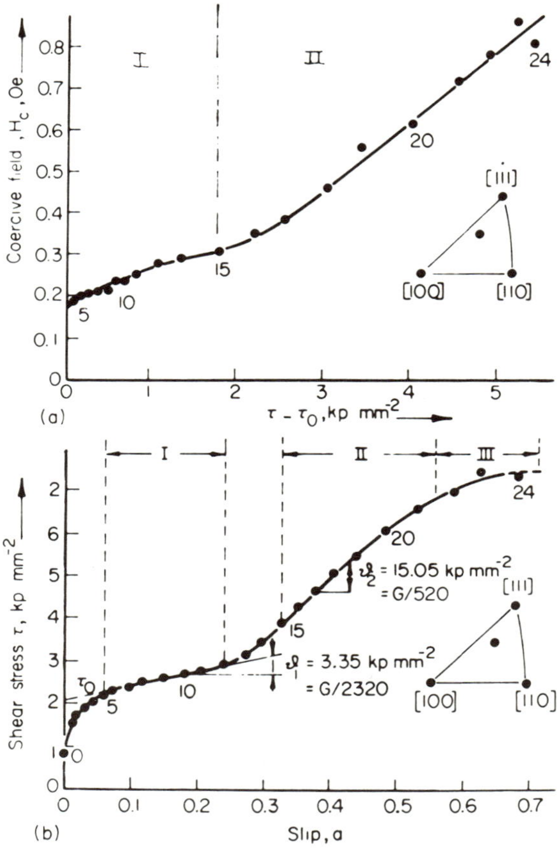

Fig. 5.42. (a) Coercive force plotted as a function of the effective shear stress and (b) the corresponding work-hardening curve measured at room temperature. Both curves were obtained from single crystals of iron with the same orientation shown by the point in the triangle. Note the striking similarity in the general shape of the curves (Traüble [1969]).

at a grain boundary would induce spikes and other complex domain patterns so as to reduce the number of free poles around the particles. An example is seen in fig. 5.46. Because the presence of impurity atoms in a boundary, especially in the inclusion form, has a strengthening effect on the boundary, dirty boundaries could effectively impede the

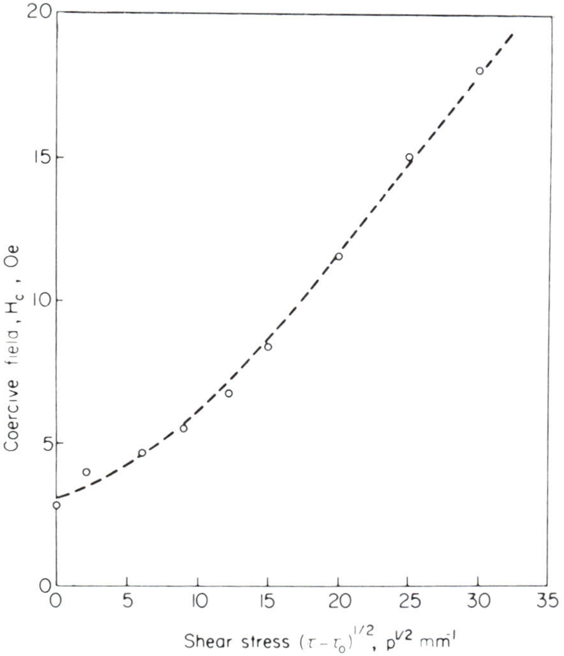

Fig. 5.43. Comparison between theory (dashed curve) and experiments (0) on the dependence of coercive force upon the state of strain. Data points were obtained from a rod-shaped Co single crystal having the *c*-axis at 50° from the rod axis (Traüble et al. [1965]).

motion of domain walls, thus adversely altering the shape of the magnetization curve and distorting and enlarging the hysteresis loop.

§2.4. *Recovery and recrystallization*; *texture*

Industrial statistics have shown that the largest tonnage of soft magnetic materials is consumed in the form of sheets or strips. Thus, rolling is an extremely important process of preparation in this industry. Normally the preparation procedure begins with the melting of raw elements into the desired alloy, followed by casting of the melt into large ingots. The ingots are then subjected to a series of rollings, initially at elevated temperatures for easy "breakdown" of the ingots. Whether intermediate annealing is required of the billets, slabs or plates

(a)

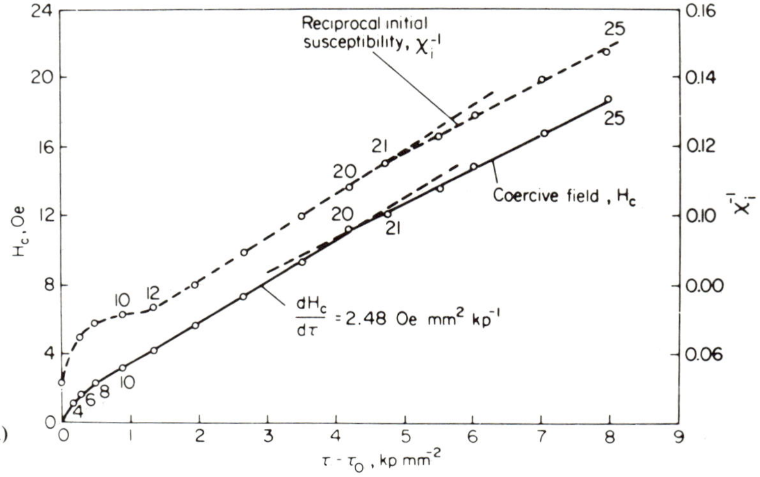

(b)

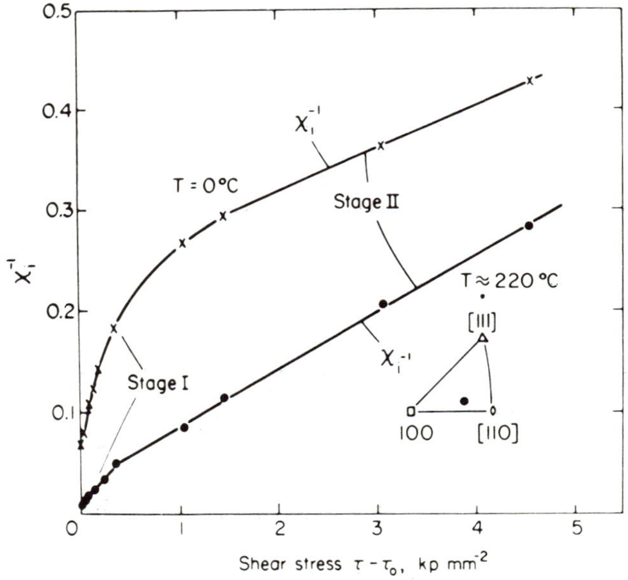

Fig. 5.44. Experimental evidence similar to that shown in figs. 5.42 and 5.43 was also obtained from Ni single crystals to support the theory of deformation dependence of (a) coercivity and (b) initial susceptibility (Traüble [1969]).

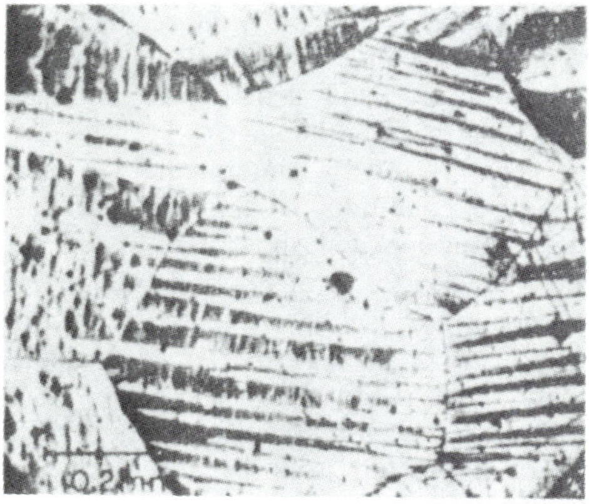

Fig. 5.45. Domain pattern showing the crossing of domains through grain boundaries (Bates et al. [1962]).

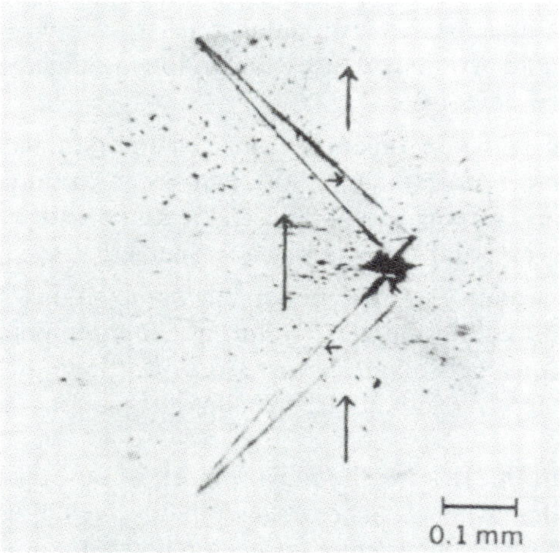

Fig. 5.46. Domain pattern around a void on the (001) surface of a Si–Fe crystal. Note that similar spikes were seen around an inclusion particle in fig. 3.20b (Chikazumi [1964]).

and whether cold rolling would be used at the end of the forming process depend on the so-called fabricability of the material, which, in turn, depends on (1) the crystal structure (fcc alloys generally show better fabricability than bcc and hcp alloys), (2) the grain structure and the mechanical behavior of the intermediate product and (3) the specified metallurgical conditions of the final product. In connection with (3), as later discussion will show, if a specific texture must be developed in the sheets, a series of hot and cold rollings coupled with intermediate annealing are usually required. By texture is meant that the majority of grains in the sheet are oriented crystallographically in a preferred orientation, not randomly.

The final sheet products prepared by cold rolling are in a severe mechanical state of stress and strain because of the heavy cold work performed on the sheets. They must be heat-treated at appropriate temperatures in a protective atmosphere. Such a heat treatment is called annealing, which often marks the final major step of a long and tedious procedure of preparation. Annealing of a cold-worked metal causes two basic phenomena, recovery and recrystallization. At moderate temperatures up to about one half of the absolute melting point, recovery occurs through one or more of the following processes:

(1) Migration of point defects. This process leads to partial removal of internal strains and the corresponding release of energy stored in the lattice during cold work. It causes the following changes in the distribution of lattice defects:

(a) annihilation between vacancies and self interstitials,
(b) trapping of vacancies or interstitial atoms at strain centers, such as grain boundaries, foreign atoms and dislocations and
(c) aggregation of point defects of each species.

(2) Plastic deformation. This occurs in some grains where residual stresses, initially within the elastic limit at room temperature, suddenly exceed the critical resolved shear stress for slip at the annealing temperature, thereby activating the slip dislocations to yield plastic strain.
(3) Polygonization. By far the most important process of recovery. In this process, edge dislocations rearrange themselves by glide (moving in the same plane) and climb (moving up or down to other planes). The term polygonization comes about because eventually edge dislocations of one sign ($\perp$ or $\top$) regroup into walls to form small-angle subgrain boundaries. Consequently each grain is broken up into several sub-

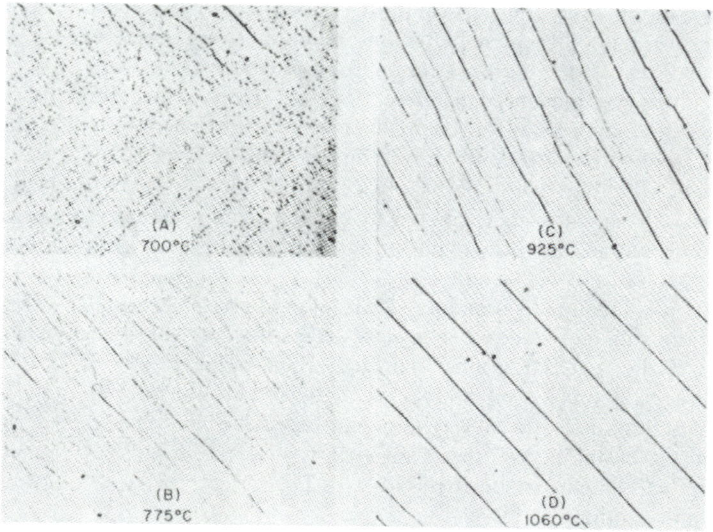

Fig. 5.47. Photomicrographs showing the process of polygonization in bent and annealed Fe–Si single crystals. All specimens were annealed for one hour at the indicated temperatures. Note that polygonization is more complete the higher the annealing temperature (Hibbard and Dunn [1956]).

grains of polygonal shapes. Figure 5.47 contains four photomicrographs to illustrate the sequential onset of polygonization in bent and annealed 3% Si–Fe single crystals as a function of temperature. In these micrographs, each fine dot is an etch pit representing a dislocation. Note that, after polygonization is completed (fig. 5.47D), most of the dislocations appear in the subgrain boundaries, leaving the interior of subgrains relatively free from dislocations.

(4) For screw dislocations, mutual annihilation of two screws of opposite sign (left- and right-handed) occurs when they run into each other. Because of the $\boldsymbol{b} \parallel \boldsymbol{l}$ property of the screws, where $\boldsymbol{l}$ is a unit vector along the dislocation line, a screw can cross slip from one slip plane to another intersecting slip plane without diffusion of atoms required for the climb process. Hence, mutual annihilation is an especially efficient mechanism of recovery for screws, even though it applies to edges as well. This mechanism, too, is held responsible for recovery in hexagonal metals, such as cobalt, deformed by easy glide.

In the latter case, the deformation is confined to the basal plane and therefore will not cause sufficient bending to produce excess edge dislocations of one sign necessary for polygonization.

Recovery is measured by detecting the changes in certain physical properties, such as electrical resistance, X-ray line broadening, hardness, yield strength and stored energy and by metallographic and transmission electron microscopy examinations. At the end of recovery, the material is expected to be free from residual stresses and, to some extent, from internal strains; but the original grains are only polygonized, not replaced.

When annealing is conducted at temperatures above $\frac{1}{2}T_m$, recrystallization takes place; whence new grains are nucleated and grown to replace the original grains. Through recrystallization, not only can strain-free grains be introduced in place of the old ones, but the size and orientation of the new grains can be controlled. The former result assures the stability of the grain structure in the application, whereas the latter changes make it possible to introduce attractive textures in the final product.

Strictly speaking, however, the development of a desired texture in a soft magnetic material relies more on the so-called abnormal grain growth in the post-recrystallization stage. Recrystallization is considered complete when all original grains are replaced by new grains. Further growth of the new grains marks the third stage of annealing known as the grain growth. While the driving force for recrystallization is provided by strain energy stored in the original grains, the driving force for grain growth is derived from the surface energy associated with grain boundaries of the new grains. The total grain boundary energy will decrease progressively during grain growth. It can be shown from arguments based on surface tension of grain boundaries that the equilibrium grain shape takes the form of a polygon of six sides with a dihedral angle of 120°. At high temperatures, polygons with fewer sides tend to shrink, whereas polygons with more sides tend to have curvatures concave outward from their centers and, on the average, will grow by grain boundary migration under the action of forces induced by surface tensions of grain boundaries. Under these conditions, the grain growth is continuous, and a uniform coarsening of grains usually occurs.

For soft magnetic materials, if textures are to be developed, the more important phenomenon of annealing is the secondary recrystallization. This is a process whereby a material that has been given an annealing treatment of primary recrystallization and normal grain growth at an elevated temperature is taken to a higher temperature to enable a few grains to grow at an abnormally rapid rate. The basic mechanism of the process remains unsettled as to whether oriented-nucleation or oriented-growth plays the key role (Beck and Hu [1966]). Meanwhile, many common features of the secondary recrystallization have been established, notably the following three: (1) Nuclei of a certain orientation dominate the process in each case; but the orientation could vary drastically under different annealing conditions. (2) Certain critically dispersed second phase precipitates, such as MnS and AlN, serve as inhibitors to impede normal grain growth of the matrix grains, thus promoting the growth of only a few grains for secondary recrystallization. And (3) the abnormal growth results in grains of unusually large sizes and sometimes converts the entire polycrystalline aggregate to a single or pseudo-single crystal. All these features contribute to the development of textures in Si–Fe. A typical example of the secondary recrystallization is shown in a 5% Si–Fe sample in fig. 5.48. If the material undergoes another abnormal grain growth after the second recrystallization, this is known as the tertiary recrystallization. Both secondary and tertiary recrystallizations have been applied to develop textures in soft magnetic materials, as the following discussion will show.

We have given a general definition of texture early in this section. More specifically, a texture developed in a rolled sheet is defined by two orientation relationships:

$(hkl) \parallel$ the rolling plane and, therefore, the surface of the sheet and
$[uvw] \parallel$ the rolling direction,

where (hkl) is the crystal plane of the majority of grains and $[uvw]$ is a crystal direction contained in the (hkl) plane (thus, $hu + kv + lw = 0$). The considerable gain from a desired texture in the performance of a magnetic material has allured some metallurgists to devote many years to developing such a texture.

Texture may be produced in an ingot by controlled solidification. Body-centered cubic metals (Fe) and alloys (Si–Fe and ferritic chrome steel) usually have their columnar grains extending in the [100] direction. The longitudinal direction of their dendrite arms is also [100]. Controlled solidification probably is the least expensive approach to produce a texture if the final product prepared by casting is acceptable. Efforts to apply this method to develop a texture in magnetic sheets, however, were mostly unsuccessful. To date, the production of a

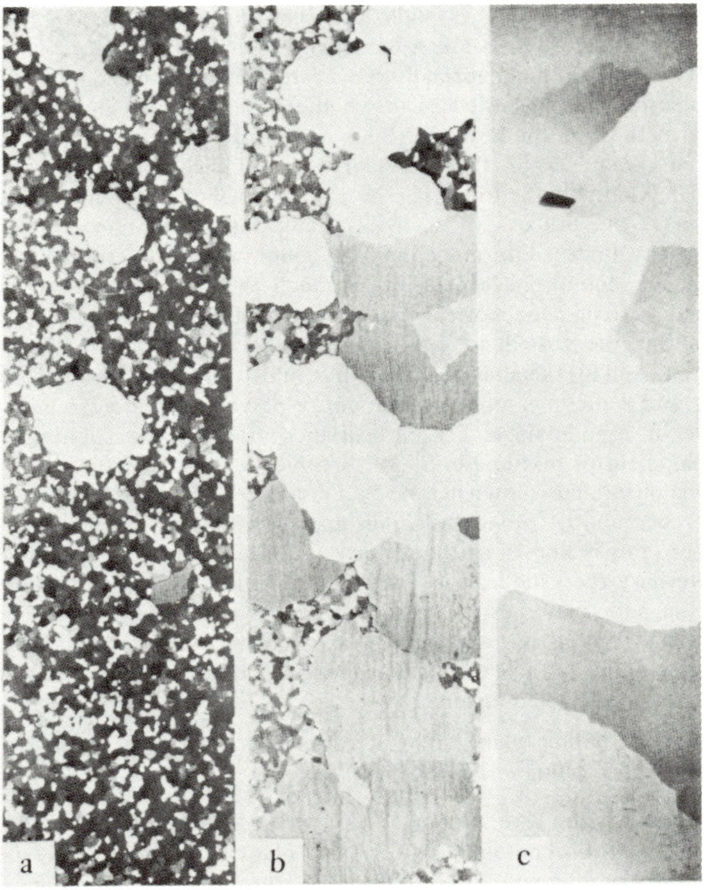

Fig. 5.48. Evidence for the preferential formation of (110)[001]-oriented grains by secondary recrystallization in 5% Si–Fe (Graham [1969]).

texture in sheet products of soft magnetic materials relies exclusively on cold rolling and subsequent annealing.

There are two important alloy systems for which textures have been developed for magnetic gains. The first and most significant achievement was made with the 3–4% Si–Fe alloys. Because the alloys are body-centered cubic, thus having the ⟨100⟩ easy axes of magnetization, and because the alloys are used largely in sheet form with the magnetizing field usually applied parallel to the plane of the sheet, it is obvious that an alignment of the ⟨100⟩ directions in the sheet plane should be highly beneficial. The first texture that could be successfully developed in Si–Fe on a massive scale was reported by Goss in Germany in 1934 (U.S. Patent 1 965 559). Goss achieved the magnificent breakthrough in producing a fairly sharp (110)[001] texture in Si–Fe by a proper combination of cold rolling and intermediate and final annealing. This texture, also known as the cube-on-edge or singly-oriented texture, is illustrated in fig. 5.49a. Since the Goss report, many industrial laboratories have modified his original process to avoid patent infringements and to improve the yield and quality of the texture. Consequently, the exact process used to produce the textured Si–Fe varies and remains confidential. Nevertheless the following procedure covers the essential steps of all commercial processes:

(1) Prepare the melt in an open-hearth furnace and cast into large moulds. The alloy usually contains 3–4% Si, 0.06–0.1% Mn, 0.03% C, 0.02% S and various minor alloying elements such as Cu, Sn, Al and P.
(2) Preheat the ingots up to 1400°C and hot roll the ingots to slabs, 25–76 mm thick.
(3) Reheat the slabs to 1300–1400°C and hot roll the slabs to plates, 1.5–2.5 mm thick.
(4) Pickle the plates in an acid solution to remove the oxide scale.
(5) Cold roll the plates in two steps of equal reduction to a final thickness of 0.25–0.36 mm (0.010–0.014″), separated by an intermediate anneal at 800–1000°C in a reducing atmosphere.
(6) Anneal briefly in wet hydrogen at 800°C to decarburize.
(7) Finally, anneal in dry hydrogen at 1200°C for texture-development.

The final sheets prepared in the above procedure are coarsely grained (1–5 mm in grain diameter) with more than 90% of the grains having the (110)[001] orientation. The texture is sharp also in another aspect that the deviation from the indicated orientation relationships is fairly small, especially in the plane alignment as shown in fig. 5.49b. The best commercial sheets with the Goss texture can reach an induction of ~1.82 T (18 200 G) under a magnetizing field of only 800 $A \cdot m^{-1}$ (10 Oe) and show a total energy loss of about 1.04 $W \cdot kg^{-1}$ (0.6 $W \cdot lb^{-1}$) at $B = 1.50$ T, 60 Hz and 0.30 mm (0.012″) thickness. More recent developments indicate that further improvements in the sharpness of the Goss texture and in the magnetic properties can be achieved by using aluminum

(a)

(b)

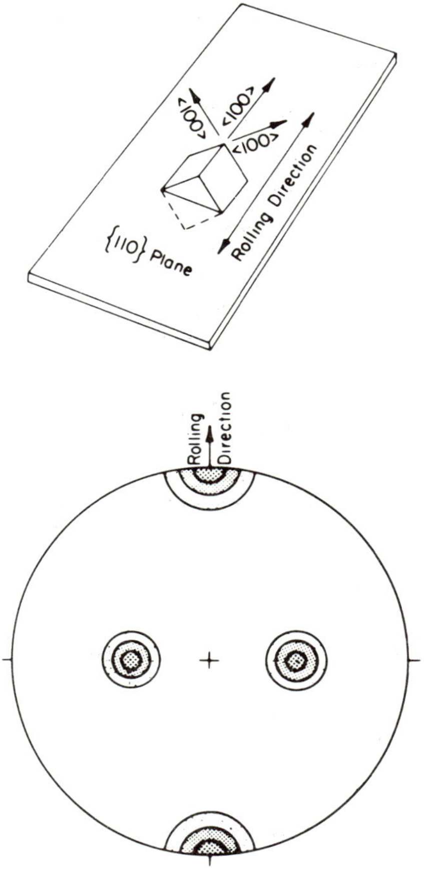

Fig. 5.49. (a) Schematic illustration and (b) {100} pole figure of the (110)[001] or Goss texture (Graham [1969]).

nitride (AlN) in addition to MnS as the inhibitors in the final anneal (Step 7), which results in the texture-forming secondary recrystallization (Tanguchi and Sakakura [1969]).

The mechanism of yielding the (110)[001] texture had not been understood until some 20 years after the work of Goss. Primary recrystallization was suspected in the early interpretation but had to be ruled out because strained Si–Fe was observed to undergo primary recrystallization in a few minutes at ≈800°C, which resulted in a number of weak textures, none being the (110)[001] orientation, in an otherwise strong background of random orientation. The correct mechanism is based on the secondary recrystallization, which selectively forms (110) grains so as to minimize the surface energy of the system by keeping the surfaces of grains in the crystal planes of the highest atomic density.

The Goss texture has made the 3% Si–Fe sheets extremely attractive for applications in power transformers and large generators, but there are some limitations inherent in this material. Having the (110) plane oriented parallel to the plane of the sheet, the Goss texture also places the $[\bar{1}10]$ and $[\bar{1}11]$ directions at 55° and 90°, respectively, to the rolling direction in the plane of the sheet. Because the $[\bar{1}10]$ and $[\bar{1}11]$ directions are harder directions of magnetization than the [001] direction, higher losses and lower permeability result when the flux must travel at an angle to the rolling direction. This occurs, for instance, at the corners of a lapped-joint transformer core or when the flux must travel in a radial direction in the rotor or stator of a motor or generator. The problem would be circumvented if there are two ⟨001⟩ easy directions lying in the sheet plane, one in the rolling direction and the other at the transverse direction. The latter situation exists with the well-known (100)[001] cube texture (fig. 5.50a) commonly found in fcc metals and alloys.

The advantages of having two easy axes of magnetization in the Si–Fe sheets had long been recognized. As early as 1935, Sixtus reported the detection of the (100)[001] component in Si–Fe. A year later, Bitter patented a method that was capable of producing a (100) [001] texture in Si–Fe. The latter texture would also have two {001} easy axes at ±45° from the rolling direction in the plane of the sheet. However, Bitter's patent was never practised because it requires cross rolling (that is, rolling the sheet alternately in two mutually perpendicular directions by turning the sheet 90° each time), which is highly impractical. Interest in the cube texture was revived in 1957 when metallurgists at Vacuum-Schmelze, A. G., Germany (Assmus et al. [1957]) announced the production of 3% Si–Fe sheets with more than 90% of the grains having the (100)[001] orientation. Their accomplishment was inspired by a discovery that Si–Fe sheets separated by a new oxide (Al_2O_3) coating during the final high temperature anneal exhibited unusually good magnetic properties. They traced these superior properties to the texture. In trying to find the reason for the production

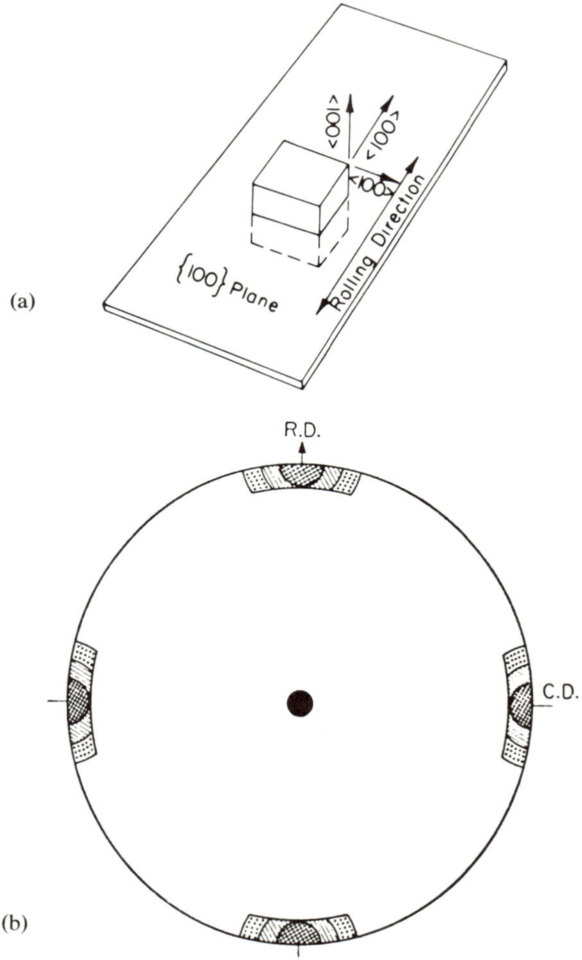

Fig. 5.50. (a) Schematic illustration and (b) {100} pole figure of the (100)[001] or cube texture ((a) Graham [1969]; (b)Wiener et al. [1958]).

of the cube texture, they discovered that the new oxide did not react with the metal surface, whereas the ordinarily used magnesium oxide was reactive.

The Vacuum-Schmelze method is now believed to base on the

anisotropy of surface energy. In surface science, it is a well-known theorem that, when the material is perfectly clean, the specific surface energy associated with the $\{hkl\}$ planes, $\gamma_{\{hkl\}}$, is the lowest in the most densely packed plane. Hence, for the bcc lattice in the clean state the lowest surface energy would be associated with the {110} planes, and $\gamma_{\{110\}} < \gamma_{\{100\}} < \gamma_{\{111\}}$. When the material is contaminated, the γ value is decreased for all planes, but to different degrees. For instance, contaminated bcc metals and alloys including Si–Fe commonly show $\gamma_{\{110\}}/\gamma_{\{100\}} > 1.0$, instead of < 1.0. Thus by orienting the (100) plane in the surface of a slightly contaminated Si–Fe sheet, the energy of the system is reduced because the surface energy is kept at a minimum.

As with the Goss texture, the original Assmus–Detert–Ibe method for production of the cube texture has been modified by various laboratories, and the actual process is kept as a trade secret. Information gathered from patents and technical publications indicates that the process generally involves the following important steps:

(1) Prepare the Si–Fe alloy as pure as economically possible in an open-hearth furnace. A state of high purity is a prerequisite for the success of the endeavor.
(2) Hot and cold roll the material to appropriate thicknesses interspersed with intermediate anneals in suitable atmospheres.
(3) Develop a primary (or secondary) texture of the (110)[001] or (120)[001] type in the anneal following the final rolling.
(4) Conduct a final anneal to develop the cube texture by secondary or tertiary recrystallization in a closely controlled atmosphere.

Besides the requirement of high purity for the alloy, the final anneal is regarded as the most crucial step because the formation of the cube texture and the sharpness of the texture both depend upon the quality and control of the atmosphere used in this step. Oxygen or (and) sulfur is considered an effective contaminant to control the surface energy such as to initiate the abnormal growth of the few pre-existing (100) [001] grains. Several metals were also suggested to help control the atmosphere. Metals prominently mentioned for this purpose are Ti, which presumably acts as an oxygen getter, and Ni, Co or Pt, which acts as a catalyst to promote the decomposition of molecular hydrogen.

The cube texture produced by the process just outlined is quite sharp. Figure 5.50b shows a (100) pole figure obtained from X-ray examination of a cube-oriented 3% Si–Fe in the early stage of the development program in a Westinghouse laboratory. The pole figure reveals that all grains have their (100) planes in the sheet plane within 6°, and deviations of the [001] direction from the rolling direction are: 63% within ±6°, 88% within ±10° and 98% within ±15°. These results are consistent with the mechanism of the formation of the texture, for which the driving force for the abnormal growth of the (100)[001] grains is derived from the anisotropy of surface energy. Hence the crystal plane, not the crystal direction, is expected to be closely controlled in the texture.

The magnetic properties of the cube-oriented 3% Si–Fe have been evaluated and compared with those of the singly-oriented and hot-rolled 3% Si–Fe. The striking advantage of the cube texture is shown in fig. 5.51, which compares the total energy loss of randomly-, singly- and cube-oriented 3% Si–Fe in various directions. The loss of the cube-oriented sheets is at a minimum in all directions. In this particular evaluation, the loss of the singly-oriented sheets in the cross direction

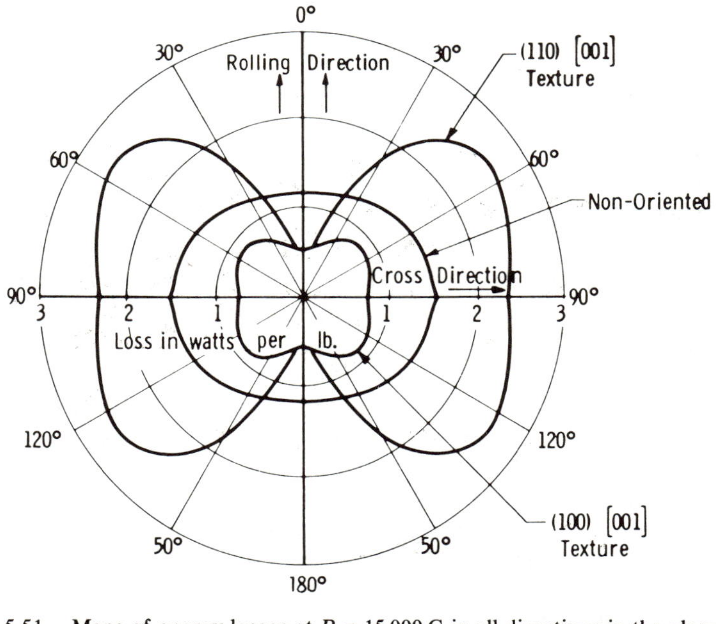

Fig. 5.51. Maps of energy losses at $B = 15\,000$ G in all directions in the plane of the sheet of three grades of 3.5% Si–Fe showing the superior performance of the cube-textured sheets (Bechtold and Wiener [1965]).

(i.e., 90° from the rolling direction) is much higher than the losses of both cube- and randomly-oriented materials.

Because of the apparent magnetic advantages of the (100)[001] texture, many rosy predictions have been made for the cube-oriented Si–Fe. The optimism was so overwhelming in the early 1960's in several industrial laboratories that one was led to believe that the replacement of the singly-oriented Si–Fe by the cube-oriented sheets would occur in a matter of a few years. Another decade has since elapsed, however, and the cube-oriented material has found hardly any applications. Its present gloomy status makes one wonder whether the promises previously made were premature. One factor that has hindered the use of the new textured Si–Fe sheets is their high cost. Even though the price of the new sheets will not be firmly quoted until a steady market is established, no doubt the cube-oriented sheets will be more expensive than the singly-oriented sheets simply because it is more difficult, and therefore more costly, to produce the cube texture. To make the situation worse, the formation of the cube texture is so dependent upon the state of impurity of the sheets and the atmosphere in the final anneal that the process can easily turn out to be unsuccessful, meaning possible low yield. Another factor is the yet unexplained finding that the cube-oriented sheets are somewhat inferior in the energy loss associated with eddy currents and perhaps with the anomalous component. Obviously, further work is needed on the reliability of the process and on the a.c. magnetic properties if the potentials of the cube-oriented Si–Fe are to be realized.

The second alloy system in which texture plays an important role is the nickel–iron alloys near the equiatomic composition. As shown in fig. 5.25, the quenched alloy FeNi has a finite value for K_1 ($\sim 0.1 \times 10^4\ \mathrm{J \cdot m^{-3}}$) as compared with the nearly zero value for the quenched alloy $Fe_{0.22}Ni_{0.78}$. Therefore, the former alloy is inherently at a magnetic disadvantage to compete with the latter. The introduction of the (100)[001] cube texture in FeNi, however, has improved the magnetic properties substantially because of the ⟨001⟩ easy axes of magnetization for this fcc alloy. An outstanding magnetic property of the cube-textured FeNi is the squareness of the hysteresis loop, which can be enhanced by applying a magnetic field during cooling after the final anneal. This, plus the fact that FeNi displays higher saturation magnetization, contains less nickel, and therefore is less expensive than $Fe_{0.22}Ni_{0.78}$ has enabled FeNi and its related alloys to be useful in some applications (see discussion in ch. 6).

The development of the cube texture in fcc metals and alloys is quite common and less difficult than with the bcc Si–Fe. For the Fe–Ni alloys, the prerequisite for the formation of the cube texture is a final severe (≈95%) reduction in thickness by cold rolling. The final anneal should be conducted in a reducing atmosphere of hydrogen at 1000–1200°C.

Although this discussion has centered on the recrystallization texture developed in sheet products by rolling and annealing, other forming processes, such as extrusion and wire-drawing, are also capable of producing very strong deformation textures, from which strong recrystallization textures of either the same or different types can, in turn, be developed upon annealing. For further information on the subject, the reader is advised to consult the book by Barrett and Massalski [1966] and an ASM seminar book edited by Margolin [1966]. Finally, it should be mentioned that little use has been made of the high saturation magnetizations of heavy lanthanides such as Gd, partly because of the large values of magnetocrystalline anisotropy constants reported for these metals. Thus, for polycrystalline Gd with random orientation, the magnetizing field required to saturate the material is much higher than $1.6 \times 10^6\ \mathrm{A \cdot m^{-1}}$ (20 000 Oe) provided by a conventional electromagnet. An effective method to solve this problem would be to develop an appropriate texture in the lanthanide. In view of the latest success with Si–Fe in the development of the cube texture, it is reasonable to forecast that such an effort on magnetic lanthanides may prove rewarding.

§2.5. *Atomic order*

A number of substitutional alloys are ordered at low temperatures and disordered at high temperatures. By "ordered" is meant that the lattice sites are grouped into sublattices, each of which is occupied predominantly by one species of atoms. The order–disorder transition in magnetic alloys interests us in a natural way because atomic or chemical ordering is strikingly analogous to magnetic ordering. Both are cooperative phenomena characterized by a sharp critical temperature and both are controlled by interactions between atoms or spins. However, our interest in atomic ordering is more than just scientific because many important soft magnetic materials base their compositions on certain stoichiometric solid solutions of Fe, Co and Ni, the majority of which are susceptible to atomic ordering. Moreover, ordering changes magnetic properties as well as other physical properties to varying degrees. Therefore, from the practical viewpoint, atomic ordering is another structural factor that we must deal with in the preparation of soft magnetic alloys.

Before discussing the effect of atomic order on properties, it is necessary to review three aspects of the ordering phenomenon; namely, the statistical theory and thermodynamics, crystallography and kinetics. The review is intended to introduce the principles underlying these aspects so that the subsequent discussion of property changes can be given in proper terms. Further basic information can be found in the articles by Muto and Takagi [1955] and Guttman [1956].

Statistical theories and thermodynamics

As for the magnetic ordering in a spin system, the theory of the formation of an ordered lattice (a superlattice) is a part of the equilibrium theory of the statistical mechanics of a cooperative system. Thus, the prominent theories of atomic order are also based on the Ising model discussed in ch. 2, §3. In their early development, the theories were confined to long-range order, which extends throughout the entire crystal or at least over many interatomic distances. The degree of long-range order is defined by a fraction W, which varies from zero at complete disorder to unity at complete order. More precisely, W may be $+1.0$ or -1.0 for complete order as illustrated in the next discussion.

Let us consider the alloy FeCo, of which the bcc structure may be resolved into two simple cubic lattices, α and β, interpenetrating each other (fig. 5.52a). Suppose there are N atomic sites in the lattice. Thus, for a perfectly pure crystal, there should be $\frac{1}{2}N$ Fe atoms and $\frac{1}{2}N$ Co

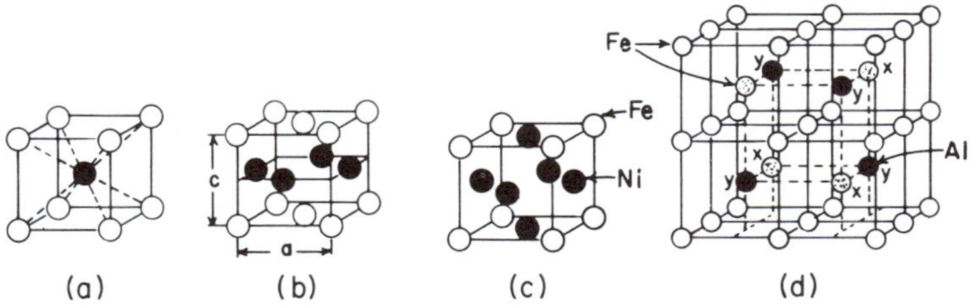

Fig. 5.52. Four types of superlattices commonly observed in ferromagnetic alloys: (a) $L2_0$ type seen in ordered FeCo, (b) $L1_0$ type seen in ordered FePt, (c) $L1_2$ type seen in ordered Ni_3Fe and (d) DO_3 type seen in ordered Fe_3Al.

atoms. Formally, the degree of long-range order is defined by

$$W = \frac{f_\alpha - F_A}{1 - F_A} = \frac{f_\beta - F_B}{1 - F_B}, \qquad (5.64)$$

where F_A and F_B are the fractions of A and B atoms in the alloy ($F_B = 1 - F_A$) and f_α and f_β are the fractions of α and β sites "rightly" occupied by A and B atoms, respectively. To verify this definition, we note that, when

(1) $W = 0$, the α and β sites will be occupied in equal numbers by Fe and Co atoms. Therefore, No. of Fe atoms in α sites = No. of Fe atoms in β sites = No. of Co atoms in α sites = No. of Co atoms in β sites $= \frac{1}{4}N$, and $f_\alpha = f_\beta = \frac{1}{4}N/\frac{1}{2}N = \frac{1}{2}$.
(2) $W = +1.0$, all α sites will be occupied by Fe atoms and all β sites by Co atoms, and $f_\alpha = f_\beta = \frac{1}{2}N/\frac{1}{2}N = 1$.
(3) $W = -1.0$, the occupancy is exactly reversed from (2) so that Fe atoms will occupy the β sites and Co atoms will occupy the α sites. Hence $f_\alpha = f_\beta = 0$. And
(4) $0 < W < +1.0$ for partial order, No. of Fe atoms in α sites = No. of Co atoms in β sites $= \frac{1}{4}N(1 + W)$, No. of Fe atoms in β sites = No. of Co atoms in α sites $= \frac{1}{2}N - \frac{1}{4}N(1 + W) = \frac{1}{4}N(1 - W)$, and

$$f_\alpha = f_\beta = \frac{\frac{1}{4}N(1 + W)}{\frac{1}{2}N} = \tfrac{1}{2}(1 + W). \qquad (5.65)$$

The numbers and fractions cited in all four cases are consistent with (5.64).

In the Ising model, the internal energy U of the alloy AB in a bcc lattice is approximated to the sum of the bond energy of all nearest neighbors. Thus we have

$$U = N_{AA}U_{AA} + N_{BB}U_{BB} + N_{AB}U_{AB}, \qquad (5.66)$$

where N_{AA} is the number of AA pairs in nearest neighbors, each of which has the bond energy U_{AA}, etc. The probability that an A atom on an α-site will form an AA pair is equal to the probability of one of its z nearest neighbors being occupied by an A atom times the coordination number z. Whence, with $z = 8$,

$$N_{AA} = \tfrac{1}{4}N(1 + W)[\tfrac{1}{4}N(1 - W)/\tfrac{1}{2}N]z = N(1 - W^2). \qquad (5.67a)$$

Similarly, $N_{BB} = N(1 - W^2)$, and

$$N_{AB} = \tfrac{1}{4}N(1+W)[\tfrac{1}{4}N(1+W)/\tfrac{1}{2}N]z + \tfrac{1}{4}N(1-W)[\tfrac{1}{4}N(1+W)/\tfrac{1}{2}N]z$$
$$= 2N(1+W^2). \tag{5.67c}$$

The internal energy in (5.66) becomes

$$U = U_0 + 2NW^2\Delta U, \tag{5.68}$$

where

$$U_0 = N(U_{AA} + U_{BB} + 2U_{AB}), \tag{5.69a}$$

and

$$\Delta U = U_{AB} - \tfrac{1}{2}(U_{AA} + U_{BB}). \tag{5.69b}$$

By applying the same procedure described in ch. 2, §3, we obtain the configuration entropy associated with atomic order

$$S_{\text{mixing}} = Nk \ln 2 - \tfrac{1}{2}N_k[(1+W)\ln(1+W) + (1-W)\ln(1-W)], \tag{5.70}$$

and an expression for atomic order under the equilibrium condition $(\partial G/\partial W = 0)$ from (5.47) at T

$$W(8\Delta U/kT) = -\ln[(1+W)/(1-W)]. \tag{5.71a}$$

This is an important result of the theory developed by Bragg and Williams (1934–5, see Cottrell [1948]), which predicts the equilibrium value for W as a function of ΔU and T. The temperature dependence of W thus predicted is depicted in fig. 5.53a. In comparison with ferromagnetism, (5.71a) may be expressed in another form (see Muto and Takagi [1955]),

$$W = \tanh[(z\Delta U/2kT)W], \tag{5.71b}$$

where z is the coordination number and has the value of eight, six or four for body-centered cubic, NaCl- or ZnS-type lattices, respectively. Note that (5.71b) resembles (2.23b), the expression for magnetization given by the Weiss molecular-field theory for the case $J = \tfrac{1}{2}$. Hence the curve shown in fig. 5.53a has the same general shape as that of a M versus T plot shown in fig. 2.13.

Another important result of the Bragg–Williams theory is the determination of the critical temperature T_c at which long-range order disappears completely ($W = 0$). With the aid of the expansion $\ln(1+x) = x - \tfrac{1}{2}x^2 + \tfrac{1}{3}x^3 - \dots$ for $-1 < x < 1$, (5.71) leads to

$$T_c = -4\Delta U/k. \tag{5.72}$$

For ordering to occur, T_c must be positive and hence ΔU must be negative. Whence the thermodynamic criterion for an alloy to be

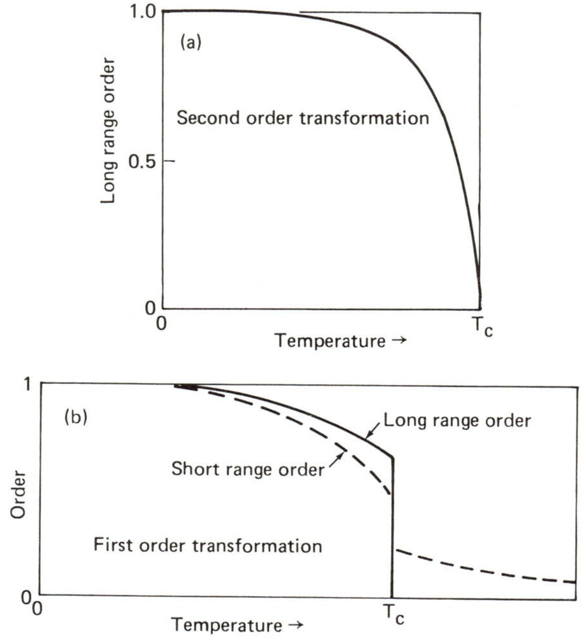

Fig. 5.53. (a) Temperature dependence of the degree of long-range order in an AB alloy. The transformation is second degree. (b) Long- and short-range order predicted for an AB_3 alloy with the long-range ordering of first degree (Kittel [1976]).

orderable is

$$U_{AB} < \tfrac{1}{2}(U_{AA} + U_{BB}). \tag{5.73}$$

The physical significance of this criterion is that atomic ordering is feasible only when the bond energy of the unlike atom pair is less than the mean bond energy of the two kinds of like atom pairs. In the same context, it can be further stated that when $U_{AB} \ll \tfrac{1}{2}(U_{AA} + U_{BB})$, the tendency for the A and B atoms to form pairs is so strong that the alloy system will eventually form one or more intermetallic compounds in the intermediate range. On the other hand, if $U_{AB} > \tfrac{1}{2}(U_{AA} + U_{BB})$, the alloy system will tend to form mixtures of two A- and B-rich solid solutions, in which like atom pairs dominate.

The Bragg–Williams theory involves the obviously oversimplified assumption that interactions between atoms of the second, third,... nearest neighbors can be neglected. Also, it fails to account for the existence of short-range order. Early specific heat measurements and X-ray studies had established the presence of antiphase domains at temperatures well above and slightly below T_c. In each domain, the atoms are ordered to a very high degree, which, however, results in $W = 0$ on average for the whole crystal. Antiphase domains are illustrated in a two-dimensional diagram in fig. 5.54a. In this diagram, there are two domains in which the atoms are perfectly ordered, but their occupancies of the α and β sites are reversed. Hence, $W = +1.0$ for the left domain, and $W = -1.0$ for the right domain. The two domains are said to be out-of-phase, and the boundary is called an antiphase domain boundary because all the atom pairs are alternately AA and BB, as compared with the AB type inside the domains. Transmission electron microscopy evidence for antiphase domain boundaries is shown in fig. 5.54b. Despite the complete order in each domain, the average degree of order over the two domains is zero.

To improve the theory, Bethe (1935, see Ziman [1964d]) treated the problem by a different method. The results of his treatment are equivalent to a quasi-chemical approximation, which invokes the chemical reaction $AA + BB \rightarrow 2AB$ to calculate the equilibrium concentrations of the three kinds of atom pairs from the relationship:

$$\frac{\langle N_{AB}\rangle^2}{\langle N_{AA}\rangle\langle N_{BB}\rangle} = 4\frac{(e^{-U_{AB}/kT})^2}{e^{-U_{AA}/kT}\,e^{-U_{BB}/kT}} = 4(e^{-\Delta U/kT})^2. \tag{5.74}$$

In his analysis, Bethe concentrated on the formation of AB pairs in a cluster of atoms and calculated the number of these "right" pairs as a function of the size of the cluster. As the temperature is increased, the number of the right atom pairs diminishes and the numbers of the wrong (AA and BB) pairs increase until a disordered state is reached in which the pairs are half right and half wrong. His analysis led to the introduction of a parameter of short-range order σ, which is defined as the probability of finding an unlike atom in the nearest neighbor of a given atom minus the probability of finding a like atom there. This definition is equivalent to stating that the probability for a certain A atom to have a B atom at a nearest neighbor is $\frac{1}{2}(1+\sigma)$, whereas the probability for the same A atom to have another A atom there is $\frac{1}{2}(1-\sigma)$. When equilibrium is reached at a temperature T, the ratio of these probabilities is governed by the Boltzmann factor

$$\tfrac{1}{2}(1+\sigma)/\tfrac{1}{2}(1-\sigma) = e^{-\Delta U/kT}. \tag{5.75}$$

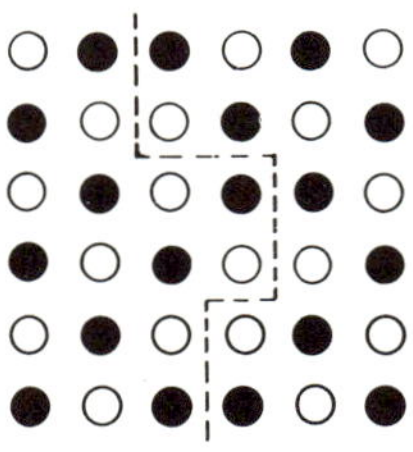

(a)

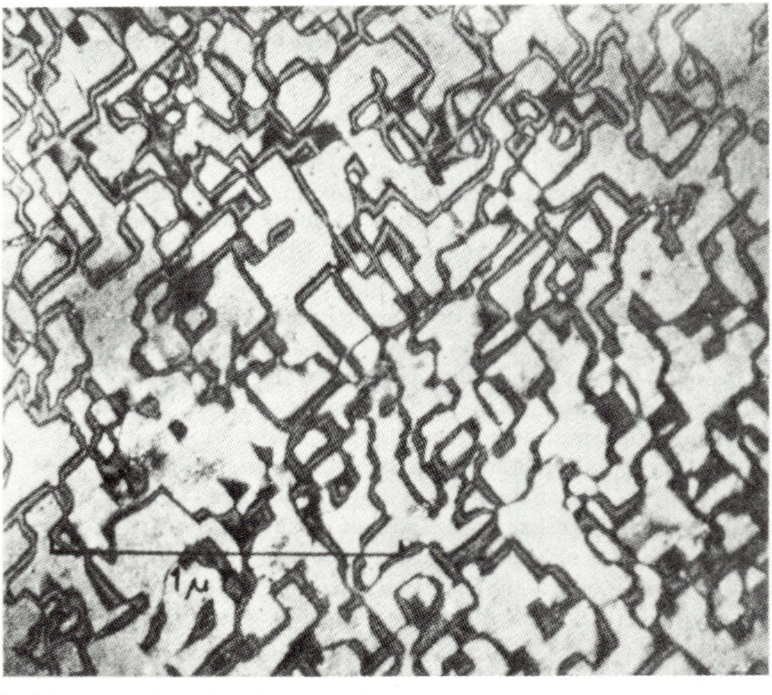

Fig. 5.54. Antiphase domain boundaries characterizing short-range order. (a) A two-dimensional diagram showing the like-atom pairs across the boundary and (b) electron microscopy evidence ((b) Fisher and Marcinkowski [1960]).

The mathematical treatment of the Bethe theory is beyond the scope of this book, and we shall, therefore, only cite the principal results. From (5.75), the theory predicts that σ will have a finite value no matter how high the temperature is. In other words, short-range order is expected to persist at all temperatures despite the effect of thermal agitation. This situation is shown in fig. 5.53b. The second important result of the Bethe theory deals with the mechanism whereby long-range order is produced. In the Bragg–Williams theory, we arrived at the conclusion that an alloy like CoFe will tend to order below T_c provided $U_{AB} < \frac{1}{2}(U_{AA} + U_{BB})$. The Bethe theory now predicts the existence of short-range order at and below T_c. How, then, can we reconcile the co-existence of the two kinds of order? To resolve the difficulty, Bethe suggested that because the nearest neighbors across the antiphase domain boundaries are atoms of similar kind, such boundaries must have high internal energy and become unstable at low temperatures. Consequently, there is a strong tendency for domains to grow so as to eliminate the boundaries. At sufficiently low temperatures, the growth of domain will be completed in the sense that all antiphase boundaries are eliminated and each crystal in the alloy is also a single domain.

From the standpoint of thermodynamics, long-range order can be classified as first or second degree* or kind according to the order of the lowest derivative of the (Gibbs) free energy showing a discontinuity at the transition temperature. For this purpose, we recall the following thermodynamic relations:

$$\left(\frac{\partial G}{\partial T}\right)_p = -S \quad \text{and} \quad \left(\frac{\partial G}{\partial p}\right)_T = V, \tag{5.76}$$

and

$$\left(\frac{\partial^2 G}{\partial T^2}\right)_p = -\left(\frac{\partial S}{\partial T}\right)_p = -\frac{C_p}{T} \quad \text{and} \quad \left(\frac{\partial^2 G}{\partial p^2}\right)_T = \left(\frac{\partial V}{\partial p}\right)_T = -\beta V, \tag{5.77}$$

where C_p is the heat capacity at constant pressure and β is the isothermal compressibility. At T_c, the ordered and disordered phases are in equilibrium; therefore the free energies of the two phases are equal. The continuity of the free energy in both degrees of order is

*More accurately, the transition should be classified as the first or second order of ordering. Such terms are obviously awkward. Thus, we prefer to use "degree" in place of "order" in ranking the transition.

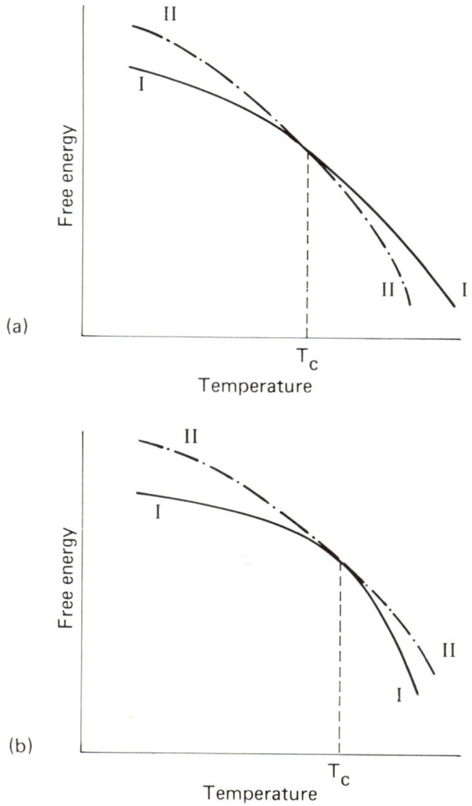

Fig. 5.55. Free energy versus temperature for (a) a first-degree and (b) a second-degree transition (Guttman [1956]).

shown in fig. 5.55. The first degree of order is characterized by a discontinuity in the S–T and V–T plots and the second degree of order by a lambda curve in the C_p–T plot (fig. 5.56). Thermodynamic arguments put forward by Stout [1948] indicate that, in a binary alloy, a second-degree transition requires not only equality of entropy and volume of the ordered and disordered phases, but also identical composition of the two phases. Thus a first-degree ordering appears on a phase diagram as two lines bounding the region where the ordered and disordered phase of different compositions coexist in equilibrium; and a

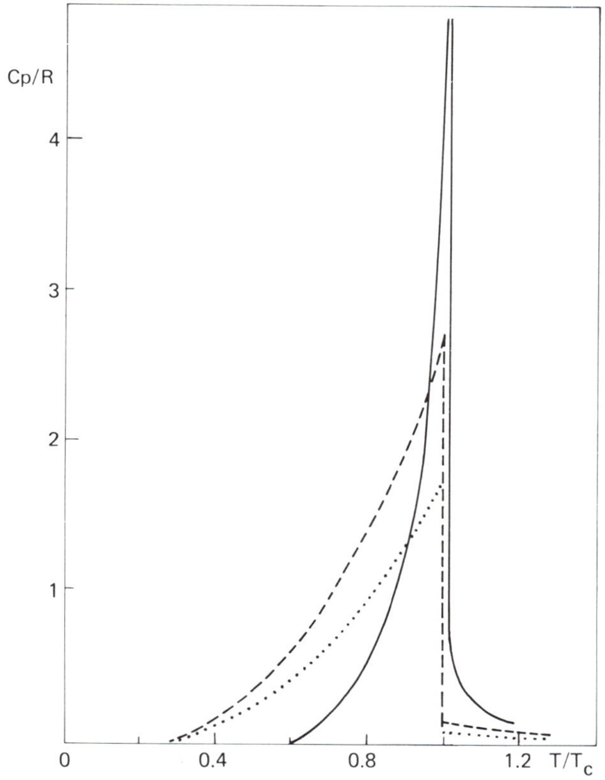

Fig. 5.56. Heat capacity versus temperature plots showing the lambda shape characteristic of the second-degree transition. Solid line – experimental results obtained from β-brass; dotted line – quasi-chemical approximation without lattice vibration; dashed line–quasi-chemical approximation with lattice vibration (Guttman [1956]).

second-degree ordering appears as a single line. Examples of the two-phase region representing the first-degree ordering are seen in the constitution diagrams of the Fe–Co (around FeCo), Fe–Ni (around Ni_3Fe) and Fe–Pt (around FePt) systems (see Hansen [1958]). The Fe–Pt system also exhibits two single lines centering around Fe_3Pt and Pt_3Fe to represent the second-degree ordering. Besides the appearance in the phase diagram, the distinction between the first- and second-degree ordering is usually made according to whether or not a latent

TABLE 5.25
Atomic order and magnetic behavior of prominent binary alloys of Fe, Co and Ni

Alloy	Atomic ordering		Crystal structure		Mode of spin coupling		Remark
	Degree	Temp. (°C)	Dis-ordered	Super-lattice	Dis-ordered	Super-lattice	
(I) Alloys of Fe, Co and Ni							
1. FeCo	1st	730	bcc	$L2_0$	Ferro.	Ferro.	Base of Permendur
2. Ni_3Fe	1st	500	fcc	$L1_2$	Ferro.	Ferro.	Base of Permalloys
3. FeNi	(a)	(a)	fcc	$L1_0$	Ferro.	Ferro.	(a) Ordering of FeNi is detectable only after the alloy has been irradiated under a magnetic field.
4. Ni_3Co	?	?	fcc	$L1_2$	Ferro.	Ferro.	
(II) Iron alloys							
5. Fe_3Al	?	~ 535	bcc	DO_3	Ferro.	Ferro.	Actual ordering sequence: bcc (disordered) $\rightarrow L2_0 \rightarrow DO_3$
6. Fe_3Si	?	~1250	bcc	DO_3	Ferro.	Ferro.	

7. Fe_3Pt	2nd	836	fcc	$L1_2$	Ferro.[b]	Ferro.[b]	(b) Atomic order causes a decrease in saturation magnetization in these five (7, 8, 9, 12 and 13) Pt-alloys, instead of an increase seen in other (1, 2, 5, 6 and 14) alloys.
8. FePt	1st	~1300	fcc	$L1_0$	Ferro.[b]	Ferro.[b]	
9. Pt_3Fe	1st	1352	fcc	$L1_2$	Ferro.[b]	Anti-ferro.[b]	
10. FePd	2nd(?)	~720	fcc	$L1_0$	Ferro.	Ferro.	
11. Pd_3Fe	2nd(?)	~800	fcc	$L1_2$	Ferro.	Ferro.	
(III) Cobalt alloys							
12. CoPt	1st	834	fcc	$L1_0$	Ferro.[b]	Ferro.[b]	Partially ordered CoPt makes an excellent permanent magnet.
13. Pt_3Co	1st	~780	fcc	$L1_2$	Ferro.[b]	Ferro.[b]	
(IV) Nickel alloys							
14. Ni_3Mn	?	~510	fcc	$L1_2$	Ferro.[c]	Ferro.[c]	(c) Most dramatic increase in M_s upon ordering. See text.
15. Ni_3Pt	2nd	580	fcc	$L1_2$	Ferro.	Ferro.	
16. NiPt	1st	645	fcc	$L1_0$	Ferro.	Anti-ferro.	

heat of transition ($\Delta H_T = \Delta S_T T_c$) is present. Both degrees of ordering are found in magnetic alloys. The classification and other descriptions of ordering in various magnetic systems are listed in table 5.25.

Crystallography. The ordering of FeCo in the bcc lattice represents a simple crystallographic case because the equiatomic composition and the subdivision of the lattice match perfectly. Thus ordering in this case only causes a slight increase in lattice parameter, as shown in fig. 5.57, with the bcc lattice remaining intact. The same situation prevails for Ni_3Fe in the fcc lattice, which can accommodate all Fe atoms at the corners of the unit cells and all Ni atoms at the face centers in the completely ordered state shown in fig. 5.52c. The ordered fcc structure of Ni_3Fe is designated as the Ll_2- or Cu_3AuI-type lattice, which is equivalent to four interpenetrating simple cubic lattices. The crystallography of other ordered magnetic alloys is more complex than FeCo

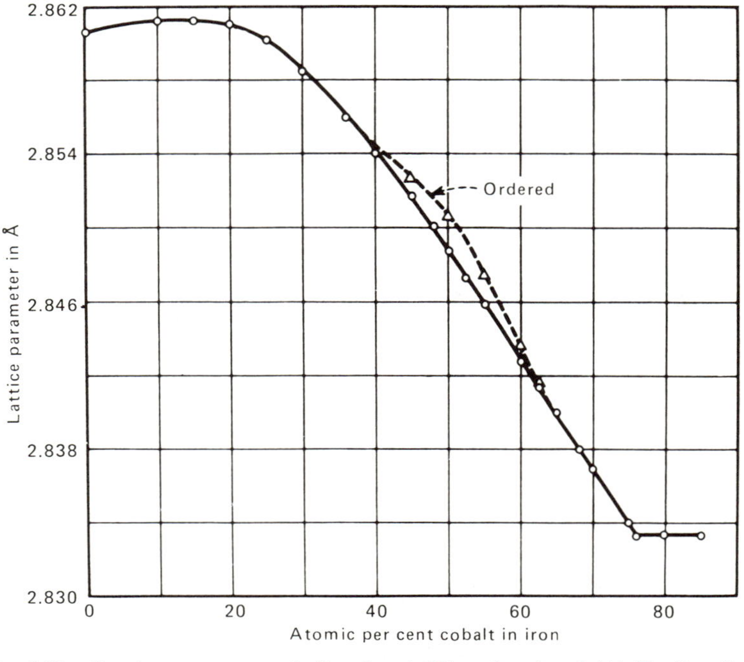

Fig. 5.57. Lattice parameter of disordered (○) and ordered (△) Fe–Co alloys (Bozorth [1951]).

and Ni_3Fe, however. For instance, the ordering of an AB alloy such as FePt in the fcc lattice necessitates the segregation of atoms of each species alternately on the (001) planes (fig. 5.52b), leading to a change in the original structure from fcc to an ordered face-centered tetragonal (fct) superlattice. (Note that fct is not one of the 14 independent space lattices. An accurate description of this ordered structure is body-centered tetragonal.) In still another case, the ordering of A_3B in the bcc lattice is accompanied by an enlargement of the unit cell by a factor of 8; that is, the unit cell of an ordered A_3B alloy such as Fe_3Al and Fe_3Si in the bcc lattice contains 8 original unit cells of the disordered structure (fig. 5.52d). In the DO_3-type superlattice of Fe_3Al, the Al atoms occupy the cube centers alternately in the [110] and $[\bar{1}10]$ directions, whereas the Fe atoms occupy the remaining cube centers as well as the corners.

Soft magnetic binary alloys that are susceptible to ordering fall mainly into the four types just described. The main features of these superlattices are summarized together with prominent examples in table 5.26.

Kinetics. To prepare an orderable magnetic alloy properly, we must know the time rate at which the equilibrium state of order or disorder is approached. If the ordered state is preferred for the product of the alloy, the temperature and time for the ordering treatment must be chosen according to the specific kinetics. On the other hand, if a disordered state is desired, we have to know first whether or not the ordering transition can be suppressed by quenching. If indeed this is the case, the critical quenching rate must still be ascertained. For either state or atomic order, the kinetics of the transition thus is extremely important in the preparation of the alloy product.

The formation of a superlattice involves the usual two processes – nucleation and growth – of many metallurgical phenomena. In atomic ordering we can envisage that, at temperatures much above T_c, small antiphase domains of partial order will constantly form and dissolve in an otherwise disordered alloy. As cooling occurs, this short-range order becomes more extensive until T_c is reached. Below T_c, some of the domains become stable and serve as nuclei. The growth of nuclei proceeds quite rapidly first until they meet. At this point, each crystal becomes an aggregate of antiphase domains. Further growth of domains will be more sluggish because there is no physical reason for any particular domain to grow at the expense of others. The driving force

TABLE 5.26

Four common types of superlattices observed in ferromagnetic alloys

Type	Crystal structure	No. of atoms per unit cell	Coordinates of atoms in the unit cell	No. of possible distinct domains	Examples
$L1_0$	Ordered face-centered tetragonal or more appropriately, body-centered tetragonal	4	2A in $(000, \frac{1}{2}\frac{1}{2}0)$ 2B in $(\frac{1}{2}0\frac{1}{2}, 0\frac{1}{2}\frac{1}{2})$	6	FePt, CoPt, NiPt, FePd
$L1_2$	Face-centered cubic	4	1B in (000) 3A in $(0\frac{1}{2}\frac{1}{2}, \frac{1}{2}0\frac{1}{2}, 0\frac{1}{2}\frac{1}{2})$	4	Ni_3Fe, Ni_3Mn, Fe_3Pt, Pt_3Fe, Pt_3Co, Ni_3Pt, Pd_3Fe
$L2_0$	Body-centered cubic	2	1A in (000) 1B in $(\frac{1}{2}\frac{1}{2}\frac{1}{2})$	2	FeCo
DO_3	Body-centered cubic with unit cells 8 times larger than those in the disordered lattice	16	4B in (000) f.c. 4A in $(\frac{1}{2}\frac{1}{2}\frac{1}{2})$ f.c. 8A in $\pm(\frac{1}{4}\frac{1}{4}\frac{1}{4})$ f.c.	4	Fe_3Al, Fe_3Si

for the coalescence of domains is provided by the surface energy associated with the antiphase boundaries, which tend to elimination during the process of growth. Bragg [1940] has shown that, for an aggregate of domains to be stable, at least four domains must meet at a corner and these domains must be crystallographically distinct. However, there are alloy systems that inherently have less than four distinct domains. For instance, the $L2_0$ superlattice formed in FeCo has only two distinct domains (Co atoms occupying α sites in one domain and β sites in the other). Hence, in FeCo and similar bcc alloys, stability cannot be attained in the group of three or more domains meeting at a corner because some of these domains are identical and will unite, leading to the eventual elimination of the corner.

The number of distinct domains in the four common superlattices are listed in table 5.26. Because the $L1_0$, $L1_2$ and DO_3 superlattices have four or six distinct domains, stability can be attained in these lattices, thereby retarding the conversion of multiple domains into a simple domain. The kinetics of ordering is thus relatively slow in these cases. Indeed, quenching has been proved effective in suppressing the transition wholly or partly in Ni_3Fe and Fe_3Al. On the other hand, the formation of the $L2_0$ superlattice in FeCo can proceed so readily that quenching is ineffective to suppress ordering. As a result, Permendur had to be introduced to replace FeCo. One purpose of adding 2% V to FeCo was supposedly to slow down the kinetics of ordering. This and other functions of V in FeCo have been discussed in §2.2.

The theory of kinetics of the order–disorder transition has been discussed in terms of the mass-action law and Eyring's theory of the rate process. In lieu of a review of the various kinetic theories, the following comments are offered:

(1) The growth of stable nuclei involves the diffusion of atoms. Hence, the kinetic theory of the phenomenon is generally characterized by the Boltzmann factor $\exp(-E_\alpha/kT)$, where E_α is the appropriate activation energy for diffusion. The time required for ordering should vary exponentially with temperature. This is verified experimentally.

(2) The actual time rate at which ordering proceeds is much slower than that dictated by atomic diffusion, however. For instance, cobalt and iron interdiffuse quite rapidly at 500°C, but the critical temperature for ordering is 730°C at the equiatomic composition. Besides the factor of stability associated with four distinct domains at a corner in the DO_3, Ll_0 and Ll_2 superlattice, there must be other reasons to explain the slow

kinetics, which would otherwise be too sluggish to produce any detectable degree of long-range order.

(3) Several special phenomena which further complicate the problem have been observed. One is the anelastic behavior induced by stress and its effect on the extent of short-range order. The other is the formation of twins in the superlattices of FePt and CoPt. Still another is the promotion of ordering in NiFe by neutron- or electron-irradiation. The latter phenomenon is particularly remarkable because NiFe had long been known as unorderable under normal conditions. Recent work conducted at the Grenoble Nucler Research Center (Paulevé et al. [1962]), however, demonstrated that, by subjecting the NiFe crystals to irradiation under a magnetic field, the alloy becomes susceptible to ordering. Evidently the production of vacancies by fast neutrons or energetic electrons and the induction of magnetic anisotropy in the lattice by the applied field are capable of increasing the ordering kinetics, which would otherwise be too sluggish to produce any detectable degree of long-range order.

The discussion in the remainder of this section deals with the effects of atomic order on various properties important in the preparation and applications of soft magnetic alloys.

Saturation magnetization. In FeCo, Ni_3Fe and other intraalloys of iron-group metals, an ordered sample possesses a slightly higher M_s value than the disordered sample. The increase is 4% in FeCo and 5% in Ni_3Fe. Figure 5.58 illustrates the increasing effect of atomic order on M_s of the alloy 2% V–FeCo. In other alloys such as FePt and CoPt, however, atomic order causes a decrease in M_s by more than 10%.

The reason for the increase in M_s in FeCo had not been explained satisfactorily until proper neutron diffraction experiments were conducted. Using a polarized beam of neutrons, Collins and Forsyth [1963] were able to detect a sensitive dependence of the Fe moment upon the environment in a series of Fe alloys containing Co up to 75 at.%. They found that, although the moment of Co atoms remained invariant at 1.8 μ_B, the moment of Fe atoms increased from 2.2 μ_B in pure Fe to just over 3.0 μ_B in alloys of 50 at.% or more Co. The increase in μ_{Fe} is ascribed to an increase in the number of Co atoms present in the nearest neighbors. As the concentration of Co increases, so will this number on a statistical basis in the disordered state, thus the steady increase in μ_{Fe} as observed. Ordering will further increase the value of μ_{Fe} because it brings Co atoms into the environment of each Fe atom at

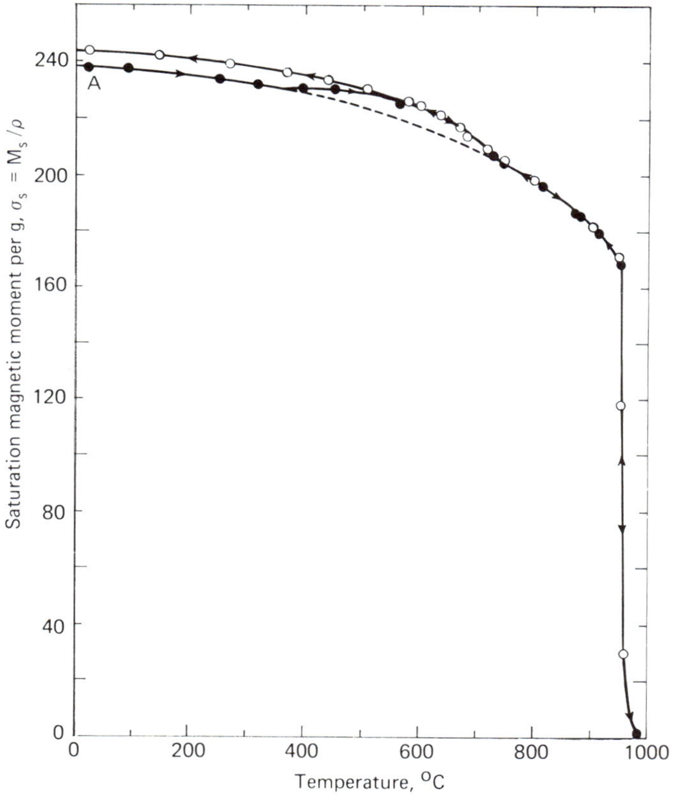

Fig. 5.58. Saturation magnetization versus temperature curves of 2% V–FeCo showing the effect of order on M_s. Measurements began at A with specimen quenched from the disordered state (Chen [1961]).

the maximum number. On this basis, the mean value of the atomic moment of the Fe–Co alloys can be calculated in both ordered and disordered states. The calculated values are shown to be in reasonable agreement with the experimental values in fig. 5.59 (Kouval [1969]).

The most dramatic increase in M_s is seen in the alloy Ni_3Mn. The disordered samples of this alloy display an average moment of 0.1 μ_B per atom. After ordering, the value jumps to 1.02 μ_B per atom. An order of magnitude increase! Neutron diffraction data obtained from powder

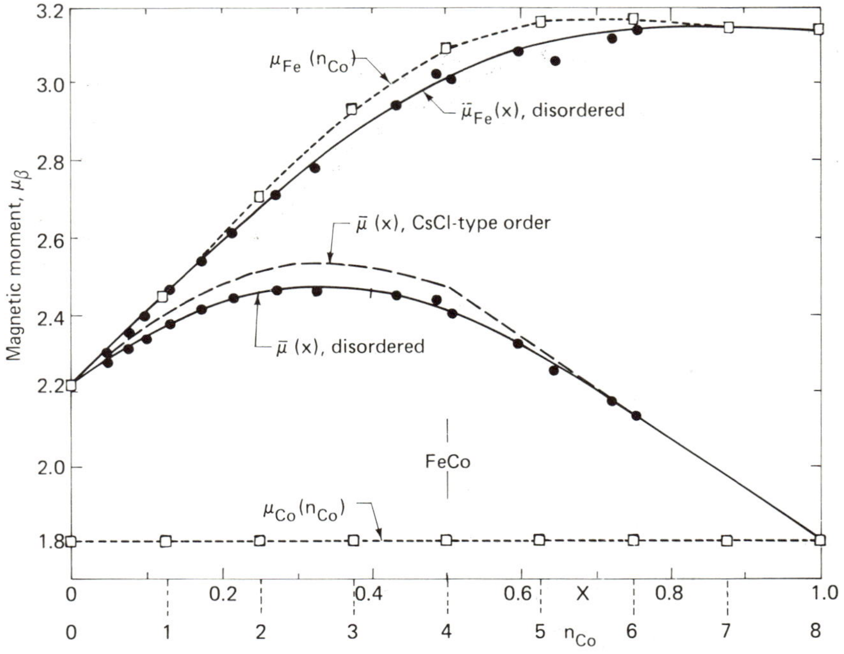

Fig. 5.59. Experimental (points) and calculated curves for atomic moments in ordered and disordered Fe–Co alloys. The average moments are plotted against alloy composition; individual moments are plotted against the number of Co nearest neighbor atoms, n_{Co} (Kouvel [1969]).

samples of the ordered ($W = 0.905$) Ni_3Mn gave the deduced values of 0.30 (±0.05) and 3.18 (±0.25) μ_B for the Ni and Mn atoms, respectively (Shull and Wilkinson [1955]). Similar Bohr magneton numbers were deduced for Ni and Mn atoms in other Ni–Mn alloys, such as NiMn, indicating an invariance of the moments with the environment in contrast to the behavior of the Fe–Co alloys. The reason for the huge increase in magnetization caused by atomic ordering in Ni_3Mn must then be related to the coupling of magnetic moments. For this reason, we assume the spin coupling to be parallel for the Ni–Ni and Ni–Mn nearest neighbor pairs, but antiparallel for the Mn–Mn pairs. In the ordered state, there are no Mn–Mn pairs; thus Mn atoms are able to make a dominant (78%) contribution to the magnetization. Meanwhile, in the disordered state, the magnetic moments of Mn atoms are cancelled whenever two Mn atoms appear in the nearest neighbors. The situation is further deteriorated by the fact that disordered Ni_3Mn has been shown to display peculiar magnetic properties, implying a strong tendency for the Ni and Mn atoms to segregate in clusters. As a result of this tendency, the matrix of the disordered alloy is broken up on a submicroscopic scale into Ni-rich, ferromagnetic and Mn-rich, antiferromagnetic regions. Such clustering makes it virtually impossible for the Mn atoms to contribute to M_s in the disordered state of Ni_3Mn.

Many interesting phenomena have been found in alloys of platinum and palladium, particularly with iron. In the Fe–Pt system, PtFe undergoes a first-degree ordering to form the Ll_0 superlattice, whereas Fe_3Pt undergoes a second-degree ordering to form the Ll_2 superlattice. Atomic ordering increases and decreases the saturation magnetization of Fe_3Pt and FePt by ≈ 10 and $\approx 15\%$, respectively. What makes the magnetic behavior of the Fe–Pt alloys even more interesting is that the Pt atoms are inherently nonmagnetic, but their presence in the iron lattice generates an appreciable moment on their own sites. According to the neutron diffraction study of Kren and Szabó [1965], each Fe and Pt atom in the ordered Fe_3Pt carried a moment of ~ 3.3 and 1.5 μ_B, respectively. The moment of the Pt atoms is induced by a polarization effect on the conduction electrons of the exchange interaction field associated with the Fe atoms. This interpretation is consistent with the observed increase in M_s of Fe_3Pt caused by ordering because all 12 nearest neighbors of a Pt atom would be Fe atoms in the completely ordered state (fig. 5.52c), thus enabling the polarization effect to reach a maximum. A much more enhanced polarization effect has been observed in dilute alloys of Pd and Pt containing less than 1% Fe, Co or Ni. In the extreme case, the dilute alloy of Pd containing 0.1% Fe features a so-called giant magnetic moment of ≈ 13 μ_B on each Fe site as a result of the polarization effect (Clogston et al. [1962]). Because of the magnetic behavior of the dilute alloys, the interpretation of the measured magnetization in the ordered Fe_3Pt before the neutron diffraction study had also assumed a "giant" moment of ≈ 3.8 μ_B carried by each Fe atom as the only source of magnetization. Obviously, such an interpretation follows the traditional thought that Pt is a non-magnetic element and thus fails to take into account the important environmental effect on the magnetic moment of both Fe and Pt atoms.

Curie temperature. Although there is a close relationship between the increasing effect of atomic order on saturation magnetization and that on the Curie temperature, there are cases in which such a relationship breaks down. For instance, Ni_3Fe and its neighboring alloys have Curie temperatures much higher than the order–disorder transition temperatures. Hence, the Curie temperatures of these Ni–Fe alloys are independent of the state of atomic order. In another case, Fe–Co alloys that contain more than 14.5 at.% Co lose their ferromagnetism mainly because of the (bcc)$\alpha \rightarrow$(fcc)β phase transformation. Hence, strictly speaking, these Fe–Co alloys have no real Curie temperatures, only virtual ones determined by extrapolation in the M_s–T plot. For the Ni–Fe and Co–Fe alloys, it is then difficult and insensible to discuss the effect of atomic order on θ_C. In still another case, the effect of atomic order on θ_C is opposite to that on M_s. For example, atomic order in FePt decreases M_s by $\approx 15\%$ as previously cited, nevertheless it raises θ_C from 540°K in the disordered state to 750°K in the ordered state.

For alloys whose Curie temperatures lie below T_c, ordering tends to raise θ_C by the amount corresponding to the increase in M_s. Typical raises are seen in Pd_3Fe from 224°C to 256°C and in Fe_3Pt from 17°C to 157°C upon ordering. In a few special cases, the increases in θ_C are very large. For example, the Curie point of Ni_3Mn lies below room temperature in the disordered state, but jumps to 450°C after ordering. This large increase in θ_C is consistent with the unusually large increase in M_s previously discussed.

Magnetocrystalline anisotropy and saturation magnetostriction. Experimental results indicate that atomic order also exerts an effect on these intrinsic magnetic properties. The changes in their coefficients, however, vary in both nature and magnitude sensitively with the composition of the alloy. For example, fig. 5.60 and fig. 5.25 show that ordering tends to increase the first anisotropy constant K_1 in the Fe–Co alloys, but tends to decrease K_1 in the Fe–Ni alloys. These figures reveal another interesting feature explained in table 5.27. Note that K_1 vanishes in the ordered state of FeCo and in the disordered state of Ni_3Fe. Consequently, ordering improves the permeability of FeCo greatly but makes Ni_3Fe a much less attractive soft magnetic alloy.

The effect of atomic order on saturation magnetostriction is shown by the following two sets of experimental data obtained for Ni_3Fe at room temperature: (A) from a single crystal by Bozorth and Walker [1953] – λ_{100} decreases from $\sim 17 \times 10^{-6}$ to $\sim 13 \times 10^{-6}$, whereas λ_{111} increases from $\sim 5 \times 10^{-6}$ to $\sim 12 \times 10^{-6}$; and (B) from polycrystalline specimens by Taoka and Ohtsuka [1954] – $\bar{\lambda}_s$ increases from $\approx 7 \times 10^{-6}$

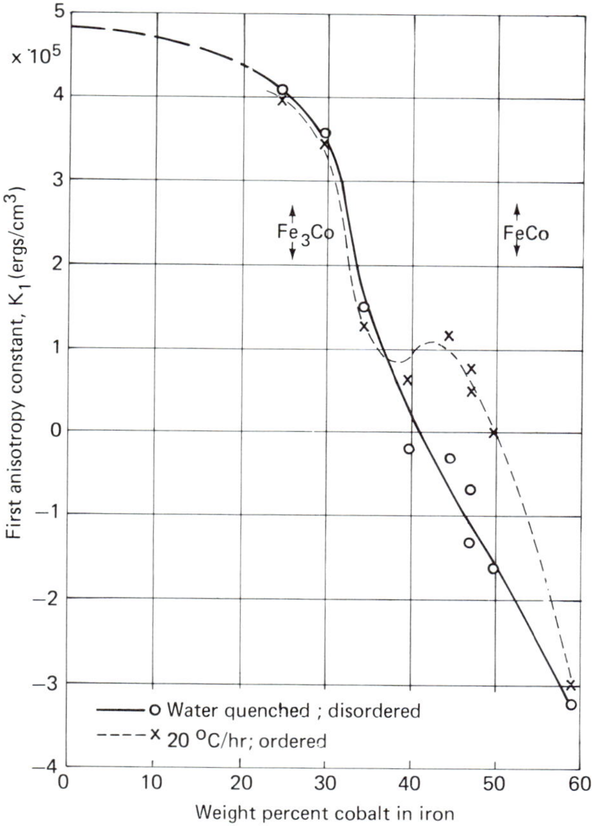

Fig. 5.60. Effect of composition and atomic order on the first constant of magnetocrystalline anisotropy K_1 of Fe–Co alloys (Hall [1960]).

to $\approx 10 \times 10^{-6}$. Using the relation $\bar{\lambda}_s = \frac{1}{5}(2\lambda_{100} + 3\lambda_{111})$ expressed in (3.19b), the data in (A) imply a change in $\bar{\lambda}_s$ from $\sim 9.8 \times 10^{-6}$ to $\sim 12.4 \times 10^{-6}$, which is in fair agreement with the data in (B). A similar increase in $\bar{\lambda}_s$ from $\sim 65 \times 10^{-6}$ to 92×10^{-6} was reported for FeCo (Smoluchowski [1951]). More recent measurements on ordered single crystals of FeCo gave $\lambda_{100} \sim 140 \times 10^{-6}$ and $\lambda_{111} \sim 40 \times 10^{-5}$, which yield $\bar{\lambda}_s \sim 80 \times 10^{-6}$, again in fair agreement with the polycrystalline datum; but λ_{100} and λ_{111} for disordered crystals were found to be the same (within $\approx 10\%$) as those for the ordered crystal (Hall [1960]). The

TABLE 5.27
The variation of the first magnetocrystalline anisotropy constant for FeCo and Ni_3Fe with the state of order

Alloy	$K_1(J \cdot m^{-3})$ deduced in the state of atomic order	
	(A) disordered	(B) ordered
FeCo	-20×10^3	~ 0
Ni_3Fe	~ 0	-2.5×10^3

discrepancy between the data reported for FeCo by the latter two workers may partly be attributed to the fast kinetics of ordering in this alloy. Because quenching is ineffective in suppressing the ordering transition in FeCo, partial long-range order might have prevailed to different degrees in the quenched specimens prepared by different workers.

Before leaving the effect of atomic order on magnetic properties, it should be mentioned that, since ordering affects M_s, θ_C, and other intrinsic properties so profoundly, its effects on the structure-sensitive properties are expected to be correspondingly appreciable. To demonstrate this point, let us refer back to the strain theory of the coercive force expressed in (4.54),

$$H_c \simeq \pi c(\lambda_0 \Delta\sigma / M_s\mu_0)(d_w/l).$$

As the isotropic longitudinal magnetostriction λ_0 and saturation magnetization M_s change with the degree of long-range order, H_c will change correspondingly. Partly because structure-sensitive properties and their changes are so complex in nature and partly because of the limited information available, we shall not discuss the subject further.

Electrical resistivity. The effect of atomic order on the electrical resistivity of alloys has been known for a long time. Upon ordering, ρ is seen to drop steadily. The change in resistivity, $\Delta\rho$, is usually so appreciable that electrical measurements have been used to determine the degree of long-range order. Although scattering of conduction electrons is more complicated in magnetic alloys because of the additional scattering caused by spins, magnetic alloys are no exceptions to the reducing effect of atomic order on ρ. A typical example is shown in fig. 5.61, which indicates the variation of resistivity of the Ni–Fe alloys with the rate of cooling after the final anneal. Curves a, b and c in this figure represent the specimen condition with $W \sim 1, 0.3–0.4$ and 0, respectively. A maximum drop of ρ appears in curve a at the 75% Ni concentration, where the ordering phenomenon is most noticeable. Note that $\Delta\rho$ at the equiatomic composition is practically nil, confirming the previous assertion that NiFe is not susceptible to ordering unless the crystals are specially treated.

Another important result shown in fig. 5.61 is the dependence of $\Delta\rho$ upon the degree of long-range order. This dependence is customarily

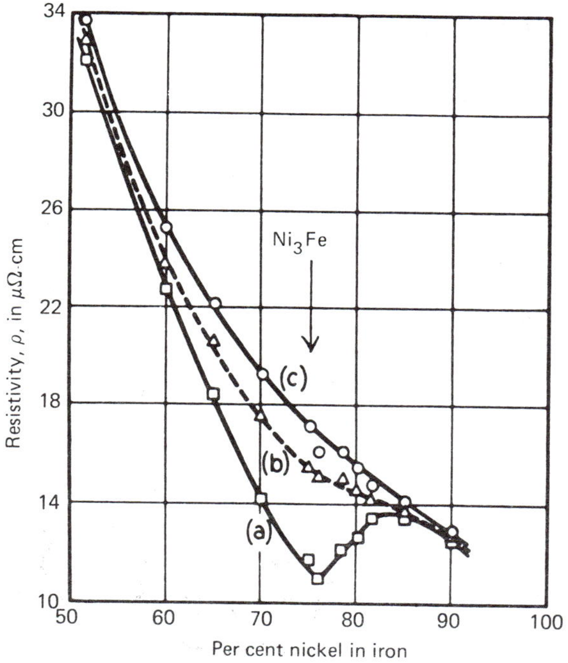

Fig. 5.61. Effect of atomic order on resistivities of Fe–Ni alloys: (a) cooled slowly, (b) cooled at 200°C per hour and (c) quenched from 600°C (Bozorth [1951]).

expressed by

$$(\rho(W)-\rho_{W-1})/(\rho_{W=0}-\rho_{W=1}) = 1-W^2 \quad \text{or} \quad \Delta\rho_W = \Delta\rho_{0-1}(1-W^2), \tag{5.78}$$

where the denominator on the left of the first expression or $\Delta\rho_{0-1}$ in the second expression is the maximum decrease in residual resistivity caused by atomic order. Equation (5.78) is the basis for the determination of W by electrical measurements. Experimentally, however, a few alloys, notably FeCo, Ni_3Mn and the nonmagnetic β-brass (CuZn), show little or no decrease in resistivity during ordering. One reason given for the absence of $\Delta\rho$ in these alloys is that their component elements have neighboring positions in the periodic table; i.e., their atomic numbers differ only by one. This explanation seems to apply well to FeCo and β-brass, but hardly to Ni_3Mn. Thus there must be other reasons, such as differences in the atomic size and electronic structural characteristics, to explain the electrical behavior of the latter alloy.

Mechanical properties. For the preparation and room-temperature applications of soft magnetic alloys, yield strength and ductility are the two most important mechanical properties for consideration in connection with atomic order. Both properties can be evaluated easily from tensile tests, and both are interrelated in the sense that a metallurgical phenomenon capable of improving yield strength will also affect (adversely) ductility. The effect of atomic order on mechanical properties typifies such an interrelationship, although caution must be exercised in discussing the individual alloys.

According to Ardley and Cottrell [1953], long-range order tends to strengthen an alloy for a geometric reason. It was shown in fig. 5.52 that a bcc and a fcc lattice in the disordered state becomes two and four interpenetrating simple cubic lattices in the ordered state, respectively. Consequently, the Burgers vectors of unit slip dislocations are changed as shown in table 5.28. Whence, a unit slip dislocation in the disordered bcc or fcc crystal is only a half-dislocation in the superlattice. More importantly, these half-dislocations must always be attached to a stacking fault, which is in the present case an antiphase domain boundary, as shown in fig. 5.62a. The fault has a surface tension due to the energy of its "wrong" atom pairs and exerts a force on the dislocation line. To move a dislocation against such a force, a shear stress must be applied of the magnitude $\tau = \gamma/b$, where γ is the specific antiphase boundary

TABLE 5.28

The change in the Burgers vectors of unit slip dislocations as a result of chemical order in the face-centered cubic and body-centered cubic metals

Disordered state		Ordered state	
Lattice	***b*** of unit slip dislocations	Superlattice	***b*** of unit slip dislocations
bcc	$\frac{1}{2}a\langle 111\rangle$	$L2_0$	$a\langle 111\rangle$
fcc	$\frac{1}{2}a\langle 100\rangle$	$L1_0$ or $L1_2$	$a\langle 110\rangle$

energy and b is the length of the Burgers vector. However, half-dislocations tend to form pairs to minimize the surface energy of the fault. Figure 5.62b shows that the formation of a pair of half-dislocations would only extend the antiphase boundary segment AB to ABA′, where the other segment BC is eliminated. Moreover, the surface tension of the faulted region between the paired half-dislocations tends to pull them together so as to further minimize the surface energy. In completely ordered crystals ($W = \pm 1$), the motion of the paired half-dislocations creates no misarrangement of atoms, thereby exerting no strengthening effect. When the long-range order is incomplete, that is, $0 < W < 1$, strengthening arises according to the amount of atomic misarrangement accompanying the passage of dislocations. This accounts for the strengthening effect of long-range order.

Short-range order may also strengthen the lattice because the motion of a dislocation through an antiphase boundary creates a new segment of boundary (Fisher [1953]). Subsequent passage of other dislocation would require higher shear stress, thus leading to a strengthening effect.

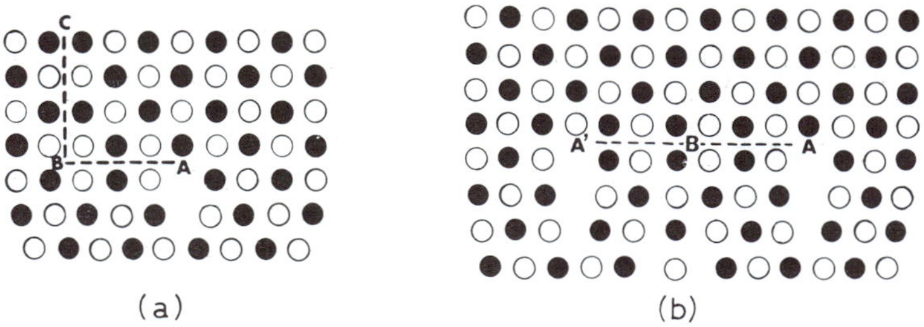

Fig. 5.62. The attachment of antiphase domain boundaries to dislocations: (a) an edge dislocation at A; (b) a superdislocation at A–A′ (Smallman [1970a]).

Chen [1959] estimated that this short-range order effect gives rise to a maximum value of critical shear stress of 6.6×10^8 dynes cm^{-2} for slip in FeCo at 27°C. In addition, Iwata [1958] proposed a third type of order as a common phenomenon in solid solutions. Based on the interaction between directional forces and inhomogeneous lattice distortion of atomic pairs, an anisotropic formation of unlike atom pairs was proposed. Such an anisotropic short-range order is accountable for the production of an induced uniaxial anisotropy in Fe–Ni alloys during magnetic anneal (§2.6). Sumino [1958] treated the combined strengthening effect of long-range order and anisotropic short-range order. Chen [1959] applied Sumino's treatment to FeCo and estimated the critical shear stress for slip to be 4×10^8 dynes cm^{-2} at 27°C.

To assess the relative importance of the various strengthening effects of atom order, Chen also estimated for FeCo at 27°C the critical shear stress needed to move an unpinned slip dislocation against the Peierls–Nabarro force at 7.5×10^7 dynes · cm^{-2}. Meanwhile, the strengthening effect of interstitial impurities by forming the Cottrell atmospheres at dislocations was estimated to yield a critical shear stress of $\sim 1 \times 10^9$ dynes · cm^{-2}. The following conclusions are thus drawn for FeCo and similar bcc alloys:

(1) Partial long-range order, isotropic short-range order and anisotropic short-range order could strengthen the alloy. At room and subzero temperatures, the critical shear stresses for slip imposed by the various strengthening mechanisms of atomic order are estimated to be less than that produced by the strengthening effect of interstitial impurities, however.
(2) In view of (1), plus the fact that it is practically impossible to completely eliminate interstitial impurities in bcc alloys, the strengthening effects of atomic order are likely to be overshadowed by that of the interstitial impurities.
(3) Even if the difficulty with the impurities can be resolved, the present theories are far from complete and reliable. Many important aspects of the ordering phenomenon are neglected. They include the influence of domain size, the structure change caused by atomic order of alloys such as FePt and CoPt and the softening effect observed in well-ordered Fe_3Al and Cu_3Au.

The adverse effect of atomic order on ductility is not only customarily expected in accompaniment with the strengthening effect, but fully exposed by, or even "unfairly" accused for, the severe brittleness of FeCo. In his extensive study on the 2%V–FeCo alloy, Chen [1961, 1964] found the accusation of atomic order as the main cause of brittleness not only misleading but, to some extent, unjustified. As was previously cited (§2.2), not all quenched specimens of the 2%V–FeCo alloy were ductile. In fact, quenching from temperatures above the order→disorder transition, but below the $\gamma \to \alpha$ transformation, failed to improve ductility. The latter result implies that the disordered α-phase is also brittle, due probably to the embrittling effect of interstitial impurities. Another reason that atomic order is not the only cause of brittleness in FeCo is that cleavage fracture caused by cracks formed on the {100} planes is only one of three modes of brittle failure. The other two modes are intergranular embrittlement caused by the segregation of trace amounts of hydrogen and oxygen at grain boundaries and interfacial cracking between grains of the retained austenite and ferrite. The latter two modes result in practically no ductility as compared with low to moderate (4 to 12% elongation) ductility in the cleavage fracture. Hence, the strengthening effect of atomic order only promotes the brittle behavior, but it should not be accused as being solely responsible for the notorious brittleness of FeCo. Further studies on the brittle behavior of FeCo and its V-doped alloys were reported by Stoloff et al. [1965].

§2.6. *Magnetic anneal, plastic deformation and allotropic transformation relating to induced magnetic anisotropy*

Magnetocrystalline anisotropy was shown (ch. 3, §3) to be a natural consequence of the spin–orbit coupling present in all ferromagnetic materials. Now we shall show that another type of magnetic anisotropy can be induced in certain materials by a magnetic field with or without the material undergoing a phase transition and by plastic deformation. The induced anisotropy is usually uniaxial in the sense that the axis of easy magnetization is unique. Thus an [*uvw*] easy axis does not imply that all ⟨*uvw*⟩ directions are easy axes as in the case of inherent anisotropy. If the easy axis turns out to be parallel to the direction of the applied field, it may not correspond to any rational crystallographic direction expressed in terms of relatively simple indices.

§2.6.1. *Magnetic anneal*

The term "magnetic anneal" is used customarily to indicate the application of a field when a material is being heat treated but does not undergo a structure change. It is by far the oldest and most widely used method to induce uniaxial anisotropy. As early as 1913, Pender and

Jones (see Chikazumi and Graham [1969]) observed a marked improvement in permeability of ~4% Si–Fe as a result of the application of an a.c. magnetic field during cooling of the alloy from 800°C to room temperature. The primary benefit of magnetic anneal, however, had not been fully recognized until Kelsall (1934, see Bozorth [1951j]) reported in detail the dramatic changes in magnetization behavior of Permalloys (55–80% Ni–Fe) caused by magnetic anneal. Subsequently many workers succeeded in applying magnetic anneal to numerous alloys in the binary systems of Co–Ni, Fe–Co, Fe–Al, Fe–Si and, of course, Ni–Fe.

The principle involved to induce uniaxial anisotropy by magnetic anneal is rather simple. It necessitates nonrandom diffusion of atoms under the influence of magnetization induced by the applied field so that a preferential alignment of like-atom pairs is developed. The preferential alignment of atom pairs constitutes a "directional order," which is responsible for the induced anisotropy. The function of the field then is threefold: (1) to induce magnetization to saturation, (2) to induce the directional order of atoms and (3) to maintain the order during cooling until atoms cease to migrate. In view of these functions, the field must be applied at a temperature just below the Curie point so that magnetization can be attained at a relatively low field strength. There is a lower temperature limit below which the application of field has little or no effect. The strength of the field required to fulfill all three functions is rather low and seldom exceeds $4000\ \mathrm{A \cdot m^{-1}}$ (50 Oe). In many cases, the field can be removed after the material has reached that temperature at which atom diffusion is likely to be unimportant. The rate of cooling is another important parameter. As a general rule, the cooling rate must be as low as economically permissible in industrial practice. If the material is an alloy susceptible to atomic ordering, the rate of cooling first must be low to temperatures just above the critical point for ordering; then the rate is increased rapidly to avoid ordering. This rate change is necessary, otherwise long-range order tends to destroy the uniaxial anisotropy induced by the magnetic field above the ordering temperature.

For a specific example, let us consider the 65-Permalloy ($Ni_{0.65}Fe_{0.35}$) whose Curie and ordering temperatures are 610 and 440–460°C*,

*Atomic ordering in $Ni_{0.65}Fe_{0.35}$ is of the first degree. Thus there is a temperature range in which ordered and disordered regions coexist in the alloy. According to the Bragg–Williams theory, the degree of long-range order at 400°C is approximately $W = 0.8$ for the alloy.

respectively. Successful magnetic anneals are conducted on this alloy under the following conditions: (A) apply a d.c. field of the order of $\approx 800\ \text{A} \cdot \text{m}^{-1}$ (10 Oe) at ≈610°C; (B) slowly cool the alloy to ≈400°C and (C) fast cool the alloy from 400°C in zero field.

Magnetic anneal can cause drastic changes in structure-sensitive magnetic properties. Moreover, these changes are permanent because they will not decay with time, nor will they be destroyed by either a magnetic field of strength considerably greater than that of the field applied during the anneal or by a moderate increase in temperature. A good example of the effect of magnetic anneal on magnetic properties is shown in fig. 5.63, which reveals a drastic change in the shape of the hysteresis loop of the 65-Permalloy. The constricted loop normally obtained from the alloy is converted to a narrow rectangular loop after the alloy has been magnetically annealed. Such a shape change is termed

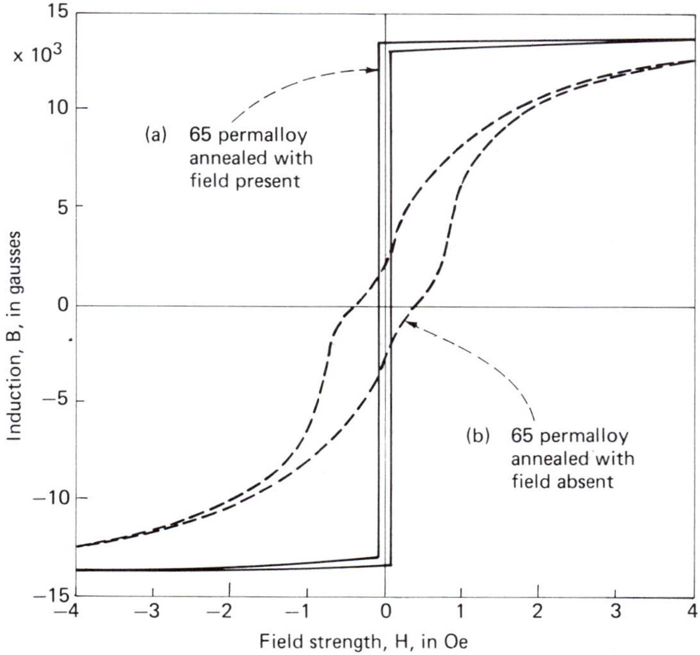

Fig. 5.63. Hystersis loops of 65 Permalloy annealed with and without a magnetic field (Bozorth [1951]).

"squaring-up" of the loop. The new loop shape imparts a large increase in retentivity and an appreciable decrease in coercivity. Also, because a square loop is capable of operating a computer memory, the magnetically-annealed 65-Permalloy thus displays attractive hysteresis characteristics for such applications. Another dramatic improvement of the magnetic properties of the 65-Permalloy is a 25-fold increase in the maximum permeability from ≈10 000 to 255 000 as shown in fig. 5.64. The initial permeability, however, remains unaffected by magnetic anneal. Other properties that may be affected severely by magnetic anneal are magnetostriction and magnetoresistance.

Because the main purpose of magnetic anneal is to induce uniaxial anisotropy in the direction of the applied field, ***H***, the material thus

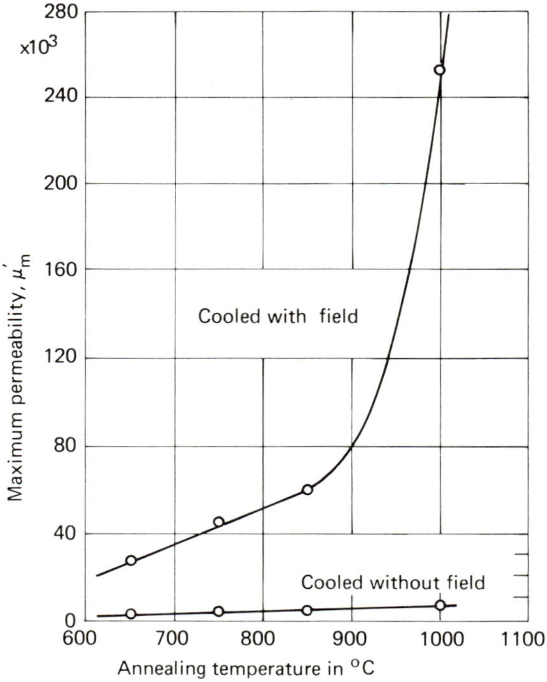

Fig. 5.64. Maximum relative permeabilities of 65 Permalloy after deformation and annealing at various temperatures for one hour with or without a magnetic field (Bozorth [1951]).

treated is expected to exhibit varying magnetic properties in different directions, particularly between directions along $\boldsymbol{H}$ and normal to $\boldsymbol{H}$. Indeed, a number of magnetically-annealed alloys display such directional dependence of magnetic properties. Figure 5.65 shows two vastly different shapes of hysteresis loops obtained from the same sample of

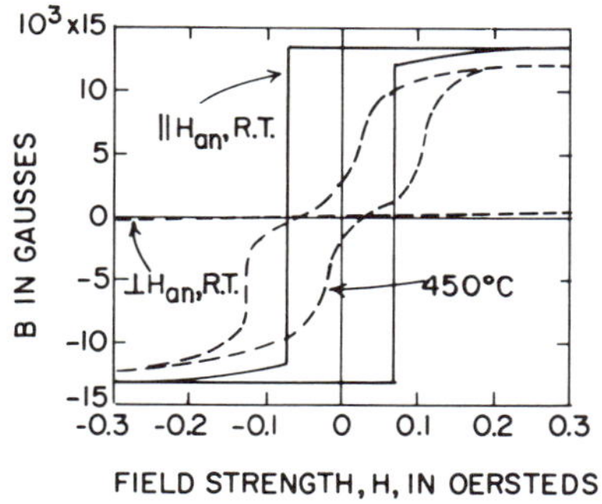

Fig. 5.65. Variation in the shape of the hysteresis loop of 45–25 Permivar showing the effect of magnetic anneal and temperature (Bozorth [1951]).

45–25 Perminvar (45% Ni, 25% Co and balance Fe) – a square loop in the $\boldsymbol{H}$ direction and an unresolvable loop in the direction normal to $\boldsymbol{H}$.

Several theories have been advanced to explain the induction of uniaxial anisotropy by magnetic anneal. An early theory suggested by Snoek [1941] attracted some attention. Néel [1952] criticized the theory by showing that the magnetostrictive interaction proposed by Snoek to account for the directional order of interstitial impurity atoms was too small in most systems. Also the Snoek theory only applies to bcc metals containing sufficient interstitial impurities. Other features of the Snoek theory will be described later. Kaya [1953] proposed another interesting theory based on a shape anisotropy, which is developed by the formation of elongated domains of an ordered phase. Again, the Kaya theory applies only to special cases. Néel [1954] and Taniguchi and Yamamoto [1954] independently proposed a theory that postulates a pseudodipole interaction between the local magnetization $\boldsymbol{M}_q$ and nearest neighbor atom pairs. As a result of further work by Chikazumi

[1956], the latter theory emerged as the preferred one partly because it applies to all alloy systems including those treated by Snoek. We shall call it the pseudodipole theory.

To describe the pseudodipole theory, consider a binary solid solution $A_{1-c}B_c$ and assume that the alloy may be ordered in long range, or directionally ordered with zero degree of long-range order. The pseudodipole interaction is assumed to vary with the type of atom pairs involved and with the orientation of the pair axis relative to the direction of $\boldsymbol{M}_q$ shown in fig. 5.66. The interaction energy of a pair of atoms at the α and β sites can be expressed as a power series in Legendre polynomials:

$$W_{\alpha\beta} = l_{\alpha\beta}(\cos^2\phi_n - \tfrac{1}{3}) + m_{\alpha\beta}(\cos^4\phi_n - \tfrac{6}{7}\cos^2\phi_n + \tfrac{3}{35}) + \cdots, \tag{5.79}$$

where $l_{\alpha\beta}$ and $m_{\alpha\beta}$ are constants characterizing the pseudodipole and pseudoquadrupole interactions, respectively, and ϕ_n is the angle between the axis of the α–β pair in the nth direction and $\boldsymbol{M}_q$. Because $l_{\alpha\beta} \gg m_{\alpha\beta}$, we need only retain the first term. The total interaction energy of the solid solution then is given by

$$E_u = \sum_n (N^n_{AA}l_{AA} + N^n_{BB}N_{BB} + N^n_{AB}l_{AB})(\cos^2\phi_n - \tfrac{1}{3}), \tag{5.80}$$

where the sum runs over all possible directions of the nearest neighbor pairs. The numbers of the three kinds of atom pairs are not independent (e.g., a change of atomic position that creates a new A–A pair also creates a new B–B pair and destroys two A–B pairs; thus 2AB → AA + BB cited in §2.5). When we count all atom pairs in the nth direction,

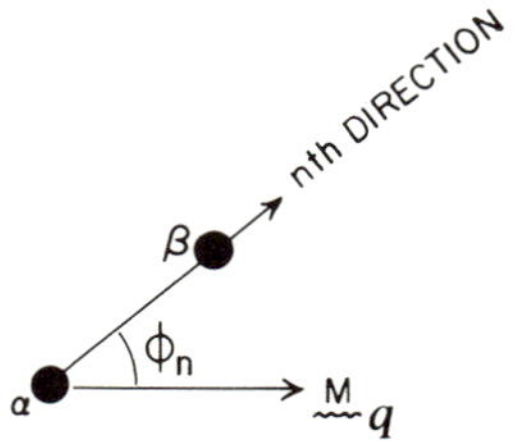

Fig. 5.66. Defining the nth direction of an atom pair with respect to local magnetization vector $\boldsymbol{M}_q$.

because each atom has two nearest neighbors along each direction n, it is readily seen that

$$2N^n_{AA} + N^n_{AB} = 2N_A, \tag{5.81a}$$

and

$$2N^n_{BB} + N^n_{AB} = 2N_B, \tag{5.81b}$$

where N_A and N_B are the total numbers of A and B atoms, respectively. Substituting (5.81) in (5.80) and using the relation $\Sigma_n (\cos^2 \phi_n - \frac{1}{3}) = 0$ gives

$$E_u = l \sum_n N^n_{BB}(\cos^2 \phi_n - \tfrac{1}{3}), \tag{5.80-1}$$

where $l = l_{AA} + l_{BB} - 2l_{AB}$, is a temperature-dependent constant measuring the energy difference between interactions of like-atom and unlike-atom pairs with local magnetization. This constant vanishes at the Curie temperature, but takes a finite value l_1 at temperature T_1, at which the field is applied. This quantity l is closely related to ΔU introduced in §2.5; its value (positive or negative) was shown to determine whether a concentrated binary alloy would be a mixture of two primary solid solutions or a single phase susceptible to ordering. In the present case, the sign of l_1 determines whether the BB pair tends to align itself parallel (for $l_1 < 0$) or perpendicular (for $l_1 > 0$) to the direction of local magnetization. The alignment of each BB pair is associated with an energy change given by $l_1(\cos^2 \phi_n - \frac{1}{3})$. The number of BB pairs in the nth-pair direction in thermal equilibrium at T_1 is thus given by

$$N^n_{BB} = N_{BB} \frac{\exp[-l_1(\cos^2 \phi_n - \frac{1}{3})/kT_1]}{\sum_n \exp[-l_1(\cos^2 \phi_n - \frac{1}{3})/kT_1]}, \tag{5.81c}$$

where N_{BB} is the total number of BB pairs present in the system at T_1. Experiments detect no anomaly in lattice parameter and elastic constants during the induction of directional order by the field, implying a weak and highly localized phenomenon. Hence it is reasonable to assume $|l_1| \ll kT_1$. So,

$$\exp[-l_1(\cos^2 \phi_n - \tfrac{1}{3})/kT_1] \approx 1 - l_1(\cos^2 \phi_n - \tfrac{1}{3})/kT_1, \tag{5.82a}$$

and

$$\sum_n \exp[-l_1(\cos^2\phi_n - \tfrac{1}{3})/kT_1] \approx z, \tag{5.82b}$$

where z is the coordination number. Putting (5.82) in (5.81c) gives

$$N^n_{\mathrm{BB}} = (N_{\mathrm{BB}}/z)\{1 - [l_1(\cos^2\phi_n - \tfrac{1}{3})/kT_1]\}. \tag{5.83}$$

If the alloy is held at T_1 long enough to attain thermal equilibrium and then cooled rapidly to T_2, where atomic diffusion cannot occur to destroy the directional order, uniaxial anisotropy is permanently induced, of which the energy is given by combining (5.80-1) and (5.83)

$$E_u = \frac{N_{\mathrm{BB}} l_2}{z}\sum_n \left[1 - \frac{l_1}{kT_1}(\cos^2\phi_n - \tfrac{1}{3})\right](\cos^2\theta_n - \tfrac{1}{3}), \tag{5.84}$$

where l_2 and θ_n are the values of l and ϕ_n at T_2, respectively. For a cubic system, we can write

$$\cos\theta_n = \alpha_1\gamma_{1n} + \alpha_2\gamma_{2n} + \alpha_3\gamma_{3n}, \tag{5.85a}$$

and

$$\cos\phi_n = \beta_1\gamma_{1n} + \beta_2\gamma_{2n} + \beta_3\gamma_{3n},$$

where α_i, β_i and γ_{in} $(i = 1, 2, 3)$ are the direction cosines of magnetization during the measurement, of magnetization during anneal and of the axis of the BB pair in the nth direction, respectively. The energy of induced uniaxial anisotropy is now expressed by

$$\begin{aligned} E_\mu &= -\frac{N_{\mathrm{BB}} l_1 l_2}{zkT_1}\sum \cos^2\phi_n \cos^2\theta_n + \text{constant} \\ &= -\frac{N_{\mathrm{BB}} l_1 l_2}{kT_1}[k_1(\alpha_1^2\beta_1^2 + \alpha_2^2\beta_2^2 + \alpha_3^2\beta_3^2) \\ &\quad + k_2(\alpha_1\alpha_2\beta_1\beta_2 + \alpha_2\alpha_3\beta_2\beta_3 + \alpha_3\alpha_1\beta_3\beta_1)], \end{aligned} \tag{5.86}$$

where

$$k_1 = \frac{1}{z}\left(\sum_n \sum_{i=1}^{3} \gamma_{in}^4 - \sum_n \sum_{i\neq j} \gamma_{in}^2\gamma_{jn}^2\right), \tag{5.87a}$$

and

$$k_2 = \frac{4}{z}\sum_n \sum_{i\neq j} \gamma_{in}^2\gamma_{jn}^2. \tag{5.87b}$$

TABLE 5.29
Theoretical values for the first and second coefficients of the induced uniaxial anisotropy

Material system	k_1	k_2
(1) Polycrystals with random orientation	$\frac{2}{15}$	$\frac{4}{15}$
(2) Interstitial impurities in bcc crystals	$\frac{1}{3}$	0
(3) fcc crystals	$\frac{1}{12}$	$\frac{4}{12}$
(4) bcc crystals	0	$\frac{4}{9}$

The constants k_1 and k_2 are the first and second coefficients of the induced uniaxial anisotropy, respectively. Their numerical values for various crystal systems are shown in table 5.29. Chikazumi [1964g] and other workers verified the theory in a variety of alloys. Their results are summarized in table 5.30. Fair agreement were attained in the fcc alloys, but no agreement was obtained in the bcc alloys of the Fe–Al system.

Some workers (Bozorth et al. [1955]) further simplified the expression

TABLE 5.30
Comparison between theoretical and experimental values for various materials of the ratio $k_1:k_2$ in the theory of uniaxial anisotropy induced by magnetic anneal

	$k_1:k_2$	
Substance	Theory	Experiment
Ni_3Fe	1:4	1:8.5
20% Co–Fe	1:4	1:3.0
12.5% Co–Fe	1:4	1:2.3–2.6
54% Ni–Fe	1:4	1:1.9–2.4
83% Ni–Fe	1:4	1:8.3–9.6
$Fe_{3.35}Al$	0:1	1:3.4
$Fe_{4.7}Al$	0:1	1:1

for E_u by converting (5.86) to the following form:

$$E_u = -F\sum_i \alpha_i^2\beta_i^2 - G\sum_{i\neq j} \alpha_i\alpha_j\beta_i\beta_j \quad (i, j = 1, 2, 3) \tag{5.88}$$

where $F = N_{BB}k_1l_1l_2/kT_1$ and $G = N_{BB}k_2l_1l_2/kT_1$. For isotropic media listed in (1) in table 5.29,

$$\begin{aligned} E_u &= \frac{-2N_{BB}l_1l_2}{15kT_1}\left(\sum_i \alpha_i^2\beta_i^2 + 2\sum_{i\neq j} \alpha_i\alpha_j\beta_i\beta_j\right) \\ &= \frac{-2N_{BB}l_1l_2}{15kT_1}\left(\sum_i \alpha_i\beta_i\right)^2 = -\frac{2N_{BB}l_1l_2}{15kT_1}\cos^2\Phi, \end{aligned} \tag{5.89}$$

where Φ is the angle between the domain magnetization and the direction of the applied field during magnetic anneal. If the alloy $A_{1-c}B_c$ has a nearly zero degree of long-range order, then $N_{BB} = \frac{1}{2}(Nc)(zc)$, where N is the total number of atoms in the specimen. In this case, we can express E_u in terms of a uniaxial anisotropy constant K_u,

$$E_u = -K_u\cos^2\Phi, \tag{5.90}$$

with

$$K_u = zNc^2l_1l_2/15kT_1. \tag{5.91}$$

The last expression discloses the composition dependence of the anisotropy constant and (5.90) is used to determine the variation of the anisotropy energy with orientation experimentally. Data obtained for the Fe–Ni alloys by Chikazumi and Oomura [1955] are shown in fig. 5.67. A good agreement is seen between the data and the predicted dependence of K_u upon concentration. The maximum values for K_u vary from $\sim 10^2\ \mathrm{J\cdot m^{-3}}$ ($10^3\ \mathrm{ergs\cdot cm^{-3}}$) in Fe–Si, through $\sim 4\times 10^2\ \mathrm{J\cdot m^{-3}}$ in Fe–Ni to $\sim 10^3\ \mathrm{J\cdot m^{-3}}$ in Fe–Co, Co–Ni and Fe–Al.

The theory is readily applied to iron containing various interstitial impurities. In a bcc lattice, although the tetrahedral interstice is larger than the octahedral interstice with respect to the radius of an inscribed sphere without causing strain by 47% ($0.126a$ versus $0.067a$, where a is the lattice constant), interstitial impurities such as carbon, oxygen and nitrogen nevertheless prefer to occupy the octahedral sites in α-Fe and some, but not all, other bcc metals. The reason is mainly geometric as illustrated in fig. 5.68a. Each octahedral site is surrounded by six host atoms of iron, of which two (1 and 2) are along the cube edge at a distance of $\frac{1}{2}a$, whereas the other four (3 to 6) are in the $\langle 110\rangle$ directions

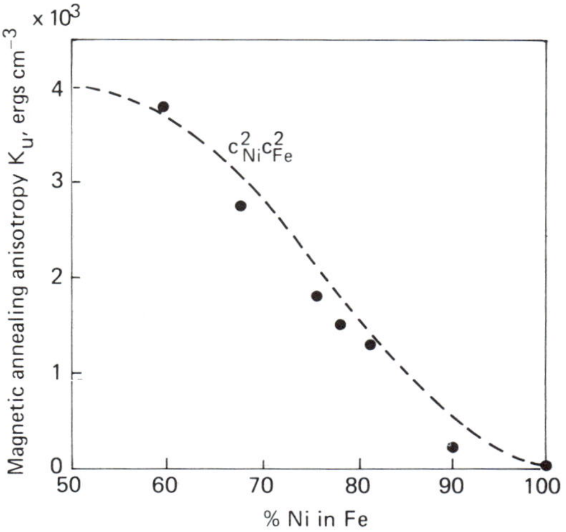

Fig. 5.67. Experimental data and theoretical curve for the constant of induced anisotropy in the Fe–Ni system (Chikazumi and Oomura [1955]).

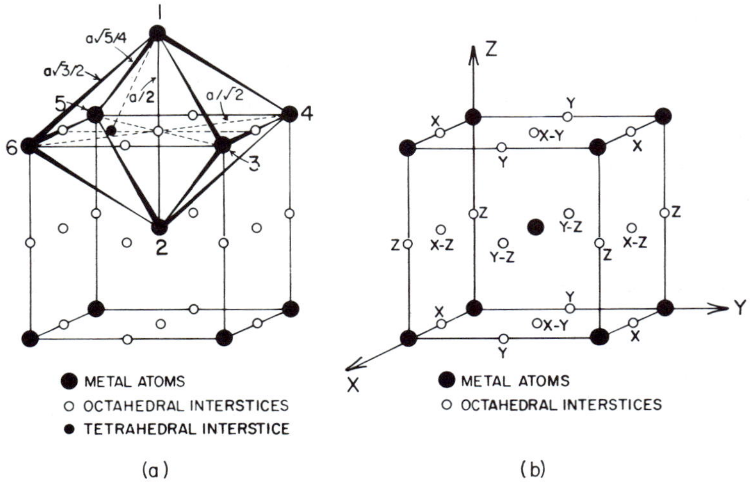

Fig. 5.68. (a) An octahedral interstitial site is surrounded by two host atoms (1 and 2) at a distance of $\frac{1}{2}a$ and four other atoms (3 to 6) at a greater distance of $a/\sqrt{2}$; (b) Classification of octahedral interstices according to their locations with respect to the crystal axes.

at a greater distance of $a/\sqrt{2}$. For Fe at 20°C, $a = 2.866$ Å; thus the open space in an octahedral interstice to accommodate an interstitial atom is $a - 2r_{Fe} \simeq 0.39$ Å along the cube edge and $2a/\sqrt{2} - 2r_{Fe} \simeq 1.57$ Å in the $\langle 110 \rangle$ directions, where $r_{Fe} \sim 1.24$ Å is the radius of an iron atom determined from the metallic bond scheme (Cottrell [1967a]). The diameters of four common interstitial atoms are

H	O	N	C
0.92	1.20	1.42	1.54 Å.

Occupancy of an octahedral interstitice by any atom of these elements would have to push *two* host atoms out along the cube edge. Similar analysis indicates that occupancy of a tetrahedral interstice by any of these impurity atoms requires pushing all *four* surrounding host atoms out. Hence the distortion caused by an interstitial atom is less at an octahedral site than at a tetrahedral site. This accounts for the preferential occupancy of octahedral sites by H, C, N and O in α-Fe.

The displacements caused by an interstitial atom of the two host atoms along the cube edge enable the octahedral interstices to be identified as the x-, y- and z-sites (fig. 5.68b), even though crystallographically they are equivalent. When an external force applied to an iron crystal produces a tensile stress in the x direction, all x-sites will be opened up, whereas all y- and z-sites will be squeezed. This situation will force the interstitial impurity atoms to move out of y- and z-sites in favor of x-sites. The well-known Snoek effect refers to the exchange-site movement among the three octahedral sites of impurity atoms under cyclic stresses. By analogy, Snoek also proposed his theory of induced magnetic anisotropy aforementioned. Because of the positive longitudinal magnetostriction in the [100] direction for Fe (table 3.2), he conceived that the application of a magnetic field to an iron crystal in [100] at a moderate temperature would not only magnetize the crystal, but also promote a redistribution of interstitial impurities to the x-type octahedral sites because these sites are opened up in accord with the positive λ_{100}. The preferential occupancy of one type of octahedral interstices by impurity atoms is equivalent to a directional order in the concentrated solid solutions. Hence uniaxial anisotropy can also be induced in α-Fe if a sufficient amount of carbon or nitrogen exists.

Many ferrites respond to magnetic anneal. A major benefit of inducing uniaxial anisotropy in magnetic oxides is to change the hysteresis loop to a square shape or to sharpen the squareness of the loop. In the case of Co-ferrites, the induced anisotropy constant K_u attains large values. For example, $K_u = 6 \times 10^4\,\mathrm{J \cdot m^{-3}}$ for $(CoFe_2O_4)_{0.75}(Fe_3O_4)_{0.25}$, as compared with $K_u = 8 \times 10^2\,\mathrm{J \cdot m^{-3}}$ for $(NiFe_2O_4)_{0.5}(Fe_3O_4)_{0.5}$. The mechanism whereby such strong anisotropy could be developed in Co-ferrites had not been settled until the role played by vacancies at the cation sites was accurately assessed. In an earlier interpretation by Iida et al. [1955, 1958], Co-ferrites were found to respond to magnetic anneal only when they were oxidized. Because oxidation of the ferrite is accompanied by an introduction of vacancies at the cation sites, these workers thus attributed the induced anisotropy to a directional order of the Co^{2+}-vacancy pairs at the octahedral sites in the inverse spinel structure (see table 5.11). Later, Penoyer and Bickford [1957] and Bickford et al. [1958] examined the variation with composition of the induced anisotropy energy in several crystals of $Co_xFe_{3-x}O_4$ $(0 < x < 0.15)$, which were grown from the melt under different partial pressures of oxygen. The coefficients F and G in (5.88) were found to increase quadratically and linearly with x, respectively. Meanwhile, the relaxation time η $(= \eta_0\,\mathrm{e}^{A/kT}$, where A is an activation energy of approximately 1 eV), which characterizes the process of magnetic anneal, was observed to increase drastically – from several minutes for crystals known to contain cation vacancies to a value too large to be determined for a crystal essentially stoichiometric with respect to oxygen content. The latter workers thus concluded that the main role played by cation vacancies is to assist the movement of Co^{2+} ions between equivalent cation sites during anneal. Even though the effect of vacancies is pronounced on the kinetics of magnetic anneal, the role played by vacancies is considered secondary as far as the development of anisotropy is concerned. The mechanism for the induction of anisotropy in ferrites is now believed to involve the formation of Co–Co ion pairs on neighboring octahedral sites along a $\langle 111 \rangle$ axis close to the direction of the applied field

§2.6.2. *Plastic deformation*

Because the mechanism responsible for the induction of anisotropy involves a directional order of BB pairs in the alloy $A_{1-c}B_c$, it is natural to expect stress either to assist magnetic anneal or to induce similar aisotropy on its own. Indeed a number of experiments that apply magnetic anneal under stress show an enhancing effect of stress on the induction of uniaxial anisotropy by a magnetic field. Even more interesting is the phenomenon in which cold rolling induces strong and varying anisotropies in various alloys. Anisotropy thus induced is called roll anisotropy, which was first detected by Six et al. [1934]. In their discovery, uniaxial anisotropy was found in FeNi sheets with the easy axis lying perpendicular to the rolling direction after a 50% reduction in thickness by rolling. Prior to the final rolling, the sheets had been severely cold-rolled and annealed for the development of a strong (100)[001] cube texture. The easy axis of magnetization, together with the retained cube texture, enables the polycrystalline sheets to adopt a lamellar domain structure as shown in fig. 5.69a. When a field is applied parallel to the rolling direction, magnetization of the sheet is effected exclusively by spin rotation indicated by arrows in the figure. Because magnetization relies on this particular process, linear relationship emerges between M and H all the way to saturation, as sketched in fig. 5.69b, leading to a constant permeability at all stages of magnetization. This outstanding property has been capitalized as the main attraction of a product of the 50% Fe–Ni alloy known as Isoperm.

Roll anisotropy is doubtless the most effective method of inducing anisotropy by plastic deformation. To understand the phenomenon, Chikazumi et al. [1957] conducted an extensive study on Ni_3Fe, which is also highly susceptible to the induction of roll anisotropy like FeNi. Their study revealed many significant features of the phenomenon. As a result, an elegant theory emerged. Their theory not only satisfactorily accounts for the complex nature of roll anisotropy, but also demonstrates the fruitfulness of applying principles of modern metallurgy to related magnetic phenomena.

Before describing the theory of Chikazumi et al., it is necessary to review the main results of their study on Ni_3Fe. The pertinent results are summarized as follows:

(1) Roll anisotropy induced in Ni_3Fe differs in both the easy axis and

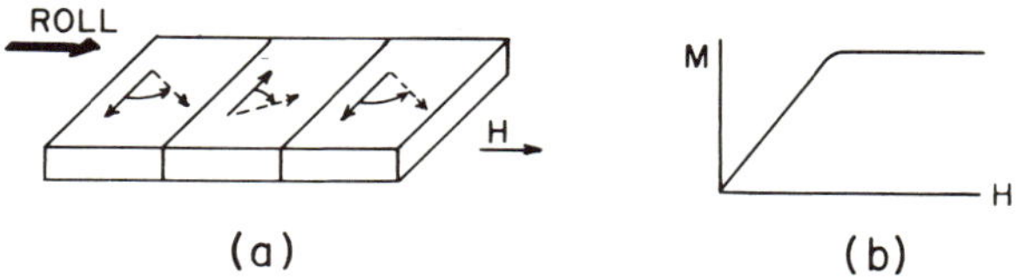

Fig. 5.69. (a) Lamellar domain structure and (b) magnetization curve of Isoperm. Dashed arrows in (a) show intermediate positions of magnetization attained in different domains upon the application of a magnetic field (Chikazumi [1964]).

the value of K_u between single crystals and polycrystalline samples. (2) For single crystals, the easy axis also varies according to the rolling plane (*hkl*) and the rolling direction [*uvw*]. Results for single crystals rolled under three (*hkl*)[*uvw*] orientations are:

(A) For the (110)[001] rolling orientation, the easy axis is developed along the $\pm[\bar{1}10]$ direction, which lies in the rolling plane and perpendicular to the rolling direction (fig. 5.70a-A). The domain structure observed on the top (110) surface consists of lamellar domains as shown in figs. 5.70b-A and c-A.

(B) In the case of (001)[110] rolling, the easy axis measured in the rolling plane first appears perpendicular to the rolling direction, then gradually changes its direction toward, and eventually parallel to, the rolling direction. However, only maze domain patterns are observed on the top and side surfaces at thickness reductions (r) of 42 and 68.5% (fig. 5.70c-B), signifying that the domain magnetization actually is at an angle to the rolling plane (fig. 5.70b-B) and the angle decreases progressively with rolling.

(C) For (001)[100] rolling, the easy axis seems to be perpendicular to the rolling direction in the rolling plane. Again, complicated maze patterns are observed on the top and side surfaces at $r = 50\%$ (fig. 5.70c-C) indicating that the easy axes in local regions neither lie in the rolling plane nor are parallel to the side surface.

(3) The constant of the induced anisotropy constant K_u varied with r and the degree of order. Evidence for these variations is shown in fig. 5.70.

(A) For (110)[001] rolling, after an initial 20% reduction, the development of anisotropy proceeds much faster in the partly ordered crystals than in the fully-ordered ones.

(B) In the (001)[110] orientation, the degree of order has little effect on K_u; but there is a change in the sign of K_u (from negative to positive) at about 20–30% reduction.

(4) For polycrystalline Ni_3Fe, the average easy axis is parallel to the rolling direction (fig. 5.70a-D), but the axis in individual crystals varies as sketched in fig. 5.70b-D and confirmed by the domain structure shown in fig. 5.70c-D.

The theory proposed by Chikazumi et al. to interpret these results is also based on a directional order of like-atom pairs, which, in this case, is produced by the movemement of single dislocations during slip. To see how the directional order comes about, let us first recall that the process of rolling is equivalent to subjecting the sheet to a radial

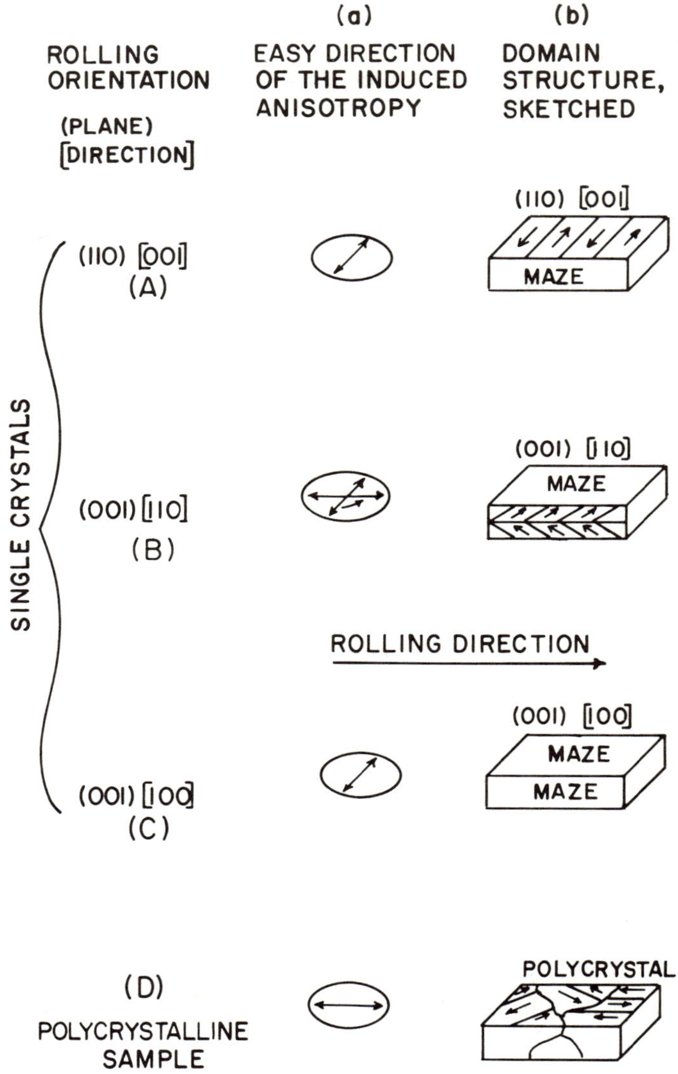

Fig. 5.70. Composite diagram showing (a) easy directions of the induced anisotropy, (b) sketched and (c) observed domain structures in Ni_3Fe single crystals subject to cold rolling in the specified orientations and in polycrystalline sample after cold rolling (Chikazumi [1964]).

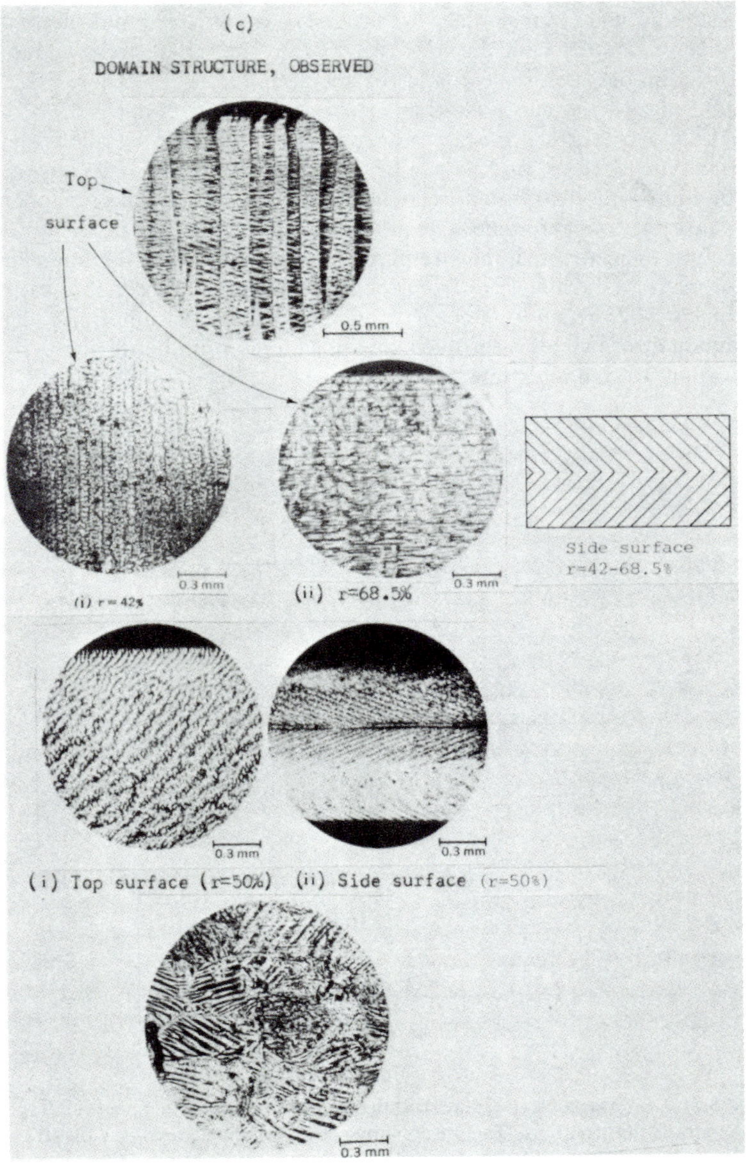
(c)
DOMAIN STRUCTURE, OBSERVED
Top surface
0.5 mm
0.3 mm
(i) r = 42%
(ii) r=68.5%
0.3 mm
Side surface
r=42-68.5%
0.3 mm
0.3 mm
(i) Top surface (r=50%)
(ii) Side surface (r=50%)
0.3 mm

Fig. 5.70(c)

pressure p_r and a tangential friction stress σ_F at any point along the surface of contact with the rolls (fig. 5.71a). The latter stress is due to the friction between the roll and the sheet, and $\sigma_F = fp_r$, where f is the coefficient of friction. In addition, as shown in the inset of fig. 5.71a, the front and rear vertical faces of an element of the sheet are acted on by normal stresses $-\sigma_x$ and $+(\sigma_x + d\sigma_x)$, respectively, which are assumed to be uniformly distributed over the faces. The horizontal components of these stresses produce a resultant tensile stress σ_X in the rolling direction to accomplish the stretching of the sheet. Many attempts have been made to calculate σ_X from the von Kármán differential equation, $d(\sigma_x h)/d\theta = 2p_r R(\sin\theta + f\cos\theta)$, under various assumptions. This resultant stress yields a resolved shearing stress τ_i in the ith slip system according to Schmid's law, which gives

$$\tau_i = \sigma_X \cos\psi_i \cos\phi_i, \tag{5.92}$$

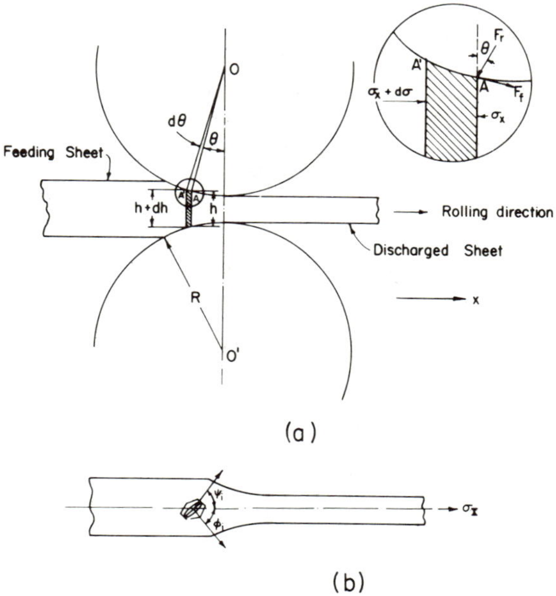

Fig. 5.71. (a) Geometry of the rolling process whereby a sheet undergoes plane-strain deformation. The inset shows the simplified analysis of stresses acting on an element of the sheet. (b) Schmid's law applied to the ith slip system in a grain under the resultant tensile stress σ_X in the rolling direction.

where ψ_i and ϕ_i are the angles between σ_X and the slip direction and the normal to the slip plane of the ith slip system, respectively (fig. 5.71b). For fcc metals and alloys such as Ni_3Fe, plastic deformation is normally performed by slip on a $\{111\}$ plane in the $\langle 1\bar{1}0\rangle$ direction. There are 12 $\{111\}\langle 1\bar{1}0\rangle$ slip systems composed of 4 $\{111\}$ planes and 3 $\langle 1\bar{1}0\rangle$ directions in each of the $\{111\}$ planes. The slip system with ψ_i and ϕ_i close to 45° attains a maximum value for τ_i; thus, that particular ith slip system will operate first to initiate slip.

In the Ll_2 superlattice of Ni_3Fe, as shown in fig. 5.72, no BB atom pairs exist across the $\{111\}$ planes in the undeformed state. Slip by a unit dislocation* with $\boldsymbol{b} = \frac{1}{2}a[01\bar{1}]$ on the (111) plane, however, will create AA and BB pairs alternately in the $[01\bar{1}]$ direction. But the passage of a second dislocation will cancel all these like-atom pairs created by the first. The creation of BB pairs in each slip system thus depends upon a number of variables, notably (1) the average number of active solitary slip dislocations per unit area in the slip plane, (2) the amount of slip produced and (3) the degree of order. Whether slip is confined to one slip system (easy glide) or occurs in two or more (multiple glide) systems depends on the orientation of the crystal with respect to σ_X. Even in the case of multiple glide, the numbers of BB pairs created across different slip planes will not be equal. Therefore, there will be an unbalanced distribution of BB pairs, leading to uniaxial anisotropy.

To calculate the energy of slip-induced anisotropy, we need to evaluate the number of BB pairs, N^i_{BB}, produced in the ith slip system per unit volume. It is easy to show in fig. 5.72 that the number of BB pairs created by the passage of an $\frac{1}{2}a[01\bar{1}]$ dislocation is given by $n_{BB} = 1/\sqrt{3}a^2$ per unit area of the slipped portion of the (111) plane for the perfectly ordered crystal, where a is the lattice constant. For partially-ordered crystals with a nonunity parameter of long-order range

*Unit slip dislocations in most fcc crystals, including Ni_3Fe, are likely to be dissociated into two Shockley partials according to the reaction:

$$\tfrac{1}{2}a[01\bar{1}] \rightarrow \tfrac{1}{6}a[\bar{1}2\bar{1}] + \tfrac{1}{6}a[11\bar{2}],$$

where vectors on the right-hand side are not unit lattice vectors connecting two regular substitutional sites, thus typifying partial dislocations. However, the net result of the motion of paired Shockley partials still creates BB pairs as shown in fig. 5.72. Thus the analysis is valid even though it does not reflect the morphology of dislocations in real crystals.

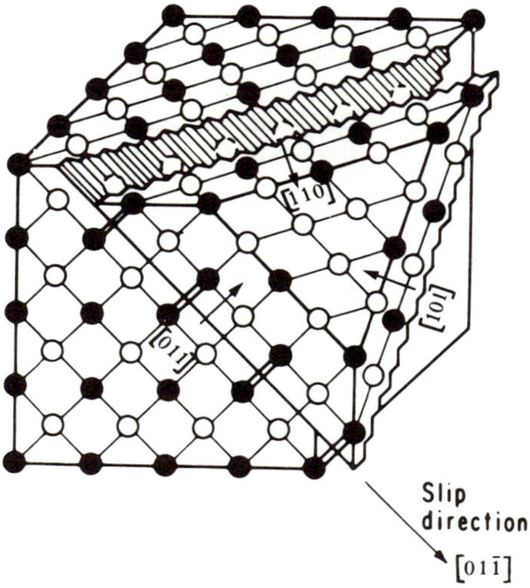

Fig. 5.72. The passage of a unit slip dislocation in the [01$\bar{1}$] direction on the (111) plane in the superlattice of an A_3B alloy produces BB pairs indicated by double lines (Chikazumi [1964]).

W, this number becomes $n_{BB} = W^2/\sqrt{3}a^2$, and hence $n_{BB} = 0$ for the completely disordered ($W = 0$) crystals. Note that in accordance with the definition of $W = +1$ (if all B atoms occupy β sites, etc.), $W^2/\sqrt{3}a^2$ also gives the excessive number of B atoms appearing at β sites in unit area of the (111) plane in comparison with that at α sites. The number of BB pairs created per unit area of the (111) plane is then given, on the average, by $\frac{1}{2}p_n W^2/\sqrt{3}a^2$, where p_n is the probability of nonoccurrence of pairing dislocation. The factor $\frac{1}{2}$ is inserted to indicate that BB pairs are created by the movement of every other dislocation. Now we define the slip density s_i as the average number of dislocations that have passed through a slip plane in the ith slip system. Since there are $\sqrt{3}/a$ atomic layers in a unit length normal to the (111) slip plane the number of slipped planes in the ith system is $\sqrt{3}p_d|s_i|/a$ per unit length in the [111] direction, where p_d is the probability of creating a dislocation on a slip plane in the system. Thus, we can write

$$N^i_{BB} = \frac{1}{2}\frac{p_n W^2}{\sqrt{3}a^2}\frac{\sqrt{3}p_d|s_i|}{a} = \frac{1}{8}npW^2|s_i|, \tag{5.93}$$

where $p = p_d p_n$ and $N = 4/a^3$ is the number of atoms per unit volume. Putting (5.93) in (5.80-1) gives, for the slip-induced anisotropy energy,

$$E_{u,\mathrm{LF}} = \frac{1}{8} NlpW^2 \sum_i |s_i|(\alpha_1\gamma_{1i} + \alpha_2\gamma_{2i} + \alpha_3\gamma_{3i})^2$$

$$= K_W \sum_i |s_i|(\alpha_1\gamma_{1i} + \alpha_2\gamma_{2i} + \alpha_3\gamma_{3i})^2, \qquad (5.94)$$

where $K_W = \frac{1}{8}NlpW^2$, α_j and γ_{ji} ($j = 1, 2, 3$) are the direction cosines of local magnetization and a BB pair with respect to the cube axes, respectively. This kind of induced anisotropy is expected for all alloys provided there exists some degree of long-range order and the slip distance is smaller than the average size of antiphase domains. It is called the long-range order-fine slip type (LF) of anisotropy. Table 5.31 lists the values of γ_{ji}, α_j^2, $\alpha_1\alpha_2$, etc., for all 12 slip systems of the fcc lattice.

Another type of anisotropy was proposed for the case where slip occurs on a coarse scale and short-range order exists in the alloy. By coarse slip is meant that the passage of dislocations is long relative to the size of antiphase domains. It is seen from fig. 5.72 that when the upper part of the crystal continues to slip downward across an antiphase domain boundary into a domain of "opposite" order, the effect is equivalent to replacing one of the three kinds of α sites (as represented by the coordinates $(\frac{1}{2}, \frac{1}{2}, 0)$, $(\frac{1}{2}, 0, \frac{1}{2})$ and $(0, \frac{1}{2}, \frac{1}{2})$; see table 5.26) by a β site in the lower part of the crystal. Consequently the direction of the BB pairs formed by the slip is changed from [011] to [110] or [101]. This result is also expected when the size of the ordered regions is small as in the case of short-range order. In either case, a directional order with BB pairs formed predominantly along one of the three ⟨011⟩ directions in each slip plane also gives rise to a uniaxial anisotropy similar to the case of LF anisotropy. Chikazumi calls it the short-range order-coarse slip (SC) type. To calculate N_{BB}^i in the SC type, we note that a $(111)[01\bar{1}]$ slip will create BB pairs in the [011], [101] or [110] direction, depending on the nature of the domain (whether A atoms occupy α or β sites) and the Burgers vectors of dislocations, but on the average the probabilities of creating BB pairs in the three directions are roughly the same and approximately equal to that for pre-existing BB pairs. Hence, the probability of finding a BB pair in a particular direction is $\frac{1}{4}$ and the number of BB pairs present in a unit area of the (111) plane after coarse slip in the ith slip system is given by $\frac{1}{4}(1-\sigma)/\sqrt{3}a^2$, where σ is the

TABLE 5.31

Direction cosines of BB pairs and local magnetization in (5.94) for long-range order fine-slip-induced anisotropy*

Slip system	Slip plane	Slip direction	Direction cosines of BB pairs			Direction cosines of local magnetization					
			γ_{j1}	γ_{j2} ($j = 1, 2, 3$)	γ_{j3}	α_1^2	α_2^2	α_3^2	$\alpha_2\alpha_3$	$\alpha_3\alpha_1$	$\alpha_1\alpha_2$
1	(111)	[01$\bar{1}$]	0	$1/\sqrt{2}$	$1/\sqrt{2}$	0	$\frac{1}{2}$	$\frac{1}{2}$	1	0	0
2		[10$\bar{1}$]	$1/\sqrt{2}$	0	$1/\sqrt{2}$	$\frac{1}{2}$	0	$\frac{1}{2}$	0	1	0
3		[1$\bar{1}$0]	$1/\sqrt{2}$	$1/\sqrt{2}$	0	$\frac{1}{2}$	$\frac{1}{2}$	0	0	0	1
4	(11$\bar{1}$)	[101]	$1/\sqrt{2}$	0	$-1/\sqrt{2}$	$\frac{1}{2}$	0	$\frac{1}{2}$	0	-1	0
5		[011]	0	$1/\sqrt{2}$	$-1/\sqrt{2}$	0	$\frac{1}{2}$	$\frac{1}{2}$	-1	0	0
6		[1$\bar{1}$0]	$1/\sqrt{2}$	$1/\sqrt{2}$	0	$\frac{1}{2}$	$\frac{1}{2}$	0	0	0	1
7	(1$\bar{1}$1)	[110]	$1/\sqrt{2}$	$1/\sqrt{2}$	0	$\frac{1}{2}$	$\frac{1}{2}$	0	0	0	-1
8		[10$\bar{1}$]	$1/\sqrt{2}$	0	$1/\sqrt{2}$	$\frac{1}{2}$	0	$\frac{1}{2}$	0	1	0
9		[011]	0	$1/\sqrt{2}$	$-1/\sqrt{2}$	0	$\frac{1}{2}$	$\frac{1}{2}$	-1	0	0
10	($\bar{1}$11)	[01$\bar{1}$]	0	$1/\sqrt{2}$	$1/\sqrt{2}$	0	$\frac{1}{2}$	$\frac{1}{2}$	1	0	0
11		[101]	$1/\sqrt{2}$	0	$-1/\sqrt{2}$	$\frac{1}{2}$	0	$\frac{1}{2}$	0	-1	0
12		[110]	$1/\sqrt{2}$	$-1/\sqrt{2}$	0	$\frac{1}{2}$	$\frac{1}{2}$	0	0	0	-1

*Reference: Chikazumi [1964g].

degree of short-range order defined on p. 313. We further note that the number of BB pairs existing in the same direction before slip is $\frac{1}{4}/\sqrt{3}a^2$. Hence, the actual number of BB pairs created by slip is $\frac{1}{4}\sigma/\sqrt{3}a^2$ per unit area. Whence

$$N^i_{\mathrm{BB}} = \frac{\sigma}{4\sqrt{3}a^2}\,\frac{p_{\mathrm{d}}|s_i|}{a/\sqrt{3}} = \frac{1}{16}Np_{\mathrm{d}}\sigma|s_i|. \tag{5.95}$$

In SC anisotropy, the (111)[01$\bar{1}$] slip creates BB pairs in three ⟨110⟩ directions normal to the (111) slip plane. The average energy of the pseudodipole interaction between local magnetization and a BB pair in any one of the three ⟨110⟩ directions is given by

$$\begin{aligned} w_{\mathrm{BB}} &= \tfrac{1}{3}l(\cos^2\phi_{[011]} + \cos^2\phi_{[101]} + \cos^2\phi_{[110]}) \\ &= \tfrac{1}{3}l(\alpha_1\alpha_2 + \alpha_2\alpha_3 + \alpha_3\alpha_1) + \text{constant}, \end{aligned} \tag{5.96}$$

where $\phi_{[011]}$ is the angle between [011] and the direction of local magnetization, etc. and α_1, α_2 and α_3 are the same direction cosines defined in (5.94).

Before calculating the energy of SC anisotropy, it should be pointed out that Néel [1954] estimated from magnetostriction data the numerical value of the important parameter l for the Fe–Ni alloys at $Nl = 3.1 \times 10^7\ \mathrm{J \cdot m^{-3}}$ ($3.1 \times 10^8\ \mathrm{ergs \cdot cm^{-3}}$). Whence, with $a \sim 3.55$ Å for Ni_3Fe at 15°C, we have $l = l_{\mathrm{NiNi}} + l_{\mathrm{FeFe}} - 2l_{\mathrm{NiFe}} \sim 3.5 \times 10^{-22}$ J (3.5×10^{-15} erg). More importantly $l > 0$ for the Fe–Ni alloys. According to the discussion connected with (5.80-1), when $l > 0$, the direction along which the largest number of BB pairs lie will be the hard axis of magnetization. In SC anisotropy, it means that the hard axis will lie normal to the slip plane, or the slip plane will become the easy plane of magnetization.

The interaction energy for the ith slip system is

$$w_i = l(n_{1i}n_{2i}\alpha_1\alpha_2 + n_{2i}n_{3i}\alpha_2\alpha_3 + n_{3i}n_{1i}\alpha_3\alpha_1), \tag{5.97}$$

where n_{1i}, n_{2i} and n_{3i} are the direction cosines of the normal to the ith slip plane. The resultant anisotropy energy is

$$\begin{aligned} E_{u,\mathrm{SC}} &= \sum_i N^i_{\mathrm{BB}} w_i = \frac{1}{16}Nlp_{\mathrm{d}}\sigma \sum |s_i|(n_{1i}n_{2i}\alpha_1\alpha_2 \\ &\quad + n_{2i}n_{3i}\alpha_2\alpha_3 + n_{3i}n_{1i}\alpha_3\alpha_1) \\ &= K_\sigma \sum |s_i|(n_{1i}n_{2i}\alpha_1\alpha_2 + n_{2i}n_{3i}\alpha_2\alpha_3 + n_{3i}n_{1i}\alpha_3\alpha_1), \end{aligned} \tag{5.98}$$

TABLE 5.32
Direction cosines of the normal to the slip plane and values for $\alpha_i\alpha_{i'}$ $(i, i' = 1, 2, 3)$ in (5.98) for short-range order coarse-slip-induced anisotropy*

Slip system	Direction cosines of normal to slip plane n_{j1}	n_{j2} $(j = 1, 2, 3)$	n_{j3}	Direction cosines of local magnetization $\alpha_2\alpha_3$	$\alpha_3\alpha_1$	$\alpha_1\alpha_2$
1, 2, 3	$1/\sqrt{3}$	$1/\sqrt{3}$	$1/\sqrt{3}$	$\frac{1}{3}$	$\frac{1}{3}$	$\frac{1}{3}$
4, 5, 6	$1/\sqrt{3}$	$1/\sqrt{3}$	$-1/\sqrt{3}$	$-\frac{1}{3}$	$-\frac{1}{3}$	$\frac{1}{3}$
7, 8, 9	$1/\sqrt{3}$	$-1/\sqrt{3}$	$1/\sqrt{3}$	$-\frac{1}{3}$	$\frac{1}{3}$	$-\frac{1}{3}$
10, 11, 12	$-1/\sqrt{3}$	$1/\sqrt{3}$	$1/\sqrt{3}$	$\frac{1}{3}$	$-\frac{1}{3}$	$-\frac{1}{3}$

*Reference: Chikazumi [1964h].

where $K_\sigma = \frac{1}{16}Nlp_d\sigma$. Values of $n_{1i}, \ldots$ and $\alpha_1\alpha_2, \ldots$ are given for the 12 slip systems in table 5.32.

In reality, both types of anisotropy are expected to be induced in the same crystal because, if the slip distance is long enough to bring one side of the crystal to another out-of-phase domain of the opposite crystal, SC-type anisotropy will be developed irrespective of the degree of long-range order. Similar reasoning applies to the onset of LF anisotropy in crystals only featuring short-range order.

To apply the theory to Ni_3Fe under different rolling orientations, we must first evaluate the quantity s_i in each case. The evaluation depends on the number of slip systems in operation and on whether or not rolling causes widening of the sheet as well as elongation in length and reduction in thickness. In all cases, the evaluation is rather tedious. Therefore, we shall only quote the results of the evaluation by Chikazumi et al., which, together with other essential features of their analysis, are given in table 5.33.

The verification of the theory with respect to the experimental results obtained from Ni_3Fe can be divided into two parts: the rationalization of the observed variation of the easy axis and the evaluation of the induced anisotropy constants. The results are summarized in table 5.33. To predict the easy axis, we must examine the direction of the induced BB pairs relative to the rolling orientation, which determines the

TABLE 5.33

Analysis of slip-induced uniaxial anisotropy by Chikazumi [1958]

Part I. Operating slip systems, density of slip and energy of induced anisotropy

Rolling orientation	Slip systems expected to operate	Value of slip density in terms of fractional reduction in thickness r	Energy of induced anisotropy	
			LF type	SC type
(1) (110)[001]	1, 2, 4 and 5	$s_1 = s_4 = -\frac{1}{2}r$; $s_2 = s_5 = \frac{1}{2}r$	$E_{u,\mathrm{LF}} = \frac{1}{2}K_w r\alpha_3^2$	$E_{u,\mathrm{SC}} = \frac{2}{3}K_\sigma r\alpha_1\alpha_2$
(2) (001)[110]	1 and 2 or 4 and 5	$s_1 = r$; $s_2 = -r$	$E_{u,\mathrm{LF}} = \frac{1}{2}K_w r\alpha_3(2\alpha_1 + 2\alpha_2 + \alpha_3)$	$E_{u,\mathrm{SC}} = \frac{2}{3}K_\sigma r(\alpha_1\alpha_2 + \alpha_2\alpha_3 + \alpha_3\alpha_1)$
(3) (001)[100]	Probably 2 and 11 or 4 and 8	$s_1 = s_8 = -r$; $s_4 = s_{11} = r$	$E_{u,\mathrm{LF}} = K_w r\alpha_2^2$	$E_{u,\mathrm{SC}} = \frac{2}{3}K_\sigma r\alpha_2\alpha_3$

Part II. The value of K_w and K_σ determined by comparison between theory and experiment for three rolling orientations

Rolling orientation	$3K_w + 2K_\sigma$	K_w	K_σ
(1) (110)[001]	23.2×10^5 (7.85×10^5)*	–	–
(2) (001)[110]	7.3×10^5 (8.0×10^5)*	1.41×10^5 (1.96×10^5)*	1.53×10^5 (1.06×10^5)*
(3) (001)[100]	–	0.94×10^5	–

*Values in parentheses are for perfect order cases.

specific slip system(s) to be activated during rolling. The slip systems expected to operate in the three rolling orientations are shown in fig. 5.73, and the easy axes or planes theoretically predicted and experimentally observed in these cases are shown in fig. 5.74. For (110)[001] rolling, the operation of four slip systems (1, 2, 4 and 5) induces LF anisotropy with a (001) easy plane of magnetization and SC anisotropy with a [01$\bar{1}$] easy axis at the intersection of two slip planes. No BB pairs are created in the (001) easy plane, whereas the two slip planes intersecting along [01$\bar{1}$] also become easy planes of magnetization. Under these circumstances, [01$\bar{1}$] should be the easy axis for induced anisotropy. This agrees with the observation that the easy axis is perpendicular to the rolling direction (results described in (2)-(A) on p. 349).

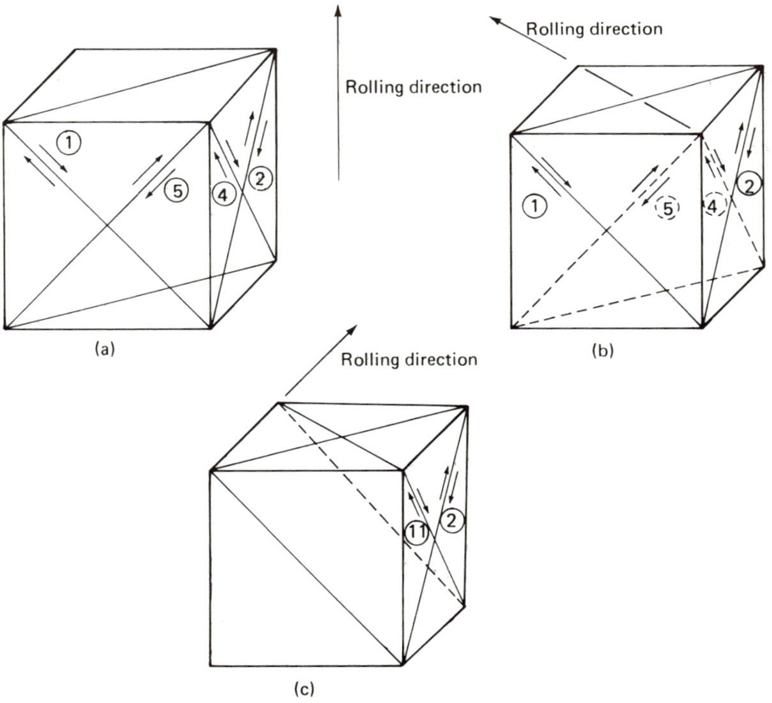

Fig. 5.73. Variation of the effective slip systems with the rolling orientation in the ordered Ni_3Fe (Chikazumi [1964]): (a) (110)[001] rolling, (b) (001)[110] rolling, (c) (001)[100] rolling.

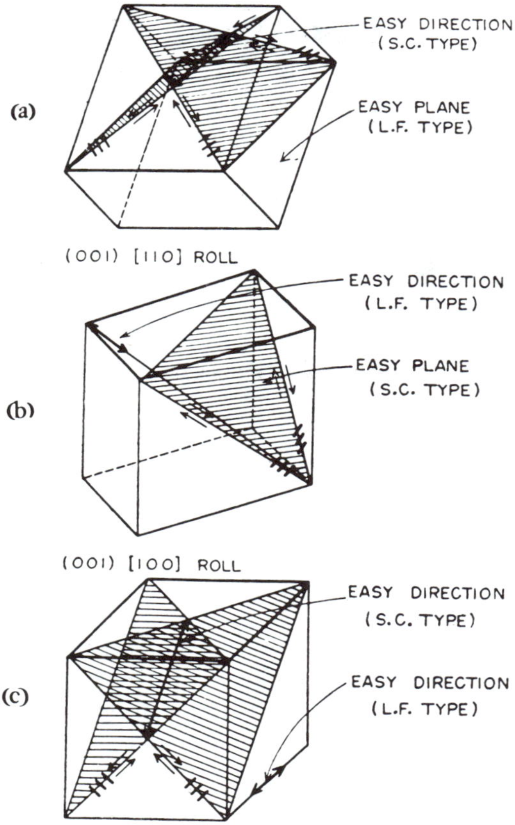

Fig. 5.74. The effective slip planes and induced anisotropy in the three rolling orientations specified. Triple short lines indicate a direction of BB pairs of the LF type (Chikazumi [1958]).

For (001)[110] rolling, the possible slip systems are again 1, 2, 4 and 5. However, slip lines observed on the side surface (fig. 5.75) indicate that only one slip plane and two slip directions are active in each half of the crystal (systems 1 and 2 in the upper half and 4 and 5 in the lower half). As far as plastic deformation is concerned, this rolling induces easy glide in contrast to the cross slip observed in the previous rolling. The reason for this difference is because the (001) and (110) rolling planes deviate from the first activated (111) slip plane at 54.7° and 35.3°,

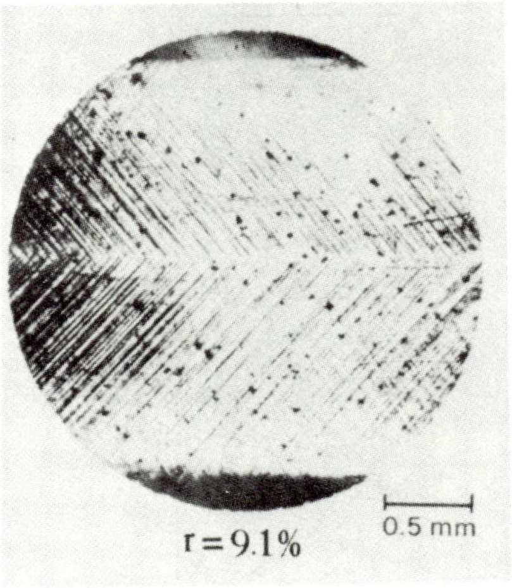

Fig. 5.75. Slip bands observed on the side surface of the rolled (001)[110] crystal of Ni_3Fe. Rolling direction is horizontal (Chikazumi [1964]).

respectively, as shown in fig. 5.76a and b. For slip to continue under constrained rolling conditions, the (111) slip plane is forced to rotate toward the rolling plane so as to bring the original slip direction into the rolling direction. In the case of (001)[110] rolling, reducing this angle from the initial value of 54.7° will bring the angle closer to 45° toward the maximum revolved shear stress according to Schmid's rule expressed in (5.92), thus enhancing the original slip plane and allowing easy glide to be maintained. In (110)[001] rolling, on the other hand, reducing the deviation angle from the initial value of 35.3° will turn the original slip plane further away from the position for the maximum shear stress at 45°, thus inducing another slip plane into operation and leading to multiple slip. Figure 5.74b shows that LF anisotropy induced by (001)[110] rolling will have the easy axis parallel to $[11\bar{1}]$, whereas SC anisotropy will have an easy plane parallel to (111). In either case, the easy axis will be at 35.3° or 54.7° from the rolling plane when viewed from the side. This agrees with the experimental results obtained at the early stage of rolling. As rolling continues, rotation of the easy-gliding plane brings the easy axis toward, and eventually into, the rolling plane. This explains why the easy axis measured in the rolling plane undergoes a change from initially perpendicular to finally parallel to the rolling direction.

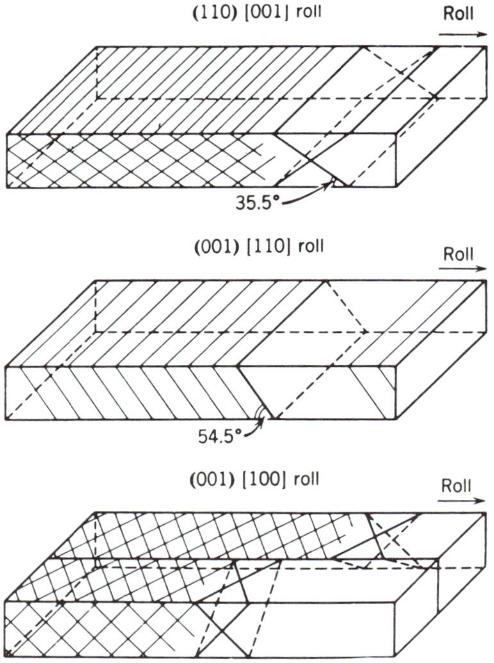

Fig. 5.76. Diagrams showing the geometry of initial slip plane relative to the rolling plane and direction in the three rolled crystals of Ni_3Fe (Chikazumi [1964]).

Slip caused by (001)[100] rolling is complicated in two respects. First, it can occur in one of two pairs (2 + 11 and 4 + 8) of slip systems in different regions, as shown in fig. 5.76c; secondly, four other systems (1, 5, 9 and 10), if activated, may widen the crystal plane, although they do not contribute to the reduction in thickness. Figure 5.74c shows that slip in systems 2 and 11 or 4 and 8 will induce LF anisotropy with an easy axis normal to the rolling direction in the rolling plane and induce SC anisotropy that does not contribute in the rolling plane because $\alpha_3 = 0$ (see table 5.33, Part I, (3)). This prediction agrees with the experiment. The SC type, however, has its easy axis parallel to the [011] direction, thus causing the resultant easy axis to make a glancing angle with the rolling plane, as was detected experimentally (results described in (2)-(C) on p. 349).

The evaluation of the anisotropy constants K_w and K_σ is based on (5.90) for the energy of uniaxial anisotropy with Φ being the angle between magnetization and the rolling direction. By conducting appropriate torque measurements, the coefficient K_u is evaluated, from which values for K_w and K_σ can be deduced by applying the following formulae:

Rolling orientation	Formula	
(110)[001]	$K_u = -\frac{1}{6}(3K_w + 2K_\sigma)r$	(5.97a)
(001)[110]	$K_u = \frac{2}{3}[3K_w + (2r-1)K_\sigma]r$	(5.97b)
(001)[100]	$K_u = -K_w r$	(5.97c)

Note in (5.97c) the zero contribution to K_u from SC anisotropy. The deduced values for K_w and K_σ are listed in table 5.33, Part II, including those in parenthesis for the case of complete order. The theory is considered satisfactory, at least qualitatively, by virtue of the fairly consistent values deduced for K_w and K_σ in various cases. The theory also gains support from the K_u versus r plot shown in fig. 5.77, in which a good agreement is seen between the calculated values (solid curve) and experimental data. Later experiments on Fe_3Al (Bünge and Muller [1957, 1958]) and analysis on anisotropy induced by wire drawing (Von Neida et al. [1968]) further demonstrate the general applicability of the theory.

§2.6.3. *Phase transition*

The third method for the induction of magnetic anisotropy involves a polymorphic transition under the influence of a magnetic field. Ideally, the phase transition for this purpose begins with a cubic lattice which transforms to an uniaxial lattice of lower symmetry. Cobalt typifies such a transformation which alters a fcc lattice to a hcp structure upon cooling below 450°C. In such phase transitions, close relationships exist between orientations of the high- and low-temperature phases. In the case of Co, the {111} planes (or ⟨111⟩ directions) in the fcc phase are equivalent to the basal planes (or the c-axis) in the hexagonal structure. Application of a magnetic field during the transition will convert one of four ⟨111⟩ directions in the cubic phase to the c-axis of the hexagonal phase.

The temperature dependence of the first and second constants of the inherent uniaxial anisotropy of Co is shown in fig. 5.78. Note that $K_1 + K_2 = 0$ at about 250°C. Thus Co shifts its easy axis from [0001]

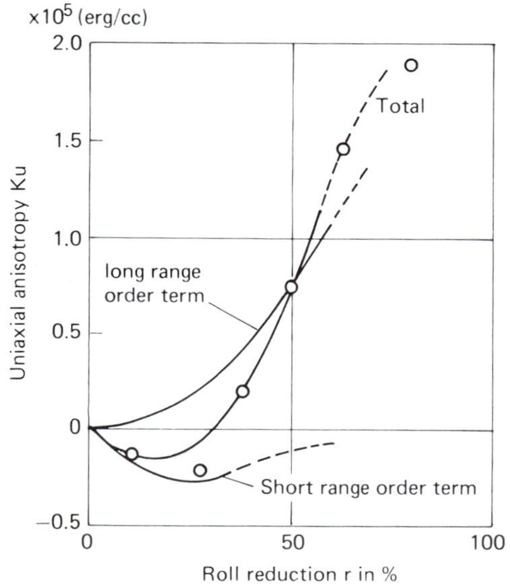

Fig. 5.77. Reasonable agreement obtained between calculated and experimental plots of K_u against the amount of thickness reduction by (001)[110] rolling (Chikazumi et al. [1957]).

along the c-axis to $\langle 11\bar{2}0\rangle$ in the basal plane as the temperature rises above 250°C. In view of this behavior, the easy axis of induced anisotropy is expected to be *perpendicular* to the direction of the applied field at room temperature provided K_u is negative. Experiments conducted by Takahashi and Kono [1960] show that cooling Co from 1000°C in a magnetic field of only $\sim 3.3 \times 10^4\ \mathrm{A \cdot m^{-1}}$ (410 Oe) at a rate of 3.3° per min induces an anisotropy with $K_u \sim -4 \times 10^3\ \mathrm{J \cdot m^{-3}}$ (-4×10^4 ergs $\cdot\ \mathrm{cm^{-3}}$) and a *hard* axis along the direction of the applied field. The same workers also examined the induced anisotropy in Co-rich alloys of the Co–Ni system with Ni up to 20 at.%. These alloys undergo the same (fcc)$\gamma \rightarrow$(hcp)ϵ transition as Co. The constant of the induced anisotropy was found to display a sensitive dependence upon composition as well as upon the strength of the field applied during the transition. For instance, K_u changes from $\sim -6.3 \times 10^3\ \mathrm{J \cdot m^{-3}}$ in pure Co through a zero value in ~4% Ni–Co to a maximum positive value of $\approx 2.3 \times 10^4\ \mathrm{J \cdot m^{-3}}$ in ~20% Ni–Co for anisotropy induced under a field of $6.4 \times 10^5\ \mathrm{A \cdot m^{-1}}$ (8000 Oe). Because of the positive value of K_u for

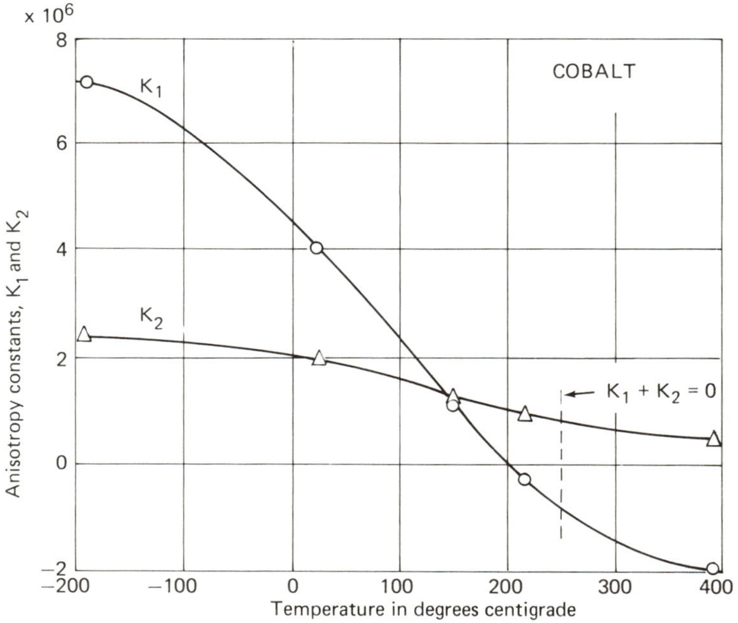

Fig. 5.78. Temperature dependence of the first and second magnetocrystalline anisotropy constants of cobalt (Bozorth [1951]).

alloys with more than ~4% Ni, the induced *easy* axis is *parallel* to the direction of the applied field in the latter alloys.

Many ferrites, too, are found to be susceptible to the induction of anisotropy during phase transition under a magnetic field. In fact, magnetite was the first material in which this phenomenon was discovered. The case of magnetite is further noteworthy for the phase transition from a cubic to a rhombohedral structure at cryogenic temperatures below −154°C. At such low temperatures, ions can hardly migrate. Atomic diffusion must be ruled out as an effective mechanism for the development of a directional order of like-ion (Fe^{2+} or Fe^{3+}) pairs. Indeed, ferromagnetic resonance measurements (Bickford [1950]) and neutron diffraction experiments (Hamilton [1958]) revealed that the directional order of cations comes about from a hopping motion of electrons from Fe^{2+} to Fe^{3+} ions on the octahedral sites, not directly involving ion migration.

The phenomenon wherein cations undergo a local rearrangement along a preferential direction through electron hopping is known as the Verwey order. When a field is applied parallel to a ⟨100⟩ direction of the cubic crystal of magnetite during the phase transition, that ⟨100⟩ axis becomes the c-axis of the rhombohedral phase as well as the easy axis of magnetization. An interesting feature of the Verwey order is that, if the direction of the applied field is suddenly changed to another ⟨100⟩ direction at a temperature just below the transition point, the original Verwey order is switched over to another configuration whose c-axis is parallel to the new ⟨100⟩ direction. This switch from one configuration to another of the Verwey order is reflected by the abrupt appearance of a new absorption line and the disappearance of the previous absorption line in the resonance spectrum. In contrast, the change in the spectrum caused by a similar change in direction of the field applied to an alloy for normal magnetic anneal is a continuous shift in position of the absorption line without changing the height of the line. The absence of the position shift of the resonance line in the former case implies that the Verwey order alters its configuration directly without going through intermediate stages as does the atomic order in alloys. Hence, magnetic anisotropy induced by an applied field during phase transition is considered different from that induced by normal magnetic anneal. This explains why phase transition is considered as a third method for inducing magnetic anisotropy. Other ferrites showing susceptibility to the Verwey order, which gives rise to uniaxial anisotropy in the low temperature phase, include Cu- and Ni-ferrites.

Suggestions for further reading

Barrett, C. S. and T. B. Massalski, 1966, *Structure of Metals*, 3rd. ed. (McGraw-Hill, New York). Chapters 10, 11, 13, 20 and 21.

Berkowitz, A. E. and E. Kneller, ed., 1969, *Magnetism and Metallurgy* (Academic Press, New York). Vol. 2, Chapters 2, 11 to 13 and 15.

Bozorth, R. M., 1951, *Ferromagnetism* (van Nostrand, New York). Chapters 3 to 8 on magnetic properties of alloys of transition metals.

Elliott, R. J., ed., 1972, *Magnetic Properties of Rare Earth Metals* (Plenum Press, London).

Reed-Hill, R. E., 1973, *Physical Metallurgy Principles*, 2nd. ed. (D. van Nostrand Co., New York).

Smit, J., ed., 1971, *Magnetic Properties of Materials* (McGraw-Hill Book Co., New York). Chapters 2 and 6 on ferrites.

Chapter 6

APPLICATIONS OF SOFT MAGNETIC MATERIALS

Soft magnetic materials have been used to perform a wide variety of magnetic functions. Some applications demand high permeability; others emphasize low energy loss at high frequencies, thus necessitating high resistivities; etc. To meet the divergent requirements of applications, numerous commercial products in the form of complex alloys or mixed ferrites were developed by metallurgical methods discussed in the preceding chapter. Hence, to facilitate our discussion on commercial soft magnetic materials, it is instructive to briefly examine the main principles that govern the applications of this class of materials.

For convenience, applications of soft magnetic materials are classified as conventional and nonconventional. We arbitrarily choose the year 1948 as the demarcation line. Applications practised before 1948 are termed conventional, and those developed thereafter are considered nonconventional. This classification, though artificial, enables us to divide conventional applications of soft magnetic materials, with a few exceptions, into four categories according to the principle or particular phenomenon involved. These categories are as follows:

(1) Electromagnetic induction. The combination of Faraday's and Lenz's laws states that, when the magnetic flux ϕ through a circuit changes, an emf ϵ is induced in the circuit such that

$$\epsilon = -\,\mathrm{d}\phi/\mathrm{d}t = -\,A(\mathrm{d}B/\mathrm{d}t), \tag{6.1}$$

where A is the area of the core carrying the magnetic induction B and the minus sign signifies the counteraction of ϵ to the flux change. The operation of electric motors, generators, transformers and chokes is based on this principle. In terms of quantity of materials consumed, applications of this category are the most important. The property requirements are low core losses, high permeability at low and medium inductions and high saturation.

(2) Magnetic force. Maxwell showed that the force per unit area between two surfaces across which magnetic induction B exists is

$$F = kB^2/\mu_0, \tag{6.2}$$

where k is a dimensionless constant having the value of $10^3/8\pi$ in SI units and $1/8\pi$ in the CGS system. Applications making use of this force include electric bells, telephone receivers, telegraph instruments, lifting magnets and moving armature relays.

(3) Magnetic shielding. When a magnetic material, in the form of a hollow sphere or cylinder, is placed in a magnetic field $\boldsymbol{H}$, the field induces magnetic poles over the free surfaces of the material, and the resultant surface density of magnetization produces a uniform field H' inside opposite to $\boldsymbol{H}$. The magnitude of the opposing field inside the sphere is given by

$$\boldsymbol{H}' \simeq \boldsymbol{H}/\mu', \tag{6.3}$$

where μ' is the relative permeability. For the hollow cylinder placed with its axis perpendicular to the field,

$$\boldsymbol{H}' \simeq \boldsymbol{H}/\mu'[1-(x^2/y^2)], \tag{6.4}$$

where x and y are the inner and outer radii, respectively. Magnetic screens are used to shield galvanometers, cathode-ray tubes, etc.

(4) Magnetostriction. Although the Joule effect (change in dimensions with magnetization discussed in ch. 3, §4) is small, large amplitude may be built up by resonance in a suitably designed core. Magnetic materials showing large constants of magnetostriction (pure nickel, 4.5% Co–Ni, 1.4% Co–2.3% Cr–Ni, Permendur and Co-ferrite) are suitable for the construction of transducers, which are used to generate sound waves for detecting submarines and ships and to produce ultrasonic vibrations for cleaning, homogenization of milk, decontamination of nuclear devices, etc.

Many of the nonconventional applications are based on the nonlinear characteristics of magnetization. Of special importance is the rectangular hysteresis loop. Examples of this group of applications include magnetic memories, d.c. converters and magnetic amplifiers. Another group of nonconventional applications uses the resonance phenomena and other ferromagnetic behavior of ferrites at microwave frequencies.

The development of soft magnetic materials for industrial applications started long before the turn of the century. Some authors

(Bechtold and Wiener [1965]) chose 1885 as the year when the history of commercial soft magnetic materials formally began. In that year, Ewing (see Bozorth [1951k]) published an important paper on the magnetic properties of iron, and George Westinghouse introduced the alternating current system of power generation and distribution. Westinghouse's innovation was made possible because of the demonstration

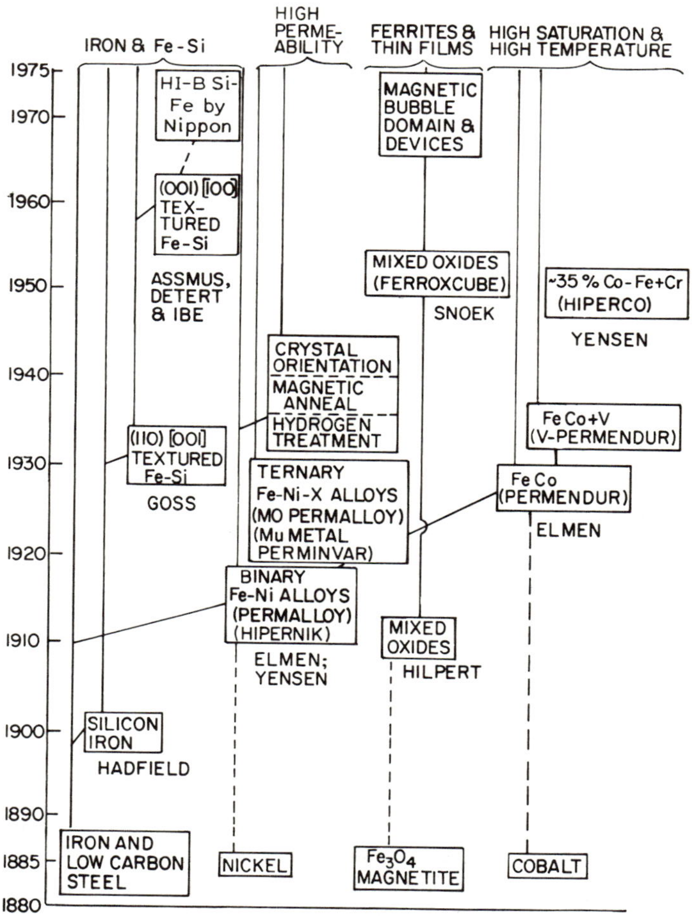

Fig. 6.1. Chronological summary of major developments of soft magnetic materials (Bechtold and Wiener [1965]).

of a transformer made one year earlier by Otto Blathey et al. in Europe (Getting [1967]). Since 1885, the history of soft magnetic materials has followed the pattern of periods of static improvements and dynamic developments as outlined in fig. 6.1. As a whole, the history is illustrious with many breakthroughs and few setbacks. The development of bubble-domain technology for memories and other devices marks the newest dynamic event. As far as metallic materials are concerned, the development of the Nippon HI-B grade with the (110)[001] texture in iron–silicon alloys (Chen [1977]) was the last breakthrough.

For a general description of soft magnetic products, it is convenient to classify them into three categories: (1) metallic materials, (2) ferrites and compounds and (3) thin films. Metallic materials are used in d.c. and low frequency (< 1 kHz) a.c. apparatus for which high electrical conductivity poses few problems. Ferrites are used at frequencies above 1 kHz and in computer memory cores. Thus the distinction between materials in the first two categories is based on both composition and applications. Thin films constitute another material category because of their thickness ($\leq 2\ \mu$m) and methods of preparation. The minimum thickness of ultra-thin metallic tapes is quoted at $\sim 3.2\ \mu$m (0.000125 in) (Bulletin TC-101B of the Arnold Engineering Company). Hence they are considered to be metallic materials, not thin films.

To compare the relative commercial importance of magnetic materials in various categories, we quote estimates of their 1967 market values by Jacobs [1969]:

I. Soft magnetic materials*		$\$315 \times 10^6$
1. Metallic materials*	$\$205 \times 10^6$	
2. Soft ferrites	$\$110 \times 10^6$	
II. Magnetic disks, drums and recording tapes		$\$280 \times 10^6$
1. Recording tapes	$\$180 \times 10^6$	
2. Magnetic drums and discs	$\$100 \times 10^6$	
III. Permanent magnets		$\$55 \times 10^6$
Estimated total for 1967		$\$650 \times 10^6$

The major magnetic media in recording tapes, discs and drums are the iron oxides, usually Fe_2O_3, but sometimes Fe_3O_4. More recently CrO_2 has been used to achieve better packing density and greater output signals. Also used as the magnetic media in tapes and drums are

*Based on electrical steels and Ni–Fe alloys produced in 1967.

thin films of Ni, Ni–Fe and Ni–Co alloys prepared by the electroplating technique. Although these metallic films are genuine magnetic materials, the oxides in fine particle form retain some of the characteristics of hard magnets. Hence the tapes, drums and discs are grouped separately from the soft and hard materials. The dominance of soft magnets ($400 million out of $650 million) in the whole field of magnetic materials is clearly shown in the above listing.

§1. METALLIC SOFT MAGNETIC MATERIALS

Materials in this category are divided into five groups according to their basic composition. They are (1) pure iron and steels, (2) iron–silicon alloys, (3) nickel–iron alloys, (4) iron–cobalt alloys and (5) miscellaneous metallic materials. The main attraction of the majority of soft metallic materials is the permeability. The development of high permeability materials is shown chronologically in fig. 6.2. Supermalloy has been exhibiting the largest values for initial (up to 120 000) and maximum (up to 1 500 000) relative permeabilities since 1947.

§1.1. *Pure iron and steels*

The term "pure iron" is used to indicate a minimum of 99.8% purity and the absence of other elements by deliberate addition. Of the three

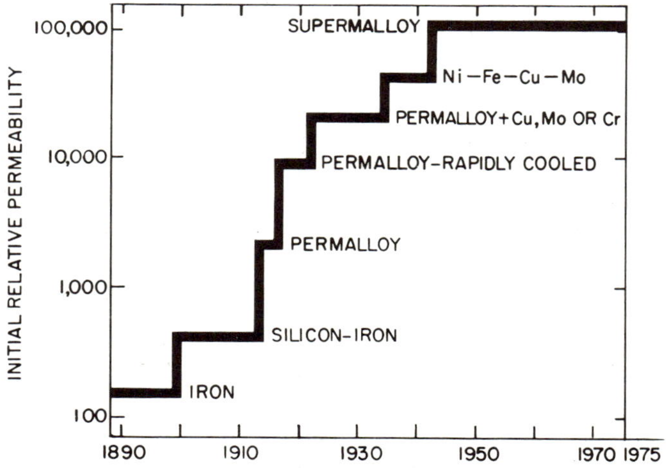

Fig. 6.2. The development of high-permeability alloys since 1890. Note the compressed vertical scale (Lee [1963]).

ferromagnetic transition metals, iron is most extensively used (nickel has some applications in transducers by virtue of its large magnetostriction, but cobalt finds no widespread use in magnetic devices). Also iron was the first soft magnetic material used on a large scale in motors, generators and power transformers before the introduction of iron–silicon alloys. Laminations in the core of the first commercial transformer built by Westinghouse in 1886 were made of iron with the following impurities: 0.046% C, 0.05% Si, 0.036% S, 0.059% N and 0.19% P, which is surprisingly high.

Pure iron exhibits quite attractive magnetic properties, such as permeabilities, saturation induction and hysteresis loss, and is relatively cheap. Its biggest disadvantage is its low electrical resistivity ($\sim 10^{-7}\ \Omega \cdot$ m or $\sim 10\ \mu\Omega \cdot$ cm at room temperature), thus making it unsuitable for most a.c. applications. Iron in the form of solid cores is used widely for telephone relays and switches, contactors, electromagnets and other d.c. devices.

Iron used for electromagnetic purpose is usually made by the basic open-hearth process from high-grade iron ore. Swedish iron and Armco iron are two high-purity commercial products prepared from Swedish iron ore, which is known for its high purity. The beneficial effect of high purity on the magnetic properties of iron is shown in fig. 6.3. An increase in purity from 99.90% to 99.99% vastly improves both the magnetization curve and the hysteresis loop.

To further improve the purity of Armco or Swedish iron, various methods discussed in ch. 4, §2.1, such as vacuum melting, electron-beam melting and zone refining, can be tried, but their applications are too expensive for extensive commercial use. A more practical method is to subject the iron product to a final hydrogen anneal (also described

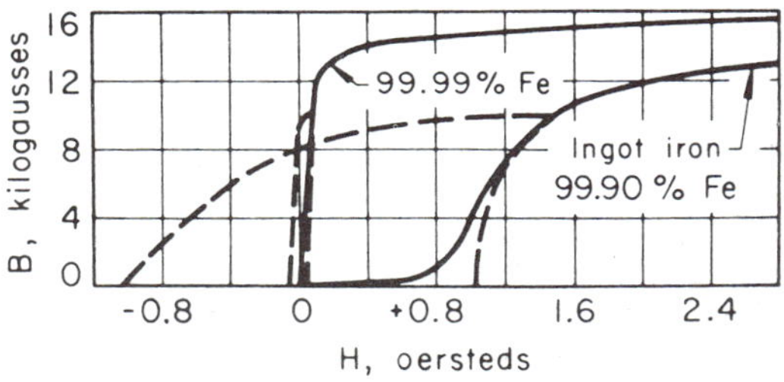

Fig. 6.3. Upper parts of hysteresis loops and B–H curves for two types of iron (Metals Handbook [1961]).

in that section). The treatment was first adopted by Yensen [1929] and later improved by Cioffi [1934]. It involves annealing the iron at between 1300 and 1500°C in an atmosphere of prepurified hydrogen in a closed box surrounded by moist hydrogen. Cioffi's treatment was capable of raising μ'_{max} from ~7000 to as high as 320 000 in the laboratory. The hydrogen anneal is effective in removing the magnetically harmful interstitial impurities such as carbon, oxygen, nitrogen and, to a lesser extent, the substitutional impurity sulfur. For pure iron, the presence of carbon and other interstitial impurities is considered especially harmful for the following reason. When these interstitial atoms are dissolved in a lattice at concentrations in excess of the equilibrium values, they tend to precipitate out slowly in fine particle form even at room temperature. This is a well-known metallurgical phenomenon called strain-aging. As a result of aging, magnetic properties are gradually worsened because the precipitates will interfere with the formation of preferred domain structures and with the motion of domain walls. Generally the carbon content is preferred to be less than 0.005% (~46 at. ppm), definitely below 0.01%.

The removal of interstitial impurities by the hydrogen treatment offers another benefit by promoting grain growth. For iron and most other metallic magnetic materials, the grains should be as large as possible because structure-sensitive magnetic properties will improve with few grain boundaries. The detrimental effect of grain boundaries on hysteresis loss of high-purity iron is shown in fig. 6.4.

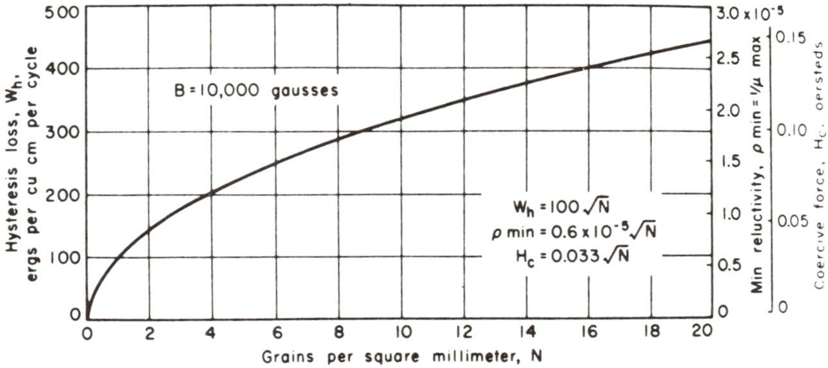

Fig. 6.4. Effect of grain size on the hysteresis loss of high-purity iron (Metals Handbook [1961]).

Iron powder has been used in the communications industry at frequencies up to 1 MHz. For this purpose, carbonyl iron powder is most often selected to take advantage of the spherical shape, the small size ($\approx 5\ \mu m$ in average diameter) and the very hard nature, mechanically, of the particles, although other grades such as hydrogen-reduced, electrolytic and mechanically-disintegrated iron powders are also used. Carbonyl iron powder derives its name from the fact that the powder is produced from the liquid iron pentacarbonyl ($Fe(CO)_5$). Briefly, the preparation procedure involves (1) the reduction of iron oxide to sponge iron in hydrogen, (2) the formation of $Fe(CO)_5$ by reacting the iron with CO and (3) the decomposition of liquid $Fe(CO)_5$ in a heated vessel under pressure through atomization. Iron powder thus prepared has a layer of insulating oxide on the surface of the individual particles, thereby satisfying an important condition for successful applications of powder. Cores made of carbonyl iron powder have surprisingly low eddy-current loss, in spite of the presence of 0.8% C and ~0.25% N.

Four prominent products of pure iron and their attractive properties responsible for the applications listed are compared in table 6.1, Part I.

Steel is a broad term used to denote all iron-base alloys. For magnetic applications, two classes of steels are involved. One is the low-carbon steels and the other the ferritic stainless steels. By low carbon is meant less than 0.2% C content. The composition of low-carbon steels is typified by AISI (Am. Iron and Steel Inst.) C1010, which contains 0.08–0.13% C, 0.3–0.6% Mn, up to 0.04% P and up to 0.05% S. Despite the presence of these elements in steels, little or no sacrifice of magnetic properties is noted in comparison with those of ingot iron (see fig. 6.5). This situation, plus the fact that low-carbon steels are inexpensive and yet exhibit far higher yield strengths than pure iron, makes the low-carbon steels attractive for many magnetic applications. Hot rolled carbon steel plates are used in the frames of rotating machinery, in motors as pole pieces or part of the pole support, whereas cold rolled steel sheets are used in fractional-horsepower motors.

Ferritic stainless steels are alloys of iron and chromium designed to resist corrosion at normal temperatures and to resist scaling and corrosion at elevated temperatures. The commercial ferritic stainless steels (AISI 400 Series) having the bcc structure are known as the magnetic stainless grades. Important examples and their properties are listed in table 6.1, Part II. Inevitably, with 12–18% Cr, the ferritic stainless steels are magnetically inferior to iron and low-carbon steels. Nevertheless, they are used in magnetic cores, solenoid switches, pole pieces and relays wherever corrosion or oxidation resistance is required while the magnetic requirements are less stringent.

§1.2. *Iron–silicon alloys*

It is estimated that the tonnage of soft magnetic materials used for ac machines outweighs that for dc applications by a factor of three. For ac applications, iron must be substantially modified by alloying to have high enough resistivity to minimize the eddy-current loss with limited sacrifice of pertinent magnetic properties. Iron–silicon alloys were

TABLE

Characterization of magnetic products

Part I. Pure

Material	Composition (% Fe–other elements)	Relative permeability μ_i'	Relative permeability μ_{max}'	B_s (T) [G]	H_c* $(A \cdot m^{-1})$ [Oe]
Ingot iron	99.8–0.02 C, 0.04 Mn, 0.03 S, 0.08 O_2 and 0.005 Si, P and N_2 each	10	1 000	2.05 [20 500]	112 [1.4]
Armco iron, Swedish iron	99.91–0.012 C, 0.018 S, 0.03 Mn and O_2 each, 0.002 Si, 0.004 P, 0.0018 N_2	200	7 000	2.15 [21 500]	80 [1.0]
Purified Armco iron (18-hr at 1475°C in dry H_2)	99.95–0.005 C, 0.003 S, 0.028 Mn, 0.003 O_2, 0.004 P and 0.0001 N_2	10 000	227 000	2.15 [21 500]	4 [0.05]
Carboxyl iron powder	99.9–≈0.05 C, Si, S and O_2 each, trace amounts of Mn and P	60	150		
Low-C steel, AISI C1010	Balance–0.08 C, 0.3 Mn, 0.08 Si, 0.07 Ni, 0.04 Cr, 0.03 P and 0.02 Mo				140 [1.74]

*At or near saturation induction.

developed under these circumstances and have been the number one soft magnetic material in both tonnage and market value since 1934 (ch. 5, §1.2). Industrial statistics show that, over the past two decades, annual production of iron–silicon alloys has averaged about 7.3×10^8 kg (720 000) tons), which amounts roughly to 40% of the total production of all magnetic materials, both soft and hard. Thus the importance of Si–Fe needs not be overemphasized. The alloys are often referred to in the industry as electrical sheets, electrical steels or silicon steels. The development of various Fe–Si alloys has already been described in ch.

6.1
of iron, steel and ferritic stainless steel
iron and steel

B_r* (T) [G]	W_h* ($J \cdot m^{-3}$/cycle) [Ergs · cm^3/cycle]	Resistivity ($\Omega \cdot m$)	Major applications
0.86 [8 600]	3 000 [30 000]	3×10^{-7}	
	500 [5 000]	1×10^{-7}	Telephone relays and switches, contractors, electromagnets and other dc apparatus
		1×10^{-7}	
		10^{-7}	Communication devices up to 1 MHz
0.8 [8 000]	≈ 4 000 [40 000]		Frames of rotating machines, pole pieces and pole supports of motors, etc.

TABLE 6.1 (cont.)
Part II. Ferritic stainless steels

AISI type	Cr (%)	C (%)	Other (%)	Condition	μ'_{max}	H_c ($A \cdot m^{-1}$) [Oe]	B_r T [G]	B_{max} T [G]	Applications
420	12	0.15 min		Annealed	670	610 [7.6]	1.06 [10 600]	1.62 [16 200]	Magnetic cores, solenoid switches, pole pieces and relays where high corrosion or oxidation resistance is required.
420	12	0.15 min		Hardened	70	4 760 [59.5]	0.65 [6 500]	1.25 [12 500]	
430	17	0.12 max		Annealed	960	488 [6.1]	1.15 [11 500]	1.62 [16 200]	
430	17	0.12 max		Hardened	278	784 [9.8]	0.51 [5 100]	1.46 [14 600]	
431	16	0.20 max	2 Ni	Hardened		1 840 [23]	0.77 [7 700]	1.53 [15 300]	

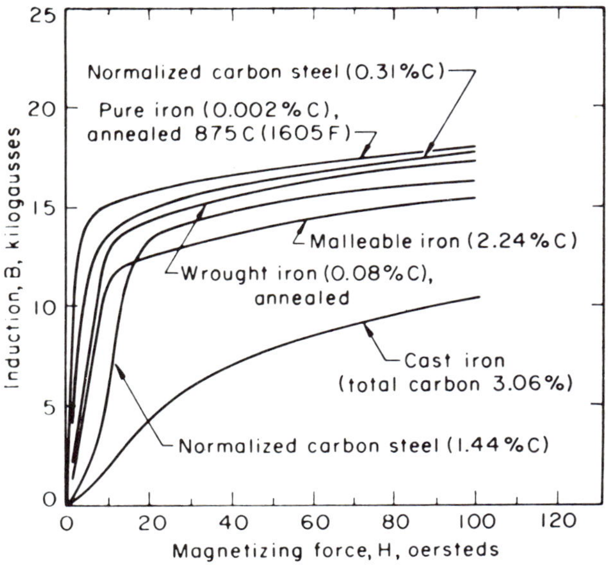

Fig. 6.5. Magnetization curves of various grades of iron and steels (Stanley [1963]).

5, §2.2 and §2.4. Also discussed in these sections are the advantages and disadvantages of alloying Fe with Si, of the (110)[001] Goss texture and of the (100)[001] cube texture. They will not be reiterated. For those interested in the detailed history of iron–silicon alloys, the article by Walter [1965] is recommended.

Although iron alloyed with up to 6.3% silicon has attractive magnetic properties with potential applications, brittleness imparted by silicon has heretofore limited the useful composition range to about 4.5%. In the AISI classification system, the 0.4–4.5% Si range is divided into nine compositions, each of which is classified by a number as well as an alternate designation, which clearly indicates the main application. Table 6.2, Part I lists these grades of Si–Fe, their comparative magnetic properties of hot-rolled sheets without texture and their outstanding features for applications. As the silicon content is increased to 4.5%, progressive property changes are noted. There is an electrical resistivity increase by a factor of 6, a coercive force decrease by ≈80%, a decrease in maximum core loss at $B = 1.5\,\mathrm{T}$ (15 000 G) by ≈75% and a density decrease by ~4%; also, there is a deterioration in saturation

TABLE 6.2

Properties and applications of iron–silicon alloys (Metals Handbook [1961]). Some characteristics and prominent applications are given in parentheses

AISI type	Alternate designation	Nominal Si content (%)	B_s (T) [G]	B_r* (T) [G]	H_c* ($A \cdot m^{-1}$) [Oe]	Max. core loss at 60 cycles ($W \cdot kg^{-1}$) [$W \cdot lb^{-1}$]	Resistivity ($10^{-7}\ \Omega \cdot m$)	Density (ρ) ($g \cdot cm^{-3}$)	Thermal conductivity at 20°C ($cal \cdot cm^{-1} \cdot {}^\circ C^{-1} \cdot s^{-1}$)
Part I. Grains oriented randomly									
M-50	Field	0.40					1.8	7.85	
	(Not subject to magnetic requirements; pole pieces and electrical apparatus for intermittent operation.)								
M-43	Armature	0.95	2.13 [21 300]	0.82 [8 200]	75 [0.94]	8.60 [3.90]	2.0 to 2.8	7.82 to 7.78	0.097
	(High core loss but good permeability at high inductions; small motors, pole pieces, and relays.)								
M-36	Electric	1.40	2.11 [21 100]	0.77 [7 700]	68 [0.85]	7.28 [3.30]	2.4 to 3.3	7.80 to 7.75	0.073
	(Excellent permeability at high inductions; rotating machines including ac and dc motors.)								
M-27	Motor	2.35	2.06 [20 600]	0.67 [6 700]	63 [0.79]	5.42 [2.46]	3.2 to 4.7	7.76 to 7.67	0.047
	(Good punching properties; continuous-duty, high-efficiency motors (10–1 000 hp) small transformers.)								
M-22	Dynamo	3.00	2.03 [20 300]	0.67 [6 700]	55 [0.69]	4.41 [2.00]	4.1 to 5.2	7.70 to 7.64	0.043
	(Good ductility, intermediate magnetic quality; stators of high-efficiency rotating machines, intermittent-duty transformers, high-reactance cores.)								
M-19	Transformer 72	3.20	2.00 [20 000]	0.63 [6 300]	44 [0.55]	3.92 [1.78]	4.1 to 5.6	7.70 to 7.62	0.039
	(Moderately high permeability at all inductions; communications equipment, high-efficiency fractional-horsepower motors.)								

M-17	Transformer 65	3.40	1.99 [19 900]	0.525 [5 250]	26 [0.33]	3.57 [1.62]	4.5 to 5.8	7.68 to 7.60	0.038
	(Intermediate properties; transformers to 1 000 kva.)								
M-15	Transformer 58	4.00	1.95 [19 500]	0.50 [5 000]	20 [0.25]	3.22 [1.46]	4.5 to 6.9	7.68 to 7.54	0.036
	(Low core loss, excellent permeability at low and moderate inductions.)								
M-14	Transformer 52	4.50	1.92 [19 200]	0.46 [4 600]	16 [0.20]	2.93 [1.33]	5.8 to 6.9	7.60 to 7.54	0.034
	(Lowest core loss of nonoriented grades, high permeability at low inductions; distribution and power transformers, high-efficiency rotating machines.)								
Part II. Grains oriented preferentially into the (110)[001] *texture*									
M-8	Oriented 80 or Tran-Cor 2X	3.00	2.01 [20 100]	–	–	1.76 [0.80]	–	7.68 to 7.64	0.043
M-7	Oriented 73 or Tran-Cor 3X	3.00	2.01 [20 100]	–	–	1.61 [0.73]	4.5 to 5.2	7.68 to 7.64	0.043
M-6	Oriented 66	3.00	2.01 [20 100]	–	–	1.46 [0.66]	4.5 to 5.2	7.68 to 7.64	0.043
M-5	Oriented 60	3.00	2.01 [20 100]	–	–	1.32 [0.60]	4.5 to 5.2	7.68 to 7.64	0.043
	(Grain-oriented sheets, M-8, M-7, M-6 and M-5, have highly directional magnetic properties, lowest core loss and highest permeability available when flux path is parallel to rolling direction; highest-efficiency distribution and power transformers and large generators.)								

*At $B = 1.0$T (10 000 G).

†At $B = 1.5$T (15 000 G) and for 0.356 mm (0.014 in) thickness.

induction by ~11%, in remanence by ~50% and in thermal conductivity (decrease) by ~80%. The randomly-oriented Si–Fe sheets (AISI M-50 to M-14) are prepared by hot rolling in three major steps: (1) preparing the alloy in a basic open-hearth or an electric furnace, (2) hot rolling the ingots into coils and (3) hot rolling the coils in packs (i.e., two or more sheets are rolled together; packs may be doubled, reheated and rerolled a number of times, depending on the gage desired) to a final thickness of from 0.318 to 0.953 mm (0.0125 to 0.0375 in, corresponding to gage 20 to 30). The finished hot-rolled sheets may have a width up to 122 cm (48 in) and a length up to 305 cm (120 in) (length is usually in multiples of the width). The sheets are pickled to remove scale and cold rolled slightly to secure satisfactory flatness.

Of the 7.3×10^8 kg (720 000 tons) of Si–Fe produced annually, about 4.6×10^8 kg (450 000 tons) are hot-rolled. The remaining 2.7×10^8 kg (270 000 tons) are hot- and cold-rolled to secure the Goss texture (Jacobs [1969]). Based on these tonnage figures, the annual production of Si–Fe thus splits into 62.5% for hot-rolled sheets and 37.5% for textured sheets. In the same estimate, however, Jacobs set the price of high-quality oriented Si–Fe at $0.20 per lb and of nonoriented sheets at $0.08–0.16 per lb. Taking the lowest price for the latter sheets, he arrived at a total market value of $100 million for the oriented sheets and $70 million for the nonoriented sheets (as compared with $25 million market value estimated for nickel–iron alloys). The prominent position of the oriented Si–Fe in the field of soft magnetic materials is clearly seen. The considerable advantage imparted to the total energy loss by the texture was shown in fig. 5.51. Other enormous improvements in properties associated with the magnetization curve and hysteresis loop by the Goss texture in 3% Si–Fe (M-6) over the nonoriented 4.75% Si–Fe (M-14) are shown in fig. 6.6.

The silicon content of the textured Si–Fe falls in a narrow range of 3 to 3.5%. No commercial Goss-textured Si–Fe sheets have been produced with more than 3.5% Si because early attempts in cold-rolling alloys containing 4% or more Si proved futile. This situation was reviewed by Getting [1967], who wondered aloud by asking, "Is 3.25% Si oriented core steel the ultimate?" He thought, and this author concurs with him, that it might pay to extend the silicon content to at least 4.5%. At this level, the brittleness problem should not be unsurmountable, while advantage can be taken of the lower magnetostriction to solve some of the remaining problems (notably noise and strain sensitivity) of power transformer cores made of 3.25% Si–Fe sheets. A

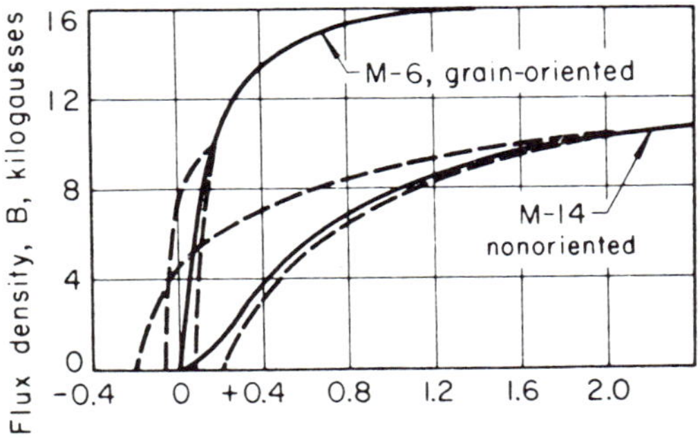

Fig. 6.6. Comparison between dc magnetization curves and hysteresis loops for grain-oriented 3.25% Si steel M-6 (flux parallel to rolling direction) and randomly oriented 4.75% Si-steel M-14 (flux half parallel and half transverse to the direction of rolling) (Metals Handbook [1961]).

recent effort (Natrita [1974]) made in Japan indicates that even 6.5% Si–Fe could be cold-rolled to less than 0.3 mm from 1 mm after the alloy had been hot-rolled by more than 70% at 600–750°C, and cold rolling of alloys containing 3.5–4.75% Si can be done easily. Magnetic properties of 6.45% Si–Fe sheets of 0.3 mm thickness are: $\mu_i' = 1460$, $\mu_{max}' = 22\,500$, $H_c = 12\ \mathrm{A \cdot m^{-1}}$ (0.15 Oe) at $B = 1.0$ T (10 000 G) and $W_h = 0.58\ \mathrm{W \cdot kg^{-1}}$ at $B = 1.0$ T and 50 Hz.

In the US market, grain-oriented Si–Fe is sold on the basis of a core-loss specification at $B = 15\,000$ G. Four grades of textured silicon-steel sheets and their properties and major applications are listed in table 6.2, Part II. Note that the only difference among them is the maximum core-loss, which varies from 0.60 to 0.80 watt per lb at 60 cycles, at the cited induction level for gage 29 (0.36 mm thick) sheets, with the nominal 3% Si content, density and resistivity remaining unchanges. Physically, the difference in the core loss is ascribed to the degree of sharpness of the texture.

Commercial products of Si–Fe, either textured or randomly oriented, are usually available as stampings (laminations) or in large sheets or long strips. The textured strips are often used to prepare closely-wound ring-shaped coils, which are particularly efficient in utilizing the texture by forming close circuits of flux in the preferred direction. The reason that silicon steels are almost exclusively used in sheet form is because they are used mainly for the construction of cores in motors, generators and power transformers. These cores must be sliced or laminated to introduce high resistances in the form of air spaces to minimize the eddy-current loss. Furthermore, the strips or sheets of Si–Fe, and, for this matter, practically all soft magnetic materials, must be insulated for high electrical efficiency. Two types of insulating materials are used: inorganic (e.g., AISI types C-4 and C-5) or organic (e.g., AISI types C-1 and C-3), depending on whether the metal will or will not be annealed afterwards for stress relief. Another important consideration of laminations is the thickness or gage number. The effect of thickness on the core loss is shown by two examples in table 6.3.

Both examples reveal the persistent decrease in core loss when the strip thickness is reduced. Thinner strips, however, are more expensive than thick ones on an equal basis of mass or volume. Therefore, the choice of the lamination thickness is a compromise between the electrical efficiency required for the particular application and the cost. A common practice is to first select a Si–Fe product in regard to the Si content and gage number and then downgrade the selection to a cheaper material of a thinner gage. In this practice, the desired electrical efficiency is retained by using less costly laminations. For example, laminations of 29 gage M-43 silicon steel may replace the original choice of 26 gage M-36 grade for equal core loss in the same stack, although the former material is less expensive.

Finally, it should be mentioned that, for Si–Fe another element, namely, carbon, must be closely controlled besides silicon. The nominal concentration of carbon is shown to vary with the Si content as follows:

	Nominal Si content				
	1%	2.5%	3%	3% (grain-oriented)	4%
Nominal % C	0.06	0.05	0.03	0.01	0.007

Excessive carbon not only adversely affects magnetic properties (e.g., increasing hysteresis loss at the rate of $\Delta W_h = 4500\ \mathrm{J \cdot m^{-3}}$/cycle per 1% C as compared with $\Delta W_h = 1000$ and $100\ \mathrm{J \cdot m^{-3}}$/cycle per 1% of S and Mn, respectively), but also causes the highly detrimental aging phenomenon previously discussed.

TABLE 6.3
Dependence of the core loss on the thickness of two sheet products

Part I. AISI type M-22 nonoriented 3% Si–Fe (ASM [1961])

Thickness (mm)[in]	Gage No.	Max. core loss, ($W \cdot kg^{-1}$)[$W \cdot lb^{-1}$] at 60 cycles, 15 kG
0.64 [0.025]	24	5.80 [2.63]
0.47 [0.0185]	26	4.98 [2.26]
0.36 [0.014]	29	4.41 [2.00]

*Measurements made on samples cut half in and half across the rolling direction.

Part II. Goss-textured (2.9–3.0)% Si–Fe prepared in laboratory (Taguchi and Sakakura [1969])

Thickness (mm)[in]	Core loss ($W \cdot kg^{-1}$)[$W \cdot lb^{-1}$] at 60 cycles, 15 kG
0.36 [0.014]	1.17 [0.532]
0.30 [0.012]	1.12 [0.510]
0.28 [0.011]	1.06 [0.480]
0.23 [0.009]	0.97 [0.440]

§1.3. *Nickel–iron alloys*

It was cited in Ch. 5, §2.2 that magnetic and electrical characteristics of the Ni–Fe system offer three attractive soft magnetic alloys: (1) ~36% Ni–Fe, known as Rhometal, for maximum resistivity; (2) 50 at.% Ni–Fe, known as Hipernik, for maximum saturation magnetization and (3) ~78% Ni–Fe, known as Permalloy, for excellent values of initial and maximum permeabilities. In addition, Isoperm, also at 50% Ni with a noted property of constant permeability, was discussed in ch. 5, §2.6. Numerous other commercial products were developed from this binary system. A summary of important commercial Ni–Fe products is given in table 6.4. Despite the large number of products, most Ni–Fe alloys are used mainly for their outstanding

TABLE
Properties and applications

Material	Nominal composition (%)			Relative permeability	
	Ni	Fe	Others	μ'_i	μ'_{max}
Rhometal	36	64	–	1 800	7 000
Sinimax	43	54	3 Si	3 000	50 000
Monimax	48	49	3 Mo	2 000	35 000
45 Permalloy	45	55	–	2 500	30 000
4750; Armco 48 Ni; Carpenter 49 alloy	47–50	50–53	–	4 000	50 000
Hipernik	50	50	–	4 500	70 000
Hipernik V; 48 Orthonik	50	50	–		55 000
Deltamax	50	50	–		150 000
Isoperm; Conpernik	50	50	–	1 500	2 000
78 Permalloy	78	22	–	8 000	100 000
4–79 Mo–Permalloy	79	16	4 Mo, 0.5 Mn	20 000	100 000
Hymu 80	79	17	4 Mo	20 000	100 000
Supermalloy	79	15	5 Mo; 0.5 Mn	75 000	800 000
Mumetal	77	16.5	5 Cu, 1.5 Cr	20 000	100 000

6.4
of nickel–iron alloys

B_s (T)[G]	B_r (T)[G]	H_c $(A \cdot m^{-1})$ [Oe]	Resistivity $10^{-7}\ \Omega \cdot m$	Some characteristics and applications
0.90 [9 000]	0.36 [3 600]	26 [0.33]	8.5	–
1.10 [11 000]		4.8 [0.06]	9.0	Above-average resistivity; high μ; high frequency coils
1.50 [15 000]		8 [0.1]	8.0	
1.60 [16 000]	0.80 [8 000]	16 [0.2]	5.0	Good overall characteristics; high μ and low losses; audio transformers, coils, relays
1.60 [16 000]	0.80 [8 000]	5.6 [0.07]	5.0	
1.60 [16 000]	0.80 [8 000]	4 [0.05]	5.0	Similar to 45 Permalloy, but more highly purified
1.60 [16 000]	1.40 [14 000]	16 [0.2]	5.0	Grain oriented in the (001)[100] texture; rectangular hysteresis loop magnetic amplifiers, coils, contact rectifiers
1.60 [16 000]	1.45 [14 500]	8 [0.1]	5.0	
1.60 [16 000]			5.0	Constant permeability; chock coils
1.07 [10 700]	0.60 [6 000]	4 [0.05]	1.6	High μ_i, low H_c and low B_r; sensitive dc relays
0.87 [8 700]	0.50 [5 000]	2.4 [0.03]	5.5	Very high μ_i; low losses; audio coils, transformers, magnetic shields
0.87 [8 700]	0.50 [5 000]	4 [0.05]	5.5	
0.80 [8 000]	0.50 [5 000]	0.48 [0.006]	6.0	Highest μ_i; low losses; high efficiency ac coils, pulse transformers and magnetic amplifier coils
0.65 [6 500]	0.30 [3 000]	4 [0.05]	6.0	Similar to 4–79 Mo–Permalloy

values of permeability. This is demonstrated in fig. 6.7, which compares the d.c. magnetization curves for five Ni–Fe products and a silicon steel (M-36). All the Ni alloys, especially Supermalloy, display much higher permeabilities than M-36.

Some other aspects of Ni–Fe discussed in Chapter 5 were:

(1) The history of and reasons for the development of the 4-79 Mo–Permalloy (§2.2);
(2) the production of the (100)[001] cube texture in NiFe and the advantages of the texture (§2.4);
(3) the effect of atomic order on saturation magnetization, Curie temperature, the first magnetocrystalline anisotropy constant, magnetostriction and electrical resistivity of Ni_3Fe (§2.5);
(4) the strong response to magnetic anneal of the 65 Permalloy and the conversion from a constricted hysteresis loop to a narrow rectangular loop (§2.5 under Magnetic Anneal) and
(5) the induction of uniaxial anisotropy by cold rolling and the benefits thereof for Isoperm and Ni_3Fe (§2.5 under Plastic Deformation).

Three commercial products of Ni–Fe alloys that have not been covered are described separately here.
Mumetal is a Ni-rich alloy with 16.5% Fe, 5% Cu and 1.5% Cr. Its composition is similar to that of the 4-79 Mo–Permalloy, with Cu and Cr replacing Mo. The attractive properties of Mumetal are (1) initial (20 000–30 000) relative permeability, (2) coercive force ($4\ \mathrm{A \cdot m^{-1}}$ or 0.05 Oe), (3) moderately high resistivity, ($6 \times 10^{-7}\ \Omega \cdot \mathrm{m}$ at 20°C), thus low total loss in low fields and (4) good ductility, which permits easy fabrication. These properties enable Mumetal to find widespread uses in the cores of current transformers, in radio apparatus at audio frequencies, for magnetic shielding, in the loading of telegraph cable and sometimes in low level magnetic amplifiers.
Sinimax is a modified 45 Permalloy with the addition of 3–3.5% Si for an increase in resistivity. Its magnetic properties, too, are much like those of 45 Permalloy with $\mu_i' = 2200$–3000, $\mu_{max}' = 50\,000$, $H_c = 4.8\ \mathrm{A \cdot m^{-1}}$ (0.06 Oe) and $W_h = 40\ \mathrm{J \cdot m^{-3}}$/cycle (400 ergs $\cdot\ \mathrm{cm^{-3}}$/cycle).
Monimax is essentially a modified Hipernik with the addition of Mo to increase the resistivity. Typically, it contains 49% Fe, 48% Ni and 3% Mo. Its magnetic properties are comparable with those of Sinimax. Both alloys are used in high-frequency coils.

Although an appropriate and closely controlled (regarding temperature, atmosphere and cooling rate) final heat treatment is essential

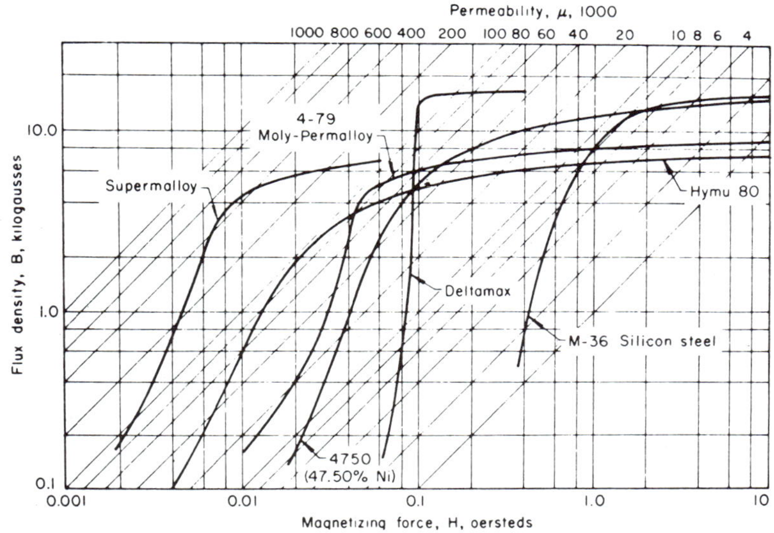

Fig. 6.7. Dc magnetization curves for various magnetic alloys, showing progressive deterioration of permeability from Supermalloy to M-36 Si–steel (Metals Handbook [1961]).

to secure desired properties for all soft magnetic properties, this requirement is especially important for the Ni–Fe alloys. A case in point is the introduction of Supermalloy, from which extraordinary values of initial (55 000 minimum and up to 125 000) and maximum (300 000 minimum and up to 1 500 000) relative permeabilities were achieved. Actually Supermalloy has a composition (79% Ni, 15% Fe, 5% Mo and 0.5% Mn) not very different from that of 4-79 Mo–Permalloy (79% Ni, 16% Fe, 4% Mo and 0.5% Mn). What makes Supermalloy so much better in permeabilities (5 times at $B = 0.002$ T to about 10 times at $B = 0.3$ T, where maximum permeability emerges) is the final heat treatment, which necessitates hydrogen of top quality for annealing and an elaborate cooling rate after the anneal, as shown by a comparison in table 6.5.

In closing this section, it is fair to state that the Ni–Fe system offers the largest number of technically important products. The useful range of alloy compositions for soft magnetic materials is roughly from 30 to 80% Ni in the face-centered cubic structure. The most attractive magnetic properties of the Ni–Fe alloys are the extraordinarily high values

TABLE 6.5
Comparison between heat treatments of Permalloy and Supermalloy

Material	Anneal		Cooling rate
	Temperature (°C)	Atmosphere	
4-79 Mo–Permalloy	≈ 1 100	Dry H_2	Ranging from furnace cooling, air cooling to quenching, depending on the permeability value desired
Supermalloy	1 300	Pure, dry H_2	Critical between 600°C and 300°C

of initial and maximum permeabilities. These properties coupled with low hysteresis losses at low inductions and reasonably high electrical resistivities (in 4-79 Mo–Permalloy and Supermalloy) make some of the Ni–Fe alloys especially suitable for applications in telecommunications engineering. The Ni–Fe alloys possess another advantage over silicon steels and Fe–Co alloys in that they are much less brittle and, therefore, can be more readily cold-rolled into thin tapes. Indeed, tapes of 4-79 Mo–Permalloy and Supermalloy are made commercially* with thickness of 3.18 μm (0.000125 in), as compared with the thinnest tapes of textures 3% Si–Fe at 25.4 μm (0.001 in) and of Supermendur at 50.8 μm (0.002 in), according to data for Tape Wound Cores in Arnold Bulletin TC-101B obtained in 1974. Against these advantages, however, must be considered their comparatively low saturation induction value (e.g., the B_s value of 78 Permalloy is just over half that of 3% Si–Fe) and their relatively high cost (5–10 times that of textured silicon steels).

§1.4. *Iron–cobalt alloys*

As far as soft magnetic alloys are concerned, the Fe–Co system differs from the Ni–Fe system in two aspects. First, there are only two outstanding Fe–Co alloys compared with the dozen Ni–Fe alloys described in the preceding section. Secondly, the main attraction of the Fe–Co alloys is the high value of saturation magnetization, not per-

*The author is indebted to the Arnold Engineering Co., Marengo, Illinois and Thomas and Skinner, Inc., Indianapolis, Indiana, for supplying technical information on their products of soft magnetic materials.

meability. In ch. 4, §2.2, we cited that Yensen developed Hiperco from the alloy 35% Co–Fe, which exhibits the highest value of M_s among all magnetic alloys, and that Elmen developed Permendur by adding 2% V to FeCo to increase the resistivity and improve the ductile behavior. Later studies led to the replacement of Permendur by Supermendur through a close control of impurities and a crucial final heat treatment featuring a fast cooling rate. Hiperco and Supermendur are the two major commercial products of the Fe–Co system in today's market. Detailed descriptions of these two materials and their applications are given in table 6.6, Part I. The Fe–Co alloys have not achieved wider application because of their high prices in comparison with other soft magnetic materials shown in tables 6.1, 6.2 and 6.4. The cost problem of Fe–Co alloys is due partly to the high cost of cobalt and partly to the brittleness, which demands more elaborate and, therefore, more costly methods to prepare the products in sheet form.

§1.5. *Miscellaneous soft magnetic alloys*

Several commercial soft magnetic alloys do not belong to any of the four groups just discussed. They are described separately as follows, and their application-related properties are compared in table 6.6, Part II.

12 Alfenol and 16 Alfenol are binary Fe–Al alloys containing 12 and 16% Al, respectively. The effect of adding Al to Fe on magnetic, electrical and mechanical properties is strikingly similar to that of Si. Thus the Fe–Al alloys exhibit much higher resistivities, but lower saturation magnetization, than pure Fe. Also, preferred orientation of the Goss or cube type can be developed in Fe–Al alloys. However, the alloys have never been developed to the same degree as the Fe–Si alloys. This uneven situation is due partly to the early difficulties encountered in melting and processing the Fe–Al alloys and partly to the success of the Fe–Si alloys. Actually, Fe–Al alloys are more attractive than Si–Fe in two respects. First, Fe–Al alloys are more ductile. For instance, alloys with up to 9% Al can be cold-rolled, and a 5% Al alloy can be cold-rolled to tapes with as much as 99.5% reduction in thickness without cracking. Secondly, Fe–Al alloys are more oxidation resistant at elevated temperatures than Si–Fe and other soft magnetic materials; thus, they may be suitable for high-temperature applications.

The 12 and 16 Alfenols were developed for high permeabilities and easy fabrication by Nachman and Buehler [1955]. Highest relative

TABLE 6.6

Properties and applications of iron-cobalt and miscellaneous iron alloys

Material	Nominal composition (%)	Relative permeability		B_s (T) [G]	B_r (T) [G]	H_c $A \cdot m^{-1}$ [Oe]	Resistivity (10^{-7} $\Omega \cdot m$)	Some characteristics and applications
	Fe–Co–Ni–Al–Others	μ_i	μ_{max}					
Part I. Iron–Cobalt alloys								
Hiperco	63.5–35–0–0–1.5 Cr	650	10 000	2.42 [24 200]	1.30 [13 000]	80 [1.0]	2.8	High μ at very high induction; high induction motors, transformers
Permendur	50–50–0–0–0	800	5 000	2.45 [24 500]	1.40 [14 000]	160 [2.0]	0.7	Very brittle; dc electromagnets and pole tips
2V Permendur	49–49–0–0–2V	800	4 500	2.40 [24 000]	1.40 [14 000]	160 [2.0]	2.7	$W_h = 1\,200$ J/M^3/cycle; similar to Permendur but ductile; receiver diaphragms
Supermendur	49–49–0–0–2V	1 000	60 000	2.40 [24 000]	2.15 [21 500]	16 [0.2]	2.7	Similar to 2V–Permendur but much higher magnetic quality

Part II. Miscellaneous soft magnetic alloys								
12 Alfenol	88–0–0–12–0	2 370	20 600	1.33 [13 300]		8 [0.1]		
16 Alfenol	84–0–0–16–0	4 000	80 000	0.80 [8 000]	0.40 [4 000]	3.5 [0.044]	15.3	High μ_{max} and good oxidation resistance
Thermenol	80.5–0–0–16–3.5 Mo	6 000	60 000	0.61 [6 100]	0.21 [2 070]	1.4 [0.018]	16.2	Modified version of 16 Alfenol with improvement in μ_i, H_c and resistivity
Sendust	85–0–0–5–10 Si	30 000	120 000			4 [0.05]		Attractive magnetic properties, but extremely poor mechanical behavior
45–25 Perminvar	30–25–45–0–0							Constant permeability at low fields; load coils
1040 alloy	11–0–72–0–14 Cu, 3 Mo	40 000	100 000			1.6 [0.02]	5.6	–

permeabilities ($\mu_i' = 4000$ and $\mu_{max}' = 80\,000$) were attained in the 16 Alfenol. Because of the tendency to atomic ordering in alloys around Fe_3Al (13.9% Al by weight), quenching from 600°C is required to avoid order in the preparation of the Alfenols. The 16 Alfenol is used as the core material in the magnetic heads for tape recording on the strength of its durability under abrasive conditions, which is five to ten times greater than that of Mumetal (Heck [1974]).

Thermenol is a modified version of 16 Alfenol with the addition of 3.5% Mo to improve the initial relative permeability (to 6000), coercive force (from 0.044–0.1 to 0.018 Oe) and resistivity at the expense of maximum permeability, remanence and saturation induction.

Sendust is a ternary alloy with a typical composition of 5% Al, 10% Si and ≈85% Fe. The name, which contains the word "dust", reflects the extreme brittleness of the alloy, which, upon crushing, disintegrates mechanically into powder. Despite its "hard" mechanical behavior, Sendust is magnetically soft, as indicated by very high values of initial (30 000) and maximum (120 000) relative permeabilities, low coercive force ($4\ A \cdot m^{-1}$ or 0.05 Oe) and low hysteresis loss. Another unusual feature of Sendust is that the permeability in fields below $80\ A \cdot m^{-1}$ (1 Oe) increases drastically by a factor of 25 at $H = 8\ A \cdot m^{-1}$ (0.1 Oe) from room temperature to a peak value at about 100°C. These remarkable magnetic properties are attributed to the nearly zero value of both magnetostriction and anisotropy constants. The extreme brittleness of the alloy, however, prohibits it from being processed into sheets. The pressed powder of Sendust finds some use in loading coils and recording heads.

Perminvars are a group of face-centered cubic ternary alloys of iron, cobalt and nickel. Their compositions center around the 45-25 Perminvar, which is composed of 45% Ni, 25% Co and the rest Fe, over wide ranges, and sometimes Mo or Cr is added to increase the resistivity. The most outstanding property of Perminvars is the constant permeability displayed at low inductions up to ~0.05 T (500 G) as the name (*Perminvar*) suggests. Another interesting magnetic property of Perminvars is that the shape of their hysteresis loops varies sensitively with composition, temperature of measurements, state of stress and the condition of final anneal (i.e., whether or not a magnetic field is applied). Three vastly different shapes of hysteresis loops observed in Perminvars are shown in fig. 5.65. Perminvar sheets have been used in loading coils.

1040 Alloy is a complex nickel alloy with 14% Cu, 11% Fe, 3% Mo and the rest Ni. Manganese may also be added up to about 1%. Despite the relatively large amount of copper present, the alloy exhibits high initial (40 000) and maximum (100 000) relative permeabilities and rather low coercive force ($16\ A \cdot m^{-1}$ or 0.20 Oe).

The d.c. magnetization curves for various prominent magnetic materials are compared in fig. 6.8. Clearly seen are the outstanding values of maximum permeability for such grain-oriented Ni–Fe alloys as Deltamax (curve 21), Monimax (curve 25) and 65 Permalloy (curve 27) and the maximum attainable values of induction for Supermendur (curve 1) and 2V-Permendur (curve 2). Detailed information on magnetic properties of various commercial products of soft magnetic materials and their applications can be found in the book by Heck [1974].

Before leaving metallic materials, a brief discussion of magnetic tapes is in order. These are thin-gage materials with thickness under ≈ 0.3 mm (0.012 in). The production of metallic tapes as thin as 3.2 μm (0.000125 in) has created many new applications for metallic magnetic materials, many of which are, otherwise, rather old. Had it not been for tapes, the use of metallic magnetic materials in transformers and magnetic amplifiers operating at moderate frequencies up to 1 MHz would have never been dreamed of. Materials for which tapes are available include singly-oriented 3.25% Si–Fe, cube-oriented Ni–Fe, nontextured 78 Permalloy and related Ni–Fe alloys (Supermalloy, Mumetal and 4-78 Mo-Permalloy), Hipernik and similar Fe–Co alloys (Deltamax and 4750) and 2% V–FeCo products (Permendur and Supermendur). Applications of thin tapes of electrical steels were surveyed by Stanley [1963] and are summarized in table 6.7. Production of the tapes involves cold rolling to the desired thickness after which the tapes are cut to the required length, coated with a layer of magnesia (MgO), wound on ceramic bobbins and subjected to an anneal in dry hydrogen, at a dew point of −40°C or less, at 1100–1200°C. The annealed cores are extremely strain sensitive, thus requiring special and careful storage and handling, and must be tested individually for magnetic characterization.

§2. FERRITES AND COMPOUNDS

When properly selected and prepared, ferrites offer several advantages over metallic materials for magnetic applications. Resistivities in the insulating range, attractive magnetic properties at high frequencies and high corrosion resistance are three notable advantages of many ferrites.

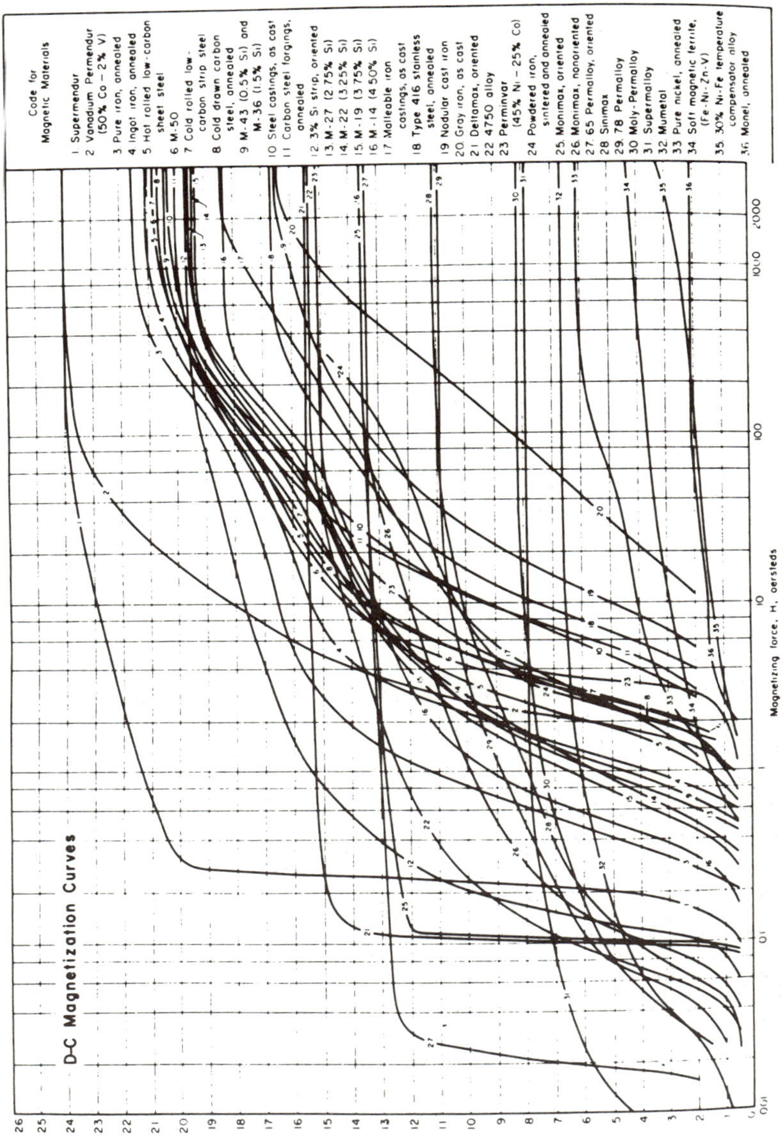

Fig. 6.8. Dc magnetization curves for various prominent magnetic products (Metals Handbook [1961]).

TABLE 6.7
Applications of sheets and tapes of metallic magnetic materials

1. Sheets of 0.356 and 0.305 mm (0.014 and 0.012 in) thickness – used at frequencies up to 400 Hz in motors, generators and transformers.
2. Tapes of 0.178, 0.127 and 0.102 mm (0.007, 0.005 and 0.004 in) thickness – used at frequencies of 400 to 2000 Hz. Prominent applications are:
   - 400-cycle motors: Automatic pilots and stabilizers and servo-mechanism motors
   - 400-cycle converters
   - Television (0.127 mm): Deflection yoke cores, flyback or horizontal-sweep transformers and vertical-sweep transformers
   - Magnetic amplifiers (including seismographs and magnetometers)
   - Other electronic devices: 400-cycle power transformers; plate and filament transformers; filter chokes; swinging chokes; charging reactors and timing pulse generators
   - Radio: Audiofrequency filters and voice-frequency modulators
3. Tapes of 0.051 and 0.025 mm (0.002 and 0.001 in) thickness – used at frequencies of 2000 Hz and higher. Prominent applications are:
   - Pulse transformers
   - High-repetition rate-charging reactors
   - Intermediate-frequency transformers
   - Blocking oscillator-type pulse transformers
   - Timing pulse generators
   - Impulse storing and memory devices (and other devices that require a high rate of change in flux with respect to time)
   - Magnetic amplifiers (where small time constants are required with a rather low power input)
4. Tapes thinner than 0.025 mm (0.001 in) – used at frequencies >1 MHz in transformers

Because the resistivities of ferrites are at least one million times greater than those for magnetic alloys, solid ferrite cores, instead of laminated cores, can be used for inductors in carrier telephony and for transformers. Solid cores often can be made in a more convenient shape than their laminated counterparts. Ferrites, however, have these disadvantages: low magnetic saturation, relatively low Curie temperature and inferior mechanical properties. In addition, because ferrite products are produced largely by ceramic technology and are hard, brittle and unmachinable, their close dimensional tolerances have to be obtained by grinding, which is an expensive process. Despite the disadvantages, ferrites are firmly established as an important class of soft magnetic materials in modern technology. They are widely used in linear, digital and microwave branches of electrical and electronic engineering. Since the successful developmental work led by Snoek in the mid 1940's, the market of ferrites has been growing at a much faster rate than that of metallic magnetic materials.

It is difficult to be as specific in the description of commercial ferrites as we have been for metallic materials because compositions of ferrites are not standardized and data are not available on the properties of specific ferrite compositions. Furthermore, design ingenuity can sometimes play a significant role in determining the functional magnetic characteristics of a ferrite device, thus tending to reduce the importance of composition and inherent magnetic properties. An example will be given in §2.1 to illustrate how design features can alter the performance of a ferrite core. The present description of commercial products of ferrites will be based on applications rather than on compositions.

In general, the principal applications of ferrites may be classified into three categories*: (1) ferrites used at frequencies up to about 300 MHz, necessitating high initial relative permeability and low energy loss; (2) ferrites used at microwave frequencies, requiring high resistivity and involving certain ferromagnetic resonance phenomena and (3) ferrites used in memory cores for displaying square hysteresis loops and reasonably fast switching time. Minor applications of ferrites include those based on large magnetostrictions in transducers and on magneto-optical effects (Faraday rotation, Kerr effects, etc.) in rotators, isolators, circulators, modulators, polarization switches and YIG lasers. Ferrites used in transducers are $CoFe_2O_4$, Fe_3O_4 and $Ni_{0.98}Co_{0.02}Fe_2O_4$, and those used in the magneto-optical devices are a number of garnets (YIG, GdIG, etc.) in pure or doped form, rare-earth orthoferrites ($MFeO_3$, where M is a rare earth ion) and barium ferrite ($BaFe_{12}O_{19}$) and divalent europium compounds described in ch. 5, §1.3.2.

§2.1. *Ferrites used at frequencies up to 300 MHz*

In this category, the spinel-type ferrites are much more extensively used than garnets, which are more expensive in the first place. The commercialization of spinel ferrites began with the introduction of four series of mixed ferrites by the Philips Laboratories in Holland. The

*Statistical data for the ferrite market in 1968 (quoted in §2.3), however, reveal that applications in categories (1) and (3) amounted to more than 90% of the total money consumed for ferrite components.

products are called Ferroxcube I for Cu–Zn ferrites, II for Mg–Zn ferrites, III for Mn–Zn ferrites and IV for Ni–Zn ferrites. Because the component single ferrites, $ZnFe_2O_4$, $MgFe_2O_4$, $NiFe_2O_4$, etc., are capable of dissolving each other to form continuous series of solid solutions, each series of Ferroxcubes has numerous mixtures. Outstanding representatives of each of the four grades of Ferroxcubes and their magnetic and electrical properties are compared in table 6.8. The quantity in the last column was defined as $\tan\delta = \mu_I'/\mu_R'$ where μ_R' and μ_I' are the real and imaginary components of relative permeability, respectively ($\mu' = \mu_R' - i\mu_I'$). In more practical terms, $\tan\delta$ measures the energy dissipated per period over the stored energy in the period; thus it is called the loss factor.

A few remarks on the Ferroxcube ferrites listed above are in order:

(1) At first glance, it seems odd to find Zn-ferrite in all four grades, because $ZnFe_2O_4$ per se exhibits no moment, as was indicated in table 5.12. However, the addition of $ZnFe_2O_4$ to other ferrites enhances saturation magnetization, as was discussed in ch. 5, §1.3.1 and shown in fig. 5.7.

(2) Although the second column indicates that Ferroxcubes are basi-

TABLE 6.8

Four commercial grades of cubic ferrites introduced by the Philips Laboratories

		Example					
Designation	Material	Composition	μ_i'	M_s $(A \cdot m^{-1})$	θ_C (°C)	$\rho(\Omega \cdot m)$	$\tan\delta$
Ferroxcube I	Cu–Zn ferrite	40 Cu, 60 Zn*	1 100		90	10^5	100
Ferroxcube II	Mg–Zn ferrite	≈50 Mg, ≈ 50 Zn	400	2.07×10^5	120	10^6	130
Ferroxcube III	Mn–Zn ferrite	45.5 Mn, 49.5 Zn, 5 Fe	1 000		110	10^2	170
Ferroxcube IV	Ni–Zn ferrite	40 Ni, 60 Zn	82		80	10^7	55

*40 Cu, 60 Zn means 40 mol% $CuFe_2O_4$ and 60 mol% $ZnFe_2O_4$, which is equivalent to $Cu_{0.6}Zn_{0.4}Fe_2O_4$.

cally mixtures of two single ferrites, a third or even a fourth ferrite may be added for certain benefits. The example listed for Ferroxcube III is a triple ferrite.
(3) Mn–Zn ferrites in grade III show attractive values of μ_i and other properties. Hence they are most widely used at frequencies up to ≈ 10 MHz. However, their resistivities ($\approx 10^2\ \Omega \cdot \mathrm{m}$) are too low to minimize the eddy-current loss at higher frequencies, for which Ni–Zn ferrites in grade IV are favored because of much higher resistivities.

Even though Ferroxcube ferrites were introduced almost three decades ago, they still dominate today's market. Snelling [1969] revealed that Ferroxcube has remained as the trade name of ferrites manufacturers in England (Mullard, Ltd.), Germany (Valvo, Gmb H), France (Coprin), the United States and, of course, Holland.

To describe ferrites in the present market, we shall follow the industrial practice, as outlined in Snelling's book, by emphasizing the requirements for the applications. Once the requirements are spelled out in terms of properties independent of the composition and geometry of the product, the selection of ferrite is merely a search for the right product through a manufacturer's catalog. For Mn–Zn and Ni–Zn

TABLE

Applications and required properties of ferrites –

Application	Nominal frequency range (MHz)	Initial relative perme-ability	$B_s^{\ddagger}$ (T) [G]	$B_{r,\infty}$ (T) [G]
Low frequency inductors	<0.2	800–2 500	0.35–0.5 [3 500–5 000]	0.08–0.14 [800–1 400]
Moderate frequency inductors, antenna rods	0.1–2	500–1 000	≈ 0.4 [4 000]	0.15–0.2 [1 500–2 000]
Wide-band and pulse transformers	up to 200	1 500–10 000	0.3–0.5 [3 000–5 000]	0.07–0.14 [700–1 400]
T.V. and power transformers, high-B_s applications	<0.1	1 000–3 000	0.35–0.52 [3 500–5 200]	0.1–0.25 [1 000–2 500]

*N. V. Phillips; **Mullard, Ltd.; $^{\Delta}$Ferroxcube Corp.; †Indiana General Corp.;

ferrites used in this category, Snelling conducted a comprehensive survey of the catalogs and trade literature of major manufacturers of the world and suggested the following classification:

(1) Mn–Zn ferrites – four areas:

(I) Low-frequency inductors up to about 20 kHz.
(II) Medium-frequency (100 kHz to 2 MHz) inductors. Antenna rods for medium- and long-wave broadcast bands.
(III) Wide-band transformers and low-power pulse transformers.
(IV) Applications requiring high saturation induction and low loss in television line scanning transformers, in deflection yokes and in power transformers at 700 Hz to 100 kHz.

(2) Ni–Zn ferrites – seven areas:

(V) $\mu_i' > 1000$ wide-band transformers at 1–300 MHz and pulse transformers for duration pulses <0.1 μs.
(VI) μ_i': 500 to 1000 for wide-band transformers at 5–300 MHz, power transformers at 100 kHz–1 MHz and antenna rods for medium- and long-wave broadcast bands.

6.9
predominantly Mn-Zn ferrites (Snelling [1969, 1972])

$H_{c,\infty}$ (A · m^{-1}) [Oe]	θ_C (°C)	Resistivity (Ω · m)	Recommended commercial products: Holland*	United Kingdom**	USA
10–30 [0.13–0.38]	140–210	0.5–7	3B, 3B3, 3B5, 3B7, 3H1	A1, A5, A13	TC9†, TC12†
40–100 [0.50–1.26]	200–280	1–20	3D3	A10	3DC$^{\Delta}$
2.8–24 [0.04–0.30]	90–200	0.02–0.5	3E1, 3E2, 3E3	A8, A7, A15	05‡, 06‡
10–30 [0.13–0.38]	180–280	0.2–1.0	3C2, 3C6, 3C8	A3, A9, A16	AK16#, AK20#, AK30#

#Arnold Engineering Co.; ‡At $H = 1000\ A \cdot m^{-1}$ (12.5 Oe).

(VII) μ_i': 160 to 490 for power transformers at 500 kHz–5 MHz and antenna rods for medium- and long-wave broadcast bands.
(VIII) μ_i': 70 to 150 for inductors at 2–20 MHz, antenna rods for short-wave broadcast bands and power transformers at 2–30 MHz.
(IX) μ_i': 35 to 65 for inductors at 10–40 MHz.

TABLE
Applications and required properties of ferrites –

Application	Nominal frequency range (MHz)	Initial relative perme-ability	B_s (T) [G]	$B_{r,\infty}$ (T) [G]
Wide-band and pulse trans-formers	1–300	>1000	≈ 0.26 at $H = 1\,000\ A \cdot m^{-1}$ [≈ 2 600]	0.85 [8 500]
High-fre-quency wide-band and power transformers	0.1–300	500–1000	0.28–0.34 at $H = 1\,000\ A \cdot m^{-1}$ [2 800–3 400]	0.15–0.19 [1 500–1 900]
Antenna rods, High-fre-quency power transformers	0.5–5	150–500	0.3–0.36 at $H = 2\,000\ A \cdot m^{-1}$ [3 000–3 600]	0.12–0.16 [1 200–1 600]
Inductors, Antenna rods, High-frequency power trans-formers	2–30	70–150	0.25–0.42 at $H = 4\,000\ A \cdot m^{-1}$ [2 500–4 200]	0.24–0.34 [2 400–3 400]
Inductors	10–40	30–70	0.24–0.28 at $H = 4\,000\ A \cdot m^{-1}$ [2 400–2 800]	0.15–0.20 [1 500–2 000]
Inductors	20–60	10–30	0.15–0.26 at $H = 8\,000\ A \cdot m^{-1}$ [1 500–2 600]	0.08–0.15 [800–1 500]
Inductors	>30	< 10	0.1–0.2 at $H = 8\,000\ A \cdot m^{-1}$ [1 000–2 000]	0.05–0.1 [500–1 000]

*Ceramic Magnetics, Inc.; **Allen-Bradley Co.; $^+$Magnetics, Inc.; ‡Indiana General

(X) μ_i': 12 to 30 for inductors at 20–60 MHz.
(XI) $\mu_i' < 10$ for inductors at frequencies >30 MHz.

Each of the 11 application areas is specified by a set of parameters on the basis of which manufacturer's recommendations for the most

6.10
predominantly Ni–Zn ferrites (Snelling [1969, 1972])

$H_{c,\infty}$ ($A \cdot m^{-1}$) [Oe]	θ_C (°C)	Resistivity ($\Omega \cdot m$)	Recommended commercial products: Holland	United Kingdom	USA
20 [0.25]	100	10		–	Pr6200‡, PR6400‡, PR6500‡, O6‡
16–48 [0.2–0.6]	90–200	$10–10^7$	Philips 4A1, 4A3, 4A4	Mullard-B1	C5N$^{\Delta}$, C7B$^{\Delta}$
80–160 [1–2]	200–370	$>10^3$	Philips-4B1	Mullard-B2	R-02**
160–500 [2–6.25]	350–490	$>10^3$	Philips-4C1, 4C6	Mullard-B10	N$^+$
300–500 [3.75–6.25]	300–500	$>10^3$	Philips-4D1, 4D2	Mullard-B4	Q2‡
500–1 600 [6.25–20]	250–510	$>10^3$	Philips-4E1, 1Z2	Mullard-B5	F6-21$^{\#}$
800–1 600 [10–20]	250–510	$>10^3$		Neosid-F29	C14A‡‡

Corp.; ‡‡Stackpole Carbon Co.; $^{\#}$D. M. Steward Mfg. Co.

suitable ferrite products are made in their catalogs. Tables 6.9 and 6.10 list the various application areas and their main property requirements for Mn–Zn and Ni–Zn ferrites, respectively. Omitted in the tables are values of several practical parameters. Readers interested in such values referred to the original tables in Snelling [1972]. All listed commercial products for the 11 areas of applications are spinel-ferrites except two cases in which hexagonal ferrites are recommended for inductors operating at frequencies >30 MHz.

In the applications of ferrites in this category, attention should be directed not just to the value of relative permeability, but also to its resolved components, μ_R' and μ_I', and the frequency dependence of both components. Hence, an important characterization of a polycrystalline ferrite product is the frequency dispersion curves for both μ_R' and μ_I', similar to those shown in fig. 4.31. Reproduced in fig. 6.9 are the relative permeability spectra obtained at room temperature for five Ni–Zn ferrites with composition ratio NiO:ZnO = 17.5:33.2 (A), 24.9:24.9 (B), 31.7:16.5 (C), 39.0:9.4 (D) and 48.2:0.7 (E). As a general trend, the real component of the complex relative permeability remains practically constant up to a frequency that depends on the type and composition of the ferrite and then falls approximately exponentially. This critical frequency is in the region of 10–100 MHz for Ni–Zn ferrites (and 1–10 MHz for Mn–Zn ferrites). The imaginary component of the relative permeability, which represents the dissipated energy leading to the so-called residual losses, rises with frequency at accelerating rates up to this critical frequency and then decreases together with μ_R'. The critical frequency is observed to be generally lower for materials that have the higher relative permeabilities at low frequencies. Snoek ascribed the latter observation to the natural resonance, or the resonance of rotation magnetization under the action of the anisotropy field H_a. The resonance frequency in this case can be calculated from the relation

$$\omega_{R,H_a} = \gamma H_a \mu_0, \tag{6.5}$$

where $\gamma = ge/2m_0 = (8.794 \times 10^{10})\, g \cdot C \cdot kg^{-1}$, is the magnetomechanical ratio. For Ni–Zn ferrites having ⟨111⟩ easy axis of magnetization, the anisotropy field is given by

$$H_a = -4K_1/3M_s\mu_0. \tag{6.6}$$

Hence, by assuming $K_1 = -5 \times 10^2\ J \cdot m^{-3}$ and $M_s = 2.39 \times 10^5\ A \cdot m^{-1}$, we obtain $H_a = 2.2 \times 10^3\ A \cdot m^{-1}$ and, for $g = 2$, $\omega_{R,H_a} \approx 5 \times 10^8$ or $f_{R,H_a} =$

$\omega_{R,H_a}/2\pi \simeq 80$ MHz, in good agreement with the critical frequency observed in Ferroxcube IV (fig. 6.9). By applying the Snoek theory, workers at Philips were able to develop another class of ferrite products designated as Ferroxplanas, which feature f_{R,H_a} values many times larger than that of the Ferroxcube IV D shown in fig. 6.9. Ferroxplanas are soft hexagonal ferrites described in ch. 5, §1.3.1. Two prominent Ferroxplanas are of the M_2Z and M_2Y types, with M being Co and Mg, respectively. Their compositions and important magnetic properties are compared in table 6.11. Experimental values for the resonance frequency are given in the last column. For hexagonal ferrites having a low anisotropy constant along the c-axis and preferred directions of magnetization in the basal plane, theory predicts the resonance frequency by the expression

$$\omega_{R,H_a} = \gamma(H_{a_1} \cdot H_{a_2})^{1/2}\mu_0, \tag{6.7}$$

where H_{a_1} and H_{a_2} are the anisotropy fields in and outside the basal plane, respectively. The observed values for these fields in Co_2Z and Mg_2Y Ferroxplanas listed in the sixth and seventh column indicate that

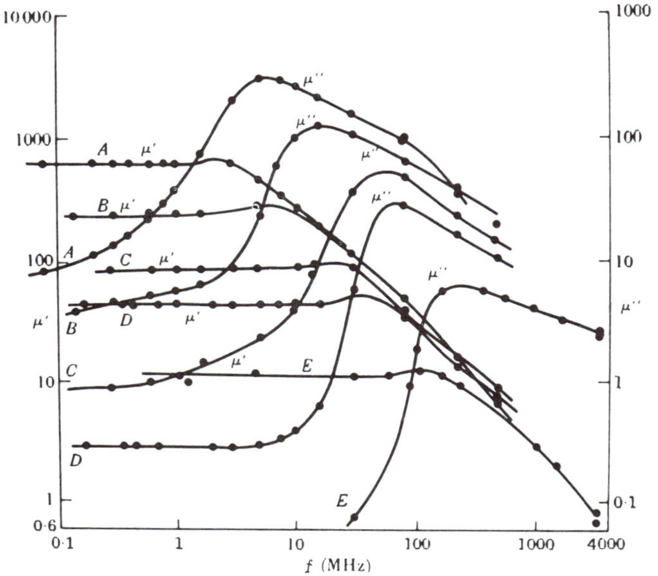

Fig. 6.9. The real (μ') and imaginary (μ'') parts of the initial relative permeability for Ferroxcube IV, A-E (Smit and Wijn [1954]).

TABLE 6.11
Composition and magnetic data for two ferroxplanas

Ferroxplana					Anisotropy field, $(A \cdot m^{-1})$		
Type	BaO:MO: Fe_2O_3	Composition	M_s $(A \cdot m^{-1})$	$\mu'_{i,R}$	H_{a_1}	H_{a_2}	f_{R,H_a} (MHz)
Co_2Z	3:2:12	$Co_2Ba_3Fe_{24}O_{41}$	2.67×10^5	~ 80	8.9×10^3	1.0×10^6	$\sim 1\,500$
Mg_2Y	2:2:6	$Mg_2Ba_3Fe_{12}O_{22}$	1.20×10^5		4.9×10^3	0.8×10^6	$\sim 1\,000$

H_{a_1} is indeed much smaller than H_{a_2}. Substituting these values in (6.7) yields 3 400 and 2 200 MHz for f_{R,H_a} for Co_2Z and Mg_2Y, respectively, in rather poor agreement with the observed values.

For low (<1 MHz) frequency applications of ferrites, in addition to high values of initial permeability, magnetocrystalline anisotropy and magnetostriction must be kept close to zero. In the single ion model, the contribution of each species of cations is approximately additive. Contributions of most cations to both anisotropy and magnetostriction are negative, except Co^{II}, which always makes a positive contribution to anisotropy, and Fe^{II}, which also makes a positive contribution to anisotropy in some cases and is the only ion known so far to make a large positive contribution to magnetostriction. Commercial products have been prepared from $Mn_{0.5}Zn_{0.4}Fe^{II}_{0.1}Fe^{III}_2O_4$ to realize the magnetic advantages of doping with Fe^{II} ions (keeping K_1 and λ small) at the expense of resistivity.

The mixed Ni–Zn ferrites in Ferroxcube IV are used primarily in the frequency range 1 to 100 MHz. For each of these ferrites, a maximum frequency limit exists, depending upon the loss allowed with respect to

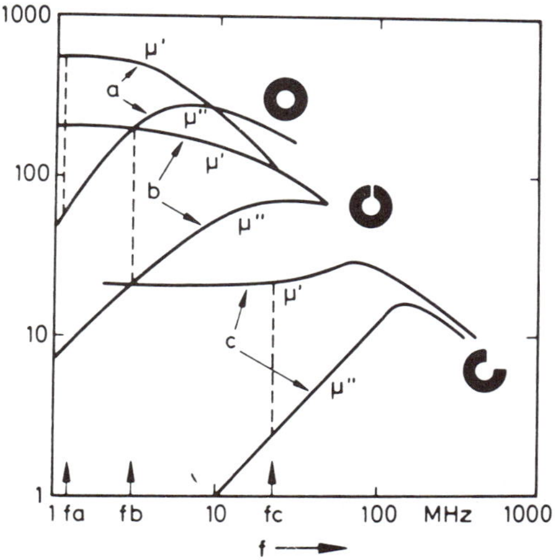

Fig. 6.10. Demonstrating how design ingenuity can improve the magnetic characteristics of ferrites in applications. The permeability spectra are widened and the peaks are sharpened by introducing a large gap in the toroidal sample (c) of $Ni_{0.36}Zn_{0.64}Fe_2O_4$ (Verweel [1971]).

the relative permeability spectra shown in fig. 6.9. However, design ingenuity could push this limit upward so as to widen the useful range of the material. For instance, it is shown in fig. 6.10 that the introduction of a small air gap in a toroidal core would shift the peak position of the μ_1' curve to a higher frequency, and a further shift is effected when the air gap is widened. The shift is also indicated by the three positions, f_a, f_b and f_c, of frequency at which $\tan\delta = 0.1$.

§2.2. *Ferrites used at microwave frequencies*

The principle underlying the usefulness of ferrites at microwave frequencies has been discussed in ch. 4, §2.2.3. It was shown in fig. 4.32b that the application of an intense steady field to a ferromagnet will cause the magnetization $\boldsymbol{M}$ to precess about the field. Meanwhile, the alternating field applied at high frequency in the $\pm x$-direction may be resolved into two rotating fields represented by two magnetic vectors of equal length rotating in opposite directions in the x–y plane. Thus one vector will rotate in the same sense as that of the precess of the magnetization vector, whereas the other vector will rotate in the opposite sense. It is then reasonable to expect that the rf susceptibility corresponding to the field rotating in the former (same) sense will exhibit the normal resonance absorption behavior in the region of the resonance frequency. On the other hand, the rf susceptibility corresponding to the rotation in the latter (opposite) sense will not exhibit resonance absorption because this rf field vector can never be "in step" with the precessing magnetization. Indeed, when a circularly polarized electromagnetic wave propagates in such a magnetic medium, a precessional resonance (and the accompanying absorption) will be produced if the rotation of the polarized wave is in the right sense with respect to the direction of magnetization precession. This phenomenon allows a variety of nonreciprocal devices to be operated at microwave frequencies in which the transmission properties in one direction are quite different from those in the other.

The gyrator developed by Hogan [1953] is a good example of the application of the gyromagnetic properties of ferrites and garnets. The construction and principle of operation of a gyrator is shown in fig. 6.11. A rod of ferrite is magnetized along the cylindrical axis in a microwave guide by a steady field generated by a solenoidal coil or a permanent magnet. A linearly polarized microwave travelling from left to right can be decomposed into two circularly polarized components whose magnetic vectors rotate in opposite directions. Since the com-

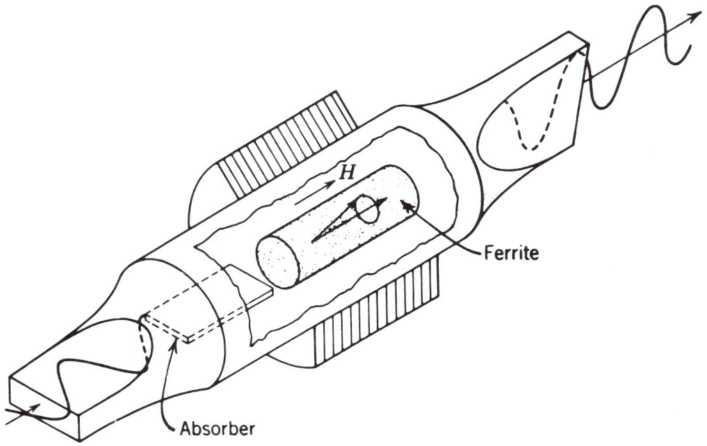

Fig. 6.11. Schematic diagram of a microwave gyrator (Chikazumi [1964]).

ponent of the same sense of rotation as the precession of magnetization will be partly absorbed, the polarization of the output wave will rotate by some angle θ. If part of the output wave is reflected back by some means into the system, the polarization will rotate another θ degrees in the same direction, making a total rotation of 2θ relative to the incident wave. Thus the reflected wave can be easily eliminated by an absorber which is so placed as to unaffect the incident beam. Hence a gyrator can be used to protect the waveguide from disturbance by reflected waves and in differential phase shift circulators. Microwave devices using ferrites and garnets include resonance or waveguide isolators, switches, tunable filters, phase shifters, circulators and amplifiers. The performance of these devices depends, among other things, on the saturation magnetization of the ferrite and the width of the resonance absorption peak. Previous discussion in ch. 5, §1.3.1 cited that yttrium–iron garnet (YIG) displays an extremely narrow ferromagnetic resonance linewidth, of the order of $\approx 50\ \mathrm{A \cdot m^{-1}}$. Thus YIG with various partial substitutions for Y finds wide application in microwave devices. Other ferrites used in these devices are spinel ferrites, such as Ni ferrite and Mg–Mn ferrite, and hexagonal ferrites, such as $BaFe_{12}O_{19}$. The full mathematical expressions relating the intrinsic properties of the magnetic materials to the parameters involved in the applications are complicated and beyond the scope of this book.

Reference should be made to the book by Lax and Button [1962] for further information on microwave ferrites and applications.

§2.3. *Square-loop ferrites* (SLFs)

Papian [1952] demonstrated that a ferrite that exhibited square hysteresis loops was capable of performing storage function as memory elements in computer systems. Since that time, SLFs have rapidly replaced metallic tapes in applications for memory cores in computers. Industrial statistics for 1968 revealed that more than 15 billions SLF cores were produced in that year, and the annual production number has been doubled every 2 to 3 years since the introduction of ferrite cores in the mid-1950's. Of the $110 million market value estimated for ferrites in 1968 by Jacobs [1969], SLF cores accounted for $55 million, ferrites used at frequencies up to ~ 100 MHz accounted for $52 million ($20 million in television receivers, $20 million used in communications and radio devices, recording heads and magnetostrictive transducers and the remaining $12 million for ferrites used in telephone communications systems) and microwave ferrites accounted for $3 million. To further illustrate the important role played by SLFs in the present high-speed computers, it is well to realize that, in the hierarchy of stores ranging from small-capacity (1–100 bits) high-speed (5–50 ns) buffer memories to very large-capacity (up to 10^9 bits), slow access (seconds or even minutes) archival stores, the main computer memory has, up to the present time, almost invariably been a medium- to high-capacity (10^5–10^7 bits), high-speed (0.25–1.0 μs) ferrite core store, leaving little room for semiconductor integrated circuit and plated-wire stores as alternatives.

Before discussing the ferrite cores, it should be remarked that the ideal shape of the hysteresis loop for storage function is a rectangle, not necessarily a square. Thus in the following discussion, it is understood that the word square is used more or less as a matter of convention and convenience, not in any way representing the actual ideal shape of the loop.

The principle of recording and storing information magnetically in a memory core is quite simple. Figure 6.12a shows a nearly ideal square loop with B_r almost equal to B_{max}. The magnetization has two stable stages, $+B_r$ and $-B_r$. These two states can be used to represent "0" and "1" of the binary number system. In a toroid, they correspond to clockwise and counterclockwise magnetization within the material. Ferrite cores are in the toroidal form to avoid self-demagnetizing effects. In principle, each core requires three windings. The "write"

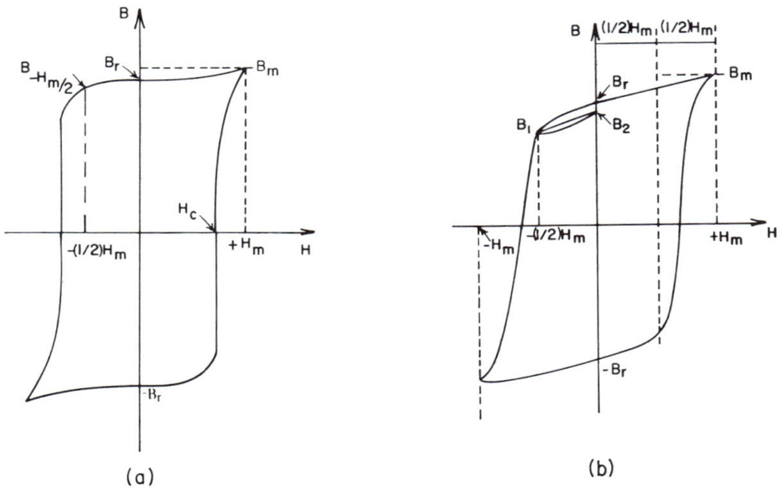

Fig. 6.12. (a) A nearly ideal square hysteresis loop recommended for performing storage function and (b) a nonideal hysteresis loop inside which a minor loop is shown for the excursion of induction between B_1 and B_2 upon the repeated application and withdrawal of a field $-\frac{1}{2}H_m$.

winding takes an input pulse to produce, say $+B_r$ for a "0". This information is now permanently recorded because the magnetization of the core material remains at $+B_r$ until a sufficiently large negative pulse is passed through the "write" winding to overcome the coercive force and to change the magnetization to $-B_r$. Now a "1" is stored in the core. To sense the state of magnetization at a later time, a reading pulse must be applied through the "read-out" winding. Whether or not an output pulse appears in the "sense" winding depends on the sense of the read-out pulse relative to that of the information stored. For a "0" stored, if the read-out pulse is of such a sense as to re-set the core to "1", an output pulse will appear in the sense winding. If the stored information is "1", no output pulse will appear. The method of read-out obviously destroys the information while it is read and is known as destructive readout (DRO). A number of nondestructive read-out (NDRO) methods were developed which automatically re-write the information after it has been read out.

In practice, the write and read-out windings may be combined, and a number of cores are arranged in an array known as a matrix within a square or rectangle comprising, say, 10 to 40 cores per side. Several matrices may be stacked to form a multi-unit store. Advanced technology has made it possible to reduce the dimensions of toroidal cores of 0.3 mm O.D., 0.18 mm I.D. and 0.3 mm thick. Such small cores not only save space, but promote fast switching.

Two basic methods of organizing magnetic cores into a computer store or memory have evolved. These are the coincident current selection system, also called the two-coordinate selection system, and the word-organized system. The latter system is used in thin film stores (to be discussed in the next section), while the former system is commonly used in ferrite-core stores. In a coincident current core store, the ferrite cores are arranged in a matrix as shown for a 4×4 matrix in fig. 6.13. In this memory, the cores are held by separate column (X) and row (Y) drive wires and a single sense wire threading through all cores. After a read operation, a selected core, say, X_2–Y_2, is left in the "0" state. During the subsequent write operation, half

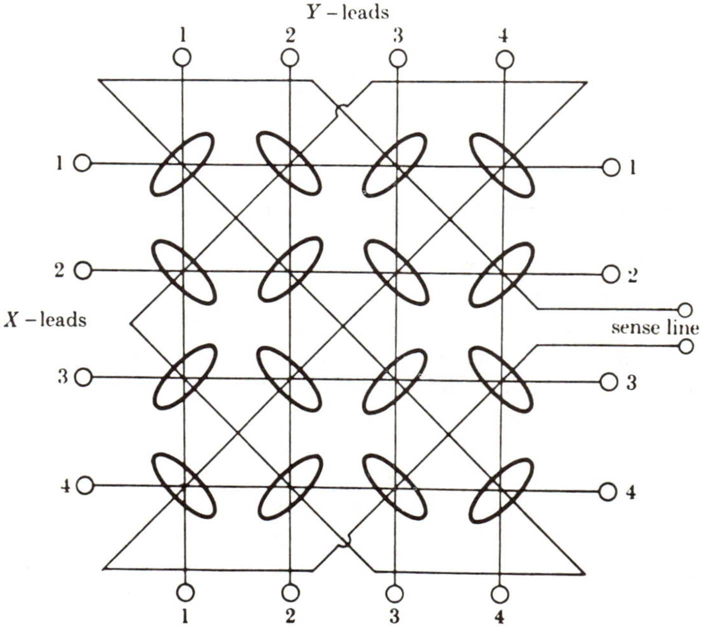

Fig. 6.13. A 4×4 ferrite core matrix operated on the coincident-current selection system by separate X- and Y-pulses and with a single sense winding (Standley [1972]).

currents are applied to both the X_2 and Y_3 wires in the opposite direction to those used during the read period so that the coincidence of these half currents at the X_2–Y_3 core will switch to the "1" state, from which a voltage pulse is induced in the sense wire. The other cores in the X_2 column and Y_3 row are subjected to only one half of the required current to switch, thus remaining in the original state. The coincident current current system has the advantage of very rapid readout of the stored information. On the other hand, it has the disadvantage of continually subjecting unselected cores to half currents that disturb their remanent state. As later discussion will show, this situation may lead to serious problems in the memory operation if the hysteresis loops of the cores deviate considerably from the square shape.

Although any ferromagnetic material with a hysteresis can be used to store a binary digit, "0" or "1", only when the hysteresis loop is sufficiently rectangular, will access to the stored information be economically gained in a large-capacity memory. Thus the most important quality of a ferrite in a memory core is the squareness of the hysteresis loop. The ratios B_r/B_m and $B_{-\frac{1}{2}H_m}/B_r$ are used to characterize the loop shape, where H_m is the switching field and B_m is the corresponding induction, as shown in fig. 6.12a. The first is called the remanence ratio and the second the squareness ratio. A good memory-core ferrite should have a value of 0.90 or greater for either ratio.

To see the adverse effect of a remanence or squareness ratio somewhat below 0.9 on the performance of a ferrite core, let us refer to fig. 6.12b. Again, suppose that the $+B_r$ state represents a "0" and the $-B_r$ state a "1". To sense the stored information, a field of $+H_m$ is applied. If the stored digit is "1", this will result in a large output voltage in the sense wire corresponding to flux change from $-B_r$ to $+B_m$, thus a total of B_m+B_r. If the stored digit is "0", the application of $+H_m$ will only induce a small voltage corresponding to a small flux change from $+B_r$ to B_m, thus a total of B_m-B_r. When the core is in the "0" or $+B_r$ state and subjected to an applied field $-\frac{1}{2}H_m$, the induction will fall to B_1. Upon removing the field, the induction will rise slightly, following the minor loop, to B_2, which is only slightly lower than $+B_r$. Subsequent applications and releases of the field $-\frac{1}{2}H_m$ will result in the excursion of B around the minor loop between B_1 and B_2. This situation corresponds to the cores in X_2 column and Y_3 row in the matrix shown in fig. 6.13, when a field of $+\frac{1}{2}H_m$ is applied simultaneously in the X_2 and Y_3 drive wires. Figure 6.14 shows the changes in induction and the corresponding induced voltages in the sense winding around the core.

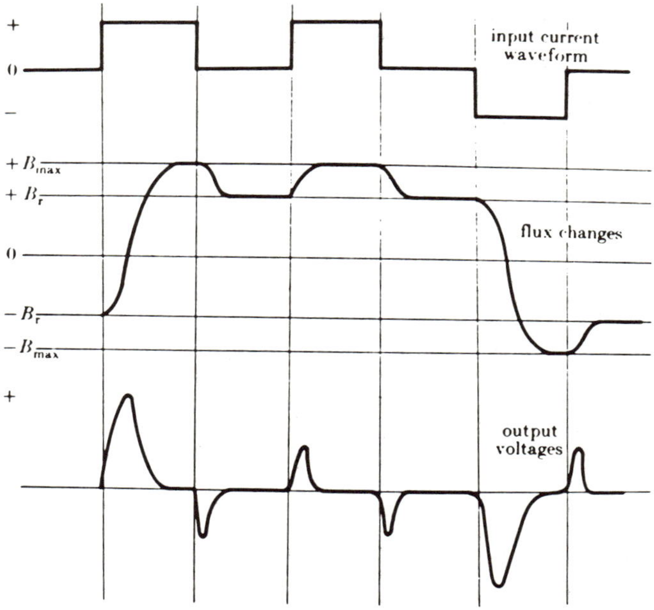

Fig. 6.14. Output voltages produced by a nonideal loop. For an ideal loop (fig. 6.12a), the small output pulses would be absent (Standley [1972]).

Note the appearance of small, unwanted pulses in several periods, which would be absent had the loop been closer to ideal squareness. The magnitude of these unwanted pulses depends on the differences between B_m and B_r. Thus if $B_r/B_m < 0.9$ appreciably, the unwanted pulses at least cause considerable noise in the sense wire and could be added together to give spurious signals.

Another important quality of SLF cores deals with the dynamics of domain wall movement. This quality is measured by the peaking time t_p and the total switching time t_s, which is determined from 10% of the current rise to the time when the voltage has dropped to 10% of its maximum value. Figure 6.15 shows the input pulse program in (a) and a typical pattern of the response voltage generated in the sense line in (b). Also defined in the figure are t_s and t_p. A useful empirical relationship is adopted between t_s and the applied field H:

$$t_s = \tau H_0/(H - H_0), \tag{6.8}$$

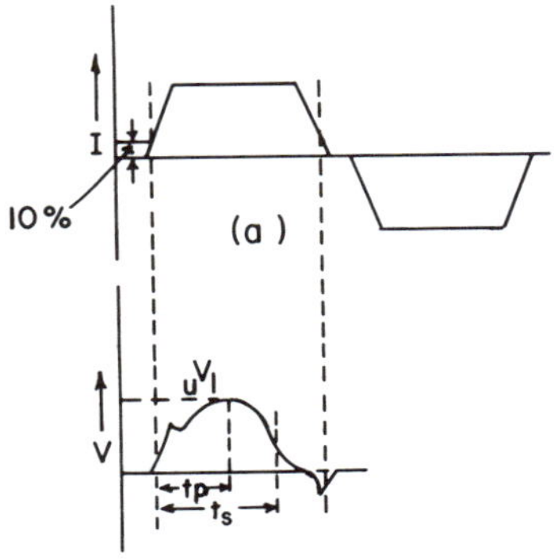

(b)

Fig. 6.15. (a) Details of the pulse program used to generate the response voltage shown in (b) (Weiss [1971]).

where τ is a constant characteristic of the material and H_0 is the threshold field for switching. The product τH_0 is called the switching coefficient S_W, and H_0 has values close to the coercive force of the material. Good memory-core ferrites should have values of t_s equal to or less than 0.5 μs, $t_p \sim \frac{1}{2} t_s$ and S_W values between 24 and 120 $A \cdot m^{-1} \cdot \mu s$ (0.3 to 1.5 Oe · μs). Because coincident-current operation limits the applied field to less than $2H_0$, the switching time cannot be decreased by increasing the applied field. However, core size is of great importance in determining the speed of a memory in terms of t_s and t_p, as shown in table 6.12. A switching time of ≈ 100 ns is attained in the smallest size core. Even at this speed, however, ferrite cores are still considered as inferior switching elements because faster switching of the order of 5–50 ns may be achieved in thin film and bipolar integrated circuit cores.

Two kinds of ferrites are available for selection as core materials: those that display square loops spontaneously and those whose loops become square after magnetic anneal. According to a phenomenological theory of Wijn et al. [1954], spinel ferrites are expected to display a spontaneous square loop with a M_r/M_s ratio of 0.87 if they satisfy three conditions: (1) pronounced magneto-crystalline anisotropy with a large negative value of K_1 and ⟨111⟩ easy directions of magnetization, (2) insignificant shape anisotropy and (3) zero or very small value of λ_{111}. The third condition is included so as to minimize the effect of stress anisotropy. Data on K_1 and λ_{111} for various single spinel ferrites are

TABLE 6.12
The variation of the peaking time (t_p), the total switching time (t_s) and other operational conditions with the outside diameter of ferrite cores

O.D. of core, (mm)	Nominal full drive current (mA)	Drive pulse rise time (ns)	t_s (ns)	t_p (ns)
2.0	385	500	3 000	1 300
1.3	450	200	900	470
0.75	475	100	600	250
0.55	750	50	200	110
0.35	1 000	25	110	60
0.30	650	25	105	55

listed in table 5.12. With respect to the three conditions, we can conclude that Co-ferrite is the least favorable ferrite because of its unusually large positive value of K_1 and the large value of λ_{111}. All other ferrites have negative K_1 values, and Mn-ferrite seems to be the most attractive because of its extremely small value of λ_{111}. Indeed, the present commercial products of SLFs are based on Mn-ferrite with various other single ferrites added to improve the properties of $MnFe_2O_4$. An effective combination is the Mg–Mn mixture in which the negative value of λ for $MgFe_2O_4$ tends to compensate for the positive value of λ_{111} for $MnFe_2O_4$, thereby leading to zero magnetostriction coefficient (presumably λ_{111}) in the proper mixture. Also, Wiechec and Kelley [1962] observed that the presence of Mn^{III} ions in the mixed ferrite enhances the squareness of the hysteresis loop. They attribute this effect to the tendency for the Mn^{III} ions to forming dsp^2 covalent bonds in a special configuration other than the octahedral (dsp^3) and tetrahedral (sp^3) bonds. These two factors have resulted in the series of mixed Mg–Mn ferrites, which are most widely used in this field. Typical compositions are $0.13Mn_3O_4{\cdot}0.87MgFe_2O_4$, which has nearly zero λ_{111} and excellent squareness of the hysteresis loop, and $x MnFe_2O_4{\cdot}(0.85 - x)MgFe_2O_4{\cdot}0.15Mn_3O_4$ with $x = 0$–10 mol %. Other commercial ferrites suitable for memory cores include (1) $xCuFe_2O_4{\cdot}yMnFe_2O_4{\cdot}zMn_3O_4$ with $x < 0.1$; (2) $xNiFe_2O_4{\cdot}yMnFeO_4{\cdot}zFe_3O_4$ with $x = 0.5$–0.70, $y = 0.05$–0.30 and $z = 0.1$–0.3 and (3) Li-ferrite mixed with a small percentage of Ni-ferrite.

Several spinel ferrites respond to magnetic anneal yielding square hysteresis loops. Commercial products of this second kind of ferrites used for memory cores include (1) $0.008CoO\cdot xZnO\cdot(0.45-x)NiO_2\cdot 0.542Fe_2O_3$ with $x = 0.13$–0.19 and (2) Ni–Fe^{II} ferrites from $0.8NiFe_2O_4\cdot 0.2Fe_3O_4$ to $0.4NiFe_2O_4\cdot 0.6Fe_3O_4$.

§3. THIN FILMS

The term "thin film" has been broadly defined as a layer of gaseous, liquid or solid matter. In the present discussion, unless stated otherwise, we use the term to denote a layer of solid, magnetic material that has a thickness in the range 10–10^4 Å and is prepared mainly by the method of vapor deposition or other related techniques such as sputtering. Although thin films had been studied long before 1955, real interest in magnetic thin films did not arise until Blois showed that films of Permalloy deposited in a magnetic field exhibited a rectangular hysteresis loop. Blois' discovery was quickly appreciated by computer-oriented researchers because such thin films could be used as bistable memory elements. Another important condition for thin films used as memory materials is a fast switching time. Dietrich and Proebster [1960] found that thin films of Permalloys were capable of reversing the magnetization from $+B_r$ to $-B_r$ in $\sim 10^{-9}$ sec. This switching time is about 100 times faster than that of a ferrite. Moreover, thin film storage elements could be prepared in batches, thus reducing handling and assembly costs. Optimism thus ran high in the early 1960's on the prospect that thin film elements might eventually replace ferrite cores in the main stores of large, high-speed computers. However, persistent technical problems and the improvement of ferrite cores have hitherto prevented thin films from achieving these objectives.

To appreciate the capability of thin film memories, it is necessary to examine the salient features of the magnetization behavior of thin films, notably the induced uniaxial anisotropy, magnetization reversal, domain walls and domain structure.

Uniaxial anisotropy. The main reason that Blois was able to obtain square loops from his thin films was that he applied a magnetic field parallel to the substrate plane during the film preparation, which induced a uniaxial anisotropy with an easy axis lying in the film plane. The origin of the induced anisotropy was initially related to the induction of a directional order of iron atom pairs in the Permalloy film by the applied field. This interpretation was substantiated by a reasonable agreement between the measured composition dependence of the anisotropy constant K_u and that predicted by the Néel theory discussed in ch. 5 §2.6.1. Later experiments, however, showed that uniaxial anisotropy could also be induced in pure metals (iron and nickel) and, in some cases, in the absence of a magnetic field if vapor deposition was carried out by an oblique incidence technique. It is now believed that, besides iron atom pairs, any deviation from the cubic symmetry of the lattice caused by lattice imperfections (vacancies, dislocations, impurity and trapped gas atoms) may also produce localized magnetic anisotropies, and an alignment of these localized anisotropies over a macroscopic region results in an average anisotropy measured in the film. The latter theory also accounts for the observation of a dispersion of the induced anisotropy described later.

In accordance with (3.9), the energy associated with the induced uniaxial anisotropy is given, to a first-order approximation for a single-domain film, by

$$E_a = K_u \sin^2 \theta, \tag{6.9}$$

where θ is the angle between the easy axis and the magnetization. As was shown in (3.13) the anisotropy field is given by

$$H_a = 2K_u/M_s\mu_0. \tag{6.10}$$

Substituting $K_u \sim 200\ \mathrm{J \cdot m^{-3}}$ $(2 \times 10^3\ \mathrm{ergs \cdot cm^{-3}})$ and $M_s = 8.36 \times 10^5\ \mathrm{A \cdot m^{-1}}$ (≈ 836 G) in (6.10) for 80-20 Permalloy yields $H_a \sim 380\ \mathrm{A \cdot m^{-1}}$ (≈ 5 Oe). Later discussion will show that this small anisotropy field plays a prominent role in the operation of a thin film storage element.

Magnetization reversal. Thin films switch faster than ferrite cores because the reversal of magnetization is effected mainly by coherent rotation instead of domain-wall motion. Thus it is important to know the switching behavior in a simple coherent rotation model. When a magnetic field is applied with components H_x and H_y in the easy and hard directions, respectively, the total magnetic energy per unit volume is

$$E = -\boldsymbol{H} \cdot \boldsymbol{M}_s\mu_0 + E_a = -M_sH_x\mu_0 \cos\theta - M_sH_y\mu_0 \sin\theta + K_u \sin^2\theta. \tag{6.11}$$

For a given $\boldsymbol{H}$, the angle θ at which E is a minimum and the magnetization is at a stable position is determined from two conditions: (1) $\partial E/\partial\theta = 0$ which gives

$$M_sH_x\mu_0 \sin\theta - M_sH_y\mu_0 \cos\theta + 2K_u \sin\theta \cos\theta = 0, \tag{6.12}$$

and (2) $\partial^2 E/\partial\theta^2 > 0$. To find the critical fields at which transitions from unstable to stable states take place, we set $\partial^2 E/\partial\theta^2 = 0$, which leads to

$$M_s H_x \mu_0 \cos\theta + M_s H_y \mu_0 \sin\theta + 2K_u(\cos^2\theta - \sin^2\theta) = 0. \quad (6.13)$$

By combining (6.12) and (6.13), we eliminate θ and obtain

$$H_x^{2/3} + H_y^{2/3} = (2K_u/M_s\mu_0)^{2/3} = H_a^{2/3}. \quad (6.14)$$

This is the equation of a hypocycloid or an astroid shown by the curved lines in fig. 6.16. The astroid delineates the critical values of field for switching; that is, there will be switching outside and no switching inside the astroid. Meanwhile, for a given θ, (6.12) represents a linear relationship between H_x and H_y, which passes through the points $(0, H_a \sin\theta)$ and $(-H_a \cos\theta, 0)$ in the figure. It can be shown that the straight line representing $\partial E/\partial\theta = 0$ is always tangent to the astroid. Thus the astroid, together with the $\partial E/\partial\theta = 0$ straight line(s), determines the magnetization direction graphically for any given field (H_x, H_y) in the following manner.

Suppose that a field is applied at an angle α with the easy direction $+H_x$, as shown in fig. 6.16. At the field strength represented by point 1, two tangents (lines A and B) can be drawn in the upper half of the astroid, thus giving two possible stable positions for the magnetization. As the field is increased from 1 through 2 and 3 (right on the astroid) to 4, which lies outside the astroid, only one tangent (line c) is possible, thus giving the only stable direction of magnetization. Various positions of such stable directions of magnetization are shown at a 15° interval of the critical angle θ_c in fig. 6.16. When the field strength is reduced, say, from point 4 to point 2 across the astroid, the change in θ_c is always continuous, whereas when H is gradually increased to pass the astroid from inside to outside, the change in θ_c may be discontinuous. This can be illustrated by moving along the dashed line from right to left, which is equivalent to having a fixed bias field of $H_y = 0.45\,H_a$ and letting H_x change from positive to negative values. As the dashed line passes from outside to inside the astroid at the crossing point a, the magnetization is directed at $\theta_c \sim 20°$. The angle increases continuously until $\theta_c \sim 45°$ at which point, b, the dashed line goes to the outside of the astroid and the magnetization rotates abruptly to $\theta_c \sim 160°$ at point c, causing a switching in the magnetization of the film. The field that causes the switching is approximately $H_x = -0.3\,H_a$ as compared with the switching field $H_x = -H_a$ if $H_y = 0$.

Another useful result of the analysis is the theoretical prediction of the hysteresis loop. Two simplest cases are for the field applied in the

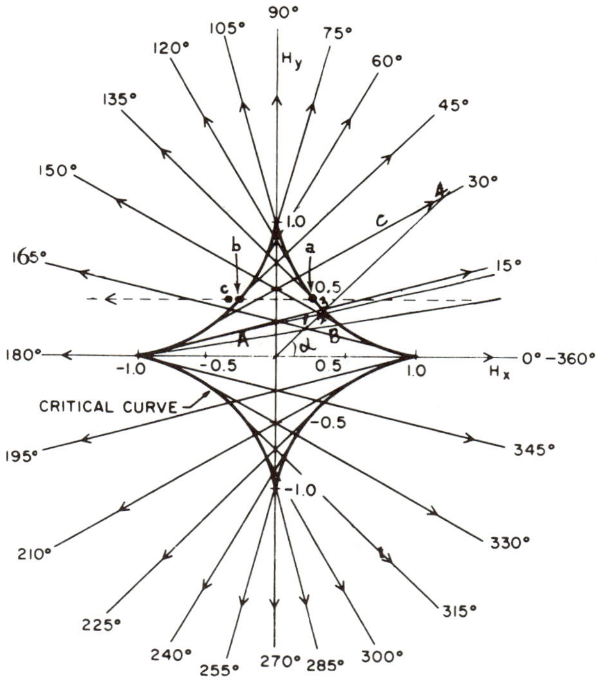

Fig. 6.16. Critical curves for single domain switching by irreversible rotation of magnetization in a thin magnetic film (Pugh and Mohr [1964]).

easy and hard directions. The analysis and results for the two cases are compared in table 6.13. An ideally square loop is predicted for the field applied in the easy direction and an unresolvable loop is predicted for the field applied in the hard direction, as shown by dotted lines in fig. 6.17a and b, respectively. Loops actually obtained in the two directions are shown by solid lines in the figures. Hysteresis loops observed in the easy direction are not ideally rectangular. They have rounded, instead of sharp, knees and the coercive forces are smaller than H_a. These differences are attributed to the involvement of domain wall movement in the actual magnetization reversal and the presence of incoherent mode of switching, thus reducing the switching threshold field.

Two other features of the magnetization behavior of thin films commonly observed should be noted. One is known as the dispersion of the induced anisotropy previously mentioned. In general, the mag-

TABLE 6.13

Comparison between processes of magnetization in the easy and hard directions for thin films

	Field applied in the easy direction along $+H_x$		Field applied in the hard direction along H_y	
1. Value of (H_x, H_y)	$(H_x, 0)$		$(0, H_y)$	
2. (6.11) becomes	$E = -M_s H_x \mu_0 \cos\theta + K_u \sin^2\theta$	(6.11.1)	$E = -M_s H_y \mu_0 \sin\theta + K_u \sin^2\theta$	(6.11.2)
3. $\partial E/\partial\theta = 0$ gives	$M_s H_x \mu_0 \sin\theta + 2K_u \sin\theta\cos\theta = 0$	(6.12.1)	$-M_s H_y \mu_0 \cos\theta + 2K_u \sin\theta\cos\theta = 0$	(6.12.2)
4. $\partial^2 E/\partial\theta^2$	$\partial^2 E/\partial\theta^2 = M_s H_x \mu_0 \cos\theta + 2K_u(\cos^2\theta - \sin^2\theta)$	(6.13.1)	$\partial^2 E/\partial\theta^2 = M_s H_y \mu_0 \sin\theta + 2K_u(\cos^2\theta - \sin^2\theta)$	(6.13.2)
5. Solutions of (6.12.1) or (6.12.2) and conditions for stable state	(I) $M_s H_x \mu_0 = -2K_u \cos\theta$; and (6.13.1) becomes $\partial^2 E/\partial\theta^2 = -2K_u \sin^2\theta$, which is always negative except $\theta = 0$ or π. This solution corresponds to an unstable state. (II) $\sin\theta = 0$ or $\theta = 0$ or π; (6.13.1) becomes $\partial^2 E/\partial\theta^2 = (M_s H_x \mu_0 + 2K_u \cos\theta)\cos\theta$ (A) For $\theta = 0$, a stable state if $H_x > -H_a$; (B) For $\theta = \pi$, a stable state if $H_x < H_a$.		(I) $M_s H_y \mu_0 = 2K_u \sin\theta$; and (6.13.2) becomes $\partial^2 E/\partial\theta^2 = 2K_u \cos^2\theta$, which is always >0. Hence this solution represents a stable state. (II) $\cos\theta = 0$ or $\theta = \pm\frac{1}{2}\pi$; (6.13.2) becomes $\partial^2 E/\partial\theta^2 = (M_s H_y \mu_0 + 2K_u \sin\theta)\sin\theta$ (A) For $\theta = +\frac{1}{2}\pi$, a stable state if $H_y > H_a$; (B) For $\theta = -\frac{1}{2}\pi$, a stable state if $H_y > -H_a$.	
6. Magnetization	$M = M_x = M_s \cos\theta = \pm M_s$		$M = M_y = M_s \sin\theta$ which has the following values: From (5.I) $M_y = (M^2\mu_0/2K_u)H_y$; and From (5.II) $M_y = \pm M_s$	
7. Hysteresis loop	Fig. 6.17a		Fig. 6.17b	

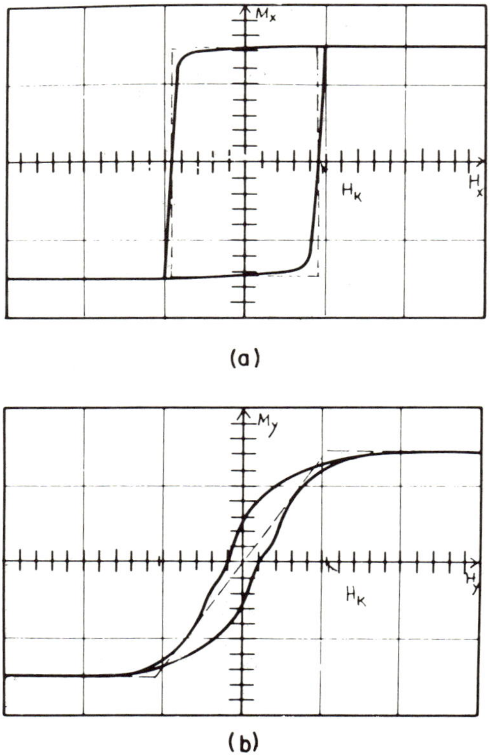

Fig. 6.17. Theoretical (dashed line) and experimental (solid line) hysteresis loops. (a) In the easy direction and (b) in the hard direction. 1 unit = 2.5 Oe, $H_c = 2.25$ Oe, $H_k = 3.0$ Oe (Middlehoek [1971]).

netization of a film does not point to one single direction but has a small spread on a microscopic scale about a mean direction. The dispersion of the film is defined as the angle that contains a certain percentage of the magnetization and is usually quoted at half the angle that contains 90% of the film magnetization. It is designated as α_{90} and has values between 0.5 and 2.5°. The second feature is called the skewness, which is the angle between the observed mean and intended directions of magnetization. (For the thin film memory element shown in fig. 6.28, skewness is defined as the angle between the word field and the hard axis.) The angle of skew also has values of a few degrees.

Domain walls. The two-dimensional geometry of thin films permits the formation of other types of domain walls than the Bloch type. When the film becomes sufficiently thin (~200 Å), the so-called Néel walls may appear. A Néel wall is defined as one in which the spins rotate about an axis perpendicular to the plane of the film. The distinction between a Néel wall and a Bloch wall is shown in fig. 6.18. Note that, for the Bloch walls, free poles are produced on the top and bottom surfaces of the sample over the width of the boundary, whereas for the Néel walls, free poles are produced over the side surfaces of the sample over the width of the boundary. By adopting an infinitely long cylinder with elliptic cross section to represent the wall, Néel [1955] arrived at the following expressions for the magnetostatic energy per unit area of domain wall:

(A) the Bloch wall $E_s = \frac{1}{2} N_{d,B} M_s^2 d_w \mu_0 = 2\pi d_w^2 M_s^2 \mu_0 / (t_f + d_w)$ (6.15)

(B) the Néel wall $E_s = \frac{1}{2} N_{d,N} M_s^2 d_w \mu_0 = 2\pi t_f d_w M_s^2 \mu_0 / (t_f + d_w)$ (6.16)

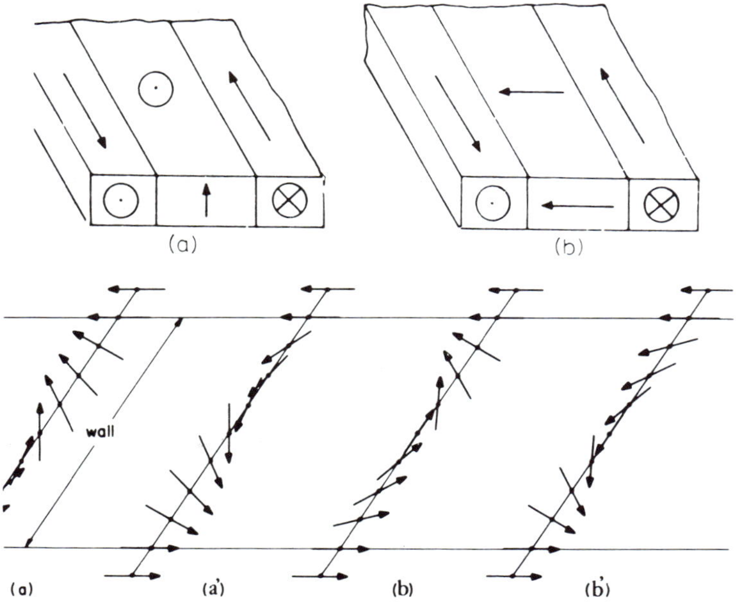

Fig. 6.18. Magnetization distribution in Bloch walls [(a) and (a')] and Néel walls [(b) and (b')] (Craik and Tebble [1965] and Middlehoek [1971]).

where $N_{d,B}$ and $N_{d,N}$ are the demagnetizing factors for the Bloch and Néel walls, respectively, d_w is the wall thickness and t_f is the film thickness. The Bloch wall has low magnetostatic energy for thick specimens ($t_f > d_w$) and the Néel wall for thin films ($t_f < d_w$). Hence the Néel walls are energetically favored when the film thickness is of the order of a few hundred angstroms.

Another special type of domain walls in thin films is the cross-tie walls. These walls are so named (Huber et al. [1958]) because they appear as short lines with tapered ends running across the main domain walls of the Néel type, as shown in fig. 6.19. The mechanism of the formation of cross-tie walls is explained in fig. 6.20. In a Néel wall, the spins at the center of the wall can be up or down, as shown in fig. 6.20a.

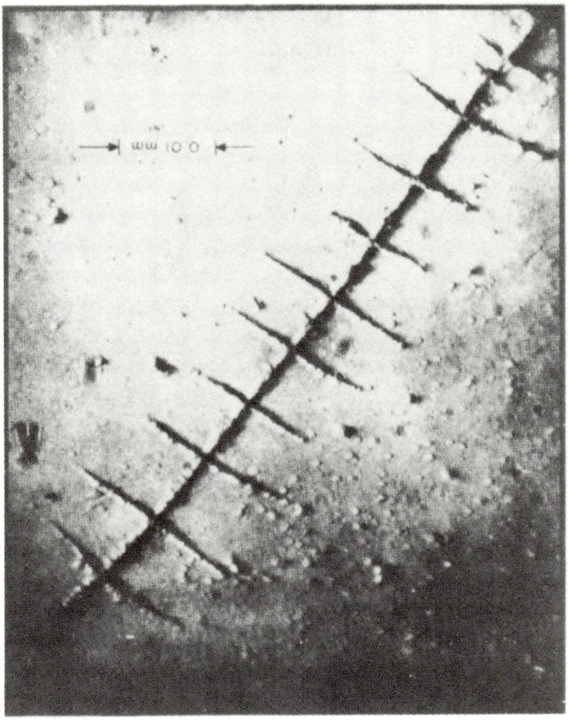

Fig. 6.19. Cross-tie domain walls observed in an 80–20 Permalloy film of 300 Å thickness. 950× (Moon [1959]).

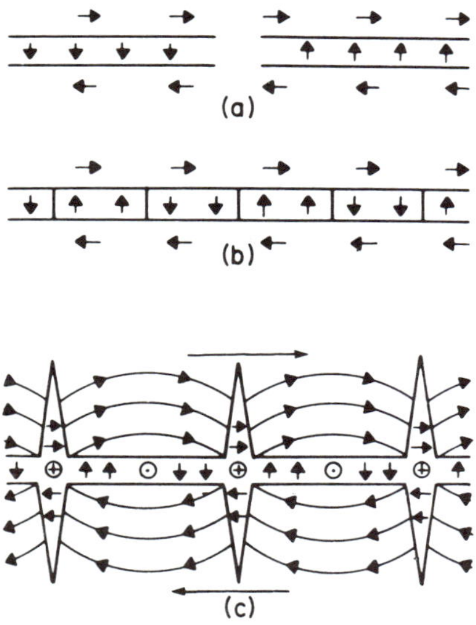

Fig. 6.20. Diagrams of possible spin arrangements explaining why cross-tie walls are formed in association with the Néel walls. (a) Unipolar Néel wall; (b) Néel wall consisting of segments of opposite polarity and (c) complete magnetization distribution of a cross-tie wall (Middlehoek [1971]).

A decrease in magnetostatic energy of the wall can be effected when segments with opposite spins alternate along the Néel wall (fig. 6.20b). In the latter arrangement, flux closure occurs between adjacent segments. This flux closure must follow the direction of magnetization in the neighboring domain; therefore, short walls perpendicular to the main wall are necessary so as to accommodate to the flux closure (fig. 6.20c). Cross-tie walls are also of the Néel type, but their complex structure makes it difficult to calculate precisely the wall energy. A rough estimate gives an energy drop of 40% in the segmented Néel wall with cross-ties with respect to an equivalent unsegmented Néel wall.

Other complex domain walls are observed on thin films. They include displaced cross-tie walls and chain walls described by Gomi and Odani [1960] and double walls described by Williams and Sherwood [1957].

Domain structure. Despite the complex domain walls just described, relatively simple domain patterns are commonly observed in thin films in the demagnetized state and under a small magnetic field applied parallel to the film plane. To appreciate this situation, it is important to realize that, because of the two-dimensional geometry, the demagnetizing factor N_d for a thin film is very small in directions parallel to the film plane. Thus, if the film is sufficiently thin (<300 Å), it can readily attain a single domain structure with the magnetization vector lying in the film plane even though the film is polycrystalline.

Observations by Bitter or optical techniques reveal that, if the domain structure of a uniaxial film contains multiple domains, the energetically more favored pattern consists of 180° domains with walls running normal to the film surface, rather than parallel to the surface. A film is termed uniaxial if it displays uniaxial anisotropy, which can be either inherent or, most likely, induced. At the edges of the film, closure domains similar to those depicted in fig. 3.17b may appear because of the finite demagnetizating field. The emergence of such a domain structure, however, may require special treatment, such as the application of an alternating field to achieve demagnetization.

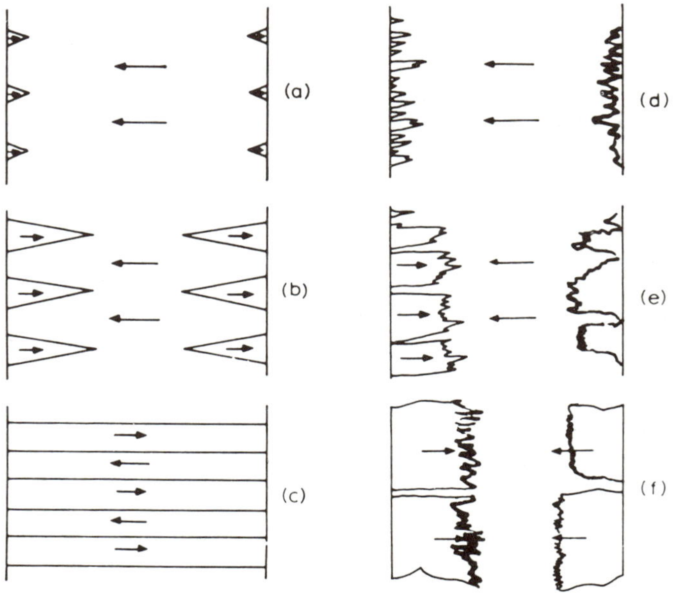

Fig. 6.21. Nucleation and growth of domains during the reversal of magnetization of ((a), (b) and (c)) a crystal and ((d), (e) and (f)) a thin film, both characterized by a uniaxial anisotropy (Craik and Tebble [1965a]).

The changes in domain structure of a uniaxial film when a field is applied in the easy direction to cause magnetization reversal are shown in fig. 6.21d–f. The reversal is accomplished by nucleation of new domains at both edges, followed by growth of the new domains. Figure 6.21f corresponds to a demagnetized state when the volume of new domains amounts to one half of the total volume, and further growth of the new domains will eventually produce a single domain structure with the magnetization reversed from the initial state. Figure 6.21a–c shows a similar reversal of magnetization observed in a bulk crystal. Two differences are noted between the two cases: (1) numerous domains are nucleated concurrently at the edges of the film and tend to fuse together as compared with few solitary domains nucleated in the crystal and (2) zig-zag domain boundaries are often produced in thin films in contrast to the spike-like appearance of partially grown domains in the crystal. Besides the formation and fusion of many domains in (1), two other effects also promote the production of zig-zag boundaries; namely, the dispersion of uniaxial anisotropy and the relatively low magnetostatic energy associated with the new domains. Experimental evidence of zig-zag boundaries is shown in fig. 6.22.

The domain structure of a thin film after a field is applied in the hard direction to attain saturation and then removed is shown in fig. 6.23. It consists of a number of 180° domains with nonuniformly spaced walls running in the easy direction. The irregular spacing of the lamellar domains is attributed to periodic spatial variation of the easy direction of the induced uniaxial anisotropy. Similar structures featuring lamellar

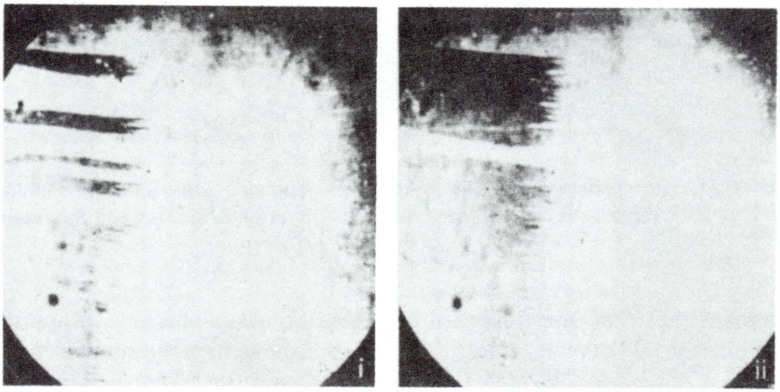

Fig. 6.22. Zig-zag domain walls in cobalt film, field increasing from (i) to (ii) (Rushton [1963]).

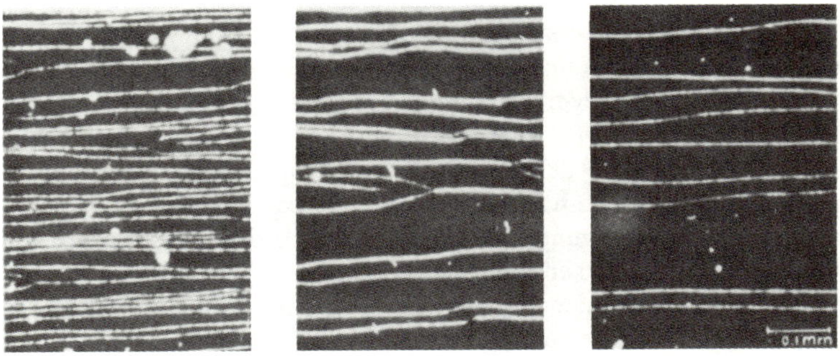

Fig. 6.23. Lamellar domains formed with $\boldsymbol{M}_s$ in the easy direction in a thin film of 80–20 NiFe after a field is applied in the hard direction (Middlehoek [1971]).

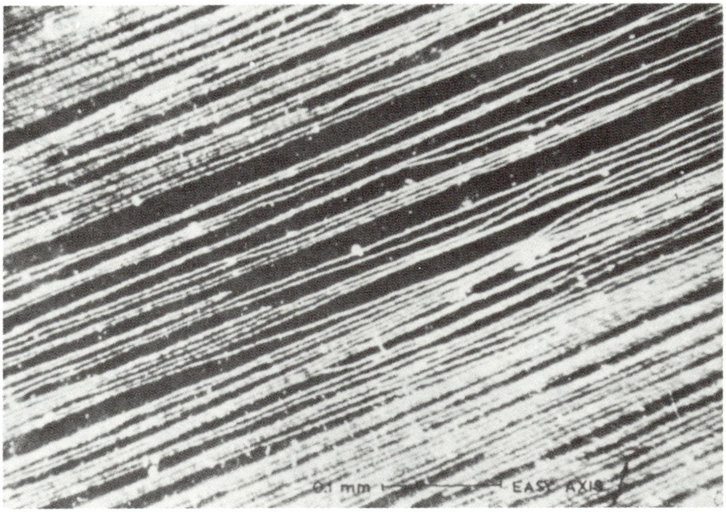

Fig. 6.24. Bitter pattern on 200 Å iron film showing long slender domains making 30° with the easy axis in the horizontal direction (Williams and Sherwood [1957]).

domains (fig. 6.24) are observed on thin films when the field is applied inclined to the easy axis. In the latter case, however, the domain walls are in a direction close to that of the applied field, and because the direction of magnetization is a compromise between the easy axis and the field direction, the domains are not 180° apart even though they bear a superficial resemblance to the 180° domain structure shown in fig. 6.23.

The foregoing discussion on domains walls and domain structure characteristic of thin films seems to imply that the switching mode might be more complex than just the coherent rotation of spins. Other processes such as incoherent rotation and domain wall motion could occur. Indeed, the actual switching behavior is quite complicated, as is shown in fig. 6.25. The astroid segment $H_a b H_a$ no longer serves as the critical curve to indicate switching for operating points (H_x, H_y) lying above the curve and no switching inside the curve. To obtain the very fast, coherent rotation mode of switching, fields much larger than H_r must be applied in region A. For fields lying outside, but close to the astroid, the magnetization is reversed by incoherent rotation in some areas (doubly shaded) and by domain wall motion in others (singly shaded). Even inside the astroid, flux reversal can still occur in region B by wall motion. Curve abc represents the threshold for wall motion, which, unfortunately, is a slow process for switching and should be avoided. Another interesting phenomenon inside the astroid is known as wall creep, for which curve ae represents the threshold. Domain wall creep comes about from small, irreversible movements of wall segments when a dc field is applied in the easy direction and a pulse field applied in the hard direction. Wall creep can exert an adverse effect on the memory operation when the ac field applied in the hard direction is continually turned on and off. After a thousand or more applications of

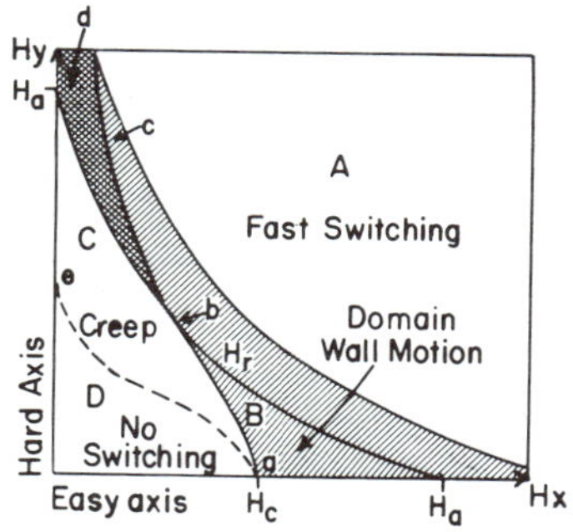

Fig. 6.25. The critical curve, $H_a b H_a$, for rotational single domain switching in comparison with curve abc representing the wall motion threshold and dashed curve ae representing the threshold for creeping. Singly shaded area-switching by wall motion and doubly shaded area-switching by incoherent spin rotation. (After Pugh and Mohr [1964].)

such an ac field, the stored information could be completely destroyed. Several methods have been suggested by Middelhoek [1971] to avoid wall creep.

Thin film memory elements and stores. Two basic forms of thin film memory elements have evolved, the flat film memory and the cylindrical plated wire memory. Because of an early conception and great market potential, the former received much greater attention and has been studied most extensively. We shall discuss this form first. Flat memories, also known as magnetic memory planes, are deposited upon a flat substrate, normally an aluminum or copper sheet covered with an insulating silica layer. The magnetic material is placed in an alumina crucible contained in a molybdenum enclosure and heated to evaporation by electron bombardment under vacuum with residual pressure at 10^{-5}–10^{-6} mm Hg. The substrate is located about 30 cm above the crucible and heated to ~400°C before deposition begins. A magnetic field of ~8000 A · m^{-1} (~100 Oe) is applied in the plane of the film for the induction of uniaxial anisotropy during evaporation. The shape of the film, either circular about 1 mm across or rectangular, is controlled by using a mask to guide the passage of vapor or by a photochemical etching technique. To complete the construction of a memory element, two or three sheets of Mylar, with conductors, are mounted on top of the film. Figure 6.26 illustrates a typical assembly of a film element.

The most widely used material for memory elements is the 80–20 Permalloy. More specifically, the composition is favored at 81.5% Ni and 18.5% Fe, for which magnetostriction is zero. Also, according to Olmen [1963], both the anisotropy dispersion in terms of α_{90} (fig. 5.27a) and coercive force (fig. 5.27b) are low at this composition. Many manufacturers add a third element to further reduce the coercive force

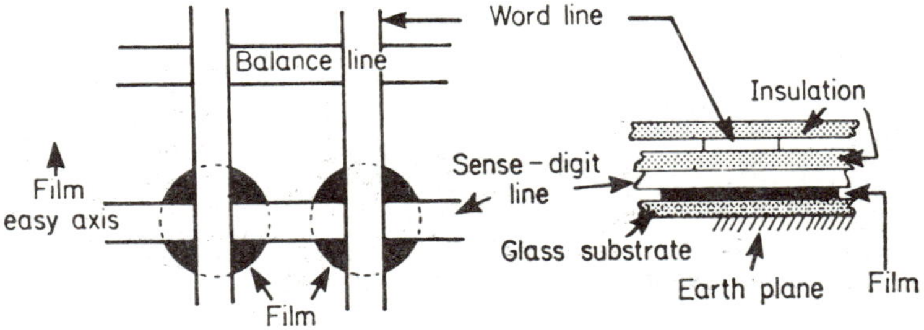

Fig. 6.26. (a) Typical planar film crosspoint and (b) the construction of the simplest type of thin film storage element (Renwick and Cole [1971]).

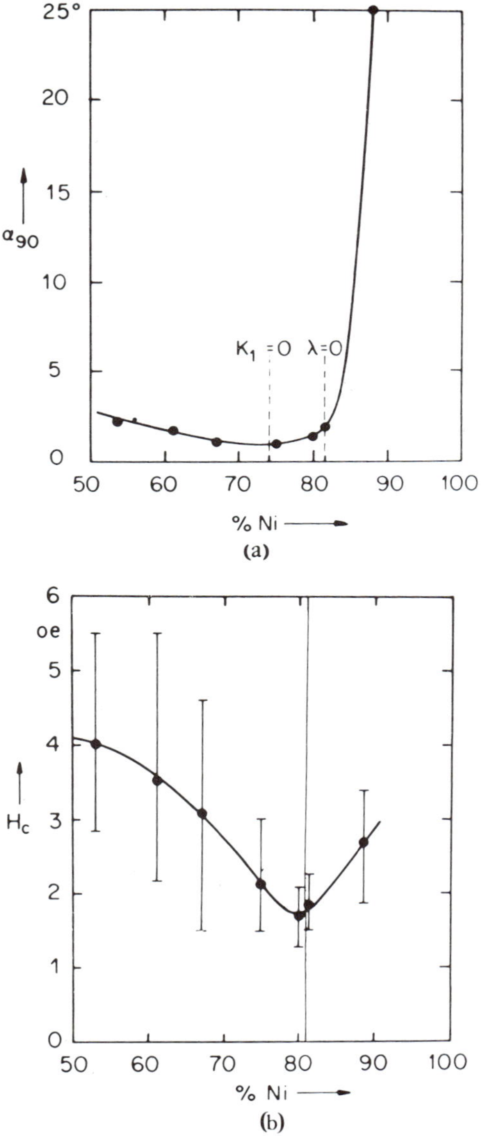

Fig. 6.27. (a) Composition dependence of angular dispersion of 1100 Å Ni–Fe films. (b) Wall-motion coercive force H_c of Ni–Fe films as a function of composition. Note both curves show a minimum at 80% Ni (Olmen [1963]).

and to improve other magnetization characteristics. Chromium has been suggested for reducing the anisotropy field H_a, for instance. Other elements mentioned in the literature include silicon and molybdenum. The composition of the film must be carefully analyzed in comparison with the nominal value. For the 81.5–18.5 composition, the original alloy before evaporation should contain 1.5% less iron, according to one study. The film thickness t_f, normally ranging between 1000 and 2000 Å, has a strong influence on the demagnetized field ($H_d \sim M_s t_f / l$, where l is the film length) and on domain wall structure. It is, therefore, measured and controlled in situ during the deposition by a sensitive device such as an ion gage monitor or a crystal oscillator.

Many proposals were made on the arrangement of film elements to build a memory store. Of these, the one by Raffel [1959] was considered as the first to satisfactorily meet the requirements of speed, drive currents and sense signals. For film stores, the coincident-current model so widely used in ferrite-core stores is considered unsuitable because the close tolerances on the coercive force are difficult to meet in thin-film elements. Thus the word-organized selection system is used for film memories. In the latter system, the film elements are arranged in a two-dimensional $n \times N$ array (hence the name 2D), where N is the number of rows, known as "words", and n is the number of film elements or bits in each word. A $16 \times N$ 2D store is shown in fig. 6.28, in which the dark rectangular patches represent the film elements with their easy axes aligned along the long side. The conductors parallel to the easy axis are called word drive, or simply word lines. Two sets of conductors run normal to the easy axis – the digit or bit (drive) lines and the sense (signal) lines. In each film core, as in the ferrite core, a "0" in the binary number system is stored by magnetizing the film along the easy direction pointing to, say, the left, and a "1" is stored by magnetization along the easy direction pointing to the right. To read information of any word, say W-2, a full current pulse I_W is sent down the W-2 line to drive all 16 bits on that line toward the hard direction. The original states of magnetization of the bits, and hence their stored information, are detected by the sense amplifiers through the 16 bit lines connected in parallel. If the pulse field associated with I_W exceeds the anisotropy field H_a, the magnetization in the cores on the W-2 line will rotate fully to the hard direction. New information is written into the memory by a combination of word pulse in W-2 line and bit pulses of selected polarity along all bit lines. The latter pulses are applied when the word pulse is dropping to zero. The polarity of the induced fields by the bit pulses determines whether the bits rotate clockwise or

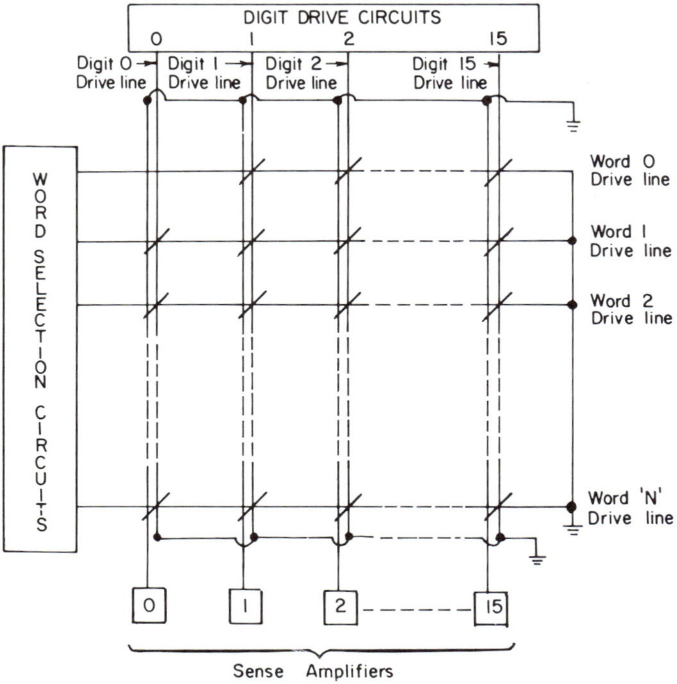

Fig. 6.28. The word-organized selection system of a $16 \times N$ thin film memory (Renwick and Cole [1971]).

counterclockwise from the hard to the easy axis, thereby storing a "0" or a "1". In this operation scheme, it is necessary to read and write all bits of a word simultaneously, thus the term word-organized system in contrast to the coincident-current, or bit-organized, system which deals with information stored in only one core. On the other hand, the reading operation in the present word-organized system does not require current-coincidence, thus causing no continual disturbance of all cores by half drive current as in the ferrite-core memory.

The early enthusiasm for magnetic film memories led to the construction of a 32-word, 10 bits per word memory installed in the TX-2 computer at MIT Lincoln Laboratories. However, despite the enormous efforts of various industrial R & D centers, thin film memories have hardly found large-scale applications in commercial computers largely because of two drawbacks. First, the manufacture of film memories turns out to be quite troublesome and expensive because one bad element renders the whole substrate useless. Secondly, the competition from ferrite core memories and semiconductor integrated circuit stores has been getting more fierce as a result of improvement of the technology for the latter kinds of memories. The continued reduction in access time and cost of the latter core systems made thin film stores uneconomic for all but very fast (1–10 ns) applications.

Plated wires are prepared by electrodepositing a thin film of the chosen magnetic material, such as the 80–20 Permalloy, to wires of a Cu alloy (Cu–Be or phosphor bronze). Wire diameter ranges from 75 to

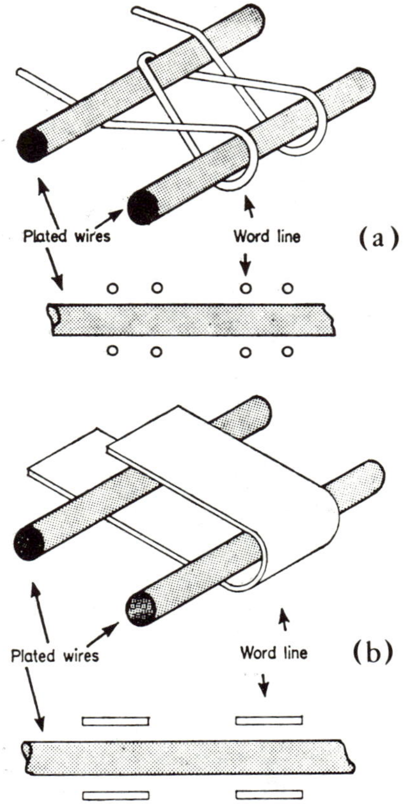

Fig. 6.29. Two methods of assembling plated wires in a memory. (a) Woven crosspoint and (b) stripline crosspoint (Renwick and Cole [1971]).

250 μm. The electroplating process is applied continuously on the wire pulled from a reel through a series of baths for cleaning, rinsing and preliminary Cu-plating before the final plating. The plated wires offer two advantages over the flat films. One is that the plated wires are less expensive and more dependable as a result of vigorous tests. The other advantage stems from the fact that the substrate wire itself serves as a digit line. The latter advantage is also utilized during electroplating in that a current passing through the wire produces a field, which then induces a uniaxial anisotropy with the easy axis around the circumference.

Plated-wire memories are constructed by arranging a number of plated wires in a matrix with orthogonal word drive lines wound around the wires. The word lines can be either fine wires used in the woven crosspoint system or strips used in the strip-line crosspoint system as shown in fig. 6.29. Memory cores are obtained at the points of intersection between the plated wire and the word line. Plated-wire memories, with capacity of 10^5–10^6 bits and cycle time between 150 and 500 ns, have been used as the main stores in some commercial computer systems (Woollons [1972]).

Suggestions for further reading

Heck, C., 1974, *Magnetic Materials and Their Applications* (Crane, Russak & Co., New York).

Snelling, E. C., 1969, *Soft Ferrites, Properties and Applications* (Iliffe Book Ltd, London).

Soohoo, R. F., 1965, *Magnetic Thin Films* (Harper & Row, New York).

Stanley, J. K., 1963, *Electrical and Magnetic Properties of Metals* (Am. Soc. for Metals, Metals Park, Ohio, USA).

CHAPTER 7

SPECIAL TOPICS

§1. RADIATION EFFECTS

The possibility that radiation may cause serious damage to solids has been recognized long before the development of nuclear reactor technology and the advent of space exploration. Events in these two fields during the past two decades, however, have dramatically demonstrated the importance of understanding the effect of radiation from both practical and scientific standpoints. Good progress has indeed been made in the formation of basic knowledge about the nature and extent of radiation damage during this period.

Although magnetic materials play only minor roles in the current thermal reactors and space vehicles, some future projects have been suggested that require a close look at the possible radiation damage to the involved magnetic materials. One project is the nuclear-electric space power system, which contains a nuclear reactor and a power generating station with necessary liquid-metal heat exchanger. Another project is the long manned space program. In the case of nuclear-space power generation, the entire electrical generating system "stands a good chance of being exposed to a high flux of nuclear radiation from the reaction," according to Beiler [1967]. The same author also speculated that, for space vehicles designed to make round trips to Mars, soft magnetic materials will be required to generate and use "megawatts of electric power not only for communication, scientific work and life support, but for the principal consumer of electric or ionic propulsion." Under these circumstances, the magnetic materials must meet the most stringent demands for efficient and economic production and utilization of electric power as well as withstand environmental conditions involving cosmic radiation.

Irradiation of a magnetic solid with energetic particles may affect magnetic and other (electrical, mechanical, thermal and radiochemical) properties. In the present discussion, however, we shall focus our attention on radiation damage to magnetic properties. At the outset of the discussion, it is worth noting that, excluding induced radioactivity, the adverse effect of irradiation has been detected only in soft materials, not on hard magnets. Radioactivity, on the other hand, is expected to be induced by neutrons, especially the thermal neutrons, in both soft and hard magnets. The induced radioactivity may present some problems in handling and storage of magnets, especially those containing appreciable amounts of cobalt because of the large cross section (37 barns) for the reaction

$$ {}^{59}_{27}\mathrm{Co} + {}^{1}_{0}\mathrm{n} \rightarrow [{}^{60}_{27}\mathrm{Co}^{*}], \qquad (7.1) $$

from which the compound nucleus undergoes the γ-decay in two isomeric forms.

For readers interested in soft magnetic materials, radiation damage deserves a consideration. This special topic is treated in two parts: the first part outlines the basic processes of radiation damage to solid lattices, and the second describes major experimental studies on the deterioration of magnetic properties after irradiation.

§1.1. *The nature of radiation damage*

When a solid is bombarded by energetic particles, it suffers damage in several possible ways. The nature of the damage varies according to (1) whether the solid is crystalline or amorphous, (2) the characteristic bonding of the host atoms and (3) the mass, charge and energy of the bombarding agents. Although neutrons, protons (in cosmic radiation) and γ-rays are the principal sources of radiation in practical problems, the list of bombarding particles used in the studies of radiation damage is rather extensive. This is partly because ion bombardment has become more important for simulation studies and industrial applications (see §2.1.2.3 on Dynamic Conversion of Magnetic Bubbles) since 1969. Practically there is no limit on the number of ions that can be used for irradiation experiments.

Among the many processes of radiation damage, four are considered of prime significance in the context of the present discussion. They are (1) atomic displacement, (2) excitation of electrons and ionization of the host atoms, (3) transmutation and (4) localized heating in both crystalline and amorphous substances. The first process is by far the most important for all types of crystalline solids. Therefore, the bulk of the discussion in this section will be on atomic displacement.

§1.1.1. *Atomic displacement*

Energetic particles traveling in a crystalline solid will inevitably collide or interact with the host atoms, thus dislodging the atoms from their normal positions. The displaced atoms will recoil through the lattice from one interstitial site to another before coming to a complete rest. A displaced atom itself is a point defect, known as a self-interstitial, whereas the original lattice site, now being unoccupied, becomes a vacancy. Hence each atomic displacement always creates a pair of point defects: a self-interstitial and a vacancy, which together constitute a Frenkel or an i–v pair (ch. 5, §2.3, fig. 5.36b).

The displaced atoms produced directly by the bombarding particles are called the primary knock-ons (PKOs), whereas those subsequently produced by the PKOs are called the secondary, tertiary displacements. (Some workers simply call all subsequent displacements "secondaries" as opposed to "primaries".) All bombarding particles, except photons and thermal neutrons, with sufficient energies are capable of producing atomic displacements directly. More importantly, the PKOs may also be highly energetic; thus they themselves can generate large numbers of displacements. For instance the PKO produced by a 1 MeV neutron in iron will gain a maximum energy of 6.9×10^4 eV, which is large enough to produce approximately 610 additional displacements. To make the matter even worse, secondary displacements are produced in a relatively small volume by closely linked chain of collision events, known as a cascade, which tends to enhance the damaging potential of radiation-induced point defects. Thus displacement of atoms is the most important process of radiation damage to metals, metallic solid solutions and intermetallic compounds.

For a quantitative assessment of the damage in metallic substances, we must know the total number of atomic displacements per unit volume n_d produced by the irradiation per unit volume. An accurate calculation of n_d is generally difficult because too many variables are involved. The usual procedure is to first compute the number of PKOs n_p produced by the bombarding particles and then to estimate the average number of further displacements $\bar{\nu}$ caused by each PKO; i.e.,

$$n_d = n_p \bar{\nu}. \tag{7.2}$$

By denoting the flux of the incident particles by ϕ ($= nv$ particles $\cdot$ m$^{-2} \cdot$ sec^{-1}, where n is the density of bombarding particles per m^3 and v the particle velocity in m $\cdot$ sec^{-1}) and the probability of collision, or the cross section for displacement by σ_d in m^2 (or barns in CGS units, where $1\,b = 10^{-24}\,cm^2$), we can express n_p by

$$n_p = \phi t n_0 \sigma_d, \tag{7.3}$$

where t is the irradiation time (thus ϕt is the irradiation fluence in particles per m^2) and n_0 is the number of atoms per m^3 in the lattice.

To calculate n_d from (7.2) and (7.3), we need to formulate σ_d and $\bar{\nu}$. Seitz [1949] made the first calculation by assuming that the displacement process features a sharp threshold energy E_d. By sharp threshold is meant and shown in fig. 7.1a that the struck atom will remain in the original site when it receives an energy Γ less than E_d and becomes displaced when $\Gamma > E_d$. This assumption thus enables us to express σ_d in terms of a differential cross section $d\sigma$ in the form

$$\sigma_d(E) = \int_{E_d}^{\Gamma_m(E)} \frac{d\sigma(E, \Gamma)}{d\Gamma}\, d\Gamma, \tag{7.4}$$

where E is the energy of the striking particle and Γ_m is the maximum energy transferred to the struck atom. The formulation of $d\sigma$ or $d\sigma/d\Gamma$ varies according to the charge and energy of the particle, which in turn determine the type of interaction of the particle with the host atom. At this point, it is instructive to differentiate two types of scattering of fast particles in a crystalline solid. One is the Rutherford scattering and the other the hard-sphere collision. The extent to which the nucleus is screened by the orbital electrons provides the basis for characterizing the interaction. Consider a charged particle of mass M_1, charge Z_1e and energy E interacting with a host nucleus of mass M_2 and charge Z_2e at a distance r, the Coulomb potential embracing the screening effect of the orbital electrons of the nucleus is given (Bohr [1948]) by

$$V(r) = (Z_1 Z_2 e^2/r)\, e^{-r/a}, \tag{7.5}$$

where $a = a_0/(Z_1^{\frac{2}{3}} + Z_2^{\frac{2}{3}})^{\frac{1}{2}}$ is the screening constant with $a_0 = \hbar^2/m_0e^2$ being the Bohr radius of the hydrogen atom. The charged particle is characterized by a collision diameter b, which is defined by

$$b = Z_1 Z_2 e^2 (M_1 + M_2)/M_2 E. \tag{7.6}$$

Physically the collision diameter is the distance at which the striking particle would approach to the atom in a head-on collision, in the absence of the screening effect. The screening is considered weak when $b \ll a$, and the Coulomb part of $V(r)$ in (7.5) dominates the interaction. Under these conditions, the scattering is well expressed by the Ruther-

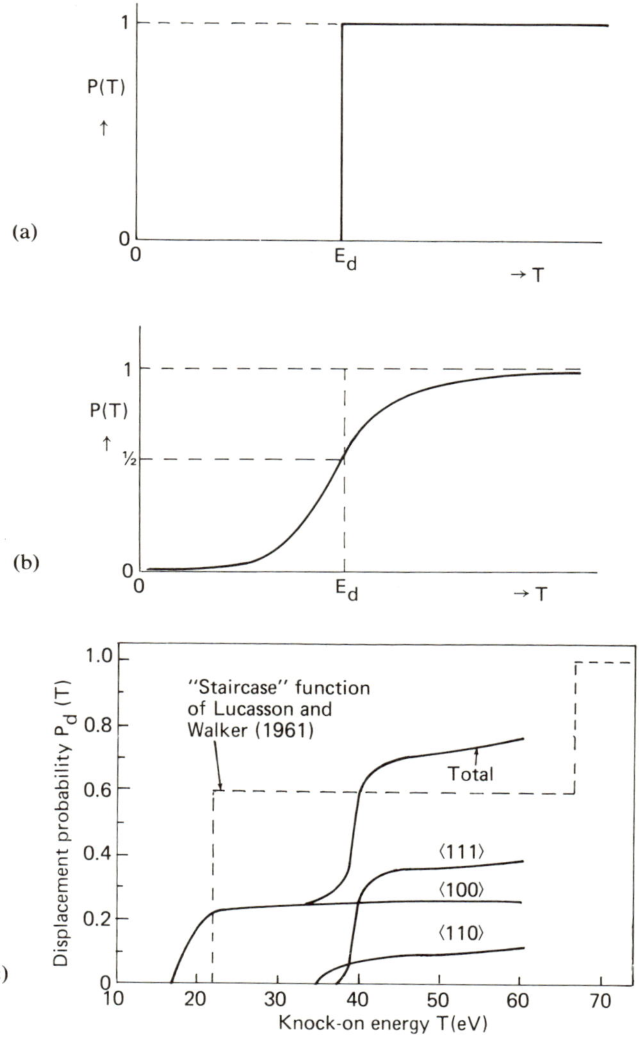

Fig. 7.1. The probability $P(\Gamma)$ of atomic displacement as a function of the kinetic energy transferred to a host atom Γ. (a) A sharp threshold energy E_d was assumed in the early analyses; (b) a curve replaces a step function when the effect of lattic crystallinity is included; (c) further revision of the $P(\Gamma)$–Γ relationship (Chadderton [1965]).

ford law, which gives the differential scattering cross section

$$\mathrm{d}\sigma(E, \Gamma)_{\text{nuclear}} = 2\pi p\ \mathrm{d}p = (\pi Z_1^2 Z_2^2 e^4/E)(M_1/M_2)(\mathrm{d}\Gamma/\Gamma^2), \qquad (7.7)$$

where p is the impact parameter shown in fig. 7.2a.

Now if $b \gg a$, the screening effect is sufficiently strong that both the striking particle and the host atom are like perfectly elastic billiard balls, giving rise to a hard-sphere collision. In this case, we have

$$\mathrm{d}\sigma(E, \Gamma)_{\text{nuclear}} = \frac{4\pi R^2(E)}{\Gamma_{\max}}\,\mathrm{d}\Gamma, \qquad (7.8)$$

where $R(E)$ is an effective hard-sphere radius. The Rutherford and hard-sphere collisions are separated approximately by setting $b = a$, at which the energy of the striking particle becomes

$$E_{\mathrm{A}} = E_{\mathrm{R}}\{(2M_1 + M_2)/M_2\}Z_1 Z_2(Z_1^{\frac{2}{3}} + Z_2^{\frac{2}{3}})^{\frac{1}{2}}, \qquad (7.9)$$

where E_{R} is the Rydberg energy (13.6 eV). Examples of both types of collisions are given later. Detailed analyses on σ_{d} can be found in the book by Thompson [1969a].

For the calculation of $\bar{\nu}$, we first replace $\bar{\nu}$ by the so-called damage function $\nu(\bar{\Gamma}_{\mathrm{p}})$, which is defined as the number of displaced atoms

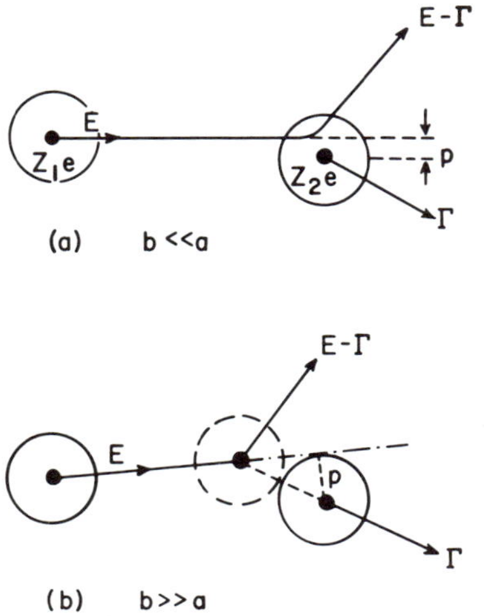

Fig. 7.2. Atomic collision of (a) the Rutherford and (b) the hard-sphere types (Nelson [1968]).

created by a PKO of an average energy $\bar{\Gamma}_p$. If nearly all PKOs have energies high enough to be in the region where $\nu(\Gamma_p)$ is a linear function of Γ_p, as in the case of fission neutrons irradiating a heavy metal, we can further assume $\bar{\nu} = \nu(\Gamma_p)$, where $\nu(\Gamma_p)$ is the damage function of a PKO of a given energy Γ_p. Whence, (7.2) becomes

$$n_d = \phi t n_0 \sigma_d \nu(\Gamma_p). \qquad (7.10)$$

Next we shall discuss the specific processes of atomic displacement by fast neutrons. For other processes of atomic displacement by fast electrons, energetic ions and gamma-rays, the reader is referred to the books by Chadderton [1965] and Thompson [1969]. The energies of the prompt neutrons produced in the fission process involving ^{235}U cover a considerable range, probably from over 10 MeV down to thermal values. Most fission neutrons, however, have energies of about 2 MeV. In regard to radiation damage, neutrons with energies greater than 0.1 MeV account for most of the damage, even though neutrons having energies greater than 100 eV can produce displacements directly. Hence neutrons with $E > 0.1$ MeV are monitored in a thermal or fast reactor when radiation damage via atomic displacements is to be examined, and these fission neutrons are generally assumed to have an average energy of 2 MeV. For the proposed fusion reactor based on the deuterium–tritium (D–T) reaction, the fast neutrons produced are monoenergetic at 14.1 MeV.

Fast neutrons interact directly with the nuclei of the atoms in the crystal. Here the neutron-atom interaction is a good example of the hard-sphere collision. In such a purely elastic collision, the maximum energy Γ_{max} transferred from the bombarding particle to a nucleus of atomic weight M_2 is given by the conservation laws of energy and momentum. Whence

$$\Gamma_{max}(E) = [4M_1M_2/(M_1 + M_2)^2]E. \qquad (7.11)$$

For neutrons with $M_1 = 1$, $\Gamma_{max}(E) = 4M_2E/(M_2 + 1)^2$. For iron and heavier elements with $M_2 > 55$, $\Gamma_{max} \sim 4E/M_2$. At the average energy of 2 MeV, fast neutrons are scattered isotropically by the majority of nuclei in the center-of-mass coordinate system of the neutron and nucleus. By isotropic scattering is meant that the struck nucleus has an equal probability of receiving energy anywhere from zero to Γ_{max}. Thus the average energy transfer is $\bar{\Gamma} = \frac{1}{2}\Gamma_{max}$. For iron with $M_2 = 55.85$, neutrons having $\bar{E} = 2$ MeV will produce PKOs with a maximum energy of $\Gamma_{max} = 1.38 \times 10^5$ eV and with the average energy of $\bar{\Gamma} \sim$

TABLE 7.1

Experimentally deduced values* of threshold energies E_d for the production of displaced atoms in metals**

Metal	E_d (eV) for the displaced atom produced in the direction indicated in single crystals			Average value of E_d for polycrystalline specimens
(I) Face-centered cubic				
Al				27
Ni				34
Cu				29
Rh	$E_d^{\langle 100\rangle} \approx E_d^{\langle 110\rangle}$			(40)
Pd				41
Ag				39
Pt	$E_d^{\langle 111\rangle} \approx 2.2 E_d^{\langle 100\rangle}$			44
Au				43
Pb				44
(II) Body-centered cubic				
Cr				60
Fe	$E_d^{\langle 100\rangle} = 20$, $E_d^{\langle 110\rangle} \sim E_d^{\langle 111\rangle} = 30$			(44)
Nb				(78)
Mo	$E_d^{\langle 100\rangle} = 35$, $E_d^{\langle 110\rangle} = 72$, $E_d^{\langle 111\rangle} = 45$			60–70
Ta	$E_d^{\langle 111\rangle} = 39 < E_d^{\langle 100\rangle}$			80–90
W				100–110
(III) Close-packed hexagonal				
	$E_d^{[0001]}$	$E_d^{\langle 11\bar{2}0\rangle}$	$E_d^{\langle 20\bar{2}1\rangle}$	
Mg				(20)
Ti				(30)
Co	40	27	40	36
Zn	19	14	20	29
Zr				(40)
Cd	$\geqslant 40$	21	19	30
Re				(60)
(IV) Tetragonal				
In				10.5

*Values in parentheses are somewhat uncertain.

**Reference: Lucasson [1976].

7×10^4 eV. Even the latter value is enormous in comparison with E_d, which ranges between 10 (for In) and 100 eV (for W) (see table 7.1). Therefore, in irradiation with fission neutrons, the PKOs are sufficiently energetic to induce many secondary displacements. In fact, the PKOs recoil at such high speeds that some of their loosely bound electrons are stripped off, making the PKOs multiply ionized at the beginning. Note that even though the incident particles, being neutrons, carry no charge, most of the PKOs are charged until they recapture the lost electrons in the later stages of their recoils.

Above a certain energy known as the ionization limit E_i, the PKOs first interact with electrons, giving rise to electronic excitation and ionization of the host atoms. Thereafter, with energies $\Gamma_p < E_i$ the PKOs first lose their energies by elastic collisions with the lattice atoms in the Rutherford mode when $\Gamma_p > E_A$, where E_A is defined in (7.9). When $\Gamma_p < E_A$, the PKOs generate further displacements by hard-sphere collisions. The regimes of energy loss of a PKO are shown diagrammatically in fig. 7.3. The production of replacements and of thermal or displacement spikes will be explained later.

The charged nature of the PKOs is one source of complications in the calculation of n_d. Another problem is in the evaluation of the neutron flux ϕ. Despite recent developments in dosimetry techniques, the determination of the integrated neutron flux from a fission spectrum still involves considerable experimental uncertainties. A third source of difficulties is connected with the threshold energy E_d. In the early crude estimate of n_d, the value of E_d was assumed to be four times the sublimation energy, which is typically 6–7 eV for metals; thus $E_d \sim$ 25 eV. Also the probability $P(\Gamma)$ for an atom to be displaced was assumed to be a step function of the transferred energy, as depicted in fig. 7.1a. Both assumptions are highly questionable. Recent studies of E_d (see Lucasson [1976]) have clearly shown that the value of E_d varies rather sensitively with (1) the target material, (2) the crystallographic direction in which the displacement occurs, or equivalently, the type of i–v Frenkel pair which is produced, and (3) the temperature. The E_d values listed in table 7.1 are given by Lucasson from a series of experimental studies. The first improvement over the step function assumed in the $P(\Gamma)$ versus Γ plot is to adopt a curve shown in fig. 7.1b so that the crystalline nature of the lattice and the effect of temperature are taken into account. In this way, $P(\Gamma) = \frac{1}{2}$, not 1, for $\Gamma = E_d$. Further improvement was suggested by Erginsoy et al. [1964], who arrived at an

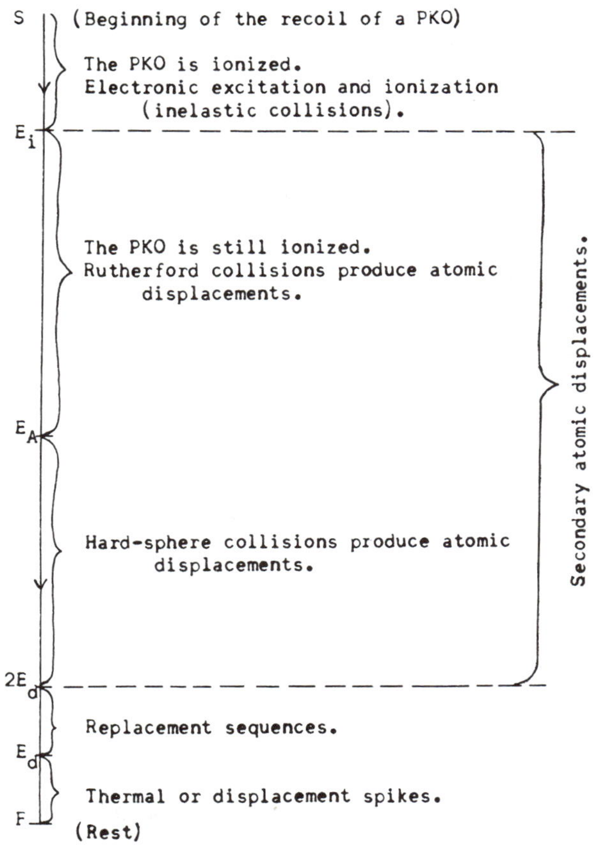

Fig. 7.3. Sequence of events for the energy loss of an energetic primary knocked-on atom in a crystalline solid.

approximately two-step function in the $P(\Gamma)$–Γ plot for the polycrystalline samples (fig. 7.1c).

Kinchin and Pease [1955] made the first serious attempt to calculate n_d. Their model still provides the basis for the current, more rigorous calculations (see Robinson [1976]). A number of assumptions were made in the Kinchin–Pease model, notably the following:

(1) The lattice atoms are stationary unless affected by collision, and those in collisions behave like hard spheres.

(2) All collisions are elastic, thus prohibiting energy loss to the lattice and in electron excitation.
(3) The collisions are independent of each other, and any spatial correlations imposed by the periodicity of the lattice are ignored.
(4) The cascade proceeds as a series of two-body collisions, in each of which $E = E' + \Gamma$, where E' is the energy retained by the striking atom and Γ is the energy transferred to the new recoil (fig. 7.4).
(5) A struck atom receiving energy less than E_d is not displaced. Nor will a recoiling atom be capable of generating a displacement if its energy is less than E_d.
(6) There exists a sharp ionization limit E_i.

Under these assumptions, Kinchin and Pease arrived at the following results for the damage function:

$$\nu(\Gamma_p) = \begin{cases} 0 & \text{for } \Gamma_p < E_d \\ 1 & \text{for } E_d \leqslant \Gamma_p \leqslant 2E_d \\ \Gamma_p/2E_d & \text{for } 2E_d \leqslant \Gamma_p \leqslant E_i \\ E_i/2E_d & \text{for } E_i < \Gamma_p \end{cases} \tag{7.12}$$

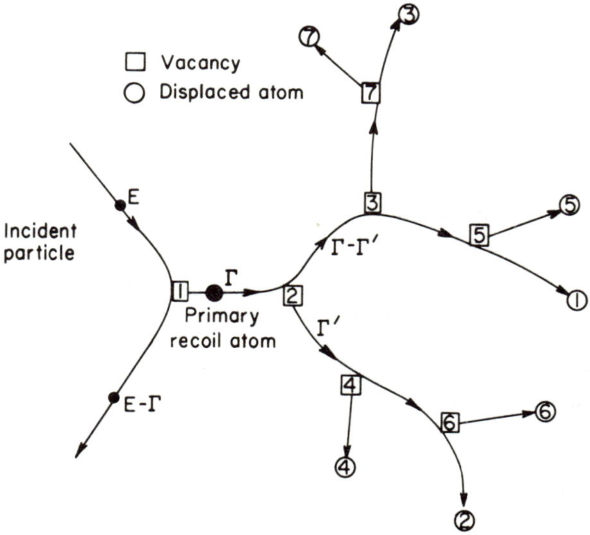

Fig. 7.4. Schematic representation of the collision cascade showing the production of vacancies and self-interstitials in equal numbers.

To calculate $\nu(\Gamma_p)$ and to compare the last two formulae, we need to know the ionization limit E_i, which is discussed in §1.1.2. Here we cite the approximate simple relation

$$E_i \text{ (in keV)} = M \text{ (in a.m.u.)}. \tag{7.13}$$

Hence for Fe, Co, Ni, and Gd, E_i has the values of 55.85, 58.93, 58.71 and 157.25 keV, respectively. In the irradiation of iron inside a thermal reactor, since the average energy of the PKOs generated by fast neutrons with $\bar{E} = 2$ MeV is ~70 keV, about 20% of the PKO energies is lost to ionization of the host atoms before secondary displacements are produced. This situation is even more prominent in the lighter metals (e.g., as much as 60% in Al) but is reversed in heavier metals (e.g., no energy of the PKOs will be lost to ionization in metals heavier than Sr).

From the E_i value given by (7.13) and by using $E_d = 24$ eV, we obtain for the neutron irradiation of iron $\nu(\Gamma_p) = 1165$ displacements per PKO from (7.12). Substituting this value and other appropriate constants for Fe ($\sigma_d = 3$b and $n_0 = 8.49 \times 10^{28}$ atoms/m^3) into (7.10) yields $n_d \approx 3 \times 10^4 \times \phi t$. Typically $\phi \sim 5 \times 10^{17}$ n · m^{-2} · s^{-1} for a thermal power reactor. Hence $n_d \sim 5 \times 10^{28}$ displacements per m^3 per year, which is equivalent to 0.55 displacement per atom (dpa) per year. At this production rate, a piece of iron placed inside the core of a thermal reactor will have all its atoms displaced roughly 11 times during the 20-year life of the reactor core. In a breeder reactor, the flux of fast neutrons will be much higher, probably reaching the value of $\phi \sim 5 \times 10^{19}$ n · m^{-2} · s^{-1}. The number of displacements will be increased to the extent that each atom in an iron piece will be displaced about 55 times per year in a fast breeder reactor.

Because of the complex nature of the displacement process, especially in the generation of cascades by PKOs, and because of the many assumptions made in the Kinchin–Pease model, the reliability of the foregoing calculation must be examined by comparing the predicted values for $\nu(\Gamma_p)$ and n_d with those deduced from sensitive experiments such as measurements of electrical resistivity or density. In all cases, the comparison indicates a rather large discrepancy (a factor of 2 to 5) between the calculated displacement number and the experimentally deduced value (see Chadderton [1965a]). The large discrepancy implies that there must be other reasons than the deficiencies of the calculation method* for the much smaller number of displacements observed. One important reason is radiation annealing, in which the induced self-interstitials and vacancies annihilate each other. This annealing comes about because self-interstitials are mobile in most irradiations except at extremely low temperatures below 20°K and because a Frenkel pair is unstable when the self-interstitial and the vacancy are within one or two interatomic distances, tending to cancel each other by recombination. The radiation annealing is enhanced by high fluences, at which the damaged regions tend to overlap, thus increasing the chances for defect annihilation. Unfortunately, there is no reliable way to predict the extent of radiation annealing.

Before leaving the discussion on atomic displacements, four phenomena must be briefly introduced. These phenomena are not only interesting, but also important in the actual production of displacements. They may play significant roles in determining the number of displacements that are produced and survive in irradiation with fast neutrons or heavy ions.

(1) *Channelling.* The anisotropic nature of the spatial arrangement of atoms in a crystal creates some open channels along certain crystal directions for easy passage of the PKOs. Atoms moving in the channels are called channelons. The preferential transfer of mass and energy of the channelons is referred to as the channelling phenomenon, which should not be confused with other channelling phenomena associated with electrons in microscopy and with dislocations in deformed metals.

A channel is usually an axis of symmetry and bordered by close-packed chains of atoms. For instance, the channels in the fcc lattice are in the ⟨110⟩ directions, each of which is surrounded by four atom chains in the same ⟨110⟩ directions. Fig. 7.5 shows a [110] channel along which the moving atoms advance alternately through a pair of atoms separated by a distance a in the [001] direction in the A, C, . . . planes and another pair of atoms separated by a distance $a/\sqrt{2}$ in the [110] direction in the B, D, . . . planes. Once a PKO enters the channel, it will travel a long distance in an almost force-free path. Experimental evidence has revealed that a PKO of 10 keV in Cu can move in a slightly curved trajectory in the ⟨110⟩ channel over some 10^3 interatomic distances. Channels in the bcc and hcp structures are in the ⟨100⟩ and

*In the refined model (Robinson [1976]), a factor $\kappa \sim 0.8$, known as the displacement efficiency, is introduced to reduce the calculated value of n_d in (7.10). This refinement hardly affects the discrepancy, however.

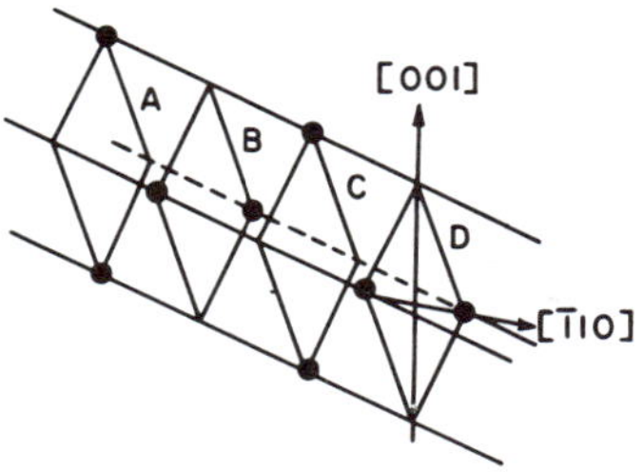

Fig. 7.5. A ⟨110⟩ channel bordered by four atomic ⟨110⟩ chains in a face-centered cubic crystal (Leibfried [1959]).

⟨11$\bar{2}$0⟩ directions, respectively. The mechanism whereby a PKO becomes a channelon is not precisely known. It is well established, however, that channelling occurs predominantly for the PKOs at high energies. Channelling obviously reduces the number of displacements, but extends the range of ion bombardment. The phenomenon has been applied to implant dopants preferentially along certain directions in semiconductors and other materials.

(2) *Focused collisions.* The anisotropy of the atom arrangement in crystals causes still another modification to the simple theory of atomic displacement. When the energy of the recoils is reduced to a sufficiently low level, the crystalline lattice will impose rigid conditions on atomic interactions such that the resulting collisions occur in a selected direction as if they are focused. A simple example of the focusing process is illustrated in fig. 7.6, where two atoms are initially at positions A and B separated by a distance D. Suppose atom A is given a momentum in a direction making an angle θ_1 with AB. A hard-sphere collision will occur when atom A reaches P at which PB $= 2R$, where R is the radius of the hard sphere. The collision will push atom B off position in an angle θ_2 to AB. Whether or not a focused collision occurs will depend on the relative value of θ_2 to θ_1, as shown by the following condition:

$$\begin{array}{llll} |\theta_2| < |\theta_1| & \text{or } D < 4R & \text{or } f < 1, & \text{focusing} \\ |\theta_2| > |\theta_1| & \text{or } D > 4R & \text{or } f > 1, & \text{defocusing} \end{array} \tag{7.14}$$

where $(-f) = \theta_2/\theta_1$ or $f = D/2R - 1$ is called the focusing parameter. Since D has the smallest value along the closest-packed directions of atoms (⟨110⟩ in fcc lattices, ⟨111⟩ in bcc lattices and ⟨11$\bar{2}$0⟩ in hcp structures), this condition clearly indicates that the focused collision is likely to proceed in these directions. It is easy to show that, in a series

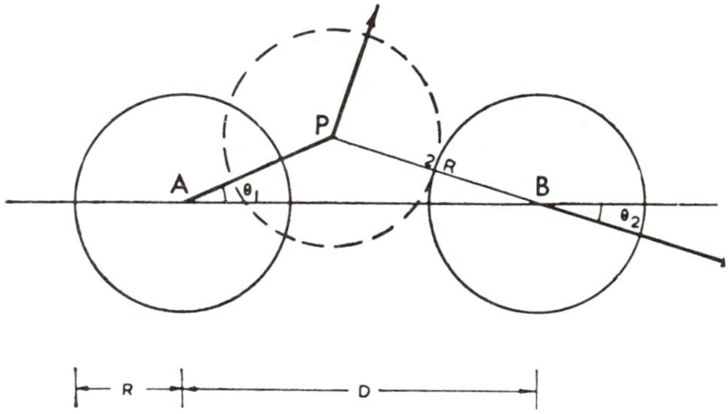

Fig. 7.6. The focusing effect on atomic collision of the hard-sphere type (Thompson [1969]).

of focused collisions, the angle for the nth collision is given by

$$\theta_n = (1-(D/2R))^n\theta_1 = (-f)^n\theta_1. \tag{7.15}$$

Hence when $f<1$ or $D<4R$, a sequence of collisions propagates from atom to atom in the focusing direction until the angle θ_n converges to zero through the factor $(-f)^n$, thereafter the collisions are of the head-on type. A focused collision sequence is shown in fig. 7.7. The physical significance of a focused collision sequence is that a bundle of energy will pass down the atom row until the energy is finally dissipated to neighboring atoms or to imperfections.

By setting $f=1$ and by choosing the appropriate interaction potential, the critical energy E_f^{hkl} below which focusing occurs can be calculated. Data compiled for fcc metals by Chadderton [1965b] give the calculated values of E_f^{hkl} in the ⟨110⟩, ⟨100⟩ and ⟨111⟩ directions at 40, 54 and 300 eV for Cu, at 87, 120 and 960 eV for Ag and at 620, 800 and 2100 eV for Au. Although the reliability of these calculated values may be questioned, they nevertheless confirm that focusing should occur predominantly in the closest-packed direction. The calculations also

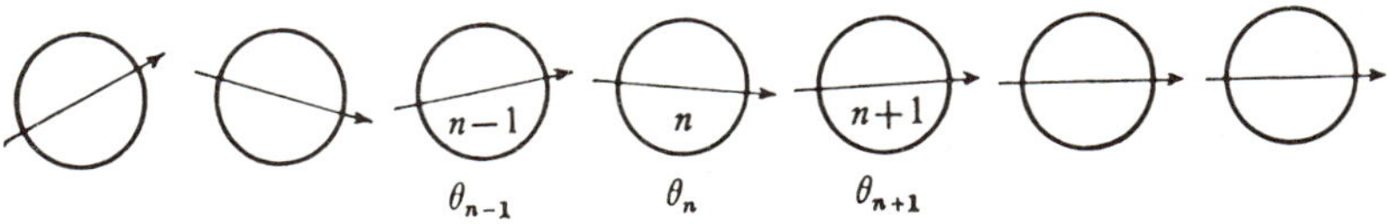

Fig. 7.7. Focused collisions in a sequence (Thompson [1969]).

reveal that focused collision sequences can be expected to exert their greatest influence on cascades in heavy metals.

(3) Replacement collisions. If a slow displaced atom A collides with a host atom B such that most of its energy is transferred to B, A will simply replace B; i.e., A comes to a rest at the site originally occupied by the B atom, which now becomes a displacement. This process is called the replacement collision, because it changes the identity of the displaced atoms, but not the total number. It may be viewed as a dynamic sort of interstitialcy interchange. In the latter view, atom A loses so much energy that it is unable to escape the potential field of the vacancy it has created. As in focusing, replacement collisions occur in a sequence, which ends when the energy of the sequence has fallen to a value E_r^{hkl} called the replacement energy. Also in the replacement process, atom B becomes a displacement with a transferred energy less than the threshold energy for displacement, E_d^{hkl}, but more than E_r^{hkl} which was believed to be not much greater than the activation energy for interstitialcy diffusion (~ 0.25 eV in copper), according to Kinchin and Pease [1955]. The ratio of replacement to displacement events is given by

$$\nu_r/\nu_d = 1.414 \ln(E_d^{hkl}/E_r^{hkl}) + 0.5, \tag{7.16}$$

which has the value 0.93 when E_d^{hkl}/E_r^{hkl} is taken to be 20. This result signified that replacements may play an important role in the production of atomic disorder in compounds, especially when the constituents have similar atomic masses. Replacement collision is also important to ordered alloys because a considerable amount of disorder may be produced by replacements during irradiation.

(4) Spikes. Spikes are regions where intense radiation damage is localized as a result of energy release in a short ($\sim 10^{-7}$ s) time. Several types of spikes have been envisaged. Here we focus our attention on the displacement spikes, leaving thermal spikes to a later discussion in §1.4 and omitting fission spikes as a special, irrelevant phenomenon. Brinkman [1962] conceived the idea of displacement spikes when considering the possibility that the mean free path $\lambda(E, \Gamma)$ between collisions of a PKO falls below the interatomic distance of the lattice. Under this condition all atoms within one interatomic distance of the trajectory of the primary are displaced as secondaries, and each of these produces, on the average, one tertiary. In this way, a large number (n_{dk}) of atoms are displaced by the PKO in a very short time ($< 10^{-11}$ s). The secondaries and tertiaries are forced away from the

primary so that n_{dk} vacancies are created in the interior of a region with n_{dk} self-interstitials at the periphery. According to the original estimate of Brinkman, the average size of a displacement spike produced by 2 MeV neutrons in copper would be 75 Å in diameter and would contain 2×10^4 atoms. The distribution of the 54 Frenkel pairs involved in a hypothetical displacement spike is shown in fig. 7.8. Such a configuration of Frenkel defects is obviously unstable because the interior is essentially hollow, whereas the density of atoms in the interstitial shell is $\sim 10\%$ greater than the normal density. Hence a pressure of about 10^{10} N/m^2 (10^{11} dynes/cm^2) will be exerted on the central region. It is therefore difficult to predict the fate of displacement spikes. Three possibilities exist. One is that the self-interstitials simply fall back into the hollow zone to recombine with vacancies, leading to the shrinkage and eventual elimination of the spike. Alternatively, the cluster of vacancies in the spike collapses to form a dislocation loop, as evidenced in a recent study of neutron-irradiated niobium by Chang and

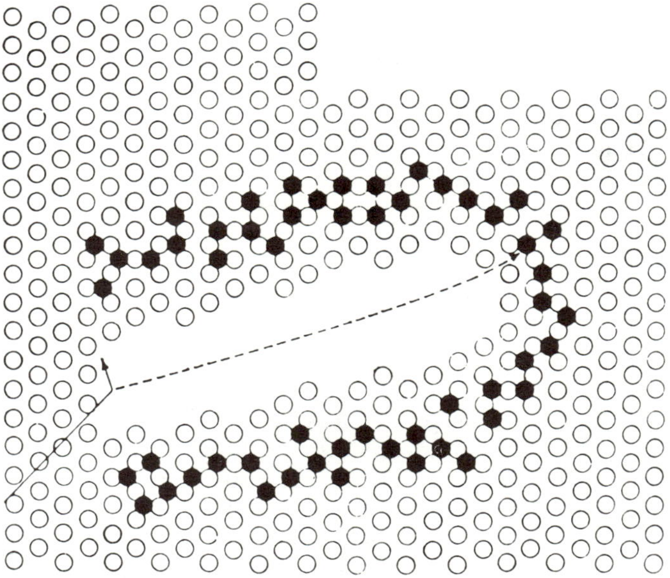

Fig. 7.8. A simple model of displacement spikes proposed by Brinkman [1962]: ○ normal atom, ● interstitial atom, —— path of primary particle, - - - path of primary knock-on.

Chen [1977]. Thirdly, if gas atoms of He, A, H, etc. migrate into the spike during the lifetime (believed to be $\sim 10^{-9}$ s) of the spike, the vacancy core is stabilized and can then act as a nucleus for the formation of a void. Recent studies using transmission electron microscopy (see Corbett and Ianniello [1972]) have furnished evidence to substantiate the nucleation of voids at the displacement spikes.

§1.1.2. *Electronic excitation and ionization*

Previous discussion in §1.1.1 has disclosed that excitation of electrons and ionization of the host atoms constitute the main process of radiation damage by energetic ions and by PKOs whose energies are greater than the ionization energy E_i. The collisions with electrons are inelastic, and the excitation likely involves the valence electrons, rather than those in the inner (K, L or M) shells. Two parameters are important in the consideration of this process. One is the stopping power of the crystal via electronic excitation. This parameter is important for irradiation with ions and other particles incapable of penetrating through the solid. According to the Rutherford law, the differential scattering cross section is given by

$$\mathrm{d}\sigma(E, \Gamma)_{\text{electronic}} = 2\pi P \frac{\mathrm{d}P}{\mathrm{d}\Gamma}\mathrm{d}\Gamma = \pi \frac{Z_1^2 e^4}{E}\left(\frac{M_1}{m_0}\right)\frac{\mathrm{d}\Gamma}{\Gamma^2}, \tag{7.17}$$

where $Z_1 e$ and E are the charge and energy of the moving particle. Applying (7.17) to the following definition of the electronic stopping power of the crystal,

$$\left(-\frac{\mathrm{d}E}{\mathrm{d}x}\right)_{\text{electronic}} = n_e \int_{\Gamma_{\min}}^{\Gamma_{\max}} \Gamma \,\mathrm{d}\sigma(E, \Gamma)_{\text{electronic}} \tag{7.18}$$

yields

$$\begin{aligned}\left(-\frac{\mathrm{d}E}{\mathrm{d}x}\right)_{\text{electronic}} &= n_e \int_{E_i}^{4m_0E/M_1} \Gamma \frac{\pi Z_1^2 e^4}{E}\left(\frac{M_1}{m_0}\right)\frac{\mathrm{d}\Gamma}{\Gamma^2} \\ &= n_e \frac{\pi Z_1^2 e^4 M_1}{m_0 E} \ln\left(\frac{4m_0 E}{E_i M_1}\right).\end{aligned} \tag{7.19}$$

Refinements of (7.19) in order to include the quantum mechanical and relativistic effects are discussed by Chadderton [1965c].

The second parameter is the ionization energy, E_i which is given by

$$E_i = \tfrac{1}{8}(M_1/m_0)E_x, \tag{7.20}$$

where E_x is equal to one-half of the Fermi energy E_f for a metal. For insulators, E_x is the lowest electronic excitation energy corresponding to the edge of the first main band in the optical absorption spectrum. For a rough estimate, the relation E_i (in keV) = M_1 (in a.m.u.) expressed in (7.13) can be used. The latter relation, together with the regimes of energy losses sketched in fig. 7.3, enables us to conclude that the importance of this process in irradiation with fission neutrons will diminish steadily as the atomic number of the target material increases. Even in light metals where $\Gamma > E_i$ by a wide margin and excitation is expected to consume most of the recoil energy, the process is of little consequence because the ionized atoms can easily capture electrons from the conduction band and regain their neutrality. The situation for insulators, however, is different, and the effect of electronic excitation lingers on.

§1.1.3. *Transmutation*

Whenever an incident particle is absorbed by a nucleus, an unstable compound nucleus is formed as an intermediate product, which decays rapidly into final products. This nuclear reaction is a transmutation because it converts a host atom into a foreign atom or another isotope. The reaction is

$$\underset{\begin{pmatrix}\text{incident}\\ \text{particle}\end{pmatrix}}{{}^{M_1}_{Z_1}\mathrm{x}} + \underset{\begin{pmatrix}\text{reactant}\\ \text{nucleus}\end{pmatrix}}{{}^{M_2}_{Z_2}\mathrm{X}} \longrightarrow \underset{\begin{pmatrix}\text{compound}\\ \text{nucleus}\end{pmatrix}}{[{}^{M_1+M_2}_{Z_1+Z_2}\mathrm{C}^*]} \longrightarrow \underset{\begin{pmatrix}\text{product}\\ \text{nucleus}\end{pmatrix}}{{}^{M_3}_{Z_3}\mathrm{Y}} + \underset{\begin{pmatrix}\text{product}\\ \text{particle}\end{pmatrix}}{{}^{M_4}_{Z_4}\mathrm{y}} + Q, \tag{7.21}$$

where $Q = [(M_1 + M_2) - (M_3 + M_4)]c^2$ is called the energy balance of the reaction. If the total mass $M_1 + M_2$ of the reactants is greater than that of the products, Q is positive and the reaction is exothermic. The energy Q released is divided between the product particle and the product nucleus. In conformity with the charge conservation, $Z_3 + Z_4$ must be equal to $Z_1 + Z_2$. For a simple example of (7.21), we cite the well-known reaction involving a thermal neutron and a boron nucleus:

$${}^{10}_{5}\mathrm{B} + {}^{1}_{0}\mathrm{n} \to [{}^{11}_{5}\mathrm{B}^*] \to {}^{7}_{3}\mathrm{Li} + {}^{4}_{2}\mathrm{He} + Q. \tag{7.22}$$

This reaction is customarily abbreviated to the form ${}^{10}\mathrm{B}(\mathrm{n},\alpha){}^{7}\mathrm{Li}$. After absorbing a neutron, the ${}^{10}\mathrm{B}$ nucleus is transmuted to a ${}^{7}\mathrm{Li}$ nucleus and an α-particle. The value of Q is readily calculated, $Q = (10.016109 + 1.008930 - 7.018222 - 4.0038727)c^2 = 2.66$ MeV, which is divided into 0.97 and 1.69 MeV associated with ${}^{7}\mathrm{Li}$ and ${}^{4}\mathrm{He}$, respectively (see Bethe [1950]).

In principle, there are many possible transmutations such as $(i, \alpha), (i, p), (i, \gamma), \ldots$ reactions, where i is the bombarding particle of the ith species. As far as radiation damage is concerned, only those transmutations with neutrons are of importance because the neutrons, being uncharged, can interact with nuclei almost exclusive of nuclear forces and therefore possess absorption cross sections comparable to, or in excess of, their elastic collision cross sections. Thermal neutrons are a special case. With energy at 0.025 eV, these slow neutrons are incapable of producing damage by direct collision, but they can rely on the (n, γ) reaction. Walker [1960] made the first serious attempt to analyze the production of atomic displacements by thermal neutrons through the (n, γ) reaction. His analysis indicated that, in some thirty elements, including Gd, Cd, Sm, Eu, Dy, Co and Ir, the number of displaced atoms produced by the recoil following capture γ-ray emission constitutes an appreciable fraction of the total number of displaced atoms, as was confirmed in a series of irradiation experiments involving nearly pure thermal neutrons by Coltman et al. [1962].

Besides the (n, fission) reaction, perhaps the most devastating transmutation induced by neutrons is the (n, α) reaction. Because helium atoms are inert and have a large atomic size close to that of copper ($r_a = 1.25$ Å), they are practically insoluble in most solids. Thus the induced α-particles tend to form gas bubbles. Moreover, these gas bubbles will not be annealed out as are voids; instead, they are easily swept to grain boundaries at elevated temperatures. Thus a preponderant occurrence of the (n, α) reaction eventually leads to a premature failure of the material. According to a study by Birss [1969], the (n, α) reaction will take place in almost any isotope, provided the kinetic energy of neutrons is high enough. The helium embrittlement resulting from the (n, α) reaction could pose a serious problem to fast and D–T magnetic fusion reactors which produce abundant high energy neutrons.

The reason that the $^{10}B(n, \alpha)^7Li$ is so notoriously well-known is twofold. First, the cross section for this reaction is at a whopping value of 3990 b. This is exceedingly unusual because the cross sections for the (n_{th}, α) reaction are generally less than 10 mb except for ^{10}B, $^6Li(\sigma = 950\ b)$, $^9Be(\sigma = 50\ b)$, $^{25}Mg(\sigma = 6.4\ b)$ and $^{27}Al(\sigma = 0.27\ b)$. Secondly, boron is often added in small amounts (0.1–0.5%) to steels for dispersion strengthening and is also present in trace amounts ($<$ 0.005%) in austenitic stainless steels. Steels are important materials in the construction of power thermal reactor cores, and type-304 and 316 stainless steels are the current choices of materials for fuel cladding in fast breeder reactors and for the construction of the vacuum wall in Tokamak-type fusion reactors. Unless preventive measures are taken, steels could develop catastrophic failures caused by the (n, α) reactions.

The (n, p) reaction is another gas-producing process. This reaction is less troublesome because hydrogen atoms can easily diffuse out of the material before they combine to form molecules or gas bubbles. Another possible type of damage by transmutation results from the reaction by which an appreciable amount of the constituent atoms is converted into foreign atoms. For instance, should a niobium-base alloy be chosen to make the first wall of a Tokamak-type fusion reactor, the irradiation of 14.1 MeV neutrons can induce the following reaction:

$$^{93}_{41}\mathrm{Nb} + {}^{1}_{0}\mathrm{n} \rightarrow [^{94}_{41}\mathrm{Nb}^*] \rightarrow {}^{92}_{40}\mathrm{Zr} + {}^{2}_{1}\mathrm{H}. \qquad (7.23)$$

According to some preliminary estimate (Stevens [1975]), (7.23) could convert 10% Nb to Zr at the end of a 10-year life of the wall. Such a contamination could lead to some undesirable consequences in the operation of a fusion reactor.

For light, charged particles (protons, deuterons and α-particles), having energies in the range 1 to 10 MeV, the absorption cross sections are extremely small, of the order of 1–100 mb, as compared with cross sections of $\approx 10^2$ b for nuclear Coulomb collisions. Hence transmutation is an ineffective process for producing damage by charged particles unless their energies reach a few hundred MeV.

§1.1.4. *Local heating in both crystalline and amorphous substances*

The heading of this subsection signifies that local heating occurs in all solids, disregarding whether the substance is crystalline or amorphous. This signification is necessary because localized heating was conceived in terms of the thermal spike for metals and other crystalline solids (Seitz and Koehler [1956]). Although the lattice nature of the amorphous materials makes the displacement process totally ineffective, there is a distinct possibility that local heating may produce nuclei for crystallization, thereby altering the basic nature of the solid from amorphous to crystalline. In the absence of other effective processes (assuming that transmutation is insignificant, which is generally true unless the fluence gets unusually high), local heating is the only process that may produce detectable radiation damage to amorphous materials.

The concept of thermal spikes emerged when Seitz and Koehler examined the situation at the end of a collision cascade. Suppose that the struck atom receives an energy Q less than the threshold energy E_d for displacement. Obviously such an energy transfer creates neither a displacement nor a replacement, according to fig. 7.3. Then Q must be released in some other fashion. A convenient way to dispose of Q is through the thermal vibration of the struck atoms by intensifying the vibrational amplitude. In the first 10^{-13}–10^{-12} s after the collision, however, the struck atom and its close neighbors are in such a violent state that the concepts of temperature and heat are inapplicable. Once this violent stage is over, a dynamical equilibrium is established in a region that can be treated by the Maxwell–Boltzmann function, and the energy distribution can now be described in terms of temperature and heat. In this context, we define a thermal spike as a region in which a sudden (within $\sim 10^{-11}$ s) release of energy causes localized heating, or even melting at the very center, of the enclosed atoms. Because of the complex nature of the event, however, it is very difficult to calculate the temperature rise in a thermal spike. Crude estimates indicate that the temperature rise probably ranges from $\sim 100°C$ to values reaching the melting point of the matrix, depending on whether the remaining energy Γ_p of the PKO to be released in the spike is small ($\sim$30 eV, Seitz and Koehler [1956]) or larger ($\sim$ 300 eV, Dienes and Vineyard [1957]).

To see the difference in local heating in thermal spikes between conductors and insulators, we cite the treatment of Seitz and Koehler, who adopted the following expressions for the temperature of spherical spikes:

$$\begin{aligned} T_{ts}(r, t) &= \frac{Q}{(4\pi)^{\frac{3}{2}}} \frac{1}{(Dt)^{\frac{3}{2}}}, && \text{for } r < (4Dt)^{\frac{1}{2}}; \\ T_{ts}(r, t) &= 0, && \text{for } r > (4Dt)^{\frac{1}{2}}, \end{aligned} \tag{7.24}$$

where $D = K/C\rho$ is the thermal diffusivity with K being the thermal conductivity, C the heat capacity and ρ the density. If we assume that the product $C\rho$ varies much less sensitively with the type of material than does K, which has typical values $\approx 10^{-1}\, \text{cal} \cdot \text{cm}^{-1} \cdot \text{s}^{-1} \cdot \text{deg}^{-1}$ for metals and $\sim 10^{-3}\, \text{cal} \cdot \text{cm}^{-1} \cdot \text{s}^{-1} \cdot \text{deg}^{-1}$ for insulators, it is readily seen from (7.24) that the heating effect in thermal spikes will be two to three orders of magnitude more intense in insulators than in metals.

Although experimental confirmation of thermal spikes encountered some difficulties in the early attempts involving phase transformations, the results obtained from sputtering experiments with 45 keV Xe^+ ions by Nelson [1965] have not only substantiated the concept of thermal spikes, but also furnished valuable information on such vital parameters as the spike temperature ΔT_{ts} (from 150°C for Zn through ~550°C for Cu, Ag and Bi to 1060°C for Ge), the radius of spherical spikes r_{ts} (~100 Å for Ag, Au, Cu, Bi and Ge and ~250 Å for Zn) and $\tau_{ts}[(1–9)\times 10^{-12}$ s]. In an earlier study, Piercy [1963] measured the size of thermal spikes by using the unusual magnetic behavior of Pt_3Fe. This stoichiometric solid solution is either antiferromagnetic or ferromagnetic, depending on whether the crystal is ordered or disordered. Thus by exposing ordered specimens to fission neutrons to fluences ranging from 3.4×10^{17} to $4.4\times 10^{18}\ n\cdot cm^{-2}$, antiferromagnetism was replaced by superparamagnetism because the heating effect of the thermal spikes introduced isolated ferromagnetic regions into the antiferromagnetic matrix. From the size of the magnetic disordered regions, Piercy concluded that each neutron collision produced an average of 2.3 thermal spikes with 3300 disordered atoms in each spike.

There is no question that local heating in thermal spikes could affect the damage function $\nu(\Gamma_p)$ by promoting the i–v recombination and defect clustering. The two- or three-fold discrepancy between the experimentally deduced number of surviving point defects and the number calculated by using (7.10) may be partly attributed to the localized heating effect of thermal spikes. Precisely how much of the discrepancy is due to this effect is not known, however.

Another effect of thermal spikes is on the atomic order of concentrated alloys. The intense heating can easily raise the temperature in the spikes well above the order–disorder transition temperature. Then the short duration of the heating and the excellent thermal conductivity of metallic systems would result in a quenching effect. Hence, a completely ordered alloy may become partially disordered wherever thermal spikes are generated during irradiation.

A third effect of thermal spikes is the destruction of crystallinity of the convertible solids whose lattices are either crystalline or amorphous, depending on the preparation procedure (most importantly on the rate of condensation from the vapor phase or the rate of freezing from the melt). Chen and Bailey [1971] reported the nearly complete destruction of the crystalline structure of a glass-forming chalcogenide ($As_{0.50}Te_{0.42}Ge_{0.08}$) as a result of reactor irradiation to the fluence of $3.4\times 10^{18}\ n\cdot cm^{-2}$ ($E > 1$ MeV) at 83°K. Although the observed crystallinity destruction could be ascribed simply to the generation of displacements, the contribution from the local heating of thermal spikes cannot be ignored.

In view of the second and third effects of thermal spikes, it seems odd to suggest that the heating effect of thermal spikes can also crystallize an amorphous substance. The suggestion is justified, however, for the fact that many amorphous materials tend to become crystalline in appropriate heat treatments. It is then possible that local heating in the thermal spikes may meet the conditions for the creation of nuclei for crystallizing the substance, especially when the solid is a thermal insulator. If the conversion process is characterized by an activation energy $\Delta F_{a\to c}$, the probability of the conversion is proportional to $\exp(-\Delta F_{a\to c}/kT_{ts})$, which in turn varies with the thermal diffusivity of the substance according to (7.24). Quantitative theory of the conversion has not yet been worked out; nor is there any experimental evidence to verify the possibility.

§1.1.5. *Annealing of radiation-induced defects*

Irradiation of crystalline solids with energetic particles has been shown to produce a variety of lattice defects, notably (1) vacancies and self-interstitials initially in equal numbers, (2) excited electrons and ionized atoms, (3) foreign atoms induced by transmutations and (4) thermal and displacement spikes. Defects in each category have vastly different fates. Those in (2) and (4) last only for very short ($\ll 1$ s) periods, whereas the transmutation products in (3) are permanent and relatively stable, although the interstitial atoms of He, H, Xe, A, etc., tend to aggregate into gas bubbles. The fate of point defects in (1) is most unpredictable. Their lives could be short ($\ll 1$ s) because vacancies and self-interstitials could recombine upon meeting each other soon after their formation. Even if they escape recombination, the surviving defects, especially the interstitials, are incessantly on the move until they are clustered, trapped or annihilated during irradiation or in post-irradiation heat treatments.

By annealing of radiation-induced defects is meant the reduction of the concentrations of vacancies and self-interstitials. This annealing phenomenon may be of two types. One type is the inherent or radiation annealing, which occurs when a new interstitial (or vacancy) is created at a location, by chance, close (a few interatomic distances) to an old vacancy (or interstitial), resulting in their spontaneous recombination. This type of annealing involves neither defect migration nor thermal effect, but is promoted and becomes important at high fluences. The second type is the thermal annealing, in which recombinations of vacancies and interstitials come about because one species of defects (most probably the interstitials) migrate close to defects of the opposite species. By virtue of the different origins of the two types of annealing, it is safe to conclude that irradiation at cryogenic temperatures, especially at and below the boiling point (20.4°K) of liquid hydrogen, will only be accompanied by radiation annealing. As the irradiation temperature is increased, thermal annealing sets in besides radiation annealing. At and above room temperature, annealing of radiation-induced defects is due more to thermal annealing than to radiation annealing unless the fluence is excessive, under which radiation annealing may contribute equally importantly.

The driving force for the annealing is due chiefly to the strain energies associated with the defects, which exist in thermal-nonequilibrium conditions. The annealing behavior of point defects is extremely complex for three main reasons. First, vacancies and self-interstitials can change their configurations, thereby altering their energetic postures and mobility without being annihilated or trapped. For vacancies, the formation of di-, tri- and multi-vacancies has been observed in annealed specimens of metals such as Al, Cu and Au, which had been quenched or irradiated (see [Smallman [1970b]). The normal configuration of solitary self-interstitials appears as dumbbells either in the ⟨100⟩ directions in fcc metals or in the ⟨110⟩ directions in bcc metals. The possibility of a crowdion configuration was advocated by Seeger [1958], and di-interstitials were examined by Johnson and Brown [1962]. Secondly, annealing can be effected by more than just (1) the recombination method. Other processes are (2) the clustering of point defects of the same species, which eventually collapse into dislocation loops, (3) the trapping at impurity atoms or precipitates, (4) the annihilation at sinks such as dislocations, grain boundaries and free surfaces and (5) the formation of voids, which are essentially aggregates of vacancies. The effectiveness of each of the five processes depends on temperature and the states of purity and strain, which control the concentration and distribution of sinks and traps. Hence it is difficult to generalize as to which process(es) would dominate the annealing of defects in irradiated solid.

The third reason for the complex annealing behavior of irradiated solids is related to the mobility of point defects. Ignoring the detailed features of their diffusion mechanisms, the frequency with which a point defect of jth species (vacancies, self- and impurity-interstitials) jumps into an adjacent similar site is approximately given by

$$\nu_j \simeq \nu_{0,j} z_j \exp(-\Delta H_j^m/kT) \tag{7.25}$$

where ΔH_j^m is the activation enthalpy for defect motion, $\nu_{0,j}$ is the vibrational frequency and z_j is the number of nearest similar sites. Although the values of $\nu_{0,j}(\sim 10^{13}–10^{14}\ s^{-1})$ and z_j $(=4$ to $12)$ for vacancies and interstitials vary within the same order of magnitude, the value of the exponential function differs greatly between vacancies and interstitials because of a substantial difference in the ΔH_j^m value. This situation has been most extensively studied for copper, for which the best values of ΔH^f and ΔH^m deduced from numerous experimental studies and theoretical analyses are listed in table 7.2, Part I. These values reveal that the ratio $\Delta H^f/\Delta H^m$ is around unity (or ≈ 1.8 according to Schilling et al. [1976]) for vacancies, but ~80 for self-interstitials. More significantly, the ratio $\Delta H_v^m/\Delta H_i^m$ is ~ 20 (or 14). Translating the physical meaning of the latter ratio into the mobility of point defects, the jump frequencies of vacancies and self-interstitials are compared at four common irradiation temperatures in Cu in table 7.2, Part II. Self-interstitials are shown to be mobile even at 4.2°K, whereas vacancies exhibit little mobility until the temperature rises to 408°K (or ~250°K, according to R. R. Coltman), which corresponds to 30% (or 18%) of the absolute melting point of copper. Despite the scarcity of data for self-interstitials, it is generally believed that the differences in ΔH_v^m and ΔH_i^m seen in Cu prevail in most other metals. For instance, the ratio $\Delta H_v^m/\Delta H_i^m$ remains to be $\sim 10–20$. Hence, self-interstitials are considered much more mobile than vacancies, irrespective of the crystal structure. This explains why most metals of the fcc (except Au), bcc (except V) and hcp structures that have been irradiated with energetic electrons or neutrons at 4.2°K have been observed to exhibit three major annealing stages in the lower part (up to $0.3\ T_m$) of the temperature spectrum. These stages are fairly well defined as regards their temperature range and mechanism, as shown together with their salient features in table 7.3.

In view of the annealing scheme depicted in table 7.3, it is reasonable to conclude that irradiation at the ambient temperature of a research or thermal power reactor around 50–100°C will have most of the induced self-interstitials annealed out concurrently in all but few refractory metals. This means that the majority of self-interstitials are recombined, annihilated at sinks, trapped or clustered to become dislocation loops; few would survive and still be free in solitary forms. Meanwhile,

TABLE 7.2
Formation and motion data for monovacancies and self-interstitial atoms in copper

Part I – Activation enthalpy in eV (Thompson [1969b])

Monovacancies		Self-interstitial atoms	
H_v^f	H_v^m	H_i^f	H_i^m
$1.1-1.2(1.29\dagger)$	$1.0-1.08(0.71\dagger)$	3.9 ± 0.5	$0.05\pm0.05(0.12^{\Delta})$
$1:1\ (1.8:1\ddagger)$		$78:1\ (33:1^{\Delta\Delta})$	
	$20:1\ (14:1;\ 6:1^{\Delta\Delta})$		

Part II – Jump frequencies calculated from (7.25) with $\nu_0\approx10^{13}$

Temperature (°K)	4.2	20	77	300	408
Monovacancies	0	0	0	$1\times10^{-3}(143\ddagger)$	$54(2\times10^{3}\ddagger)$
Self-interstitial atoms	0	30 $(0^{\Delta\Delta})$	6.4×10^{10} $(1.7\times10^{6\Delta\Delta})$	1.7×10^{13} $(1.2\times10^{12\Delta\Delta})$	2.9×10^{13} $(4.0\times10^{12\Delta\Delta})$

*Equivalent to $0.3T_m$ of Cu.
†Values given by Schilling et al. [1976].
‡Values subsequently calculated for the ratios and ν_j from the data of Schilling et al.
$^{\Delta}$Value quoted by R.R. Coltman (private communication).
$^{\Delta\Delta}$Values subsequently calculated for the ratios and ν_j from the data given by Coltman.

the induced vacancies are hardly annealed for a lack of mobility during irradiation, except those lost by recombination due to the motion of self-interstitials. Examination by transmission electron microscopy of the as-irradiated specimens has thus revealed defect clusters, which vary in nature (vacancy or interstitial), geometry (the Burgers vector and the habit plane), size and distribution mode sensitively with the temperature and fluence of irradiation, the species of metal and the purity of specimen. The diffraction contrast of defect clusters has appeared in the form of black dots, which are unresolved clusters of point defects, simple loops and complex loops, each of which contains

TABLE 7.3

Three stages in the lower part of the temperature spectrum of the annealing behavior of irradiated metals and alloys (Schilling and Sonnenberg [1973])

Stage	Temp. range (°K)	Main mechanism(s)	Special features
I	Up to 30–50°K	1. Interstitial–vacancy recombination in close Frenkel pairs for substages I_A, I_B and I_C. 2. Further recombination of migrating interstitials with vacancies, or 3. Migrating interstitials combine to form clusters, or 4. Trapping of migrating interstitials by impurity atoms in substages I_D and I_E.	1. Many well-defined substages, I_A, I_B, I_C, I_D and I_E. 2. Stage I is absent in irradiated Au.
II	From 30–50°K to 200°K	1. Growth or rearrangement of interstitial clusters formed in Stage I for high-purity metals. 2. Detrapping of interstitials from impurity atoms for impure metals and dilute alloys.	1. Continuous recovery throughout the entire stage in many metals whereas several substages appear in others. 2. The characteristics of Stage II are extremely sensitive to the state of purity.
III	From 200°K up to 20% of the absolute melting temperature	1. Mainly migration of single vacancies to (A) sinks and surviving interstitials (Al, Pt, and Au) and (B) to form their own clusters (Cu, Ni and Ag) or to join the existing vacancy loops nucleated from the collapsed collision cascades (Nb and Mo). 2. Possibly migration of surviving interstitials of the dumbbell configuration when mechanism III-1 fails (Seeger [1976]).	1. Annealing is more or less complete at the end of this stage in Pt and Al; but remains incomplete in Cu, Ni and Ag. 2. Strong dependence of annealing characteristics upon irradiation dose.

many subclusters. (See Chang and Chen [1977].) Upon post-irradiation annealing, the interstitial clusters generally shrink and disappear, whereas the vacancy aggregates tend to grow until the annealing temperature reaches high values at which vacancies existing in thermal nonequilibrium are eventually eliminated. Irradiation at higher temperatures will have the same annealing effect on the induced point defects as the post-irradiation heat treatments except when the irradiation temperature exceeds $0.3T_m$. In the latter case, void formation occurs in metallic materials provided that the fast fluence is sufficiently high. This phenomenon comes about because dislocations display a slight preference for absorbing self-interstitials, thereby enabling vacancies to establish an appreciable supersaturation, leading to void formation. It is indeed unfortunate that the conditions favoring void formation are found in the core of the reactor because the current choice of cladding materials, 304- or 316-type stainless steels, of the fuel elements has been shown to swell as much as 10–18% caused by void formation at the fluence of 10^{22} neutrons ($E > 0.1$ MeV) per cm^2.

§1.2. *Radiation damage to soft magnetic materials*

Knowing the nature of radiation damage and the annealing behavior of radiation-induced point defects, we are now in a position to examine the effect of irradiation on soft magnetic materials. As might be expected, the most damaging process is atomic displacement. The induced point defects may directly impair the structure-sensitive magnetic properties or indirectly affect the intrinsic magnetic properties by causing some changes (disorder, short-range directional order or lattice-site switching) in the arrangement of atoms or ions. There also exists the special case in which the exposure of amorphous magnetic substances to fast neutrons may convert the lattice to crystalline in isolated regions as a result of local heating in the thermal spikes. In this connection, it is of interest to find out whether or not the amorphous films of Gd–Co alloys developed for the bubble-device applications are radiation resistant to avoid such a possible lattice conversion in a nuclear environment.

§1.2.1. *Radiation effects on the intrinsic magnetic properties*

Because the intrinsic magnetic properties are insensitive to the conditions of crystal- and grain-structures, irradiation will not affect such properties as M_s and θ_C in pure metals and dilute solid solutions. For ordered, concentrated alloys and for compounds, however, irradiation may cause some changes in the arrangement of atoms or ions, thus altering M_s, and other intrinsic magnetic properties. Three studies are quoted here to illustrate such effects:

1. Reduction of saturation magnetization of γ-Fe_2O_3 *and* Fe_3O_4 *by neutron irradiation.* Henry and Salkovitz [1959] exposed samples of γ-Fe_2O_3 (90.9% purity with 6.52% Fe_3O_4, 1.63% Al_2O_3, 0.65% H_2O and 0.53% SiO_2) and a mixture of 77.22% Fe_3O_4 and 21.76% γ-Fe_2O_3 in the Brookhaven Pile at the flux of 2.12×10^{12} thermal neutrons per $cm^2 \cdot s$ and 6.2×10^{12} fast neutrons per $cm^2 \cdot s$ for 555 h. The fluences were estimated at 5.4×10^{18} thermal and 1.2×10^{17} fast neutrons per cm^2. The irradiation resulted in a reduction of the saturation magnetization by 3% in γ-Fe_2O_3 and 15% in Fe_3O_4. The reducing effect on M_s of both compounds is attributed to a transfer of the Fe^{3+} ions from octahedral sites to tetrahedral sites (see fig. 2.14), but the much larger reduction of M_s in Fe_3O_4 was not explained satisfactorily.

2. Changes in the first anisotropy constant K_1 *and magnetostriction.* Hall et al. [1959] measured K_1, $\lambda_{s,[100]}$ and $\lambda_{s,[110]}$ in single crystals of a number of soft magnetic materials before and after a neutron irradiation conducted in the same Brookhaven reactor to a fluence of 1.7×10^{18} total neutrons per cm^2 at 50°C. Their results are summarized in figs. 7.9 and 7.10. The irradiation did not change either K_1 or the two λ_s in Fe, Ni, (1.5–6.4)% Si–Fe and (3 and 5.7%) Al–Fe alloys. However, it caused detectable changes (decrease or increase) in K_1 and λ_s in both [100] and [110] directions in the crystals of compounds or alloys that undergo the order–disorder transition. The latter materials include Fe_3O_4, (8.1–17%) Al–Fe, (40–80)% Ni–Fe, (29–44)% Co–Fe, 10% Co–Ni, 2% Mo–81% Ni–Fe, 5% Mo–79% Ni–Fe, 3% Mo–15% Al–Fe and 45% Ni–25% Co–Fe. The irradiation-induced changes are consistent with a decrease in atomic order for all alloys except those of the Al–Fe system. The behavior of the Al–Fe alloys seemed to imply the operation of an additional mechanism, which was not specified by Hall et al.

3. Evidence for short-range directional order induced by irradiation. A sensitive shape change of the hysteresis loops of several industrial products: Monimax, Deltamax and Allegheny 4750 of NiFe and Permalloy, Supermalloy and Mumetal of $Ni_{0.75}Fe_{0.25}$ was detected after a similar irradiation in the Brookhaven nuclear facility by Schindler et al. [1959] at 115°C. The shape changes, especially the appearance of the constricted loops (fig. 7.11b), are taken as evidence for the onset of short-range directional order of the Taniguchi–Néel type (ch. 5, §2.6.1).

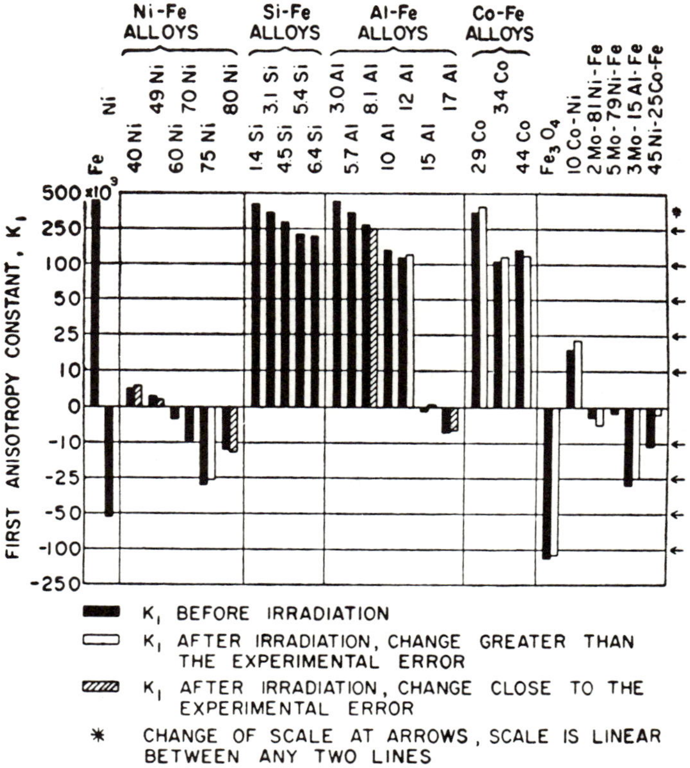

Fig. 7.9. Effect of neutron irradiation to a moderate fluence (1.7×10^{18} total neutrons/cm^2) on the first anisotropy constants of various soft magnetic metals and alloys (Hall et al. [1959]).

The change in the basic type of magnetism of Pt_3Fe from antiferromagnetism in the ordered state to superparamagnetism in the mixed state as a result of neutron irradiation described in §1.1.4 is another example of the effect of irradiation on the intrinsic magnetic properties of alloys involving atomic order. Chen et al. [1974] observed a similar effect of neutron irradiation on the mode of spin coupling in NiPt, which together with Pt_3Fe are two of the isomorphous series of Pt–M alloys, where M = Cr, Mn, Fe, Co and Ni and $c_{Pt}/c_M = 3:1$, $1:1$ and $1:3$.

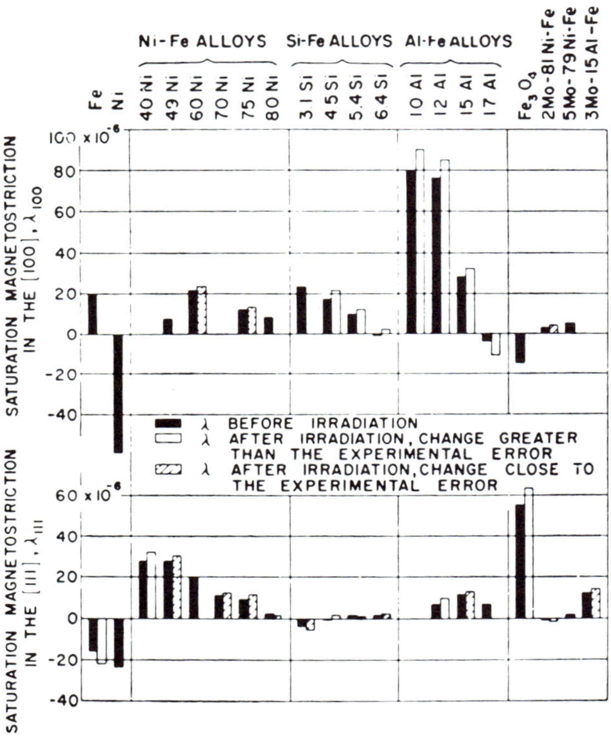

Fig. 7.10. The same neutron irradiation was seen to affect the saturation magnetostriction coefficients in the [100] and [111] directions in many soft ferromagnets (Hall et al. [1959]).

§1.2.2. *Deterioration of the structure-sensitive magnetic properties*

Many magnetic properties such as μ_i, μ_{max}, $H_{c,\infty}$, $B_{r,\infty}$ and the hysteresis loop are sensitive to the presence of defects in the lattice. Hence irradiation could adversely affect these properties. To inhibit the motion of domain walls, point defects must be clustered into aggregates comparable in size to the wall thickness (≈50 to 2000 Å). Point defects initially induced in the form of Frenkel pairs will not impair even the most structure-sensitive magnetic properties. For this reason, it is reasonable to state that radiation-induced point defects per se cause no damage, but their "daughter" products – dislocation loops, black dots and defect complexes involving interstitial impurity atoms – are detrimental to the structure-sensitive magnetic properties.

Workers at the US Naval Ordinance Laboratory (NOL) extensively investigated the effect of neutron irradiation on ferromagnets, both soft and hard (Gordon and Sery [1964]). The purpose of their study was to evaluate a wide variety of practical materials with respect to the Environmental Requirement Guide for Electronic Component Parts adopted by the Director of Defense Research and Engineering, US Department of Defense. The Guide defines a "hyperenvironment" as a combination of temperature at 500°C and a neutron fluence up to 1×10^{16} "fast" (actually epicadmium with $E>0.5$ eV) neutrons per cm^2 in reactor irradiation or up to 1×10^{13} neutrons per cm^2 in nuclear pulsed irradiation. Unfortunately, therefore, the NOL experiments were performed at relatively low levels of neutron fluxes ($\sim1\times10^{11}$ epicadmium neutrons per cm^2 per sec) and fluences ($\sim2\times10^{18}$ neutrons per cm^2) at 75–90°C. Despite these shortcomings, the experiments yielded interesting results, which are partly reproduced in figs. 7.11–7.15 and in tables 7.4 and 7.5. The results are summarized as follows:

(1) Two popular soft materials – Supermalloy and 4-79 Mo–Permalloy – were found to be the most sensitive materials to radiation damage by neutrons. At the fluence of $2\times10^{18}\,\mathrm{n\cdot cm^{-2}}$ ($E>0.5$ eV), both $\mu'_{H=20\,\mathrm{Oe}}$ and $\mu'_{\max}$ were reduced by 93% and $H_{c,\infty}$ increased more than eightfold (table 7.4). Figure 7.12 shows the progressive deterioration of the dc magnetization curve of Supermalloy. Starting from the fluence of $1.3\times10^{18}\,\mathrm{n\cdot cm^{-2}}$, however, the deterioration proceeds at a decelerating rate, implying that there is probably a saturation level of fluence beyond which no further damage to the magnetization curve will occur.

(2) While the nickel alloys display high susceptibilities to radiation damage, several important core materials of the iron-base alloys are "surprisingly" radiation-resistant. The latter materials include the 2V-Permendur, randomly oriented 3.5% Si–Fe and 3% Si–1% Al–Fe (table 7.4). Even for grain-oriented 3% Si–Fe, only $\mu'_{H=20\,\mathrm{Oe}}$ suffered an 18% drop at $2\times10^{18}\,\mathrm{n\cdot cm^{-2}}$; all other properties – $\mu'_{\max}$, $H_{c,\infty}$, B_r and $B_r/B_{\max}$ – are hardly affected. Note, too, the essentially invariant behavior of $\mu_{\max}$ and $H_{c,\infty}$ for a number of materials in figs. 7.13 and 7.14.

(3) As shown in table 7.5, nickel ferrite, 2–81 Mo–Permalloy dust and Sendust-flake cores exhibit significant increases in high frequency energy losses. Within experimental error, however, carbonyl iron-dust cores showed none of the increases. In contrast to the laminated and tape-wound cores, the former cores showed some degree of recovery toward their pre-irradiation properties after 10-month storage at room temperature. In a separate study, Alley [1959] also reported that various

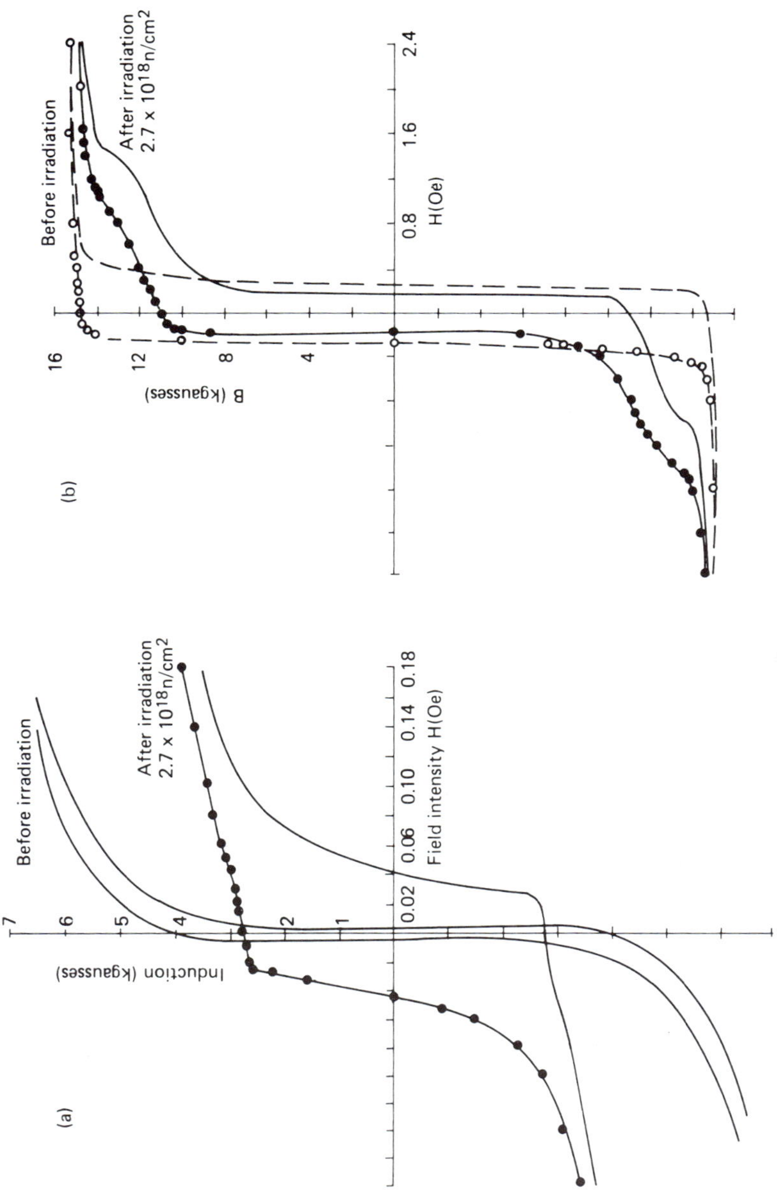

Fig. 7.11 Effect of neutron irradiation on the hysteresis loop. (a) The increase in coercivity and the decrease in retentivity cause a considerable shape change towards squareness; (b) the emergence of constricted loops. (Gordon [1960]).

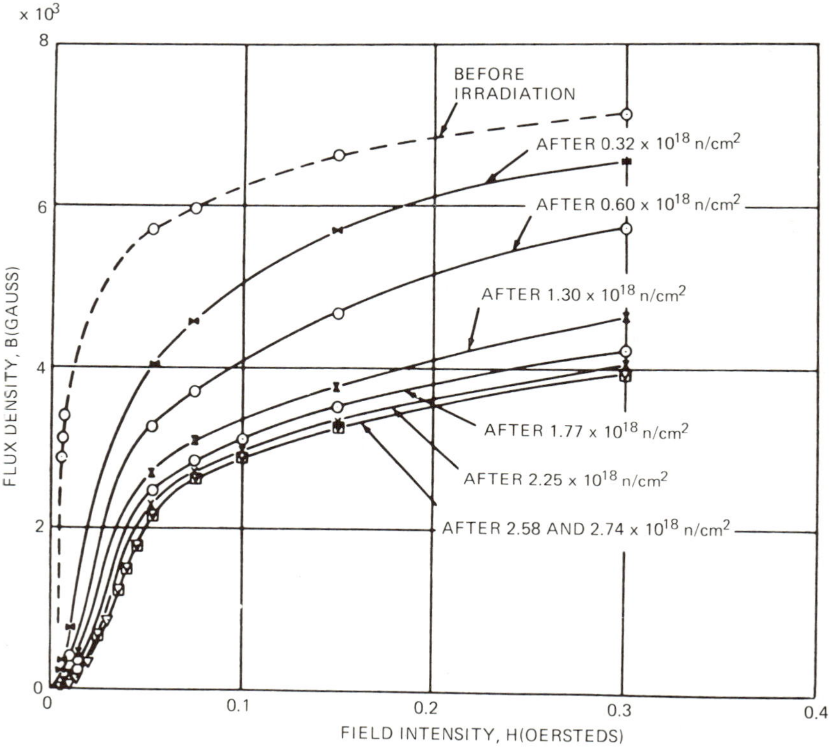

Fig. 7.12. Progressive deterioration of the magnetization curve during neutron irradiation (Gordon and Sery [1960]).

ferrites showed negligible permanent damage caused by reactor irradiation up to $\sim 10^{17}\ \mathrm{n \cdot cm^{-2}}$ ($E > 1$ MeV).

(4) Neutron irradiations were seen to seriously distort the hysteresis loops of Supermalloy, grain-oriented 50%Ni–Fe and other Ni-base alloys. The distortion either results in the appearance of constricted portions above the shoulders of the loop (fig. 7.11b) or is reflected by severe degradation of both remanence and coercive force (fig. 7.11a). This effect is partly attributed to changes in the degree of atomic order, which has been discussed in §1.2.1.

Several experimental studies were conducted to investigate the effect of pulsed neutron irradiation. The experiments involved the exposure

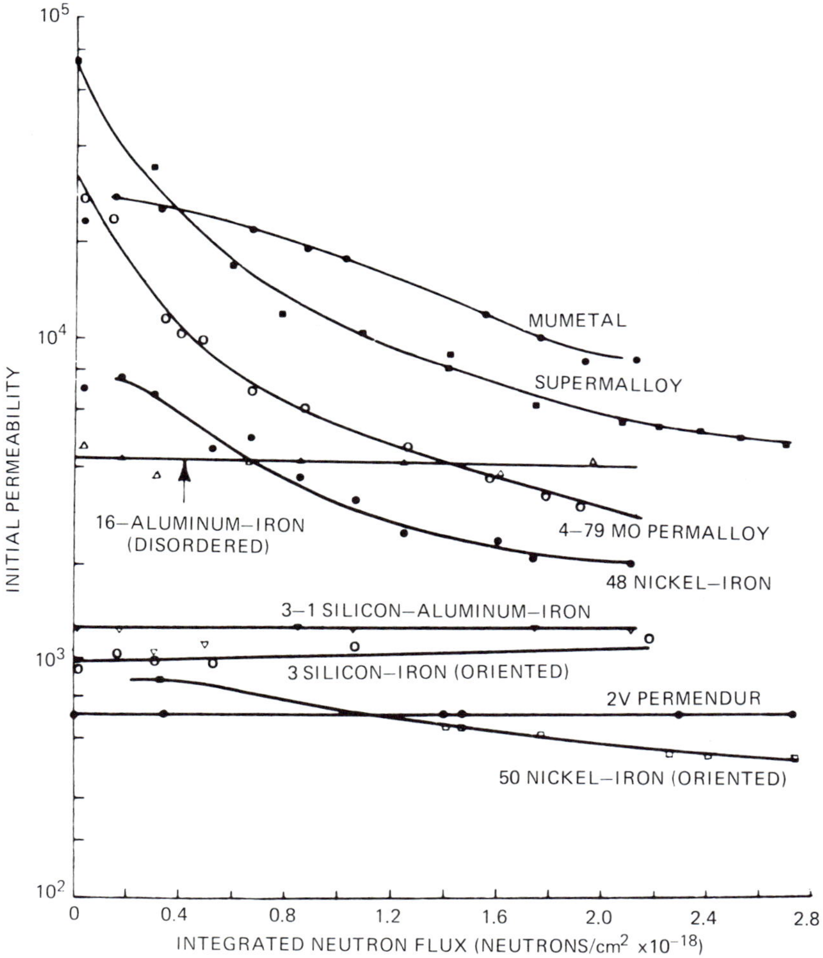

Fig. 7.13. The initial permeabilities of Permalloy and other nickel alloys are shown to be highly susceptible to radiation damage by fast neutrons, in sharp contrast to strong resistances of many iron-base alloys (Gordon [1960]).

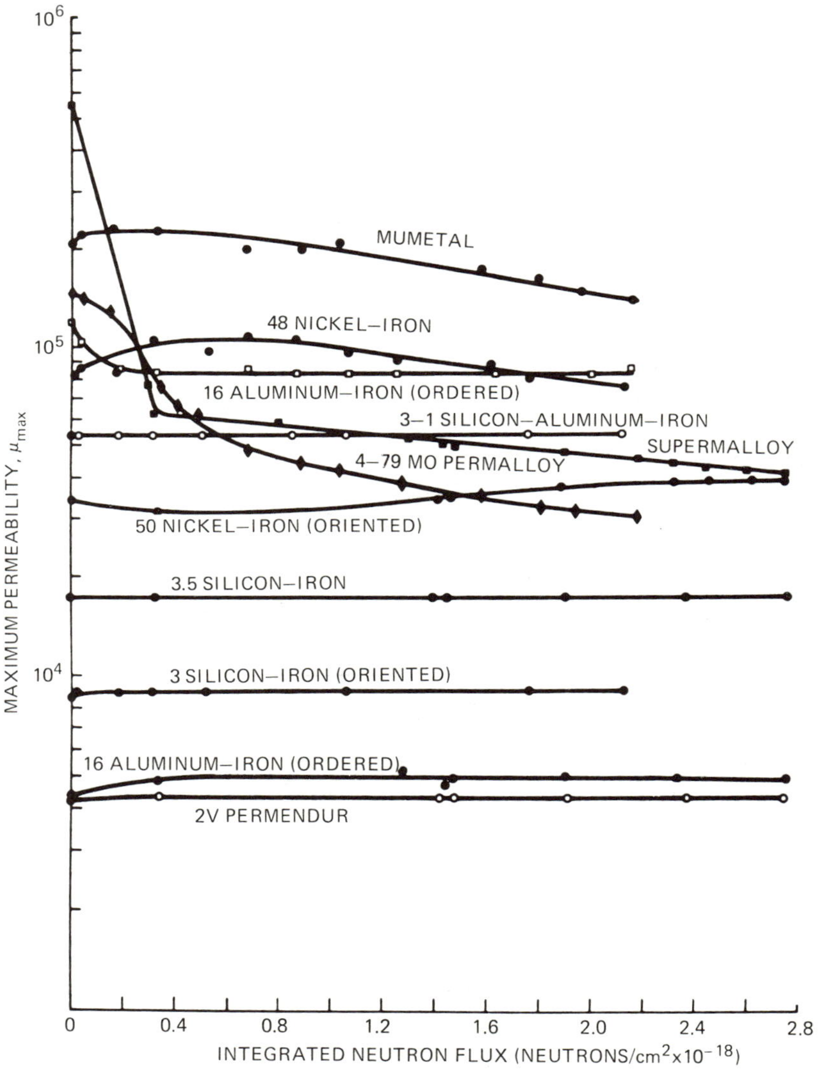

Fig. 7.14. Neutron irradiation data showing the severest damage to maximum permeability of Supermalloy (Gordon [1960]).

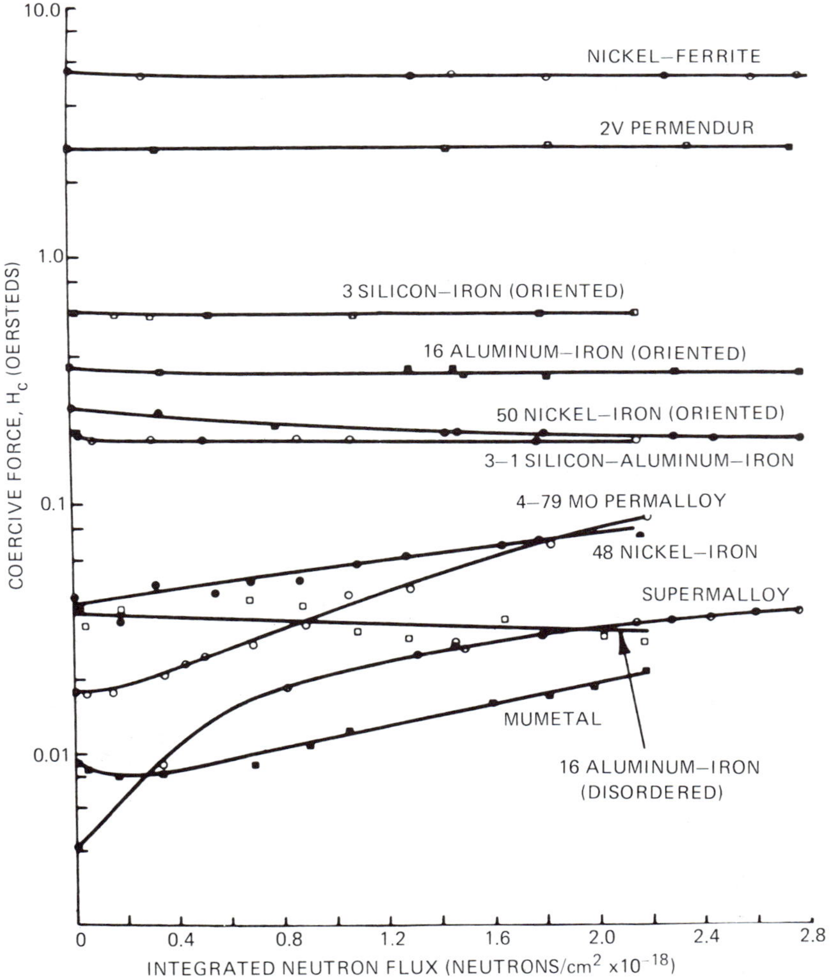

Fig. 7.15. Moderate increases in coercive force of some soft magnetic alloys caused by neutron irradiation (Gordon [1960]).

TABLE 7.4

Magnetic core materials: percentage change in dc magnetic properties as result of irradiation with $\approx 2 \times 10^{18}$ total neutrons/cm^2 (Gordon and Sery [1964])

Materials	$\frac{\Delta\mu'_{20}}{\mu'_{20}}$	$\frac{\Delta\mu'_{max}}{\mu'_{max}}$	$\frac{\Delta H_c}{H_c}$	$\frac{\Delta B_r}{B_r}$	$\frac{\Delta B_m}{B_m}$	$\frac{\Delta(B_r/B_m)}{B_r/B_m}$
Supermalloy	−93	−93	815	−38	−3	−36
4–79 Mo–Permalloy	−89	−79	403	−44	1	−44
Mumetal	−65	−38	158	−26	−3	−23
48 Nickel–iron	−70	−10	99	−26	−2	−26
50 Nickel–iron (oriented)	−31	15	−28	−24	−4	−21
3.5 Silicon–iron	8	−1	6	1	0	1
3 Silicon–iron (oriented)	18	1	−2	−3	−1	−1
3–1 Silicon–aluminum–iron	1	1	−2	−2	1	−1
16 Aluminum–iron (ordered)	34	15	−8	8	−5	13
16 Aluminum–iron (disordered)	−4	−4	−2	–	–	–
2V Permendur	3	2	−2	−1	−1	0

μ_{20} – permeability at induction of 20 G
μ_{max} – maximum permeability
H_c – coercive force
B_r – residual induction or remanence
B_m – induction at field intensity of 30 oersteds
B_r/B_m – loop rectangularity or squareness ratio.

TABLE 7.5

Powder and ferrite cores: effects of irradiation with $\approx 2 \times 10^{18}$ total neutrons/cm^2 (Gordon and Sery [1964])

Material		μ'	a	e	c	$R/\mu Lf$	
Nickel ferrite	Before	14	450	–	–	9 000	(50 Kc, 1 G)*
	After	14	400			13 000	
2–81 Mo–Permalloy	Before	123	1.5	20	10	500	(25 Kc, 20 G)
	After	122	1	40	10	1 000	
Carbonyl iron (plastic binder	Before	9	–	2	–	3 500	(500 Kc, 1.7 G)
	After	9		2.5		2 500	
Carbonyl iron (glass binder)	Before	9	–	2.5	–	3 000	(500 Kc, 1.4 G)
	After	9		1.5		3 500	
Sendust flake (flakenol)	Before	175	10	5	180	500	(25 Kc, 20 G)
	After	195	11	30	20	1 100	

*$2\pi/\mu Q = R/\mu Lf = aB + ef + c$,
where $R/\mu Lf$ = core loss factor (Legg),
c = residual loss coefficient, a = hysteresis loss coefficient and e = eddy current loss coefficient.

of magnetic materials to nuclear explosions at various distances, thus yielding varying doses of total (fast and thermal) neutrons up to $3 \times 10^{14}\,\mathrm{n \cdot cm^{-2}}$ (Linden and Long [1958]). The results, however, were inconclusive concerning the effect of nuclear explosions on magnetic materials.

From the results obtained in the studies conducted at NOL, Gordon [1960] concluded that neutron irradiations (up to $2 \times 10^{18}\,\mathrm{n \cdot cm^{-2}}$ with $E > 0.5\,\mathrm{eV}$) affect only the *softest* magnetic materials, whereas the harder ($H_0 > 0.5\,\mathrm{Oe}$) core materials (and the permanent magnets) showed no appreciable decays. As a further conclusion of the NOL studies, the radiation sensitivities of the softest magnetic materials are compared with those of other engineering materials in fig. 7.16. The comparison clearly indicates that even the softest magnetic materials

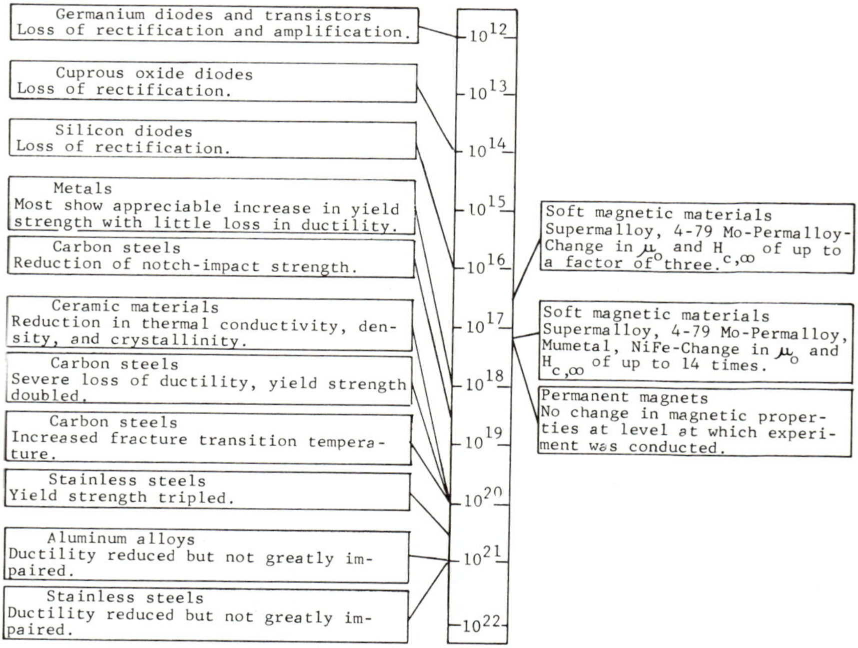

Fig. 7.16. Comparison of the lowest levels of neutron fluence at which various classes of engineering materials are affected. Note that only the softest magnetic alloys such as Supermalloy, 4-79 MO-Permalloy and Deltamax are susceptible to neutron damage at fluences less than $10^{16}\,\mathrm{n/cm^2}$ (Gordon and Sery [1964]).

fare reasonably well relative to the electronic properties of semiconductors.

§2. MAGNETIC BUBBLES AND DEVICES

It has been revealed in fig. 6.1 that magnetic bubble domains and the associated device technology represent the latest dynamic development in the long history of soft magnetic materials. Indeed since the paper by Bobeck [1967], who first discussed the great potential of magnetic bubbles, there has been a flurry of laboratory activities on this type of materials and their prototype devices that has seldom been witnessed in the entire field of magnetics and electronics. Bubble domains have generated this interest because they make possible drastically new solid-state devices for data storage with unparalleled packing density up to 10^{11} bits per cm^2. They also inspire considerable scientific interest because new theory and new experimental techniques are required for the analysis of complex domain walls and domain behavior.

To define magnetic bubbles, we must first refer to the cylindrical domains, which feature a closed wall in the shape of a cylinder. Magnetic bubbles are cylindrical domains with a circular cross section. A third kind of this special group of domains is the strip domains, which are the branches in a meandering quasi-parallel structure of domains commonly seen in certain ferromagnetic materials in the demagnetized state (figs. 7.17a and b). Strip and cylindrical domains were first observed on the polished surface of yttrium orthoferrite ($YFeO_3$) by the Faraday magneto-optic technique by Sherwood et al. [1959]. Their work was largely regarded as only scientifically interesting, however, concerning some "unusual" domain structure (see Craik and Tebble [1965b]). The practical value of bubble domains remained unrealized until a paper by Michaelis [1967], which inspired Bobeck to foresee the tremendous device potential of magnetic bubbles. In his paper, Michaelis described the ease with which isolated domains could be activated to propagate in thin films. Bobeck thus remarked that such outstanding dynamic properties of magnetic bubbles are ideal for solid-state devices such as memories and logic circuits.

The present discussion on bubble domains is divided into three parts. §2.1 deals with the basic properties of magnetic bubbles with emphasis on their static and dynamic behavior. §2.2 covers the classification and characterization of bubble materials. And §2.3 discusses the principles underlying the operation of bubble devices and the promiment features of some prototype devices.

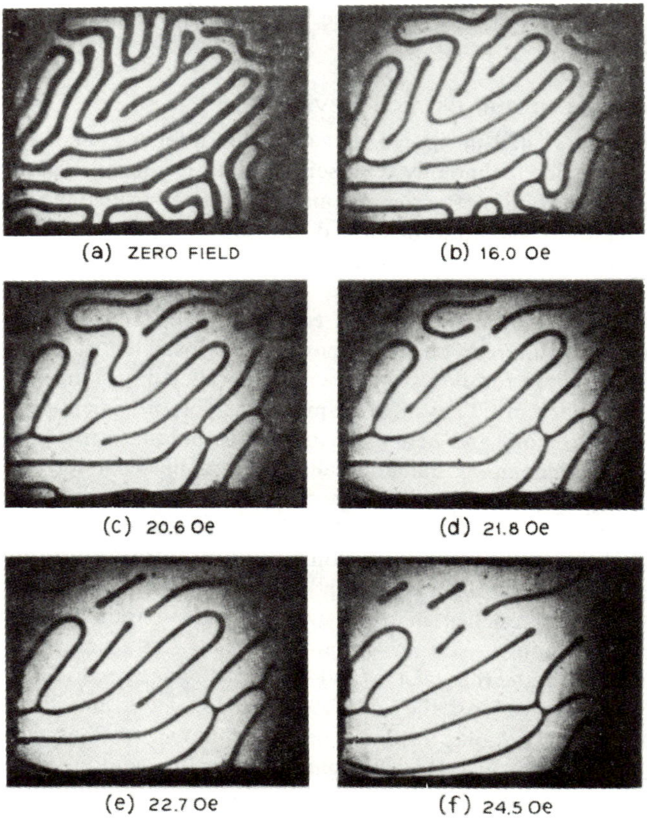

Fig. 7.17. Magneto-optic Faraday observations of domain structures on an "as-grown" $TmFeO_3$ platelet, 2.3 μm thick. Starting from (a) the demagnetized state, the crystal is progressively magnetized to (j) by an increasing bias field applied perpendicular to the platelet surface. After reaching 37 Oe, the field is slowly withdrawn, (k) to (l), and the crystal regains its domain structure of strips,

§2.1. *Properties of bubble domains*

The first requirement for a material to produce and support bubble domains is to have a net moment μ in an outstanding direction. By this is meant μ lies in a crystal direction for crystalline materials or in the direction normal to the thin film of an amorphous material. It is immaterial whether the substance is ferromagnetic, ferrimagnetic or

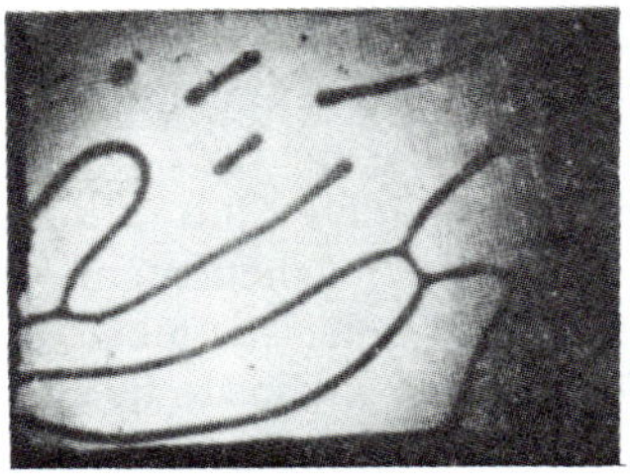

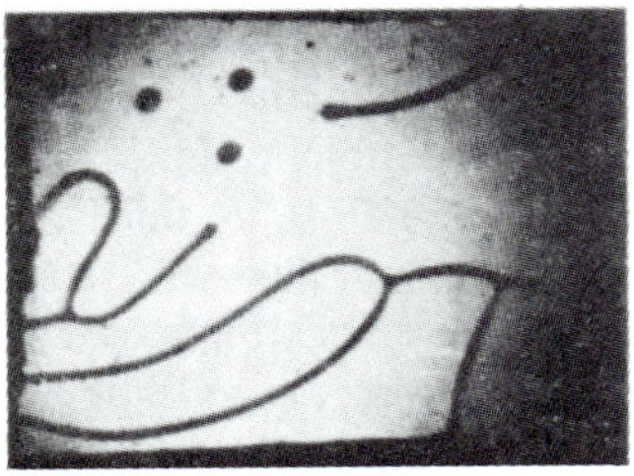

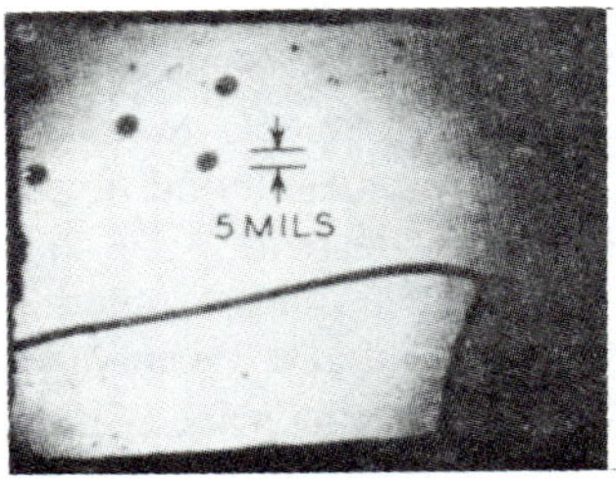

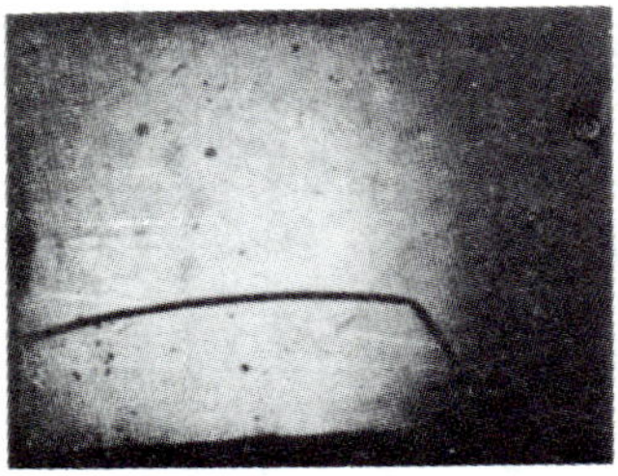

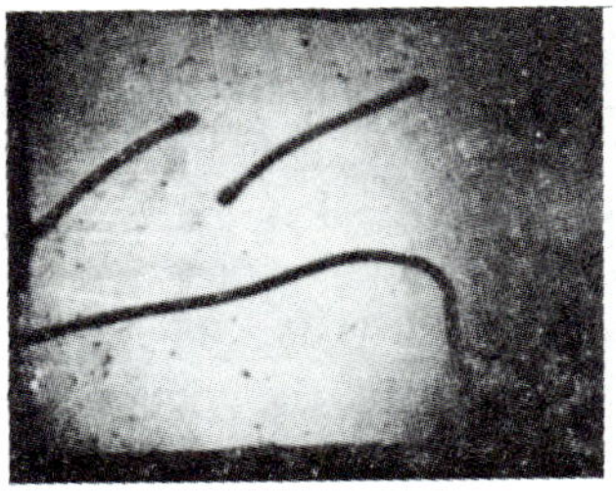

which are evenly distributed in opposite directions. Note in (g), (h) and (i) the conversion of strip domains to magnetic bubbles and in (j) the shrinkage and eventual disappearance of bubbles, at different stages of magnetization (Bobeck [1967]).

even antiferromagnetic*. Another requirement is that the material must exhibit uniaxial anisotropy such that the easy axis of magnetization is

*As will be shown in §2.2.2.1, the rare-earth orthoferrites display canted antiferromagnetic spin arrangement of Fe^{3+} ions. Many of these ferrites are capable of producing typical bubble domains.

normal to the crystal platelet or to the surface of the thin film. It does not matter whether the uniaxial anisotropy exists inherently or is induced under certain conditions not necessarily involving a magnetic field (Richards [1976]). A third requirement involves the ratio of the uniaxial anisotropy field H_k to the saturation magnetization M_s. In terms of bubble-device technology, this ratio is called the quality factor $Q \equiv H_k/M_s$ ($H_k/4\pi M_s$ in CGS units). For bubble domains to be stable, Q must be greater than unity. Cobalt is a good example of the usefulness of this factor. Being ferromagnetic with magnetocrystalline anisotropy along the c-axis, cobalt satisfies the first two requirements. But its Q value is $\sim\frac{1}{10}$, thus Co is incapable of supporting stable bubble domains. Various ferromagnets are compared with respect to the third requirement in fig. 7.18. The solid line represents the condition $Q = 1$. Materials whose Q values are less than unity lie below the solid line and are thus ruled out as bubble materials. These include such prominent metallic ferromagnets as Fe, Co, Ni and Permalloy. Ferromagnets situated above the solid line are, in principle, bubble materials, because their Q values are greater than unity. They include MnBi, hexagonal ferrites, garnets, orthorferrites and amorphous Gd–Co alloys. Other aspects of the figure will be discussed later.

When all three requirements are met, the domain structure of a thin specimen in the demagnetized state will be composed of strip domains (fig. 7.17a) with a rare appearance of cylindrical domains. Whether magnetization of the specimen would convert some of the strip domains into cylindrical or bubble domains (figs. 7.17b to j) depends on how the magnetic field $\boldsymbol{H}$ is applied. This implies that the generation and stability of bubble domains correspond to certain favorable energy conditions under the influence of $\boldsymbol{H}$. Therefore the energy considerations are of prime importance and should be discussed first.

§2.1.1. *Strip domains in an applied field*

Consider a magnetic platelet of infinite dimensions in the x–y plane and thickness h (fig. 7.19). In the presence of a magnetic field $\boldsymbol{H}$ applied in the z-direction, a strip of "reverse" magnetization appears with width W between two strips magnetized in the opposite direction to $\boldsymbol{H}$. The value of W depends on the magnitude of $\boldsymbol{H}$, but is independent of the wall energy because the widening or narrowing of the strip by wall movement will not change the wall area or the specific wall energy, provided the walls are rigid. The rigidity of the wall is then assumed in

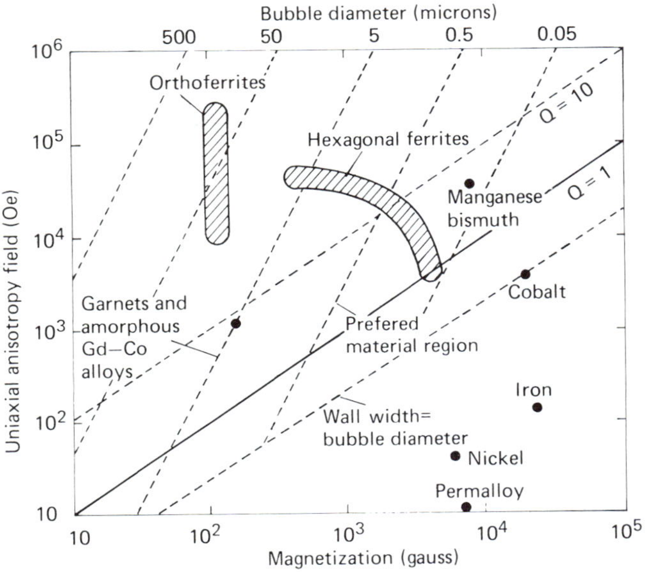

Fig. 7.18. Characterization of bubble materials in terms of the quality factor $Q(=H_k/M_s)$ and the bubble diameter. The diamond-shaped region between $Q = 1$ and 10 and between bubble diameters of 0.5 and 5 μm represents optimum values for Q and the bubble diameter in applications (Bobeck et al. [1975]).

the present analysis. By applying Ampere's rule [$\nabla \times \boldsymbol{H} = \boldsymbol{J} + \partial \boldsymbol{D}/\partial t$, where $\boldsymbol{J}$ is the electric current density and $\boldsymbol{D}$ is the electric displacement (see Menzel [1960])], Kooy and Enz [1960] and Bobeck [1967] arrived at the relation

$$\frac{\bar{H}_z}{M_s} = \frac{2}{\pi}\left[\tan^{-1}\left(\frac{h}{W}\right) - \frac{W}{2h}\ln\left(1 + \frac{h^2}{W^2}\right)\right], \tag{7.26}$$

where $\bar{H}_z$ is the average z-component of the magnetostatic field acting on the strip domain walls by neighboring magnetic poles. The quantity on the left-hand side may be considered as a normalized field with respect to M_s, and it becomes $\bar{H}_z/4\pi M_s$ in CGS units. Equation (7.26) has simplified forms in two extreme cases, as shown below:

$$\bar{H}_z/M_s = 1 \qquad \text{for } W \ll h \text{ or } W \to 0 \tag{7.26a}$$

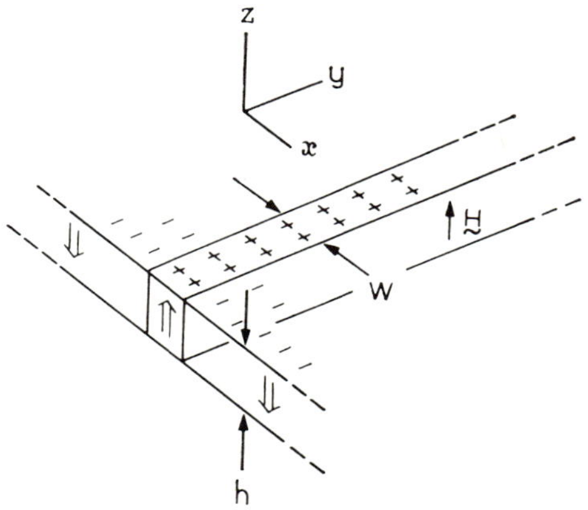

Fig. 7.19. A straight strip domain with rigid walls is assumed to appear between regions magnetized in the opposite direction for a simplified analysis (Bobeck [1967]).

and

$$\bar{H}_z/M_s = 1/\pi(W/h), \qquad \text{for } W \gg h. \tag{7.26b}$$

Consequently, (7.26) may be approximated to the form

$$\bar{H}_z/M_s = 1/[1 + \pi(W/h)]. \tag{7.27}$$

Figure 7.20 plots the normalized field against W/h with the normalized applied field $\boldsymbol{H}/M_s$ superposed. The strip is in stable equilibrium at $W/h \approx 0.95$ when the $\bar{H}_z/M_s$ curve intersects at A with the horizontal line representing the applied field. If the strip is narrower than this, $\bar{H}_z$ exceeds H and the strip tends to expand towards A. If the strip is too wide, H predominates and the strip tends to shrink towards A.

§2.1.2. *Magnetic energy of bubble domains*

The geometric features of bubbles create a basic difference in the length of the domain walls between bubble and strip domains. For the strip domain shown in fig. 7.19, the length of the two rigid walls is constant irrespective of the width. For bubble domains, the wall length is along the circumference and thus equal to $2\pi r$, where r is the bubble radius. As the bubble grows, so will the wall length of the bubble. Consequently, a wall-energy term must be added to the other two terms in (7.26). Figure 7.21 shows a bubble with magnetization opposite to the rest of the plate.

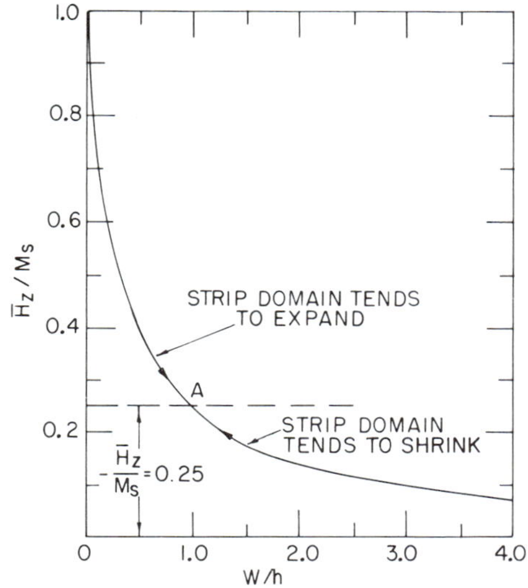

Fig. 7.20. The stability of the strip domain is secured at the intersection of the curve plotted from solutions to (7.26) and the horizontal line representing the applied field.

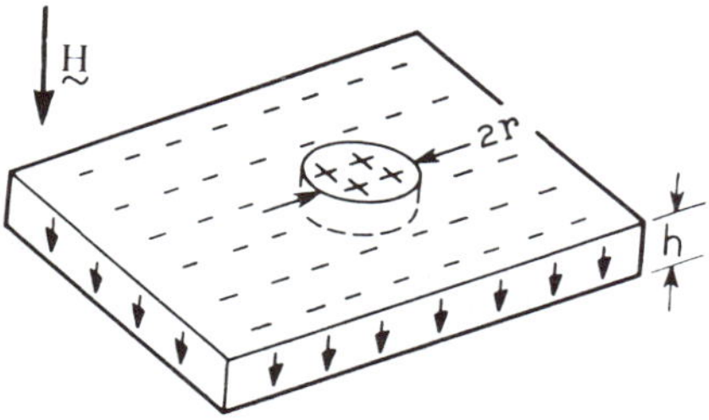

Fig. 7.21. A cylindrical domain is introduced with diameter $2r$ in a plate of thickness h by a bias field $\boldsymbol{H}$ (Bobeck [1967]).

The bubble has a total magnetic energy E_{total} relative to other portions of the plate, and E_{total} can be resolved into three components:

$$E_{\text{total}} = E_\gamma + E_H + E_s. \tag{7.28}$$

The wall energy E_γ is given by

$$E_\gamma = 2\pi r h \gamma_w, \tag{7.29}$$

where γ_w is the specific surface energy of the bubble-domain wall in $J \cdot m^{-2}$. The energy associated with the applied field is given by

$$E_H = -\mu_0 \int_V \boldsymbol{M}_s \cdot \boldsymbol{H} \, dV = \mu_0 M_s H \pi r^2 h. \tag{7.30}$$

The general expression for the magnetostatic energy has been derived in appendix A3. From (A3.14), we have

$$E_s = -\tfrac{1}{2}\mu_0 \int \boldsymbol{M}_s \cdot \boldsymbol{H}_d \, dV, \tag{7.31}$$

where $\boldsymbol{H}_d$ is the demagnetizing field. The actual calculation of E_s is complex and difficult to be rigorous. Several attempts were made, of which the expression obtained for H_d by Callen and Joseph [1971] is quoted here:

$$H_d = M_s \bigg/ \left[1 + \frac{\pi}{K}\left(\frac{2r}{h}\right)\right], \tag{7.32}$$

where $K = 1 + \ln[2\pi(2r/h)]$. For the likely range of bubbles at $0 < 2r/h < 10$, Callen and Joseph chose $\pi/K = 0.75$, which was later reduced to 0.726 by O'Dell [1974a] to attain a better agreement with the exact solution. Note the similar form between (7.32) and (7.27). Combining (7.31) and (7.32) yields

$$E_s = -\frac{1}{2}\left[\frac{\mu_0 M_s^2}{1 + (\pi/K)(2r/h)}\right] \pi r^2 h. \tag{7.33}$$

§2.1.3. *The stability of magnetic bubbles*

A bubble is in an equilibrium state under a magnetic field when its total magnetic energy is at a mimimum, or equivalently $\partial E_{total}/\partial r = 0$. Furthermore, to ensure static stability, the condition $\partial^2 E_{total}/\partial r^2 > 0$ must be imposed. Partial-differentiating (7.28) with respect to r and rearranging gives

$$\frac{\partial E_{total}/\partial r}{4\pi r h \mu_0 M_s} \simeq \frac{\gamma_w}{2r\mu_0 M_s} + \tfrac{1}{2}H - \tfrac{1}{4}H_d. \tag{7.34}$$

This equation is only approximate because the dependence of H_d on r has been ignored. But it serves a useful purpose in that two of the three energy terms appear naturally as magnetic fields. Thus the first term on the right-hand side may also be treated as an effective wall field and the term on the left-hand side as the net effective field acting on the domain wall. Before discussing the effect of the three component fields, we recall from the relation $F_i = -\partial E_i/\partial r$, which equates the negative of the first-order partial derivative of energy with respect to r to the force acting on the domain by the ith field. Injecting this interpretation into (7.34), we can state that the force exerted by the applied field on the domain wall tends to contract the volume of the bubble. In other words, E_H is an inward radial force. The wall field also tends to reduce the wall area to zero. The magnetostatic field H_d, on the other hand, tends to equalize the overall magnetic surface charge by an increase in bubble volume. In this context, we rewrite (7.34) to obtain

$$\frac{F_{\text{total}}}{4\pi rh\mu_0 M_s} \simeq -H_\gamma - \tfrac{1}{2}H + \tfrac{1}{4}H_d, \tag{7.35}$$

where $H_\gamma = \gamma_w/2r\mu_0 M_s$ is the wall field. Figures 7.22(a)–(c) show the variation in the three component fields with r, (d)–(f), their corresponding physical situations and (g) the combination of H_γ, H and H_d into the net effective field in (g). The dashed curve representing H_{total} in (g) crosses the abscissa twice, indicating that there are two r values at which $\partial E_{\text{total}}/\partial r = 0$. However, the lower value at (a) represents unstable equilibrium because $\partial^2 E_{\text{total}}/\partial r^2 < 0$; hence only the second r value at (b) is in stable equilibrium with $\partial^2 E_{\text{total}}/\partial r^2 > 0$. We conclude from the shape of the dashed curve that (1) the bubble in fig. 7.21 is in stable equilibrium when $r = r_b$, (2) it tends to expand when $r_a < r < r_b$ and (3) it tends to shrink when $r < r_a$ or $r > r_b$.

The foregoing discussion deals with the stability conditions of bubbles for a given plate thickness. In the design of a bubble device, it is desirable to keep the bubble radius as small as possible. The conclusions we have just drawn imply that the desired minimization of r must be achieved by adjusting the plate thickness. Indeed, the theory of static stability of cylindrical domains developed by Thiele [1969, 1970] has shown just that. Before discussing Thiele's theory, we shall refer to an important analysis of the strip domains by Kooy and Enz [1960], whose work led to the introduction of another material parameter known as the characteristic length $\lambda \equiv \gamma_w/\mu_0 M_s^2$ ($\lambda \equiv \gamma_w/4\pi M_s^2$ in CGS units). Consider a thin plate of thickness h, which contains strip domains as shown in fig. 7.23. For a simple analysis, the domain structure is assumed to have these idealized features under an applied field $\boldsymbol{H}$ in the z direction: (1) parallel strip domains, (2) straight walls and (3) division of the plate into regions of equal width W, each of which consists of two strip domains with "+" and "−" magnetizations of width W_+ and W_-, respectively. (Obviously, $W_+ = W_-$ for $H = 0$

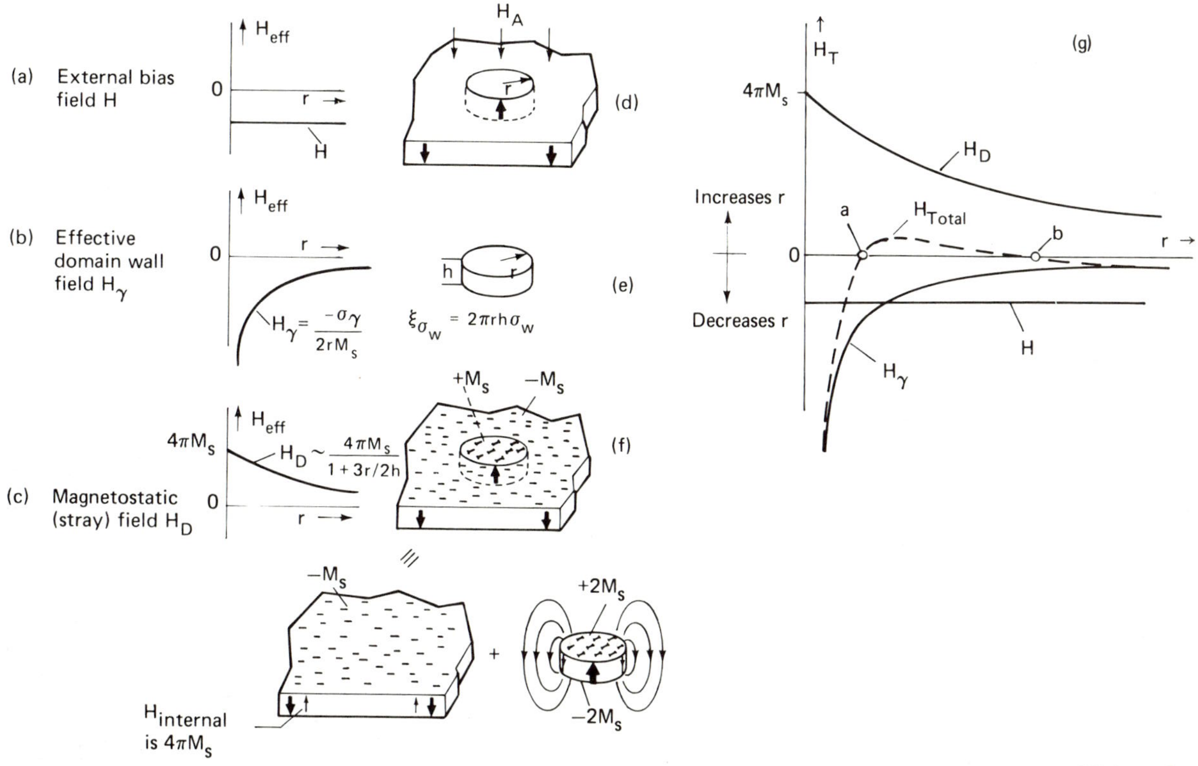

Fig. 7.22. The bias field (a) and the effective domain-wall (b) and magnetostatic (c) fields are combined to establish static stability (g) of a bubble domain. (d), (e) and (f) show the physical situations of (a), (b) and (c), respectively (Bobeck et al. [1975]). All drawings are based on CGS units.

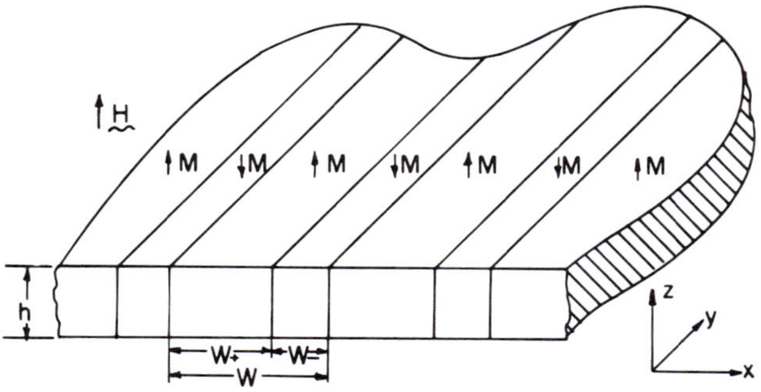

Fig. 7.23. A simplified model of strip-domain structure (Bobeck and Torre [1975]).

and $W_+ > W_-$ for nonzero fields.) In this "ideal" structure, there are on the average $2/W$ walls per unit length. Hence the wall energy per unit square is

$$E_\gamma = 2hr/W. \tag{7.36}$$

The energy associated with the applied field in a unit square is

$$E_H = -\mu_0(\Delta W/W)HM_sh = -\mu_0 HMh, \tag{7.37}$$

where $\Delta W = W_+ - W_-$ and

$$M = (\Delta W/W)M_s. \tag{7.38}$$

The magnetostatic energy E_s can be calculated from (7.31) (see Bobeck and Torre [1975a]), which leads to

$$E_s = \tfrac{1}{2}\mu_0 M^2 h + \frac{2\mu_0 M_s^2 W}{\pi^3}\sum_{n=1}^{\infty}\frac{1}{n^3} \times \sin^2\left[\tfrac{1}{2}n\pi\left(1+\frac{M}{M_s}\right)\right]\left[1-\exp\left(-\frac{2\pi nh}{W}\right)\right]. \tag{7.39}$$

Combining (7.36), (7.37) and (7.39) into (7.28) gives the total magnetic energy of the plate per unit square. It is readily seen that E_{total} now varies with M and W. Therefore, to establish stable equilibrium, we must obtain the minimum total energy by differentiating E_{total} with respect to both M and W; that is,

$$\frac{\partial E_{\text{total}}}{\partial M} = 0 \quad \text{and} \quad \frac{\partial E_{\text{total}}}{\partial W} = 0. \tag{7.40}$$

These two conditions give

$$\frac{M-H}{M_s}+\frac{W}{h\pi^2}\sum_{n=1}^{\infty}\frac{1}{n^2}\sin\left[n\pi\left(1+\frac{M}{M_s}\right)\right]\left[1-\exp\left(-\frac{2\pi nh}{W}\right)\right]=0, \tag{7.41a}$$

and

$$\frac{\gamma_w}{\mu_0 M_s^2 h}+\frac{W^2}{\pi^3 h^2}\sum_{n=1}^{\infty}\frac{1}{n^3}\sin^2\left[\tfrac{1}{2}n\pi\left(1+\frac{M}{M_s}\right)\right]$$
$$\times\left[\left(1+\frac{2\pi nh}{W}\right)\exp\left(-\frac{2\pi nh}{W}\right)\right]=0. \tag{7.41b}$$

From (7.41b), the characteristic length is formally introduced from the first term

$$\lambda \equiv \gamma_w/\mu_0 M_s^2 (\equiv \gamma_w/4\pi M_s^2 \text{ in CGS units}), \tag{7.42}$$

as the second bubble-material parameter. This parameter is so called because it has the dimension of length. Actually, it is not a real length, nor can it measured directly. The desired values for λ in connection with bubble devices will be discussed in §2.2.1. Equations (7.41a) and (7.41b) can be solved simultaneously for M and W by numerical methods. Duijvestijn and Boonstra [1960] conducted a numerical analysis for $BaFe_{12}O_{19}$ ($\gamma = 2.8\times10^{-3}\ \mathrm{J\cdot m^{-2}}$ and $M_s = 345\,000\ \mathrm{A\cdot m^{-1}}$) and confirmed the condition $W_+ = W_-$ in a zero field. They also found that λ has a very small value, $\approx h/80$ and, as H/M increases above 0.4, W_+ starts to increase rapidly until it approaches infinity for a value of H/M at ≈ 0.77. Their findings were in good agreement with experimental results obtained from a single crystal plate of $BaFe_{12}O_{19}$, 0.3 μm thick, by Kooy and Enz [1960] provided that a demagnetizing factor of 0.95 was used to account for the finite size of the plate, instead of the value of 1.0 for an infinite plate. The theory developed for the role played by the magnetostatic field in the stabilization of cylindrical domains is quite complex. Readers who are interested in the detailed derivation are referred to the original papers by Thiele [1969, 1970]. Here we shall discuss some of the important results of his theory. The first condition for stable equilibrium, viz., $\partial E_{\text{total}}/\partial r = 0$, leads to

$$(\lambda/h) + a(H/M_s) - F(a) = 0, \tag{7.43}$$

where the three terms represent the wall field, the applied field and the magnetostatic field with

$$a = 2r/h, \tag{7.44}$$

called the aspect ratio and

$$F(a) = \frac{2}{\pi} a^2 \left[\frac{(1+a^2)^{\frac{1}{2}}}{a} G\left(\frac{a^2}{1+a^2}\right) - 1 \right]. \tag{7.45}$$

In the last equation,

$$G\left(\frac{a^2}{1+a^2}\right) = \int_0^{\frac{1}{2}\pi} \left(1 - \frac{a^2 \sin^2 \alpha}{1+a^2}\right)^{\frac{1}{2}} \mathrm{d}\alpha, \tag{7.46}$$

which is the complete elliptic integral of the second kind. The second condition for stable equilibrium, viz., $\partial^2 E_{\text{total}}/\partial r^2 > 0$, leads to

$$\frac{H}{M_s} \frac{\mathrm{d}F}{\mathrm{d}a} > 0. \tag{7.47}$$

From (7.43), we obtain a necessary, but not sufficient, condition for the bubble to have a certain aspect ratio under the applied field:

$$H/M_s = [F(a) - \lambda/h]/a. \tag{7.48}$$

Figure 7.24 plots H/M_s against h/λ, which is the normalized plate thickness with respect to the characteristic length. A band emerges in this plot to define the region of stable bubbles. Below the band, the bubble tends to run out into strip domains; above the band, the bubble will

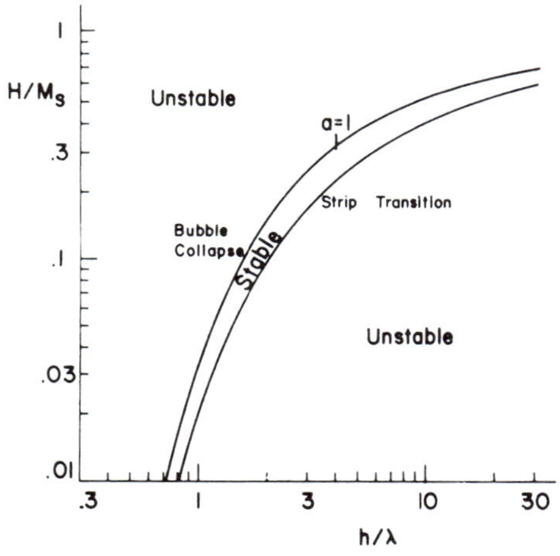

Fig. 7.24. Conditions for bubble stability in the $Q(=H/M_s)$ versus h/λ plot (Bobeck and Torre [1975]).

collapse. Another graphical presentation of the stability condition is shown in fig. 7.25, which plots the bubble diameter against the plate thickness with both quantities normalized to the characteristic length. Again, a stable band exists between two unstable regions similar to those seen in fig. 7.24. Of practical significance is the revelation in fig. 7.25 that the size of the stable bubble reaches a minimum value of $2r = 3.9\lambda$ when the plate thickness is $h = 3.3\lambda$.

The Thiele theory also deals with the variation of bubble stability with the normalized plate thickness when the bubble is deformed. As far as the wall energy is concerned, the round shape of a bubble represents the minimum value, thus reflecting the stable state. When the bubble is deformed, say to an elliptic cross section, there will be an increase in energy ΔE proportional to $(\Delta r)^2$, where Δr is the change in domain size. Despite the energy increase, the domain will remain deformed by virtue of the coercivity $H_{c,\infty}$ of the material until ΔE exceeds the work required to overcome the coercivity. The maximum fractional change in bubble radius is given by

$$\frac{\Delta r}{r} = \frac{a}{S_0(a) - (\lambda/h)} \frac{H_{c,\infty}}{M_s}, \tag{7.49}$$

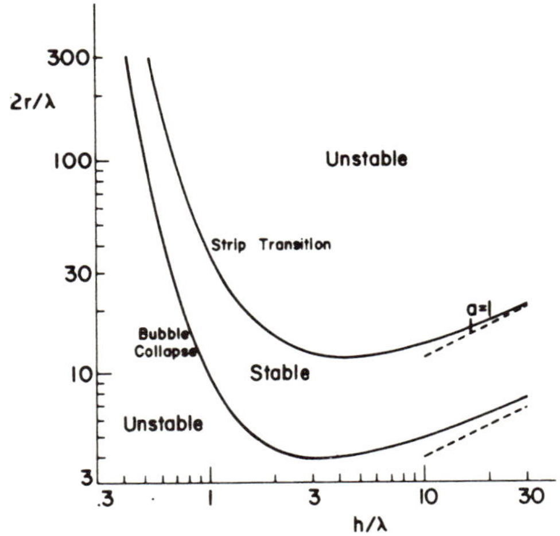

Fig. 7.25. Conditions for bubble stability in the bubble diameter versus plate thickness plot, both normalized to the characteristic length (Bobeck and Torre [1975]).

where

$$S_0(a) = F(a) - a(\mathrm{d}F/\mathrm{d}a). \tag{7.50}$$

This result is shown by plotting $(\Delta r/r)/(H_{c,\infty}/M_s)$ against h/λ in fig. 7.26. A stable bubble band is seen to lie between a region of elliptical deformation below the band and a region of radial deformation above the band. The bulging of the stability band in the middle signifies that, for plate thicknesses that minimize the bubble size, the bubbles are also most resistant to deformation. Specifically, Thiele refers to $h = 4\lambda$ as the optimum plate thickness. At this thickness, the value of the bubble aspect ratio varies from approximately 1 near bubble collapse to 3 near the bubble-to-strip transition. The optimum values for the aspect ratio and the bubble diameter are thus chosen to be 2λ and 8λ, respectively.

§2.1.4. *Bubble dynamics*

The dynamic behavior of magnetic bubbles is of great importance for the obvious reason that the propagation of bubbles is an essential operation of all bubble devices. Hence the success and advance of bubble devices depend to a large extent on how well we understand

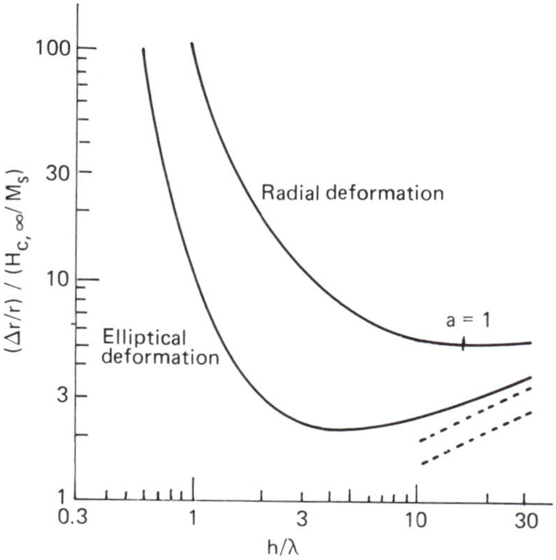

Fig. 7.26. Variation of deformation stability with normalized plate thickness (Bobeck and Torre [1975]).

bubble dynamics. Another reason for the study of bubble dynamics is that the motion of bubbles has some interesting features absent in conventional domain structures. These features will be dealt with in §2.1.4.2. It will be shown in §2.1.4.3 how the study of bubble dynamics has led to the discovery of hard bubbles and a phenomenon known as dynamic conversion.

§2.1.4.1. *Wall mobility.* It was shown in ch. 4, §2.2.2 that the motion of a planar domain wall at velocity v induced by an applied field H is described by

$$V = \mathrm{d}x/\mathrm{d}t = \xi(H - H_t), \tag{4.93}$$

where $\xi = 2M_s\mu_0/\beta$ is the wall mobility with β being the damping coefficient of the medium. The threshold field H_t required to activate the wall has the value close to the coercivity of the matrix. Hagedorn [1971] obtained an expression for the mobility from the magnetization variation inside the wall by first simplifying (4.93) to

$$v = \xi H, \tag{7.51}$$

under the assumption that the matrix exhibits zero coercivity and, therefore, zero threshold field. The power dissipated per unit area during the planar wall motion is given by

$$P = 2\mu_0 M_s H v. \tag{7.52}$$

Combining (7.51) and (7.52) yields

$$\xi = 2\mu_0 M_s v^2/P. \tag{7.53}$$

To calculate the power dissipation as a function of magnetization variation from the rear face to the front face of the wall, Hagedorn invoked the Gilbert equation expressed in (4.128) and attained from (7.53) the following expressions of mobility

$$\xi = \frac{4\mu_0\gamma A_{ex}}{\beta\gamma_w} = \frac{\mu_0\gamma d_w}{\beta\pi} = \frac{\mu_0\gamma}{2\beta}\left(\frac{\lambda}{Q}\right), \tag{7.54}$$

where γ is the gyromagnetic ratio and d_w is the wall thickness. The last two expressions are obtained from the first one by using two relations previously described in ch. 3:

$$d_w = \pi(A_{ex}/K_u)^{\frac{1}{2}}, \tag{3.35}$$

and

$$\gamma_w = 4(A_{ex}K_u)^{\frac{1}{2}}; \tag{3.36}$$

and two other relations,

$$\lambda = (8\, A_{ex}Q/\mu_0 M_s^2)^{\frac{1}{2}}, \tag{7.55}$$

and

$$\lambda/d_w = (2/\pi)Q. \tag{7.56}$$

Note from (7.54) that high mobility of the domain wall necessitates a thick wall or a high value of the ratio λ/Q. In other words, to attain high mobility for high data rates in memory devices, the characteristic length of the material must be kept large and the quality factor kept small.

§2.1.4.2. *Three modes of bubble motion.* For a conventional domain, the displacement of the domain wall leads only to either expansion or contraction of the domain. For a bubble, however, wall displacements can result in three modes of domain motion; namely, (1) expansion or contraction, (2) translation and (3) rotation. To see how these bubble motions come about, let us consider an isolated bubble induced at a stable diameter D_0 by a bias field $\boldsymbol{H}$. A pulsed field $\boldsymbol{H}_p$ is applied in three modes. First, H_p is uniform and in the same direction as $\boldsymbol{H}$ so that its application simply increases the total bias field by that amount, leading to a shrinkage of the bubble by an inward radial wall motion at a uniform velocity ξH_p. Now, if the uniform pulsed field is applied in the opposite direction to $\boldsymbol{H}$, the wall motion will remain radial, but becomes outward, resulting in an expansion of the bubble. In either way, the bubble motion is not very useful in the operation of bubble devices except when a bubble is required to be annihilated by the collapse process.

For a bubble to have a translational motion, the bias field must not be uniform but must vary spatially. Figure 7.27 illustrates the behavior of a bubble in a bias field that varies sinusoidally with the distance in the reference direction. The minimum position of the field provides the stable location for the bubble where the pressures exerted on the wall at both sides by the field are exactly balanced out. At other locations, the pressure will be higher on one side than on the other. Hence, the bubble will move to the right or left, depending on whether the bubble is initially situated on the left or right of the field minimum. This second mode of bubble motion can be realized by applying a pulsed field $\boldsymbol{H}_p$

which varies linearly with distance in the direction of translation. The superposition of $\boldsymbol{H}_p$ to $\boldsymbol{H}$ creates a field gradient such that the change in the field in the z direction, H_z, across the bubble diameter is ΔH_z. Consequently the local field acting on the left-hand side of the bubble is initially increased from H to $H+\frac{1}{2}\Delta H_z$, thus tending to make the wall move inwards into a weaker field. The situation with the right-hand wall is exactly reversed – the local field is reduced from H to $H-\frac{1}{2}\Delta H_z$ and the wall will move outwards into a stronger field. The net result is a translation of the bubble. Because the front wall tends to accelerate, the bubble cannot remain circular, however.

To let the bubble have translational motion without destroying its circular shape, we need to superimpose to the original bias field a parabolic field well. Again, there will be a field gradient ΔH_z across the bubble diameter; however, the local field seen by the whole wall remains at H, whereas the total bias field is decreased inside the bubble and increased outside. With such a distribution of the bias field, the bubble will have a stable translational motion to the right without distorting the size and shape. Also along with the bubble motion, the field well moves so that the bubble remains to lag behind the field well by a constant distance. This mode of bubble motion corresponds to the ideal propagation in bubble devices. Detailed analyses of the bubble motions are discussed by O'Dell [1974b].

The third mode of bubble motion is a modification of the second mode. This modification uses the versatile function of an overlay. It is shown in fig. 7.32a that an overlay is a thin ferromagnetic material such as Permalloy deposited on an insulating spacer, which had been deposited on the top surface of the bubble-crystal plate. The Permalloy

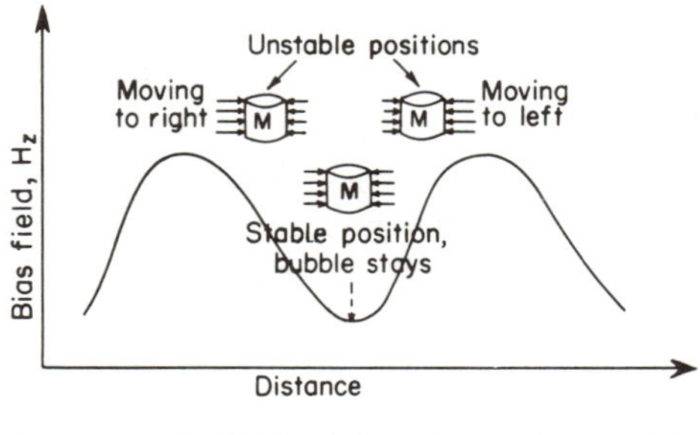

Fig. 7.27. One stable (at the field minimum) and two unstable positions of a magnetic bubble in an oscillating bias field.

overlay used here is in the form of a round disc and magnetized to saturation by an in-plane drive field H_D. Suppose that an isolated bubble is attached to the overlay on the side where the magnetization of Permalloy counteracts against $\boldsymbol{H}$ so as to produce a bias field minimum at this location. Now, if the in-plane drive field is rotated, the bubble will rotate with, but remains lagging behind by a small angle, the field. In this way, a local field will remain acting on the wall in order to maintain the rotational motion of the bubble. This mode of bubble motion was observed stroboscopically by Rossol and Thiele [1970].

§2.1.4.3. *Bubble translation in a viscous medium.* In the second mode of bubble motion, a bubble moving to the right with a translational velocity v_x in a nonideal material will experience a viscous drag, which acts normally on the moving wall. The magnitude of the drag force is given for a wall segment $\mathrm{d}l$ by

$$\mathrm{d}F_{x,\xi} = -(2Mh/\xi)v_x \cos\theta \,\mathrm{d}l, \tag{7.57}$$

where $v_x \cos\theta$ is the velocity component normal to the wall. The total drag force is then given by

$$F_{x,\xi} = \int_l \mathrm{d}F_{x,\xi} = -2\int_{-\frac{1}{2}\pi}^{+\frac{1}{2}\pi} (2Mh/\xi)v_x \cos^2\theta \, r\,\mathrm{d}\theta = -\pi r(2Mh)v_x/\xi. \tag{7.58}$$

This drag force must be equal to the driving force $F_{x,H}$ exerted on the bubble by the bias field gradient causing the bubble translation. The required bias field gradient is equivalent to having $\boldsymbol{H}$ at the center of the bubble. The force acting normal to the bubble wall at P is given by

$$\mathrm{d}F_{x,H} = \left(H + \frac{\partial H_z}{\partial x} r\cos\theta \pm H_{c,\infty}\right)(2Mh)(r\,\mathrm{d}\theta), \tag{7.59}$$

where h is the plate thickness. Note that the effect of coercivity $H_{c,\infty}$ of the material is to simply add to, or substract from, the normal force on the wall, depending on whether $\frac{1}{2}\pi < \theta < \frac{3}{2}\pi$ or $-\frac{1}{2}\pi < \theta < \frac{1}{2}\pi$. Whence,

$$F_{x,H} = 2Mh\left\{\int_{-\frac{1}{2}\pi}^{+\frac{1}{2}\pi} [H + (\partial H_z/\partial x)r\cos\theta - H_{c,\infty}]r\cos\theta\,\mathrm{d}\theta \right.$$
$$\left. + \int_{+\frac{1}{2}\pi}^{+\frac{3}{2}\pi} [H + (\partial H_z/\partial x)r\cos\theta + H_{c,\infty}]r\cos\theta\,\mathrm{d}\theta\right\} \tag{7.60}$$

If $\partial H_z/\partial x$ is kept constant, then (7.60) is simplified to

$$F_{x,H} = 2Mh[\pi r(\partial H_z/\partial x) - 4H_{c,\infty}]r. \tag{7.61}$$

Equating (7.61) to (7.60) and solving for v_x gives

$$v_x = \xi[r(\mathrm{d}H_z/\mathrm{d}x) - (4/\pi)H_{c,\infty}]. \tag{7.62}$$

Because $2r(\mathrm{d}H_z/\mathrm{d}x) = \Delta H_z$, (7.62) becomes

$$v_x = \tfrac{1}{2}\xi[\Delta H_z - (8/\pi)H_{c,\infty}]. \tag{7.63}$$

This expression of bubble-wall velocity was obtained by Thiele [1971], who also showed a reasonably good agreement between the calculated values of v_x and the observed results reported by Rossol and Thiele [1970] provided that the size and circular shape of the bubbles were kept unchanged during motion.

When the applied field gradient attains high values as is the case in device operation, v_x reaches a maximum value, which, according to Slonczewski [1973], is given by

$$v_{x,\max} = 24\gamma A_{ex}/h(\mu_0 K)^{\frac{1}{2}}. \tag{7.64}$$

This limiting value of v_x is of great importance in applications because it limits the maximum data rate of bubble devices. For most favorable bubble materials for which $\gamma = e/m_0$ because $g = 2$ for spins only and in which the bubble height equals the bubble diameter, $v_{x,\max}$ is in the range 20–30 m · s^{-1}. For certain garnets such as $(Eu_{1.45}Y_{0.45}Si_{0.6}Ge_{0.5})(Ca_{1.1}Fe_{3.9})O_{12}$, the Landé factor was observed to be greater than 30; hence the associated gyromagnetic ratio $\gamma (= ge/2m_0)$ has high values. Le Craw et al. [1975] thus measured unusually high bubble velocities (>600 m · s^{-1}) in the latter materials.

§2.1.4.4. *Dynamic conversion; hard bubbles.* In a dynamic study of epitaxial films of $(GdTmY)(Ga_{0.8}Fe_{4.2})O_{12}$ grown on a (111)-oriented substrate of GdGa-garnet, Tabor et al. [1972] detected anomalous behavior of some bubbles, notably the following:

(1) The translational motion of some bubbles proceeded at an angle, not parallel, to the direction of the applied pulse-field gradient.
(2) The distances over which these abnormal bubbles were translated under a fixed pulse field were erratic because they varied by as much as a factor of 10.

(3) The abnormal bubbles displayed much lower mobilities than those predicted from (7.54).
(4) Bubbles were seen to collapse over a range of the applied bias field, instead of all collapsing at exactly the same collapse field H_l defined in the expression

$$v = \xi[H_p - (H_l - H)(d_0 - d)/(d_0 - d_l)], \tag{7.65}$$

where v is the velocity of the moving wall, H is the bias field applied in the negative direction, d_0 is the stable diameter of the bubble under H, H_p is the pulsed field applied in the same direction as H and d_l is the diameter of the collapsing bubble.

These "anomalous" properties were attributed to a dynamic conversion of bubbles, from a normal to a hard type. A normal bubble is defined as having only the Bloch walls, whereas a hard bubble has a mixed wall of Bloch and Néel segments. The conversion is termed dynamic because it occurs during the motion of a bubble at velocities typically between 1–2 m/s and the bubble returns to the normal state when motion ceases. The simplest Bloch wall of a normal bubble is shown in fig. 7.28a and b, in which the magnetization vector rotates in a plane parallel to the wall in the clockwise (cw) and counterclockwise (ccw) direction, respectively. These simple Bloch wall structures probably never exist; otherwise two kinds (left- and right-handed) of normal bubbles would have been recorded. The absence of such experimental evidence implies that the Bloch walls of normal bubbles actually have more complex structures. Slonczewski et al. [1972] showed in fig. 7.28c that the Bloch wall of a normal bubble may be partly left-handed and partly right-handed so that the bubble is capable of moving in the direction parallel to the drive-field gradient. A simple wall structure of a hard bubble is illustrated in fig. 7.28d, which consists of cw and ccw Bloch segments separated by Néel segments. In this structure, each Bloch segment is terminated by a small transitional region, called a Bloch line, in which the magnetization vector rotates cw or ccw in a plane perpendicular to the wall. An enlarged, and probably more realistic, representation of the wall structure of a hard bubble is shown in fig. 7.29. More Néel segments are shown in the latter structures. Moreover, each Néel segment has a finite width in which the spin distribution at the midpoint of the wall is sketched in the lower diagram. The Bloch lines have a higher energy density than the Bloch segments because the spins must turn through a greater angle in going from lattice site to lattice site. For this reason, it is readily seen from fig.

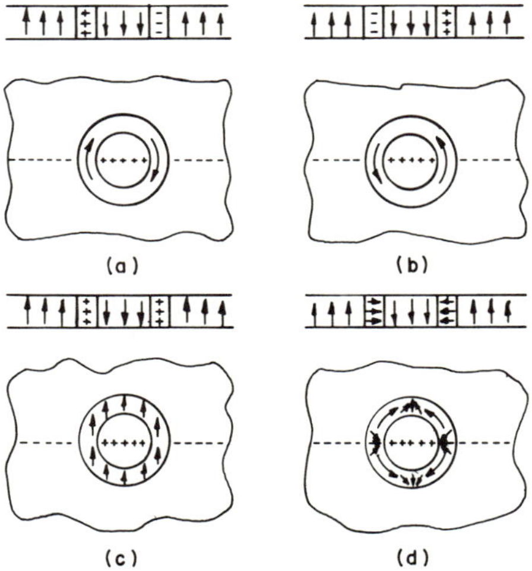

Fig. 7.28. Three [(a) to (c)] normal and one (d) hard bubble-wall structures. (a) A simple Bloch wall in the cw direction; (b) a similar Bloch wall in the ccw direction; (c) a wall structure partially (a) and partially (b); (d) four Bloch segments separated by Néel segments, which are also known as the Bloch lines (Bobeck et al. [1975]).

7.28d that, as the bubble contracts, a larger proportion of its wall volume is occupied by the Bloch lines. Consequently, after the bubble diameter reaches a certain value during contraction, the total wall energy will no longer decrease, but may even increase. This explains why hard bubbles are more resistant to collapse and display an extended range of stability with respect to the bias field strength.

Results obtained from subsequent dynamic studies have indicated (Bobeck and Torre [1975e]) that hard bubbles are characterized by a diameter variation of roughly 10:1 over a stable bias range of approximately $0.23M_s$ in comparison to a 3:1 diameter variation over a stable bias range $0.1M_s$ for normal bubbles. For example, for a typical bubble material with 52 μm thickness and 0.67 μm characteristic length, normal bubbles collapse at 3 μm when the bias field is equivalent to $0.43M_s$, whereas hard bubbles persist to a diameter of 0.7 μm when the bias field reaches the value equivalent to $0.71M_s$.

The dynamic behavior of hard bubbles has been analyzed on the basis of the Landau–Lifshitz–Gilbert equation expressed in (4.128) by Vella-Coleiro et al. [1972], Malzemoff and Slonczewski [1972] and

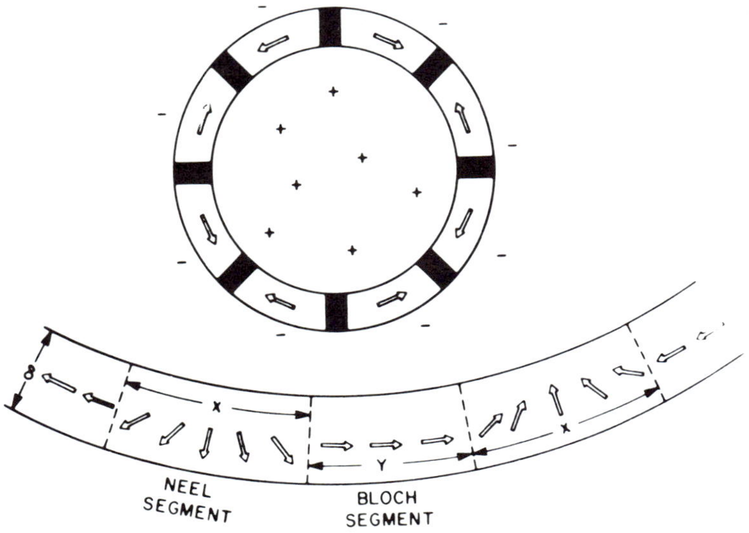

Fig. 7.29. A more complex hard-bubble wall structure than that shown in fig. 7.37d. The lower diagram shows the spin distribution at midpoint of wall (Bobeck and Torre [1975]).

Thiele [1973]. Because of the skew propagation of hard bubbles, it is necessary to resolve the velocity of their translations into two components, $v_\perp$ and $v_\parallel$ in the directions perpendicular and parallel to the field gradient, respectively. According to Vello-Coleiro et al., the ratio of $v_\perp/v_\parallel$ becomes large and the ratio of $v_\parallel$ to the drive field becomes very small at large values of the drive field. Expressions for the calculation of $v_\perp$ and $v_\parallel$ are derived by Thiele. Meanwhile, Hagedorn [1974] proposed a model for the dynamic conversion based on the nucleation and propagation of Bloch lines within the moving domain wall. The temporary immobile state of a hard bubble is believed to contain both moving horizontal Bloch lines and stationary Bloch lines, the latter of which are parallel to the cylindrical axis of the bubble domain. Material imperfections and composition inhomogeneities are postulated to play an important role in nucleating the Bloch lines, thus accounting for the erratic behavior of the propagating bubbles.

Hard bubbles have been widely observed in garnet films grown epitaxially, but they were rarely detected in garnet platelets cut from bulk crystals grown from flux; nor have they been found in orthoferrites. The appearance of hard bubbles seems to depend on the intrinsic characteristics of the material as well as on the gradient of such parameters as composition, stress and uniaxial anisotropy as a result of thickness nonuniformity. Three techniques have been developed for suppressing the formation of hard bubbles, which are highly undesir-

Fig. 7.30. Evidence for suppressing the formation of hard bubbles in a LPE film of $(YGdTm)_3(FeGa)_5O_{12}$ by proton implantation. The micrograph shows two drastically different domain structures under a bias field of $5.6\,kA \cdot m^{-1}$ (70 Oe). Many of the domains in the unimplanted area on the left-hand side are still strips, which form hard bubbles as the bias field is increased to $8.4\,kA \cdot m^{-1}$ (105 Oe). All of the bubbles in the implanted side have normal characteristics with collapse fields near $6\,kA \cdot m^{-1}$ (75 Oe) (Bobeck and Torre [1975]).

able from the viewpoint of device operation. One technique uses ion implantation of protons (Wolfe and North [1972]). The effectiveness of this technique is demonstrated in fig. 7.30. The second technique is to prepare the garnet films on multilayers, instead of on single layers (Bobeck et al. [1973]). The third technique is based on the direct deposition of a very thin (≈300 Å) Permalloy film formed on the garnet film (Lin and Keefe [1973]).

§2.2. *Bubble materials*

It has been stated in §2.1 that, for a solid to be a bubble material, three requirements must be fulfilled: (1) a net moment in a prominent crystal or geometric direction, (2) uniaxial anisotropy of which the easy axis of magnetization coincides with the prominent direction cited in (1) and (3) an appropriate (>1) ratio of the anisotropy field to saturation magnetization. For the bubble material to be attractive in device applications, however, there are further requirements, which will be discussed in §2.2.1. A general description of various classes of bubble materials will be given in §2.2.2.

§2.2.1. *Material requirements for bubble devices*

Before discussing other material requirements, let us elaborate on the third requirement concerning the quality factor Q. This factor must be greater than unity in order to stabilize the bubbles by keeping the spins aligned in the direction of the anisotropy field and normal to the plate surface. If $Q<1$, M_s will be aligned in the plate surface so as to reduce the magnetostatic energy E_s in (7.28). Another reason for requiring $Q>1$ is seen from the static stability condition for bubbles in terms of the characteristic length λ. The model used to derive this λ-related condition assumes that the magnetization lies only along the plate normal and that the domain walls have zero width. The validity of both assumptions increases with increasing Q value, because the degree of rotation of M_s due to the effect of demagnetizing field and the ratio of the domain diameter in the preferred range to the wall width are both proportional to H_k/M_s. A third reason for favoring high values of Q is related to the nucleation of bubble domains. From the device viewpoint, spontaneous nucleation of undesired bubbles must be avoided. To achieve this, the nucleation field H_n must be greater than the sum of the bias field and the demagnetizing field; otherwise domains will be self-nucleating upon the application of the bias field. Since the bias field is less than M_s, the latter condition may be written as $H_n > M_s$. For a "defect-free" crystal*, H_n is taken to be equal to the anisotropy field. Hence we have the approximate condition $H_k > M_s$ for the prevention of self-nucleation of domains.

Despite these three reasons, the value of Q must be kept at a reasonably low level so that the bubble diameter is small and the wall mobility is high [see (7.54)]. The optimum value of Q, therefore, is a compromise of these considerations. A general range recommended for Q is $1 < Q < 10$, as shown by the diamond-shaped area in fig. 7.18. The third requirement practically rules out all orthoferrites and all hexagonal ferrites for device applications, even though they display bubbles. Enclosed in this diamond area are garnets and amorphous Gd–Co films. A more elaborate plot of H_k against M_s is shown in fig. 7.31, which maps out the attractive device materials in specific areas for different classes of bubble materials with respect to the Q requirement. It is worth repeating that to meet the Q requirement, M_s must be made as low as possible for a given anisotropy. This is a rare case in the application of ferromagnetic materials because usually saturation magnetization is preferred to be as high as possible in order to improve the efficiency of the application.

Other material requirements are:

(4) A small characteristic length λ. Discussions in §2.1.3 have shown that the greatest bubble stability is attained when the thickness of the crystal plate or film is around 4λ and the bubble diameter about 8λ. Hence a small value of λ is necessary to restrict both the plate thickness and the bubble diameter.

(5) High mobility of bubble walls to attain high bit rate and high storage density. From the expression for bubble mobility in (7.54), the wider the wall or the higher the ratio of λ to Q, the higher the mobility.

(6) Low coercivity. The pertinent equation is (7.62), which indicates clearly that low $H_{c,\infty}$ enables the translational motion of bubbles to proceed at high velocities. This requirement means that the material must be as free from "defects" as possible. The presence of "defects" also promotes the nucleation of undesirable domains.

(7) Capability of being prepared into a crystal platelet or a thin film, of

*The term "defect-free" is used only in the sense commonly practised in bubble-device technology. Here, the defects refer to macroscopic or semi-microscopic imperfections in the lattice which are capable of acting as nuclei. In terms of materials science, such defects are of the volume type discussed in ch. 5, §2.3.

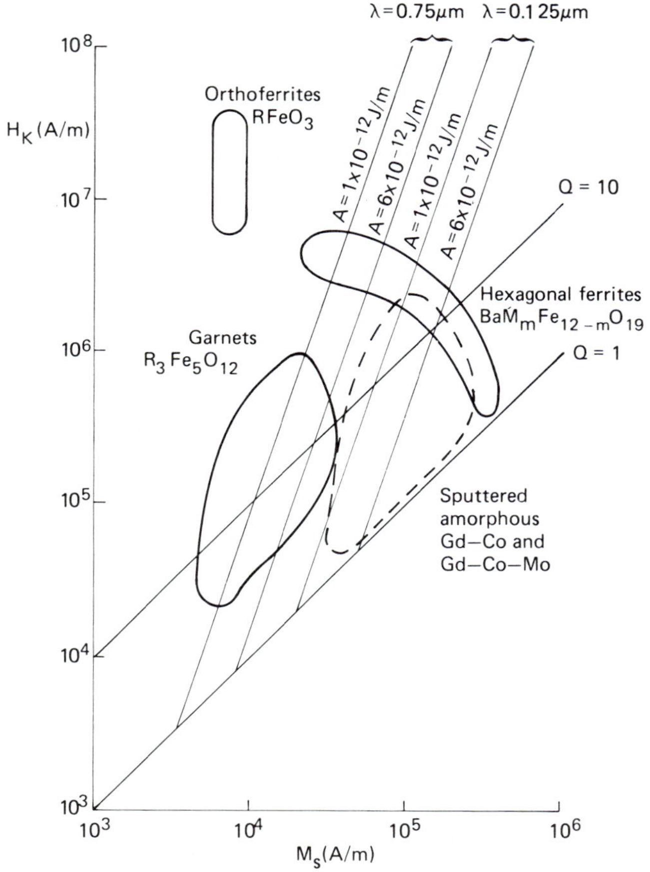

Fig. 7.31. Further characterization of bubble materials in terms of the quality factor $Q(=H_k/M_s)$, the characteristic length λ and the exchange stiffness A. Note that only the material groups of garnets and sputtered amorphous Gd–Co films are partially enclosed in the preferred, diamond-shaped region bordered by $Q = 1$ and 10 and $A = 1 \times 10^{-12}$ and $6 \times 10^{-12}\,\mathrm{J \cdot m^{-1}}$ (Covault [1975]).

the order of 4λ. Also the thickness must be uniform within 5%. Otherwise bubbles could be trapped in the thicker regions, causing unreliability in the device operation.

(8) A compensation temperature of the material well below room temperature, with the magnetization as nearly constant as possible in

the operation range of temperature of $\approx 100°C$. The relative independence of magnetization and other magnetic properties on temperature is particularly important for bubble materials used in such devices as repertory dialers (see Michaelis and Danylchuk [1971]) and desk calculators.

(9) Low coefficients of magnetostriction so that the coupling mechanism between the bubble and the lattice responsible for magnetostriction will be ineffective in retarding the motion of bubbles.

(10) Finally, composition homogeneity and low concentration ($<5/cm^2$) of magnetic defects to ensure reliability in the device operation. Magnetic defects are defined as structural features that impede or deflect a bubble as it moves in its path in the device. They include inclusions, dislocations, cavities in, and pits, scratches and magnetic dusts on the surface of, the bubble platelet or film (Nielsen [1976]).

§2.2.2. *Crystalline bubble materials*

§2.2.2.1. *Crystalline bubble materials.* There are three classes of crystalline substances capable of supporting magnetic bubbles. They are rare-earth orthoferrites, uniaxial garnets and hexagonal ferrites. The rare-earth orthoferrites, such as $YFeO_3$ (Sherwood et al. [1959]), $TmFeO_3$ and $LuFeO_3$ (Bobeck [1967]) were the first bubble materials discovered. Garnets, however, possess the most attractive properties for applications, thus promising to be the best device materials. For this reason, our emphasis in this section will be on garnets, leaving discussion on other crystalline materials somewhat brief.

§2.2.2.1.1. *Rare-earth(RE) orthoferrites.* These compounds are of the general formula $RFeO_3$, where R is a rare earth. They have an orthorhombic structure with the lattice parameters a and b nearly equal and smaller than c by about 30%. The direct exchange between spins of Fe^{3+} ions is negligible, but the superexchange (fig. 2.18) involving the intervening O^{2-} ions leads to an antiferromagnetic spin coupling of Fe^{3+} ions in alternating a-directions. Moreover, because of a lattice distortion, the Fe^{3+} spins are slightly canted by 0.5° to the a-axis. The canted antiferromagnetic spin arrangement gives rise to a weak ferromagnetism ($M_s \sim 8000\ A \cdot m^{-1}$ or $4\pi M_s \sim 100\ G$) in the c-axis. One exception is $SmFeO_3$, which has its net moment along the a-axis at room temperature. Also, the easy axis of magnetization in many of the RE-orthoferrites can be shifted from the c- to the a-direction by decreasing the temperature. The Néel temperature for all RE-orthoferrites is about 400°C. Pertinent data on the crystal structure and magnetic properties for RE-orthoferrites are compiled in table 7.6. Two RE-orthoferrites can dissolve each other to form a continuous series of mixed orthoferrites. Two mixed compounds, $Sm_{0.6}Er_{0.4}FeO_3$ and $Sm_{0.55}Tb_{0.45}FeO_3$, are included in table 7.6. The mixing technique was found valuable in attaining desirable bubble characteristics and the required temperature dependence for device applications.

RE-orthoferrites are optically transparent, especially to the red spectrum. The optical technique based on the Faraday effect is thus applicable for studying the domain behavior of this class of magnetic materials. For example, domains in a $TmFeO_3$ platelet, 58 μm thick, are readily observed by using the Faraday rotation of transmitted light. This optical property of RE-orthoferrites coupled with many other attractive magnetic properties led to the discovery of cylindrical and bubble domains and to the early promise of new devices. Many of the basic analyses on the domain behavior and device technology were conducted on RE-orthoferrites. In the final overall appraisal, however, garnets are considered far superior to RE-orthoferrites as bubble device materials.

§2.2.2.2. *Uniaxial rare-earth garnets.* A general discussion of the rare-earth garnets (REG) has been given in ch. 5, §1.3.1(2). We recall that these ferrimagnetic compounds have the same complex structure as the mineral garnet. Their general formula is $R_3Fe_5O_{12}$, where R is a rare-earth ion. This formula is better expressed with respect to the distribution of the cations among three kinds of lattice sites as $\{R_3^{III}\}[Fe_2^{III}](Fe_3^{III})O_{12}$, where { }, [ ] and ( ) designate the dodecahedral, octahedral and tetrahedral sites, respectively. Being cubic crystallographically, REG are intrinsically nonaxial, as evidenced by their easy directions of magnetization along the cubic diagonals. However, uniaxial anisotropy can be induced in mixed REG with two or more species of rare earth ions, as was first discovered by Bobeck et al. [1970] in $Gd_{2.34}Tb_{0.66}Fe_5O_{12}$ platelets cut in certain orientations from beneath a {112} facet of the bulk crystal grown from the fiux-controlled melt. Later, Le Craw et al. [1971] also found bubble domains in $(Eu_2Er)(Ga_{0.7}Fe_{4.3})O_{12}$ cut from beneath the {110} facets of the bulk crystal. The phenomenon of inducing the uniaxial anisotropy features (1) a dependence of the symmetry and magnitude of the uniaxial anisotropy on the orientation of the facet beneath which the platelets were cut, (2) the requirement of two or more species of rare-earth ions and (3) the variation of the sign and magnitude of the anisotropy with the species of the rare-earth ions. Callen [1971] and Rosencwaig et al. [1971] attributed the uniaxial anisotropy to a growth-induced preference in site occupancy of the rare-earth dodecahedral sublattice. Subsequently, Rosencwaig and Tabor [1972] worked out a phenomenological theory based on the nonrandom distribution of rare-earth ions in the dodecahedral sites. Their theory invokes two parameters defined in terms of (1) site preferences determined by the relative size of the

TABLE
Data for rare-earth orthoferrites

Rare earth	Lattice parameter* (Å)			Saturation magnetization**		Néel temperature* (°K)
	a	*b*	*c*	M_s $(A \cdot m^{-1})$	$4\pi M_s$ (emu)	
Y	5.283	5.592	7.603	8 350	105	643
La	5.556	5.565	7.862	6 600	83	738
Pr	5.495	5.578	7.810	5 650	71	707
Nd	5.441	5.573	7.753	4 930	62	693
Sm	5.394	5.592	7.711	6 680	84	673
Eu	5.371	5.611	7.686	6 660	83	663
Gd	5.346	5.616	7.668	7 480	94	661
Tb	5.326	5.602	7.635	10 900	137	652
Dy	5.302	5.598	7.623	10 180	128	645
Ho	5.278	5.591	7.602	7 240	91	643
Er	5.263	5.582	7.591	6 440	81	641
Tm	5.251	5.576	7.584	11 140	140	631
Yb	5.233	5.557	7.570	11 370	143	632
Lu	5.213	5.547	7.565	9 460	119	622
$Sm_{0.6}Er_{0.4}$					83	
$Sm_{0.55}Tb_{0.45}$					108	

*D. Treves, J. Appl. Phys. **36** (1965) 1033.
**Bobeck and Torre [1975f].
***Bobeck et al. [1969].
†NA: not available.

rare-earth ions and (2) magnetic interactions involving the anisotropic exchange fields acting on the R-ions.

A summary of the measured results on bubble-related properties of flux-grown crystals of various mixed garnets is given in table 7.7. Despite the success in growing large-size single crystals of mixed REG, the application potential of garnets in bubble devices had been still somewhat limited until the epitaxial growth techniques were developed. The latter techniques provide the efficiency, economy and ease with

7.6
evaluated as bubble materials

Material parameter***				
Experimentally deduced value			Calculated value	
Bubble diameter (mils)	Bias field (Oe)	Platelet thickness (mils)	$l_d(\equiv \gamma_w/4M_s^2)$, (mils)	γ_w (ergs/cm^2)
3.0	33	3.0	2.5	1.8
		NA†		
		NA†		
7.5	3.2	2.0	4.4	1.1
6.0	3.0	1.1	2.9	1.3
5.5	10.5	2.0	3.7	1.6
3.7	16	2.4	2.9	1.7
1.7	51	2.2	1.4	1.7
2.0	32	1.6	1.7	1.8
4.5	12	2.1	3.3	1.7
6.0	8	2.0	3.9	1.6
2.3	37	2.3	1.9	2.4
3.8	41	3.0	3.0	3.9
7.5	10.5	2.0	4.3	3.9
1.0	33	1.8	0.80	0.35
0.75	61	2.0	0.40	0.30

TABLE 7.7

Magnetic and bubble characteristics of flux-grown mixed garnets, some of which were doped with aluminum or gallium to raise the value of the quality factor Q.*

Material	M_s ($kA \cdot m^{-1}$)	$4\pi M_s$ (G)	$(TC)_m$ (%/°C)	λ (μm)	γ_w ($mJ \cdot m^{-2}$)	ξ ($m^2 \cdot A^{-1} \cdot s^{-1}$)	$[\xi]$ ($cm \cdot Oe \cdot s^{-1}$)
$Er_2Tb_1Al_{1.1}Fe_{3.9}O_{12}$	10.8	136	− 1.8	1.27	0.19	0.007	55
$Gd_{2.34}Tb_{0.66}Fe_5O_{12}$	10.9	137	+ 3.2	1.53	0.23	0.015	120
$Gd_{0.95}Tb_{0.75}Er_{1.3}Al_{0.5}Fe_{4.5}O_{12}$	14.4	181	+ 1.2	0.35	0.083	0.008	100
$Eu_2Er_1Ga_{0.7}Fe_{4.3}O_{12}$	19.7	247	+ 0.1	0.64	0.31	0.021	165
$Eu_2Er_1Ga_{0.7}Fe_{4.3}O_{12}$	15.6	196	+ 0.1	0.73	0.22	0.025	200
$Y_2Gd_1Al_{0.8}Fe_{4.2}O_{12}$	26.1	328	+ 0.38	0.21	0.18	0.023	2 000
$Y_{1.8}Eu_{0.2}Gd_{0.5}Tb_{0.5}Al_{0.6}Fe_{4.4}O_{12}$	35.8	450	0	0.22	0.36	0.024	186
$Eu_{1.5}Gd_{1.5}Al_{0.5}Fe_{4.5}O_{12}$	17.4	219	+ 0.9	1.78	0.29	0.023	1 200
$Eu_{1.9}Gd_{1.1}Al_{0.5}Fe_{4.5}O_{12}$	12.7	160	+ 0.5	0.67	0.14	0.254	1 200
$Ca_2Bi_1V_1Fe_4O_{12}$	4.4	55	− 0.18	4.77	0.13	0.138	2 000
$Pr_1Gd_2Ga_{0.5}Fe_{4.5}O_{12}$	12.1	152	− 2.0	1.50	0.27	0.053	425
$(TC)_m = \Delta M_s/(M_s \Delta T)$ near RT							
$\lambda = \gamma_w/4\pi M_s^2$							

*Bobeck and Torre [1975g].

which single-crystalline films of mixed garnets can be prepared having superior properties with respect to the ten requirements listed for device applications. Because of these outstanding accomplishments, coupled with the fact that amorphous bubble films have some inherent drawbacks, it is reasonably safe to conclude that bubble device technology will heavily rely on the epitaxially-grown films of mixed garnets. Further efforts on the epitaxy-growth approach have led to the development of two methods for the preparation of garnet films; the chemical vapor deposition (CVD) and the liquid phase epitaxy (LPE). Details of the CVD method are reviewed by Nielsen [1971] and Varnerin [1971], and of the LPE method by Stein [1974]. Although both methods have been successful, the LPE method is favored because the garnet films grown from the liquid phase contain fewer "defects" than those from the vapor phase.

The phenomenological theory of Rosencwaig and Tabor is applicable to the uniaxial anisotropy induced in the epitaxially-grown garnet films. Uniaxial anisotropy may also be induced under stress in garnet films prepared by the epitaxy-growth techniques, however, when the lattice parameters of the crystal mismatch those of the substrate. Because of the latter possibility, it has been suggested that appropriate stresses should be applied to augment growth-induced anisotropy, thus achieving as nearly as possible the optimum value of the quality factor Q by properly choosing the substrate. With all these advantages of epitaxial growth techniques, we are able to specify the garnet films in meeting the ten requirements for device applications. An example of material specifications for devices operated by field-access circuits is shown in table 7.8. These specifications are intended for 10^5 bits/cm^2 and a 2×10^5 bits/sec data rate. The density of 5 defects per cm^2 obviously refers

TABLE 7.8
Device specifications of bubble materials (Bobeck and Torre [1975h])

Bubble diameter ($2r$)	$6\ \mu$m
Bubble collapse field (H_l)	$8\ \mathrm{kA \cdot m^{-1}} \pm 1\%[100\ \mathrm{Oe} \pm 1\%]$
Magnetization ($M_s[4\pi M_s]$)	$14.3\ \mathrm{kA \cdot m^{-1}}[180\ \mathrm{G}]$
Thickness (h)	$6\ \mu\mathrm{m} \pm 1\%$
Mobility (ξ)	$>0.025\ \mathrm{m^2 \cdot A^{-1} \cdot s^{-1}}[>200\ \mathrm{cm \cdot s^{-1} \cdot Oe^{-1}}]$
Coercivity ($H_{c,\infty}$)	$<23.9\ \mathrm{A \cdot m^{-1}}[<0.3\ \mathrm{Oe}]$
Defect density	$<5\ \mathrm{cm}^2$
Anisotropy field (H_k)	$>1.5 M_s(1.5 \cdot 4\pi M_s)$

to macroscopic imperfections such as inclusions, cavities and surface faceting. Bubble properties observed on epitaxial films of prominent mixed garnets are compared in table 7.9.

§2.2.2.3. *Hexagonal ferrites.* At first glance, hexagonal ferromagnets such as MnBi and ferrimagnets such as barium ferrites (ch. 5, §1.3.1(3).) are natural materials to satisfy the first and second requirements for bubble materials. The prominent barium ferrites in this group include several magnetoplumbites, notably $(Sr, Ba, Pb)Fe_{12}O_{19}$ and $BaFe_{18}O_{27}$. Hence Van Uitert et al. [1970] explored the suitability of this class of ferrites for bubble-device applications. Although MnBi and most of the hexagonal ferrites have Q values greater than unity (fig. 7.31), they fall in a region just outside the favored diamond-shaped region bound by $Q = 1$ and 10 and $A_{ex} = 6 \times 10^{-12}\ J \cdot m^{-1}$ for $\lambda = 0.125\ \mu m$ and $A_{ex} = 1 \times 10^{-12}\ J \cdot m^{-1}$ for $\lambda = 0.75\ \mu m$. Moreover, magnetoplumbites were found to display wall mobilities 3 to 4 orders of magnitude smaller than those of orthoferrites. Unless their bubble characteristics are substantially improved, hexagonal ferrites are considered unattractive for device applications. An effective method to improve the bubble characteristics of magnetoplumbites is to modify the composition by substitution. For instance, substitution of Al, Ga, Cr, Rh or $Ti_{0.5}Co_{0.5}$ for Fe tends to reduce M_s and the anisotropy energy of the Sr- or Ba-magnetoplumbite, thereby shifting their positions in figs. 7.18 and 7.31 to the left because of an increase in the characteristic length.

§2.2.3. *Amorphous bubble materials*

A drastically different class of bubble material was discovered by Chaudhari et al. [1973], who applied the sputtering technique to prepare thin films from various Gd–Co intermetallic compounds. The films were found to exhibit (1) an amorphous structure, (2) a composition deviation from the target material and (3) a single, though metastable, phase. The composition deviation is referred to that of the target, which is usually an intermetallic compound, or sometimes a mixture of two compounds.* The film usually has a nonstoichiometric composition related to the stoichiometric composition of one compound. The actual composition of the film is controlled by the magnitude of the negative bias

*Because of a much greater difference in atomic size than Hume-Rothery's 15% and for other chemical and electronic reasons, Gd and Co only form seven intermetallic compounds (Gd_4Co_3, $GdCo_2$, $GdCo_3$, Gd_2Co_7, $GdCo_5$ and Gd_2Co_{17}) in the binary system in lieu of any intermediate solid solutions (Lihl et al. [1969]).

TABLE 7.9

Magnetization and bubble properties of garnet epitaxial films prepared by the LPE method (Bobeck and Torre [1975i])

Material	M_s ($kA \cdot m^{-1}$)	$4\pi M_s$ (G)	h (μm)	λ (μm)	γ_w ($mJ \cdot m^{-2}$)	Bubble diam. (μm)	ξ ($m^2 \cdot A^{-1} \cdot s^{-1}$)	$[\xi]$ ($cm \cdot s^{-1} \cdot Oe^{-1}$)
$Eu_2Er_1Ga_{0.7}Fe_{4.3}O_{12}$ [(a)]	13.8	173	6	0.17	0.17	7	–	–
$Eu_1Er_2Ga_{0.7}Fe_{4.3}O_{12}$	9.5	120	14	2.2	0.20	22	–	–
$Eu_{0.6}Y_{2.4}Fe_{3.85}Ga_{1.15}O_{12}$	22.3	280	4	~0.4	–	4	–	–
$Gd_{0.47}Y_{2.53}Fe_{3.95}Ga_{1.05}O_{12}$	11.9	150	4	~0.4	–	4	–	–
$Er_2Eu_1Ga_{0.7}Fe_{4.3}O_{12}$	23.5	295	5.6	0.5	0.35	5	0.006	45
$Er_{1.99}Gd_{1.01}Ga_{0.22}Fe_{4.78}O_{12}$	11.8	148	5	0.9	0.173	9	0.011	89
$Y_{0.94}Gd_{1.07}Yb_{0.57}La_{0.42}Al_{0.7}Fe_{4.3}O_{12}$	19.1	240	7.3	0.46	0.21	5	0.035	275
$Y_{1.03}Gd_{1.29}Yb_{0.68}Al_{0.7}Fe_{4.3}O_{12}$	13.9	175	2.1	0.51	0.125	5	0.251	2 000
$Y_{1.5}Eu_{1.5}Al_{0.7}Fe_{4.3}O_{12}$	37.8	475	–	0.17	0.29	2	–	–
$Y_{1.3}Eu_{1.7}Al_1Fe_4O_{12}$	19.9	250	–	0.42	0.21	4	~0.060	~480
$Y_1Gd_1Tm_1Fe_{4.2}Ga_{0.8}O_{12}$	16.3	205	–	0.75	0.27	8	0.063	500
$Gd_{0.86}Y_{1.52}Yb_{0.62}Ga_{0.89}Fe_{4.11}O_{12}$	11.9	150	3.6	0.80	–	8	0.188	1 500
$Eu_2Y_1Fe_5O_{12}$ [(b)]	124.0	1 559	1.51	0.066	–	0.7	–	–
$Sm_{0.4}Y_{2.6}Ga_{1.2}Fe_{3.8}O_{12}$	15.9	200	6	0.60	–	6	0.044	350
$Y_{1.88}Lu_{0.2}Ca_{0.92}Ge_{0.92}Fe_{4.08}O_{12}$	13.6	171	11	0.47	0.11	6	0.251	2 000

Note: Unless otherwise indicated films were grown on {111} $Gd_3Ga_5O_{12}$ substrates.

[a]Grown on {110} $Gd_3Ga_5O_{12}$ substrate.

[b]Grown on {100} $Sm_3Ga_5O_{12}$ substrate.

voltage applied to the substrate. The films not only retain the ferrimagnetic spin coupling as in the compound, but also display the so-called perpendicular anisotropy, that is, uniaxial anisotropy with the easy axis of magnetization perpendicular to the film surface.

Data for spontaneous magnetization obtained at 4.2 to 600°K by Tao et al. [1974] indicate that amorphous $Gd_{1-x}Co_x$ films with $0.36 < x < 0.96$ display the same collinear antiparallel coupling between spins of Gd and Co atoms as that established in the pertinent compounds. There exist three regimes of magnetization; namely, Gd-dominating, compensated and Co-dominating. To see these three regimes, it is useful to express the magnetization M_x of a $Gd_{1-x}Co_x$ film in the form

$$M_x = |M_{Gd}(1-x) - M_{Co}x|. \tag{7.66}$$

The critical composition x_c at which M_x vanishes is called the compensation composition. For films with $x < x_c$, $M_{Gd}(1-x) > M_{Co}x$, therefore, Gd dominates the magnetization of the film. The reverse is true for films with $x > x_c$, for which the magnetization is dominated by Co. An equivalent and more convenient form of (7.66) is based on the atomic moments:

$$\bar{\mu} = |\mu_{Gd}(1-x) - \mu_{Co}x|, \tag{7.67}$$

where $\bar{\mu}$ is the mean moment per atom in the film. According to Tao et al., however, the atomic moments of Gd and Co are not constant, and again, there are noticeable differences between values for the compounds used in the target for sputtering and their derived amorphous films. The deduced values for μ_{Co} at 4.2°K in Bohr magnetons per Co atom are: (A) Compounds – 0.65 in $GdCo_3$, 1.0 in $GdCo_2$ and 1.6 in Gd_2Co_7, $GdCo_5$ and Gd_2Co_{17} and (B) Films – 1.45 in $Gd_{0.33}Co_{0.67}$ and tending to 1.4 at higher Co concentrations, Assuming $\mu_{Gd} = 6.40\ \mu_B$ and $\mu_{Co} = 1.65\ \mu_B$, (7.67) gives $x_c = 0.795$ and $\bar{\mu} = 0.04\ \mu_B$ for the $Gd_{0.2}Co_{0.8}$ film, which is equivalent to $M_s = 250\ A \cdot m^{-1}$ (2.5 G). Such a low value of M_s can easily make the films satisfy the third requirement for bubble device materials mentioned on p. 480. Hence the amorphous $Gd_{1-x}Co_x$ films with $x \approx 0.8$ are taken as the optimum composition for bubble-device applications.

The binary Gd–Co system has an inherent drawback in relation to the device applications. Magnetizations of the binary films in the composition range ($0.75 < x < 0.85$) of interest have a strong temperature dependence at room temperature, thus in contrast to requirement No. 8. A simple way to remove this drawback is to add a nonmagnetic element to Gd–Co so as to reduce the net magnetization. Elements thus far investigated for this purpose include Au, Cu, Cr, Mo, Si and V. Early studies seem to favor Mo, which tends to decrease μ_{Co} most efficiently as shown by the following expression for the ternary film $(Gd_{1-x}Co_x)_{1-y}Mo_y$:

$$\mu_{Co}^{y} = \mu_{Co}^{y=0} - (5.0)y[y + (1-y)x]^{-1}. \tag{7.68}$$

whence, (7.67) is modified to

$$\bar{\mu} = (1-y)|(1-x)\mu_{Co}^{y} - x\mu_{Co}^{y}|. \tag{7.69}$$

Chaudhari et al. [1975] reported that the presence of 16 at.% Mo in $Gd_{0.2}Co_{0.8}$ reduced the saturation magnetization from 1.0×10^5 A · m^{-1} (100.3 G)* to 0.38×10^5 A · m^{-1} (38.2 G), a decrease of 62%. Meanwhile, Hasegawa et al. [1975a] observed that a decrease in the Mo concentration from 16 to 10 at.% would decrease T_p by ~80°K, but would increase θ_C by ~230°K, increasing the gap between T_p and θ_C by 310°K. Hence by adjusting the composition of the ternary film, that is, by varying both the Mo content and the Co/Gd ratio, the intrinsic magnetic properties such as M_s, T_p and θ_C can be "tuned" to the device requirements.

After preliminary examination of the bubble behavior of the sputtered depositions, the amorphous films of ternary Gd–Co alloys were considered possible device materials to challenge the garnets. It was claimed that the amorphous films of Gd–Co are even superior to the epitaxially grown films of garnets. For instance, in the preparation of amorphous films, the substrate conditions are much less stringent. Both crystalline and amorphous (glass) substrates can be used, and the substrate temperature is at room temperature, thus requiring none of the special equipment for heating as in the case of epitaxial growth of garnet films. Moreover, being amorphous, the films show less sensitivity to "defects" of their bubble characteristics. These advantages make it possible to integrate not only various bubble circuits on the same chip as has been practised in the microcircuitry technology, but also to combine bubble circuits with semiconductor circuits to attain maximum efficiency and versatility. The amorphous films, however, are metallic in nature; therefore, they are not electrical insulators like the garnets. This means that a dielectric of high breakdown strength must be used

*Note the appreciable difference between the M_s value reported for the undoped $Gd_{0.2}Co_{0.8}$ by the latter workers and that adopted in the selection of the optimum composition. The reason for the discrepancy is unknown.

between a conductor line and the film. Also, because of the short history of the amorphous materials, considerable work still lies ahead on detailed and more critical film characterization connected with the device operations. The fate of morphous films in the bubble-device technology remains uncertain at present.

The foregoing discussion on the amorphous bubble films has concentrated on the Gd–Co and Gd–Co–Mo systems. Although they are currently considered the most promising amorphous materials for bubble-device applications, by no means are they the best choices. It is possible that material systems of amorphous films finally used in the devices is yet to be found. Other systems suggested for exploration in this endeavour include (Ni, Fe or Co)–Si (Klokholm and Plaskett [1974]) and Ni–Au (Lee and Suits [1975]).

§2.3. *Application of magnetic bubbles*

The impetus of the great application potential of bubble materials has inspired tremendous interests in the fields of magnetics and electronics. It seems appropriate and necessary to devote this last section to a description of how the bubble materials can be used in devices. Because the applications are still in the developing stage and because novel techniques and new devices are being proposed in large numbers, it is futile to discuss specific application techniques and devices here. Instead, only principles underlying the functions of bubble devices will be described along with a brief discussion on a few basic bubble devices. The future of the bubble-device technology will be commented on in the concluding remarks.

§2.3.1. *Operational principles of bubble devices*

In general, the function of a bubble device depends on four principal operations: generation, propagation, detection and annihilation of bubbles. An auxiliary operation, which involves the transfer of bubbles, may be considered as a special operation of propagation.

(I) Generation. The appearance of strip domains in fig. 7.17 implies that magnetic bubbles exist only by accident in the demagnetized state of orthoferrites and garnets. Therefore bubbles must be generated by the application of a proper bias field. The earliest method of bubble generation was to shrink a strip domain to a point, as is shown in figs. 7.17g to i. The bias-field strength at which the strip-to-bubble conversion occurs is called the transition field H_t. A further increase in the bias field causes the bubbles to decrease in diameter. Eventually at a

second critical field H_l, known as the collapse field, all bubbles disappear and the material is saturated. The latter field H_l is higher than H_t anywhere up to ≈ 63%, depending upon the thickness of the film (Bobeck and Torre [1975b]).

A single bubble can be nucleated in a saturated film by applying a large magnetic field opposite to the saturation magnetization. The critical field required for the nucleation of a bubble is termed the nucleation field H_n. For materials with zero exchange stiffness,

$$H_n = H_k - M_s, \tag{7.70}$$

which is based on the fact that the demagnetizing field aids nucleation of bubble domains. For most real bubble materials with nonzero A_{ex}, however, the magnitude of H_n is increased by a term based on the domain-wall energy. Bobeck et al. [1971b] set $H_n \geqslant 0.5\, M_s$ from the observation that $H_k > 1.5\, M_s$ for "defect-free" materials. Because H_n has relatively large values, this applied reverse field must be limited in a localized region; otherwise the nucleated domain will quickly expand to cover the entire film. This generation method thus has the drawback that the current required to produce the local nucleating field is excessive even for materials with low M_s values.

A more convenient method is to pinch a portion of the strip domain off or to slice an existing bubble into halves by using current or Permalloy circuits in various schemes. This method offers the advantage that the field required for pinching off a bubble is much smaller than H_n. Also, because the new bubbles are generated from the existing strip- or bubble-domains, the generation is better controlled insofar as there is no danger of accidentally creating a bubble. This method was applied by Perneski [1969], whose basic design of a bubble generator is shown in fig. 7.32. The generator consists of a disc overlay situated to the left of a series of T-bars forming a propagation track. The disc and the bars are Permalloy thin films deposited on an insulating film, which had been deposited on the top surface of the bubble film. The disc is perhaps five times the diameter of an isolated bubble in the material. It is magnetized to saturation by an in-plane drive field. A source domain is permanently trapped by the generator disc at the edge where the + poles lie. As the in-plane drive field H_D rotates from → to ↓, the trapped domain is forced to stretch into the "+" end of the adjacent bar. In continuing the rotation of the drive field from ↓ through ← to ↑, the "−" poles near the stretched portion of the domain give rise to a pinching-off action, dividing the source domain into two parts. Every

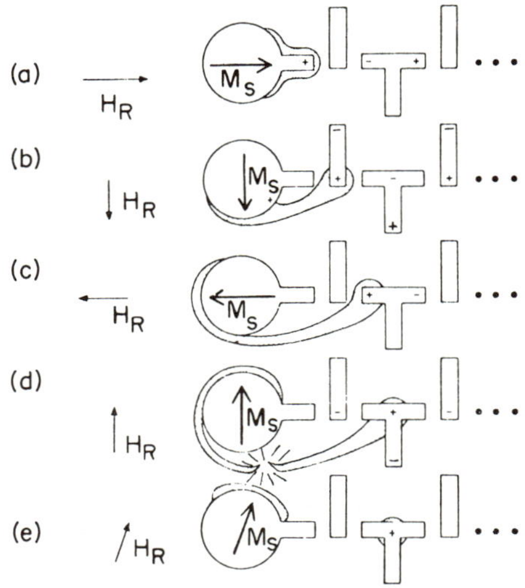

Fig. 7.32. Perneski's method of generating bubbles from a source domain, which is stretched and divided during its rotation during its rotation around the generator.

time the drive field completes one revolution, one bubble is launched onto the propagation track, and all the bubbles already on the track move ahead one pattern period.

New bubble domains may also be generated by splitting existing bubbles. A simple splitter may be an overlay consisting of two T's arranged in a modified H-shape like ⊣T. The overlay is located at the end of a closed T-bar propagation track, as illustrated by O'Dell [1974c]. Such a generation of bubbles is equivalent to signal duplication, or amplification, in data storage and retrieval.

(II) Propagation. In principle, bubbles can be propagated by creating a gradient in the applied field or in other parameters such as the wall energy, film thickness and magnetization. Figure 7.27 has illustrated that a bias field that varies sinusoidally in the x-direction will exert an unbalanced pressure on the bubble at positions other than the minimum. Hence a bubble situated on the left- (or right-) hand side to a minimum will be pushed to the right (or left). Thiele and Morrow [1967] and Ashkin and Dziedzic [1972] have demonstrated a controlled bubble movement under a laser beam, whose temperature profile produced sufficient gradients in both wall energy and magnetization.

A reliable and easy bubble propagation required for operation of devices can be accomplished by a travelling wave-like gradient of an additional field to the bias field. The bias field is usually produced by a permanent magnet, whereas the additional field is provided by either current-carrying conductors or a Permalloy overlay. The former is called the current-access method and the latter the field-access method. In either method, bubbles are made to travel along certain tracks made of conductors or magnetic overlays in certain patterns.

Propagation tracks involving conductors are easy to design, but they are difficult to fabricate for small ($\sim 1\,\mu$m) bubbles found in the uniaxial garnet platelets and in the amorphous Gd–Co films. Moreover, the current required may be excessive, thus causing instability of the bubbles. For these reasons, the field-access method is presently favored.

There are two types of magnetic overlays: the asymmetric keeper type and the transverse field type. The first is typified by the angel-fish propagation track, which consists of small triangles of Permalloy overlays in a nose-to-tail configuration. On such a track, a bubble makes inchworm motions by expanding and contracting its size when the drive field is modulated (Bobeck et al. [1969]). This type of propagation overlay has three disadvantages (1) narrow operating margins, (2) low speed and (3) difficulty in performing other operations, such as detection.

The transverse field-access method is used most extensively for bubble propagation as well as other operations. The first design of the propagation track based on this method was described by Perneski [1969], and consists of a series of T-bars (fig. 7.32). The operational principle of such a track is explained in fig. 7.33. Because of the geometric dependence of the demagnetizing field and the shape of the overlay, magnetic poles are generated only at the ends, not along the sides by the in-plane field. As this field rotates clockwise, the minimum of the field normal to the surface of the bubble film is made to move from left to right in accordance with the long dimensions of the T and the bars arranged relative to one another. Correspondingly, the magnetic poles at the ends of the overlay bars are systematically redistributed, leading to the orderly propagation of the bubble from left to right. Each revolution of the field rotation causes the bubble to advance one period of the track. The propagation will reverse its direction when the in-plane field rotates counterclockwise.

The design of a T-bar circuit is difficult owing to the complex

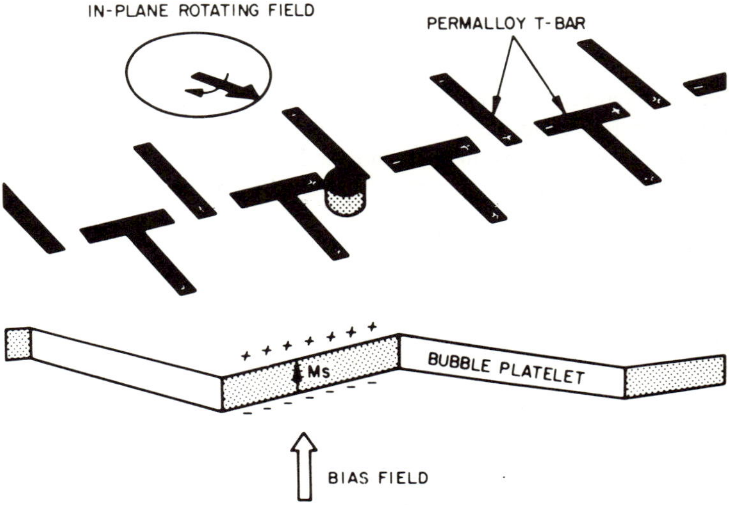

Fig. 7.33. Schematic diagram showing the propagation of a bubble under a rotating in-plane field on a T-bar track (Bobeck and Torre [1975]).

nonlinear interaction between the Permalloy drive field and the bubble. Analyses of the three-dimensional distribution of the field potential were performed on the circuits of T-bars, Y-bars and chevrons by George and Archer [1972–4] and by Almasi et al. [1975]. Despite their efforts, dimensions of the circuit elements and specifications of their assembly are determined partly by results obtained from empirical calculations and partly by the trial-and-error method. Bonyhard et al. [1973] cited the following dimensions of a typical T-bar circuit overlayed on a high mobility ($\xi = 0.025\ \mathrm{m}^2 \cdot \mathrm{A}^{-1}$ or $200\ \mathrm{cm} \cdot \mathrm{s}^{-1} \cdot \mathrm{Oe}^{-1}$) garnet platelet: (1) width of T's and bars: $a = 3.6\ \mu\mathrm{m}$, (2) center-to-center spacing between two T's: $8a$, (3) total height of each T or bar: $8a$, (4) circuit period: $8a$, (5) Permalloy overlay thickness: $\frac{1}{10}a$, (6) Permalloy to garnet spacing: $\frac{1}{4}a$ and (7) average bubble diameter: $\frac{3}{2}a$. Note that all dimensions are in the microscopic range in the sense that the fabrication of the circuits necessitates the use of a microscope. This circuit is capable of operating up to 400 Hz.

Despite the wide acceptance of the T-bar circuits in the design of bubble devices, other circuits involving such configurations as Y-bars, symmetric and asymmetric chevrons, X-patterns and asymmetric half disks have been suggested for additional advantages or for special reasons. George and Archer [1973] analyzed the T-bar, Y-bar and symmetric chevron patterns and showed that the T-bar pattern is capable of producing a good travelling wave for bubble propagation because the amplitude of the gradient of the field-well remained reasonably constant during the cycle of the drive-field roation. The X pattern described by Parzefall and Littwin [1973] offers the advantage that the bubbles need not propagate from end to end under the Permalloy element. The T-bar track with two gaps per propagation period has now been replaced in most laboratories by more efficient single-gap structures, which consist of asymmetric half disks or asymmetric chevrons (Bobeck [1977]).

The chevron pattern produces a smoother bubble motion than the T-bar pattern and also allows the elements to be packed more closely. These advantages coupled with the versatile capabilities of chevrons have enabled these circuits to replace the T-bar patterns in current device-technology. The principle of operating chevrons in a propagation track is similar to that of the T-bar track. Figure 7.34 shows the motion of two bubbles from left to right as the in-plane drive field rotates clockwise. Note that the bubble "wets" all the Permalloy overlays that form the propagation path. In this way, the possibility for bubbles to enter secondary paths is reduced. This is a further advantage of the

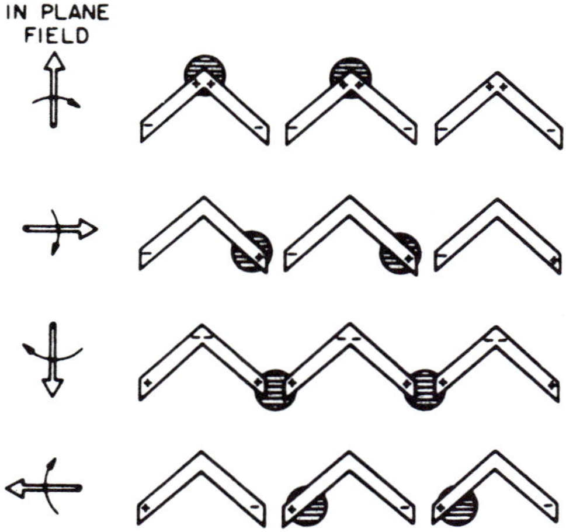

Fig. 7.34. Propagation of a pair of bubbles from left to right under a clockwise-rotating in-plane drive field on a symmetric chevron track (Bobeck and Torre [1975]).

chevron circuits over the T-bar circuits, which tend to cause diversion into the secondary paths of moving bubbles. Still another advantage of the chevrons is that they can be connected at alternate ends in the  pattern to form a magnetoresistance detector, which, together with other chevron propagation tracks, will build a "guard rail" that rings the active circuit area, from which unwanted bubbles are screened out.

(III) Detection. The methods just described for the generation and propagation of bubbles enable us to construct a simple data store. For instance, putting a disc generator (fig. 7.32) and T-bars together in a closed loop would allow data to be fed onto the track and circulated continuously. To make such a store useful, however, we need some means to read out the stored information in a later time. Hence the third major component of the store is a detector, which normally consists of a sensor and some electronic accessories for signal passing, amplification and registration. The sensor is placed close to the propagation track at a strategic point so that it can detect the passage of a bubble through that point by an electrical pulse or by an optical image. In this way, the stored datum is read out without being destroyed. In other cases, a destructive read-out may be used, which offers the advantage that all the energy contained in the bubble can then be coupled into the sensor.

Many techniques are available for bubble detection, among which those based on (1) magnetoresistance, (2) the Hall effect, (3) galvanomagnetics and (4) the optical effect are prominent. The magnetoresistance sensors are by far the most successful, and our discussion will be focused on this technique.

In the presence of a magnetic field, electrical resistivity of a ferromagnetic material is either increased or decreased, depending on whether the material is magnetized to saturation in the direction parallel ($\rho_\parallel$) or perpendicular ($\rho_\perp$) to that of the current flow. This effect is expressed by

$$E_x = j[\rho_\perp + (\rho_\parallel - \rho_\perp)\cos^2\phi], \qquad (7.71)$$

where E_x is the electrical field in the x-direction, j the current and ϕ the angle between M_s and the x-direction shown in fig. 7.35a. Because $\rho_\parallel > \rho_\perp$ and the difference $\Delta\rho = \rho_\parallel - \rho_\perp$ amounts to a few percentages of the resistivity (ρ_0) in zero field, the magnetoresistance effect can be used to detect the bubble in the following manner. The Permalloy sensor is placed near a bar-overlay in the propagation track. In the absence of a bubble, the in-plane drive field H_D will saturate the sensor in the

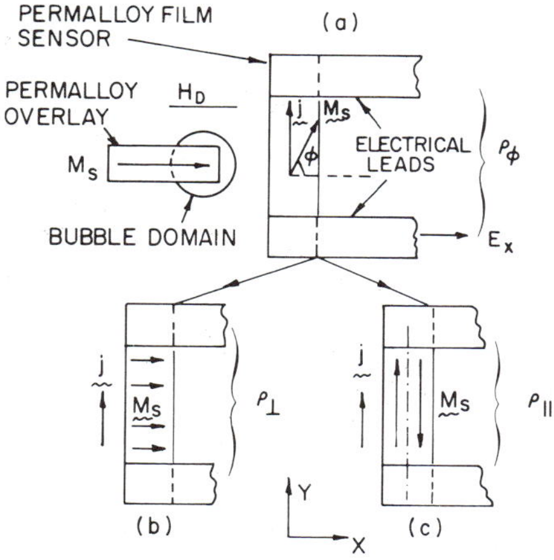

Fig. 7.35. (a) Construction of a magnetoresistance sensor. (b) "Normal" magnetoresistance occurs in the absence of a bubble at the end of a Permalloy-bar overlay. (c) "Parallel" magnetoresistance occurs when a bubble is at the end of the overlay. The stray field of the bubble cancels out the in-plane field, thereby letting shape anisotropy dominate the direction of magnetization of the sensor along its long dimension.

direction normal to that of the current (fig. 7.35b), thus corresponding to $\rho_\perp$. When a bubble appears at the end of the bar, the stray field associated with the bubble will cancel out the drive field, leaving the sensor in a demagnetized state with domains equally distributed in opposite directions parallel to the current by virtue of the shape anisotropy (fig. 7.35c). Because magnetoresistance is an even effect, the resistivity of this demagnetized state is the same as the state in which the material is saturated in the same direction as the current, thus corresponding to $\rho_\parallel$.

Magnetoresistivities of Ni–Fe alloys have been measured as a function of composition and film thickness by Mitchell et al. [1964] and by Battarel and Galinier [1969]. For 80% Ni–Fe films prepared by vapor deposition, $\Delta\rho/\rho_0 \sim 0.03$ and $\Delta\rho = 6 \times 10^{-9}\ \Omega$-m (0.6 $\mu\Omega$-cm). Applying this $\Delta\rho$ value and $j = 10^{10}\ \text{A} \cdot \text{m}^{-2}$, we get $\Delta V = j\Delta\rho = 60\ \mu\text{V}/\mu\text{m}$, which means, in the case of an "ideal" detector, an output voltage of 60 μV times the sensor length in microns in the y-direction in fig. 7.35. A signal of the order of $10^2\ \mu$V may be barely detectable against amplifying and system noise, however. Therefore the detection signal must be intensified by various methods, which are discussed by Bobeck and Torre [1975c].

(IV) Annihilation. The method of bubble annihilation varies according to the type of circuits used. With the conductor circuits, annihilation is accomplished through the collapse of bubbles by simply pulsing the particular conductor loop at which the bubble is positioned. Similarly, bubbles on a T-bar or chevron propagation track can be annihilated by energizing a current loop in the locality so as to locally increase the bias field. For example, a current pulse of 0.2 A of 1 μs duration opposite in polarity to that used for generation suffices to effect the bubble annihilation. Another method involves the use of passive annihilators, which resemble generators shown in fig. 7.32, except that the associated in-plane field rotates in the reverse direction. Also, guard rails may be

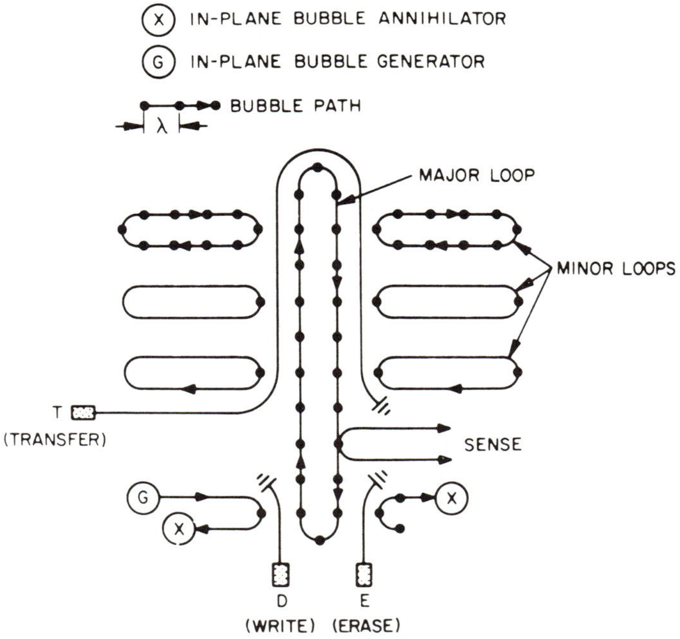

Fig. 7.36. A simple shift register, which consists of a major loop, six minor loops for data storage, a generator, an annihilator, a sensor and a transfer port (Bobeck and Torre [1975]).

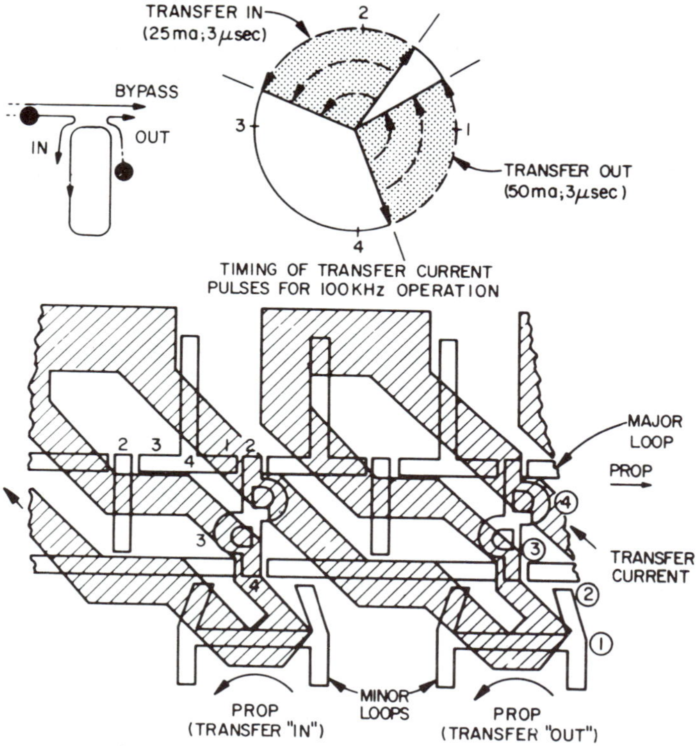

Fig. 7.37. Two S-shaped overlays are inserted in the major loop for the transfer of bubbles between major and minor loops (Bobeck and Torre [1975]).

used to direct the unwanted bubbles into an area of the bubble chip, which is separated from the active device area.

(V) Transfer. In some devices, transfer of bubbles in a controlled manner is essential to reduce retrieval time. For example, the operation of a shift register shown in fig. 7.36 necessitates the transfer of bubbles between the major and minor loops so as to store and retrieve data conveniently and efficiently. The transfer of bubbles is performed by the S-shaped gate shown in fig. 7.37. This figure contains (1) a portion of the major loop made of T's and bars, (2) two S-shaped transfer gates, (3) two distorted H-shaped overlays located next to the transfer gates and (4) a portion of the S-shaped conductor circuit. The H-shaped

overlays are needed to connect the major and minor loops. The numbers in fig. 7.37 indicate the positions of bubbles with respect to the involved in-plane rotating drive field. Suppose a bubble enters from the upper left bar and continues its propagation $2 \rightarrow 3 \rightarrow 4 \rightarrow 1$ on the major loop. Further propagation of the bubble, however, is blocked by the current at position 2, which is at the upper end of a transfer gate. Instead, the gate will first deflect the bubble to the lower end through positions 3 and 4 of the gate and eventually transfer the bubble to the minor loop on the left. This operation of transferring a bubble from the major loop to a minor loop is termed "transfer-in". To transfer-out a bubble, the reverse procedure applies, as is illustrated by the route ① → ② → ③ → ④ for a bubble being transferred from a minor loop on the right to the major loop. Nelson [1974] used a different design of the transfer gate for the chevron circuits.

§2.3.2. *Bubble devices*

The most attractive applications of bubble materials doubtless lie in the digital memories. Advantages are taken of the small size (down to less than 1 μm in diameter), reasonably high mobility ($>$ 1500 cm $\cdot$ sec^{-1} $\cdot$ Oe^{-1}) and interactions of bubbles to build memories with superior capabilities such as (1) high packing density up to 10^9 bits/cm^2, (2) less than 1 ms access time even for large files and (3) versatile switch and logic functions. Bubble chips successfully prepared and tested at the Bell Laboratory (Bobek and Torre [1975d]) in 1974 have these features: (1) storage capacity – 200 000 bits stored in a 0.5×0.5 cm area (or 80 K bits per cm^2), (2) bubble diameter – 5 μm which can be manipulated at a 100 kHz stepping rate, (3) circuit period $\sim$25 μm, (4) bubble material – garnet epitaxial films in wafer form $\sim$2 cm in diameter and (5) bubble circuits fabricated on 2 mask levels.

In comparison with the magnetic drums, disks and tapes currently in use at one end of the spectrum of memory technology (fig. 7.38), the extreme compactness of bubble memories is a distinct advantage. Moreover, unlike the drums, disks and tapes, bubble devices have no moving parts, thus eliminating the problems of cost and noise, unreliability and other engineering considerations caused by moving parts. At the other end of the spectrum, semiconductor random-access memories are gradually replacing ferrite core memories. The non-mechanical nature of the access and circuit-processing techniques used so prominently for semiconductor devices are also applicable to bubble memories. The bubble technology, however, is superior to the semi-

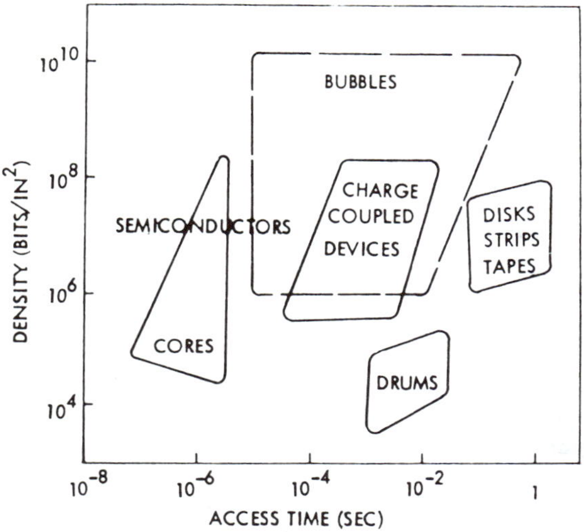

Fig. 7.38. Appraisal of various storage devices in terms of packing density and access time. It is projected that bubble devices will eventually bridge the gap between semiconductor cores and disks, strips, tapes and drums (Covault [1975]).

conductor technology in three aspects. (1) The preparation of the bubble films, even in the case of single-crystalline garnets, is easier. (2) Only single mask level and no alignment is required in processing bubble modules. (3) The first and second advantages together with the high density of bits would make the per-bit memory cost much lower than that of integrated silicon memories. The cost of bubble devices will eventually be at a level competitive with many mechanically accessed memory devices.

The major applications of bubble materials are expected to be for general-purpose mass storage. In such applications, bubble chips with capacities of $\geq 10^5$ bits are required, and the basic chip organization is either a simple shift register or, for improved access performance, a selectively accessible array of storage loops. As shown in fig. 7.36, a shift register is an assembly of bubble generators, propagation tracks, sensors and other necessary circuits capable of performing four basic functions of data: writing by bubble generation, clearing by annihilation, selection by transfer and reading by detection. In addition, it is advantageous to detect replicas rather than the original bubbles of the stored data; hence a fifth function, bubble replication, is required for a shift register. This name is adopted to reflect the fact that the bubbles are both holders and conveyors of information and can easily be shifted from position to position in the assembly. A large number of shift registers can be assembled into various matrix configurations similar to the word organization shown in fig. 6.28. Besides striving for high data density in bits per cm^2, another objective of the matrix design is to achieve both high efficiency in the use of accessory circuits and reasonable access and latency times. The maneuverability in the design of bubble memories is clearly illustrated in the two examples compared in table 7.10 (Chang [1974]). To attain a moderate data capacity, say ≈ 16 K bits in the first example, 128 shift registers, each with 128 bits forming a minor loop, are linked by a major loop. This memory needs only six interconnection pins and four circuits (one each for writing, clearing, sensing and reading), and the maximum access and latency times are both $128t_R$, where t_R is the field rotation period. To attain a much higher data capacity, say $\approx 268M$ bits in the second example, instead of arranging the major–minor loops in a single array, the chip is divided into smaller arrays arranged into 16 rows and 16 columns. Each array consists of 1024 shift registers, each with 1024 bits, organized in major–minor loops as in the first example. There are $(1024)^2 = 1\,048\,576$ bits in each array. The access and latency times are $1024t_R$, which is 8 times that in the first example, but would still be considered fast in comparison with the mechanically accessed magnetic disk file. The 256 arrays are arranged in a 16×16 coincident selection matrix, which has 268 455 456 bits altogether. The 16 arrays in each column are connected by a single transfer line, whereas the 16 arrays in each row are connected by double sensing, single clearing and single writing lines, thus requiring 81 interconnection pins and $4 \times 16 = 64$ circuits. The second memory has a capacity 1.6×10^4 times greater than the first memory at a slight ($8\times$) expense of the access time and with a moderate ($16\times$) increase in the number of circuits and interconnections.

In addition to general purpose mass-storage applications, bubble memories can be constructed for special functions. Repertory dialers and digital switching networks are two examples (Bobeck et al. [1975]). Unlike other magnetic memory devices, such as ferrite cores and Permalloy films, bubble memories may have logic capabilities. A simple example of bubble logic is seen in fig. 7.39. From the bubble logic point of view, the subsystem shown in fig. 7.39 is an inverter; the output is $\bar{X}$

TABLE 7.10

Two examples of bubble-memory design (Chang [1974])

Design example 1	Design example 2
16 384 bits ($\overline{128^2}$)	268 435 456 bits ($\overline{1024^2} \times \overline{16^2}$)
	Coincident selection from 16×16 arrays
Major/minor loops	Each array: Major/minor loops
128 S.R. each with 128 bits	1024 S.R. each with 1024 bits
$T_{ACC} = 128\ T_R$	$T_{ACC} = 1024\ T_R$
$T_{LAT} = 128\ T_R$	$T_{LAT} = 1024\ T_R$

		Pins	Semiconductor circuits		Pins	Semiconductor circuits
Generator	(write)	1	1	Generator	16	16
Annihilator	(clear)	1	1	Annihilator	16	16
Transfer	(select)	1	1	Transfer	16	16
Sensor	(read)	2	1	Sensor	2×16	16
Ground		1		Ground	1	
Total			4	Total	81	64

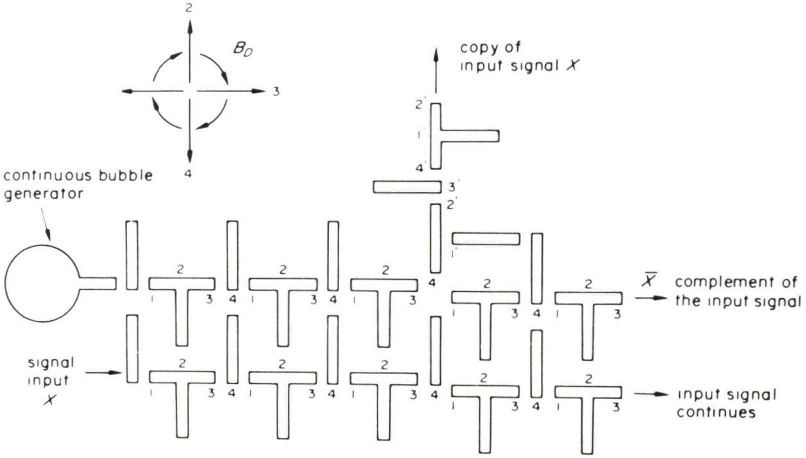

Fig. 7.39. Illustrating bubble logic capability through the interaction between bubbles propagating in two parallel tracks. When the bubble domain generator is replaced by a binary bubble signal Y, the output leaving the top of the subsystem is $X \cdot Y$ and that leaving upper row is $\bar{X} \cdot Y$, thus forming a logical AND gate (O'Dell [1974]).

for an input signal X. Now if the continuous bubble generator is replaced by a second bubble signal Y, then the output leaving the top of the subsystem is $X \cdot Y$ and that leaving the upper row output is $\bar{X} \cdot Y$. Thus we have a logic AND gate. The bubble logic would make it possible to construct integrated-circuit bubble chips just like the integrated-circuit semiconductor chips.

Another application of bubble materials is in display devices. By virtue of the magnetic Faraday effect on the garnet films and of the magnetic Kerr effect on the amorphous GdCo films, the possibility of producing a numerical display on an oscilloscope of the numbers counted by using bubbles was demonstrated by several workers, notably Shoji [1972] and Ashkin and Dziedzic [1972]. By incorporating a laser beam in the bubble display, the latter authors were able to write their signatures in script with characters only $\sim 100\ \mu$m high.

§2.3.3. *The future*

Early in 1975, workers at the Bell Laboratory succeeded in making bubble circuits with capacities around 65 K bits (Bobeck et al. [1975]). Such success has prompted the interested industries to intensify their efforts in producing some less sophisticated bubble memories in the pilot plants, according to a recent announcement in Electronics (March 4, 1976). In March 1977, a bubble machine was installed in one of the Bell Telephone switching offices in Detroit. The machine, called the 13A Voice Announcement System, contains four serial shift register garnet chips, each with a storage capacity of 68 121 bits. It has 32 pins for external connections and measures 3.05 cm × 5.59 cm × 1.52 cm high. The 272 484 bit machine is used to record and announce "call–assist" messages (Chen [1977]).

Despite the rapid and successful development of the bubble technology, a great deal still needs to be done in the areas of material selection and characterization and circuit design and fabrication. For instance, to make the bubble devices competitive with existing memories, cost and reliability problems must be solved. To realize the potential compactness,

TABLE 7.11

Present status and future predictions for circuit processing, magnetic bubble materials and circuit performance (Bobeck et al. [1975])

	PRESENT	FUTURE
Circuit processing	Photolithography	Electron beam or X-ray lithography
Definition		
Min. feature size [microns]	≈ 1.5	< .5
Implied ckt. period (microns)	≈ 15	< 4
Perfection		
Mask (defects · cm^{-2})	≈ 2.5	–
Metallization (defects · cm^{-2})	≈ 2.5	–
Magnetic bubble film	EPI garnet	EPI garnet amorphous film
Bubble size [microns]	3	< 1
Mobility [cm · s^{-1} · Oe^{-1}]	1.500	> 1.500
Implied bit rate [M bits · s^{-1}]	2.5	> 10
Propagation achieved [M bits · s^{-1}]	2	–
Req. major functions [M bits · s^{-1}]	0.4	–
Perfection		
Magnetic pt. defects [defects · cm^{-2}]	2.5	–
Uniformity [% · cm^{-1}]	< 5	< 5
Operating temp. range	−25 to +100°C	−50 to +100°C

finer ($<0.5\ \mu m$) feature size must be achieved in the fabrication of circuits. Therefore electron-beam or X-ray lithography must be adapted for convenient applications in place of the current technique based on photolithography. Also, to keep the bubble size less than 1 μm in diameter for ultimate compactness, higher quality bubble films must be sought through the improvements in material composition and processing. Data listed in table 7.11 exemplify the present status and future predictions of bubble-device technology.

There is no question that bubble devices will eventually play a vital role in the storage, retrieval and analysis of information. The prognosis of this emerging technology is perhaps best depicted in fig. 7.38, which shows that bubble devices will eventually bridge the capacity-data retrieval time between semiconductor memories and ferrite cores on one extreme of the spectrum and drums, disks, strips and tapes on the other.

Suggestions for further reading:

(I) On radiation effects

Chadderton, L. T., 1965, *Radiation Damage in Crystals* (Methuen, London).

Dautreppe, D., 1969, *Radiation Effects in Magnetic Materials*, in: *Studies in Radiation*, Vol. III, G. J. Dienes, ed. (Gordon and Breach, New York).

Thompson, M. W., 1969, *Defects and Radiation Damage in Metals* (Cambridge University Press, London).

(II) On magnetic bubbles and devices

Bobeck, A. H. and E. Della Torre, 1975, *Magnetic Bubbles* (North-Holland, Amsterdam).

O'Dell, T. H., 1974, *Magnetic Bubbles* (The McMillan Press Ltd., London).

APPENDIX 1

Conversion table of magnetic and select physical quantities

Quantity	Symbol	SI unit	Multiplying factor used to convert: One SI unit to CGS units	Multiplying factor used to convert: One CGS unit to SI units	CGS unit
Length	L	Metre (m)	10^{2}	10^{-2}	Centimetre (cm)
Mass	M	Kilogram (kg)	10^{3}	10^{-3}	Gram (g)
Time	T	Second (s)	1	1	Second (s)
Current	I	Ampere (A)	10^{-1}	10	Abampere
Force	F	Newton (N)	10^{5}	10^{-5}	Dyne
Work and energy	W and E	Joule (J)	10^{7}	10^{-7}	Erg
Power	P	Watt (W)	10^{7}	10^{-7}	$\mathrm{Erg \cdot s^{-1}}$
Electric charge	q	Coulomb (C)	10^{-1}	10	Abcoulomb
Magnetic pole	m	$\mathrm{A \cdot m}$	$10^{8}/4\pi$	$4\pi \times 10^{-8}$	
Magnetic flux	Φ	Weber (Wb)	10^{8}	10^{-8}	Maxwell (Mx)
Magnetic moment	μ	$\mathrm{A \cdot m^{2}}$	10^{-9}	10^{9}	
Magnetization	M	$\mathrm{A \cdot m^{-1}}$	10^{-3}	10^{3}	Gauss (G)
Magnetic induction	B	Tesla (T) or $\mathrm{Wb \cdot m^{-2}}$	10^{4}	10^{-4}	Gauss
Magnetic field	H	$\mathrm{A \cdot m^{-1}}$	$4\pi \times 10^{-3} = 1.257 \times 10^{-2}$	$10^{3}/4\pi = 7.958 \times 10$	Oersted (Oe)
Magnetic susceptibility	χ	(dimensionless)	$1/4\pi = 7.958 \times 10^{-2}$	4π	(dimensionless)
Relative susceptibility	χ'	$(\mathrm{H \cdot m^{-1}})^{-1}$	1.984×10^{-13}	5.040×10^{12}	(dimensionless)
Permeability	μ	$\mathrm{H \cdot m^{-1}}$	$10^{7}/4\pi$	$4\pi \times 10^{-7}$	(dimensionless)

APPENDIX 1 (cont.)

Quantity	Symbol	SI unit	Multiplying factor used to convert: One SI unit to CGS units	Multiplying factor used to convert: One CGS unit to SI units	CGS unit
Relative permeability	μ'	(dimensionless)	–	–	(dimensionless)
Permeability of free Space	μ_0	$4\pi \times 10^{-7}$ H · m^{-1}	–	–	1.0
Anisotropy constant	K	J · m^{-3}	10	10^{-1}	Erg · cm^{-3}
Electric potential	V	volt (V)	10^8	10^{-8}	Abvolt
Inductance	L	Henry (H)	10^9	10^{-9}	Centimetre
Electric resistivity	ρ	Ohm · metre (Ω · m)	10^2	10^{-2}	Ohm · centimetre (Ω · cm)

APPENDIX 2

Values of select physical constants

Constant	Value in SI unit	Value in CGS unit
Speed of light in free space, c	2.997925×10^{8} m · s^{-1}	2.997925×10^{10} cm · s^{-1}
Rest mass of the electron, m_0	9.1091×10^{-31} kg	9.1091×10^{-28} g
Charge of the electron, e	1.60210×10^{-19} C	1.60210×10^{-20} cm$^{\frac{1}{2}}$ · g$^{\frac{1}{2}}$
Charge to mass ratio for electron, e/m_0	1.758796×10^{11} C · kg^{-1}	1.758796×10^{7} cm$^{\frac{1}{2}}$ · g$^{-\frac{1}{2}}$
Planck's constant, h	6.6256×10^{-34} J · s	6.6256×10^{-27} erg · s
$\hbar$	1.05450×10^{-34} J · s	1.05450×10^{-27} erg · s
Permeability of free space, μ_0	$4\pi \times 10^{-7}$ H · m^{-1}	1.0
Bohr magneton, μ_B	9.2732×10^{-24} A · m^{2}	9.2732×10^{-21} erg · G^{-1}
Boltzmann's constant, k	1.38054×10^{-23} J · °K^{-1}	1.38055×10^{-16} erg · °K^{-1}
Avogadro's constant, N_0	6.02252×10^{23} g-mole^{-1}	6.02252×10^{23} g-mole^{-1}
Gas constant, R	8.3143 J · °K^{-1} · g-mole^{-1}	8.3143×10^{7} erg · °K^{-1} · g-mole^{-1}
Magnetomechanical or gyromagnetic ratio ($eg/2m_0 =$)	(8.7941×10^{10}) g kg^{-1} · s · A^{-1}	(8.7941×10^{7}) g cm$^{\frac{1}{2}}$ · g$^{-\frac{1}{2}}$

APPENDIX 3

DEMAGNETIZING FACTOR AND MAGNETOSTATIC ENERGY

DEMAGNETIZING FACTOR

The magnetization of a ferromagnetic body of finite size will produce magnetic free poles at the surface (internal or external) under one of two conditions (1) the normal component of the magnetization changes ($\boldsymbol{n} \cdot \boldsymbol{M} \neq 0$) or (2) the magnetization is nonuniform ($\operatorname{div} \boldsymbol{M} \neq 0$). Whenever this situation exists, a magnetic field will be induced inside the body in a direction opposite to that of the magnetization. The induced field is called the demagnetizing field $\boldsymbol{H}_\mathrm{d}$ because it always tends to oppose the applied field $\boldsymbol{H}$. Hence for a nonzero $\boldsymbol{H}_\mathrm{d}$, the effective field $\boldsymbol{H}_\mathrm{eff}$ acting on the body is always less than the applied field because

$$\boldsymbol{H}_\mathrm{eff} = \boldsymbol{H} - \boldsymbol{H}_\mathrm{d}. \tag{A3.1}$$

A simple example of the demagnetizing effect is shown in fig. A3.1, in which a magnetic field $\boldsymbol{H}$ is applied normal to the two surfaces of a magnetizable plate of thickness t, large width and infinite length. The magnetization $\boldsymbol{M}$ induced in the direction of $\boldsymbol{H}$ will produce free poles

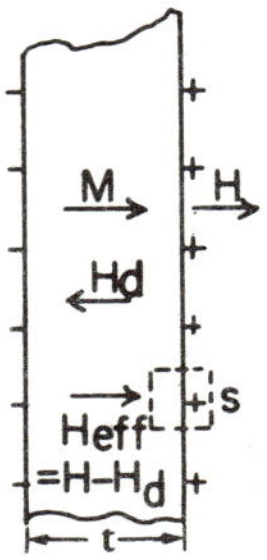

Fig. A3.1. The demagnetizing effect in a plate magnetized in the transverse direction (Chikazumi [1964]).

on the surfaces as shown. A demagnetizing field $\boldsymbol{H}_d$ is produced by the free poles in the direction opposite to $\boldsymbol{M}$, and hence to $\boldsymbol{H}$. As might be expected, the magnitude of $\boldsymbol{H}_d$ is proportional to that of $\boldsymbol{M}$. Thus, we can express in SI units

$$H_d = N_d M, \tag{A3.2}$$

where N_d is the dimensionless constant of proportionality called the demagnetizing factor. The value of N_d varies from zero to unity, depending upon the shape of the material and the direction of magnetization. To calculate N_d, we apply the Gauss theorem, which states that the surface integral of the normal component of the field, H_n, over a region is given by the total free pole strength inside that region. For the plate shown in fig. A3.1, we have

$$\int\int H_n \, dS = m, \tag{A3.3}$$

for a volume whose surface area S is indicated by the dashed lines. If the magnetization is assumed to be uniform throughout the plate, so will be the distribution of the free poles. Consequently, the left-hand side of (A3.3) becomes $H_d S$ and the right-hand side becomes MS using (1.5). Hence,

$$H_d = M. \tag{A3.4}$$

Comparing with (A3.2) gives $N_d = 1$ in this case. Now, if the plate is magnetized in any of the lateral directions parallel to the surfaces, the free poles will appear at the edges at an infinite distance and the

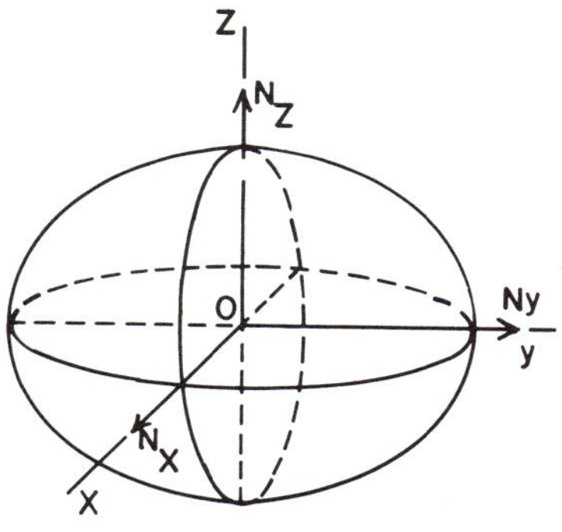

Fig. A3.2. Three demagnetizing factors along the principal axes of an ellipsoid.

demagnetizing field in the plate would be zero; that is, $N_d = 0$. Zero demagnetizing field also exists in an infinitely long rod magnetized longitudinally. For laboratory studies and in production, however, the method widely used to avoid the demagnetizing effect is to prepare the specimen or product in a ring form with the magnetizing field produced circumferentially by a toroidal winding. In this way, no magnetic poles are produced on the free surface.

In general, the calculation of N_d is quite complex unless the magnetization is uniform, or even impossible if N_d varies from point to point. Maxwell showed that a uniform field would induce a uniform magnetization in an isotropic body only if the body has the shape of an ellipsoid, thereby having a finite surface of the second degree. For a regular ellipsoid, whose three principal axes are chosen as the axes of rectangular coordinates (fig. A3.2), there exists a simple relation,

$$N_x + N_y + N_z = 1, \qquad \text{(A3.5)}$$

where N_x, N_y and N_z are the three demagnetizing factors in the Ox, Oy and Oz directions, respectively. Thus, for a sphere, $N_x = N_y = N_z = \frac{1}{3}$. For an infinitely long rod of circular cross section with its axis along Ox and magnetization in any of the lateral directions, $N_x = 0$ and $N_y =$

TABLE A3.1.
Demagnetizing factors for rods and ellipsoids magnetized parallel to the long axis (Bozorth [1951k])

Dimensional ratio (length/ diameter)	Rod	Prolate ellipsoid	Oblate ellipsoid
	1.0	1.0	1.0
1	0.27	0.3333	0.3333
2	0.14	0.1735	0.2364
5	0.040	0.0558	0.1248
10	0.0172	0.0203	0.0696
20	0.00617	0.00675	0.0369
50	0.00129	0.00144	0.01472
100	0.00036	0.000430	0.00776
200	0.000090	0.000125	0.00390
500	0.000014	0.0000236	0.001567
1 000	0.0000036	0.0000066	0.000784
2 000	0.0000009	0.0000019	0.000392

$N_z = \frac{1}{2}$. For an infinite plate in the yz plane as shown in fig. A3.1, $N_y = N_z = 0$ and $N_x = 1$ as has been derived. Numerical values for the demagnetizing factor in rods and prolate and oblate ellipsoids are given in table A3.1 for various length-to-diameter ratios. Values for demagnetizing factors of a finite circular cylinder magnetized longitudinally and of an infinite rectangular bar magnetized transversely are found in the first volume of this Series of Monographs by Brown [1962].

Production of the demagnetizing field at the internal surface can occur in many different modes. One mode involves the domain wall at which $\boldsymbol{n} \cdot \boldsymbol{M} \neq 0$ as in the case shown in fig. 3.12d. Another mode involves such internal surfaces as grain boundaries, interfaces between two phases and boundaries of inclusion particles. At these surfaces, neither $\boldsymbol{n} \cdot \boldsymbol{M}$ nor div $\boldsymbol{M}$ vanishes. A third mode operates at the perimeter of a cavity, where $\boldsymbol{n} \cdot \boldsymbol{M} \neq 0$.

The practical significance of the demagnetizing field is based on the appreciably large values of H_d for all ferromagnetic materials. Take iron for instance. With $M_s = 1.71 \times 10^6\ \mathrm{A \cdot m^{-1}}$ at 17°C (table 4.1), a spherical iron specimen will produce a demagnetizing field of the magnitude

$$\begin{aligned} H_d &= N_d M_s = \tfrac{1}{3} \times 1.71 \times 10^6 \\ &= 5.7 \times 10^5\ \mathrm{A \cdot m^{-1}} (\approx 7120\ \mathrm{Oe}). \end{aligned}$$

Thus, to magnetize this specimen to the saturation value, the applied field must exceed $5.7 \times 10^5\ \mathrm{A \cdot m^{-1}}$. Another area where the demagnetizing field is important is in the application of soft magnetic materials in magnetic circuits. The design of such circuits must take the effect of H_d into careful consideration.

MAGNETOSTATIC ENERGY

The magnetic energy, or more precisely the magnetostatic energy, is another important consideration of magnetization that the reader should comprehend. This energy appears in several forms, of which two will be discussed. The first is the potential energy, E_{mag}, of a magnetized body in an applied field $\boldsymbol{H}$. It is equivalent to the energy of interaction of a permanent magnet having the same magnetization $\boldsymbol{M}$ with the field; that is,

$$E_{mag} = -\boldsymbol{M} \cdot \boldsymbol{H} \mu_0 = -MH\mu_0 \cos\theta, \tag{A3.6}$$

where θ is the angle between $\boldsymbol{M}$ and $\boldsymbol{H}$. For convenience we often set

$E_{mag} = 0$ at $\theta = 0$ by simply rewriting the expression as $E_{mag} = MH\mu_0(1 - \cos\theta)$.

Another form of the magnetostatic energy is the self-energy, E_s, of a magnetized body. This energy is so named because it arises from the interaction of the magnetization with the field created by the magnetization itself. Two examples may be cited. One is the self-energy of a permanent magnet in its own field and the other is the magnetic energy associated with the demagnetizing field. While the first example does not concern us here, the second example was shown in ch. 3, §5 which plays a vital role in determining the domain structure in equilibrium.

The magnetostatic self-energy is given for unit volume by

$$E_s = -\tfrac{1}{2}\boldsymbol{M}\cdot\boldsymbol{H}\mu_0, \tag{A3.7}$$

where the factor one-half enters to prevent counting the magnetic dipoles twice. To derive (A3.7), let us consider a magnetizable body of unit volume. This body is assumed to contain n elemental permanent magnets, and therefore n positive and n negative free poles. Let the strengths of these free poles be $m_1, m_2, \ldots, m_{2n}$ and the magnetic potentials of the poles at their final positions be $\phi_1, \phi_2, \ldots, \phi_{2n}$. In building up the magnetization to the final value M, it is envisaged that the $2n$ poles are brought together from infinity and their strengths are gradually increased by small increments at the same rate. At any instant during the process, the poles have the strengths $\alpha m_1, \alpha m_2, \ldots, \alpha m_{2n}$, where α varies from 0 at zero magnetization at infinity to 1 in the final state. In a single step, the strength of each pole will be increased by the amount $m\,\mathrm{d}\alpha$ from αm to $(\alpha + \mathrm{d}\alpha)m$. Correspondingly, when the strength of the ith pole is αm_i, the potential at the momentary position at that pole will be $\alpha\phi_i$. The work done in each step of such a magnetization process is given by

$$\mathrm{d}E_s = \mu_0 \sum_{i=1}^{2n} m_i\,\mathrm{d}\alpha \cdot \alpha\phi_i. \tag{A3.8}$$

The total work done in magnetizing the body will be

$$E_s = \mu_0\left(\int_0^1 \alpha\,\mathrm{d}\alpha\right)\sum_{i=1}^{2n} m_i\phi_i = \tfrac{1}{2}\mu_0 \sum_{i=1}^{2n} m_i\phi_i. \tag{A3.9}$$

Now refer to the individual magnets. If the length of the qth magnet is l_q, this magnet has the magnetic moment

$$\mu_q = m_q l_q, \tag{A3.10}$$

whereas the difference in magnetic potential between its two poles is

$$\phi_q^+ - \phi_q^- = -H_\parallel l_q, \tag{A3.11}$$

where $H_\parallel$ is the component of the magnetic field parallel to the moment μ_q. Combining (A3.10) and (A3.11) gives

$$\begin{aligned} m_q\phi_q^+ - m_q\phi_q^- &= -(\mu_q/l_q)(H_\parallel l_q) \\ &= -(\boldsymbol{\mu}_q \cdot \boldsymbol{H}_q), \end{aligned} \tag{A3.12}$$

where $\boldsymbol{H}_q$ is the magnetic field acting on the qth magnet. Applying (A3.12) to (A3.9) yields

$$E_s = -\tfrac{1}{2}\mu_0 \sum_{q=1}^{n} \boldsymbol{\mu}_q \cdot \boldsymbol{H}_q. \tag{A3.13}$$

For the magnetostatic energy associated with the demagnetizing field, $\boldsymbol{H}_q$ becomes H_d and $\Sigma_{q=1}^{n}\, \mu_q$ per unit volume is precisely the magnetization $\boldsymbol{M}$ of the body. Consequently, we have

$$E_s = -\tfrac{1}{2}\mu_0 \boldsymbol{M} \cdot \boldsymbol{H}_d = \tfrac{1}{2}\mu_0 N_d M^2. \tag{A3.14}$$

This result was used in the analysis of domain structures in various metals in ch. 3, §5. Another application of (A3.14) is illustrated in fig. A3.3. Magnetic free poles are shown to be induced at the interface of an inclusion particle. The demagnetizing field associated with the free poles is in the direction opposite to that of $\boldsymbol{M}$. Because $N_d = \frac{1}{3}$ in this case, the particle has a total magnetostatic energy of the magnitude

$$E_s = \tfrac{1}{6}\mu_0 M^2 V, \tag{A3.15}$$

where V is the volume of the particle.

Fig. A3.3. The demagnetizing effect in a spherical inclusion particle.

REFERENCES

Alley, R. E. Jr., 1959, J. Appl. Phys. **30**, 284S.

Almasi, G. S., Y. S. Lin, E. Muvro and M. Slusarczuk, 1975, AIP Conf. Proc. No. 24, 635.

Alstad, J. and S. Legvold, 1964, J. Appl. Phys. **35**, 1752.

Anderson, J. C., 1968, Magnetism and Magnetic Materials (Chapman and Hall, London).

Anderson, P. W., 1950, Phys. Rev. **79**, 350.

Anderson, P. W., 1961, Phys. Rev. **124**, 41.

Anderson, P. W., 1963, Magnetism, Rado, G. T. and H. Suhl, eds. (Academic Press, New York) p. 25.

Ardley, G. and A. H. Cottrell, 1953, Proc. Royal Soc. **A219**, 328.

Argyle, B. E., S. Charap and E. W. Pugh, 1963, Phys. Rev. **132**, 2051.

Arnold, H. D. and G. W. Elmen, 1923, J. Franklin Inst. 195, 621.

Ashkin, A. and J. M. Dziedzic, 1972, Appl. Phys. Lett. **21**, 253.

Assmus, F., K. Detert and G. Ibe, 1957, Z. Metállk. **48**, 344.

Bakish, R. (ed.), 1962, Int. to Electron Beam Technology (Wiley, New York).

Bardell, P. R., 1960, Magnetic Materials in the Electrical Industry, 2nd ed. (MacDonald, London).

Barkhausen, H., 1919, Physik, Z. **20**, 401.

Barrett, C. S. and T. B. Massalski, 1966, Structure of Metals, 3rd. ed. (McGraw-Hill, New York) a. 626, b. 367, c. 247 and d. 249.

Bates, L. F., 1963, Modern Magnetism (Cambridge Univ. Press, London) a. 14 and b. 40.

Bates, L. F., D. F. Craik and E. D. Isaac, 1962, Proc. Phys. Soc. **79**, 970.

Battarel, C. P. and M. Galinier, 1969, IEEE Trans. Magnetics MAG-**5**, 18.

Beaudry, B. J. and K. A. Gschneidner, 1978, in: Handbook on the Physics and Chemistry of Rare Earths, K. A. Gschneidner and L. Eyring, eds. (North-Holland, Amsterdam).

Bechtold, J. H. and G. W. Wiener, 1965, in: The Sorby Centennial Symposium on the History of Metallurgy, Smith, C. S. ed. (Gordon and Breach, New York) p. 501.

Beck, P. A. and H. Hu, 1966, Recrystallization, Grain Growth and Textures, Margolin, H. ed. (Am. Soc. for Metals, Metals Park, Ohio, U.S.A.).

Becker, R., 1951, J. Phys. Radium **12**, 332.

Becker, R. and W. Döring, 1939, Ferromagnetismus (Verlag. Julius Springer, Berlin).

Beiler, A. C., 1967, J. Appl. Phys. **38**, 1161.

Berkowitz, A. E. and E. Kneller (ed.) 1969, Magnetism and Metallurgy, vols. 1 and 2 (Academic Press, New York).

Bertaut, F., 1951, J. Phys. Radium **12**, 252.

Bethe, H., 1935, Proc. Royal Soc. (London) **A150**, 552.

Bethe, H., 1950, Rev. Mod. Phys. **22**, 213.

Bickford, L. R., Jr., 1950, Phys. Rev. **78**, 449.

Bickford, L. R. Jr., J. M. Brownlow and R. F. Penoyer, 1958, J. Appl. Phys. **29**, 441.

Birss, I. R., 1969, J. Nucl. Mat. **34**, 241.

Bitter, F., 1931, Phys. Rev. **38**, 1903.

Bitter, F., 1932, Z. Phys. **74**, 295.

Blank, J. M., 1961, J. Appl. Phys. **26**, 975.

Bloch, F., 1930, Z. Physik **61**, 206.

Bloch, F., 1932, Z. Physik **74**, 295.

Blois, M. S., Jr., 1955, J. Appl. Phys. **26**, 975.

Blum, S. L. and P. C. Li, 1961, J. Am. Ceram. Soc. **44**, 611.

Bobeck, A. H., 1967, Bell Sys. Tech. J. **46**, 1901.

Bobeck, A. H., 1977, paper presented at Electro 77, April 1977.

Bobeck, A. H. and E. Della Torre, 1975, Magnetic Bubbles (North-Holland, Amsterdam) a. 23, b. 5, c. 191, d. 202, e. 147, f. 98 and g. 121.

Bobeck, A. H., R. F. Fisher, A. J. Perneski, J. P. Remeika and L. G. van Uitert, 1969, IEEE Trans. Magnetics, **MAG-5**, 544.

Bobeck, A. H., E. G. Spencer, L. G. Van Uitert, S. C. Abrahams, R. L. Barns, W. H. Grodkiewicz, R. C. Sherwood, P. H. Schmidt, D. H. Smith and E. M. Walters, 1970, Appl. Phys. Lett. **17**, 131.

Bobeck, A. H. and I. Danylchuk, 1971a, International Conf. on Ferrites, Hoshino, Y., S. Iida and M. Sugimota, eds. (Univ. Tokyo Press) p. 361.

Bobeck, A. H., L. G. Van Uitert and E. M. Walters, 1971b, IEEE Trans. Magnetics MAG-7, 461.

Bobeck, A. H., S. L. Blank and H. J. Livingston, 1973, AIP Conf. Proc. No. 10, 498.

Bobeck, A. H., P. I. Bonyhard and J. E. Gensic, 1975, Proc. IEEE **63**, 1176.

Bohr, N., 1948, Kgl. Dan. Vid. Selsk., Mat.-Fys. Medd. **18**, No. 8.

Bonyhard, P. I., J. E. Geusie, A. H. Bobeck, Y. S. Chen, P. C. Michaelis and J. L. Smith, 1973, IEEE Trans. Magnetics, Vol. **MAG-9**, 285.

Bozorth, R. M., 1951, Ferromagnetism (van Nostrand, New York) a. 621, b. 486, c. 506, d. 778, e. 86, f. 51, g. 60, h. 571, i. 503, j. 904 and k. 849.

Bozorth, R. M. and J. H. van Vleck, 1960, Phys. Rev. **118**, 1493.

Bozorth, R. M. and J. G. Walker, 1953, Phys. Rev. **89**, 624.

Bozorth, R. M., E. F. Tilden and A. J. Williams, 1955, Phys. Rev. **99**, 1788.

Bragg, W. L., 1940, Proc. Phys. Soc. **52**, 105.

Bragg, W. L. and E. J. Williams, 1934, Proc. Roy. Soc. (London) **A145**, 699.

Brailsford, F., 1966, Physical Principles of Magnetism (van Nostrand, London) a. 123, b. 216, c. 212 and d. 90–91.
Braun, P. B., 1957, Phillips Res. Repts. **12**, 491.
Brillouin, L., 1927, J. Phys. Radium **8**, 74.
Brinkman, J. A., 1954, J. Appl. Phys. **25**, 961.
Brinkman, J. A., 1962, Societa Italian Di Fisica, XVIII Corso (Academic Press, New York).
Brown, W. F., Jr., 1941, Phys. Rev. **60**, 139.
Brown, W. F., Jr., 1951, Phys. Rev. **82**, 94.
Brown, W. F. Jr., 1962, Magnetostatic Principles in Ferromagnetism (North-Holland, Amsterdam) pp. 190–192.
Bünge, H. J. and J. G. Müller, 1957, Wiss. Z. Hochsch. Verk. Dresden **5**, 327.
Bünge, H. J. and J. G. Müller, 1958, Z. Metallk. **49**, 40.
Bunshah, R. F. (ed.), 1968, Techniques of Materials Preparation and Handling, Vol. 1, Part 2 (Interscience, New York).
Bunshah, R. F. and M. A. Cocca, 1968, Techniques of Materials Preparation and Handling, Vol. 1, Part 2, Bunshah, R. F., ed. (Interscience, New York).
Burke, J. E., 1959, Kinetics of High-Temperature Processes, Kingery, W. D., ed. (Wiley, New York).
Burton, J. A., R. C. Prim and W. P. Slichter, 1953, J. Chem. Phys. **21**, 1987.
Callen, H., 1971, Appl. Phys. Lett. **18**, 311.
Callen, H. and R. M. Joseph, 1971, J. Appl. Phys. **42**, 1977.
Campbell, C. C. M. and W. Leiper, 1974, AIP Conf. Proc. No. 18, 319.
Carr, W. J., Jr., 1959, Magnetic Properties of Metals and Alloys (Am. Soc. for Metals, Metals Park, Ohio, USA).
Carr, W. J., Jr., 1966, Handb. Phys. **18-2** (Springer-Verlag, Berlin) p. 294.
Carter, R. E., 1961, J. Am. Ceramic Soc. **44**, 116.
Chadderton, L. T., 1965, Radiation Damage in Crystals (Methuen, London) a. 30, b. 61 and c. 143.
Chambron, W., D. Dautreppe, L. Néel and J. Paulevé, 1962, Compt. Rend **255**, 2037.
Chang, C. P. and C. W. Chen, 1974, Script Met. **9**, 349.
Chang, C. P. and C. W. Chen, 1977, Crystal Lattice Defects, in press.
Chang, H., 1974, Proc. National Computer Conf. 847.
Chaudhari, P., J. J. Cuomo and R. J. Gambino, 1973, IBM Res. Dev. **17**, 66.
Chaudhari, P., J. J. Cuomo, R. J. Gambino, S. Kirkpatrick and L. J. Tao, 1975, AIP Conf. Proc. No. 24, 562.
Chaudron, G., 1962, Ultra-High Purity Metals, Smith, R. L., ed. (Am. Soc. for Metals, Metals Park, Ohio, USA).
Chen, C. W., 1958, J. Appl. Phys. **29**, 1337.
Chen, C. W., 1959, J. Appl. Phys. **30**, 1995.
Chen, C. W., 1961, J. Appl. Phys. **32**, 348S.
Chen, C. W., 1962a, Phil Mag. **7**, 1753.
Chen, C. W., 1962b, Phys. Rev. **129**, 121.

Chen, C. W., 1963, Physics Letters **7**, 16.
Chen, C. W., 1964, Cobalt, No. 22.
Chen, C. W., 1965, Solid State Communications **3**, 231.
Chen, C. W., 1977, Proc. Int. Conf. on Magnetic Alloys and Oxides, J. Magnetism and Magnetic Materials, **7**, 308.
Chen, C. W. and D. M. Bailey, 1971, Radiation Effects **10**, 65.
Chen, C. W., J. D. Greiner and R. W. Buttry, 1974, AIP Conf. Proc. No. 18, 432.
Chikazumi, S., 1956, J. Phys. Soc. Japan **11**, 551.
Chikazumi, S., 1958, J. Appl. Phys. **29**, 346.
Chikazumi, S., 1964, Physics of Magnetism (Wiley, New York) a. 182, b. 189, c. 277, d. 275, e. 129, f. 250.
Chikazumi, S. and C. D. Graham, Jr., 1969, Magnetism and Metallurgy, Vol. 2, Berkowitz, A. E. and E. Kneller, eds. (Academic Press, New York) p. 617.
Chikazumi, S. and T. Oomura, 1955, J. Phys. Soc. Japan, **10**, 842.
Chikazumi, S., K. Suzuki and H. Iwata, 1957, J. Phys. Soc. Japan **12**, 1259.
Chiswell, B. and E. C. M. Grigg, 1971, **SI Units** (Wiley, Australasia, Sydney).
Cioffi, P. P., 1934, Phys. Rev. **45**, 742.
Clogston, A. M., 1962, Phys. Rev. **125**, 439.
Clogston, A. M., B. T. Matthias, M. Peter, H. J. Williams, E. Corenzwit and R. C. Sherwood, 1962, Phys. Rev. **125**, 541.
Collins, M. F. and J. B. Forsyth, 1963, Phil. Mag. **8**, 401.
Coltman, R. R., C. Klabunde, D. L. McDonald and J. K. Redman, 1962, J. Appl. Phys. **33**, 3509.
Coltman, R. R., 1975, private communication.
Corbett, J. W. and L. C. Ianniello (eds.), 1972, Radiation-Induced Voids in Metals (National Technical Information Service, Springfield, Virginia, USA).
Cottrell, A. H., 1948, Theoretical Structural Metallurgy (Arnold, London) p. 208.
Cottrell, A. H., 1958, Trans. AIME, **212**, 192.
Cottrell, A. H., 1967, An Introduction to Metallurgy (Arnold, London) a. 35 and b. 200.
Covault, M. F., 1975, unpublished work.
Craik, D. J. and R. S. Tebble, 1965, Ferromagnetism and Ferromagnetic Domains (North-Holland, Amsterdam) a. 204 and b. 109.
Crawford, J. H., Jr and L. M. Slifkin (ed.), 1972, Point Defects in Solids, Vol. 1 – General and Ionic Crystals (Plenum Press, New York).
Cullity, B. D., 1972, Int. to Magnetic Materials (Addison–Wesley, Reading, Massachusetts, USA).
Cuomo, J. J., R. J. Gambino and E. S. Kirkpatrick, 1974, IBM Tech. Disclosure Bull. **17**, 2202.
Danan, H., A. Herr and J. P. Meyer, 1968, J. Appl. Phys. **39**, 669.
Darken, L. S. and R. W. Gurry, 1946, J. Am. Chem. Soc. **68**, 798.
Dautreppe, D., 1969, Studies in Radiation, Vol. III, Dienes, G. J. ed. (Gordon and Breach, New York) p. 1.

De Barr, A. E., 1953, Soft Magnetic Materials used in Industry (Inst. Phys., London).

De Gennes, P. G., 1966, C. r. Acad. Sci. **247**, 1836.

De Vries, G., D. W. van Geest, R. Gersdorf and G. W. Rathenau, 1959, Physica **25**, 1131.

Dieter, G. E., Jr., 1976, Mechanical Metallurgy, 2nd ed. (McGraw-Hill, New York) p. 210.

Dietrich, W. and W. E. Proebster, 1960, J. Appl. Phys. **13**, 281S.

Dillon, J. F., Jr. and H. E. Earl, Jr., 1959, J. Appl. Phys. **30**, 202.

Dinhut, J. F., J. P. Eymery and P. Moine, 1972, Phys. Stat. Sol. (a) **12**, 153.

Dirac, P. A. M., 1929, Proc. Roy. Soc. **A123**, 714.

Domb, C., 1965, Magnetism, Vol. IIA, Rado, G. T. and H. Suhl, eds. (Academic Press, New York).

Döring, W., 1948, Z. Naturforsch. **3a**, 373.

Duijvestijn, A. J. W. and B. P. A. Boonstra, 1960, Philips Res. Repts. **15**, 390.

Dykstra, L. J., 1969, Magnetism and Metallurgy, Berkowitz, A. E. and E. Kneller, eds. (Academic Press, New York) p. 513.

Economos, G., 1955, J. Am. Ceram. Soc. **18**, 241.

Economos, G., 1958, M.I.T. Lab. for Insul. Research Tech. Rept. No. 129.

Elliott, R. J., 1968, Comments Solid State Physics **1**, 85.

Elliott, R. J. (ed.), 1972, Magnetic Properties of Rare Earth Metals (Plenum Press, London).

Elmen, G. W., 1929, J. Franklin Inst. **207**, 583.

Elmen, G. W., 1931, Bell Lab Record, **10**, 2.

Erginsoy, C. G., H. Vineyard and A. Englert, 1964, Phys. Rev. **133**, A595.

Evans, R. D., 1955, The Atomic Nucleus (McGraw-Hill, New York).

Fisher, J. C., 1953, Phys. Rev. **91**, 232.

Fisher, R. F. and R. H. Morrow, 1970, unpublished work quoted by A. H. Bobeck and E. Della Torre (1975).

Fisher, R. M. and M. J. Marcinkowski, 1960, Phil. Mag. **6**, 1385.

Freeman, A. J., 1972, Magnetic Properties of Rare Earth Metals, Elliott, R. J., ed. (Plenum Press, London) p. 248.

Friedel, J., 1954, Phil. Mag. Supplement **3**, 446.

Friedel, J., 1958, Nuovo Cimento Supplemento **7**, 287.

Galt, J. K., B. Matthias and J. P. Remeika, 1950, Phys. Rev. **79**, 391.

Geller, S. and M. A. Gilleo, 1957, Acta Crystallog. **10**, 239.

George, P. K. and J. L. Archer, 1973, J. Appl. Phys. **44**, 444.

George, P. K. and J. L. Archer, 1974a, AIP Conf. Proc. No. 18, 116.

George, P. K. and J. L. Archer, 1974b, IEEE Trans. Magnetics **MAG-10**, 821.

Getting, M. P., Jr., 1967, J. Appl. Phys. **38**, 1074.

Gilbert, T. L. and J. M. Kelley, 1955, Proc. Conf. Magnetism and Magnetic Materials (A.I.E.E. Publ.) p. 253.

Glasstone, S. and M. C. Edlung, 1952, The Elements of Nuclear Reactor Theory (van Nostrand, New York) p. 71.

Gomi, Y. and Y. Odani, 1960, J. Phys. Soc. Japan, **15**, 535.
Goodenough, J. B., 1954, Phys. Rev. **95**, 917.
Gordon, D. I., 1960, Proc. Inst. Environmental Sci. Annual Tech. Meeting, 205.
Gordon, D. I. and R. S. Sery, 1964, J. Appl. Phys. **35**, 879.
Gorter, E. W., 1950a, C. r. Acad. Sci., Paris **230**, 192.
Gorter, E. W., 1950b, Nature (London) **167**, 798.
Gorter, E. W., 1954, Philips Res. Repts. **9**, 321 and 403.
Gorter, E. W., 1955, Proc. Inst. Radio Engrs **43**, 1945.
Goss, E. P., 1934, USP 1 965 559.
Graham, C. D., Jr., 1961, J. Phys. Soc. Japan **16**, 1481.
Guttman, L., 1956, Solid State Physics, Vol. 3, Seitz, F. and D. Turnbull, eds., Academic Press, New York).
Gyorgy, E. M., M. D. Sturge, L. G. Uitert, E. J. Heilner and W. H. Grodkiewicz, 1973, J. Appl. Phys. **44**, 438.
Hagedorn, F. B., 1971, AIP Conf. Proc. **5**, 72.
Hagedorn, F. B., 1974, AIP Conf. Proc. **18**, 222.
Hagedorn, F. B., 1974, J. Appl. Phys. **45**, 3129.
Hall, R. C., 1960, J. Appl. Phys. Suppl. **31**, 157S.
Hall, R. C., W. S. Byrnes and R. G. Crawford, 1959, J. Appl. Phys. **30**, 288S.
Hamilton, W. C., 1958, Phys. Rev. **110**, 1050.
Hansen, Max, 1958, Constitution of Binary Alloys (McGraw-Hill, New York).
Hasegawa, H. and J. Kanamori, 1971, J. Phys. Soc. Japan **31**, 382.
Hasegawa, H. and J. Kanamori, 1972, J. Phys. Soc. Japan **33**, 1945.
Hasegawa, R., B. E. Argyle and L. T. Tao, 1975a, AIP Conf. Proc. No. 24, 110.
Hasegawa, R., R. J. Gambino and R. Ruf, 1975b, Appl. Phys. Lett. **27**, 512.
Hausner, H., 1973, Handbook of Powder Metallurgy (Chemical Publishing Co., New York).
Heck, Carl, 1974, Magnetic Materials And Their Applications (Crane, Russak and Co., New York) p. 589.
Heikes, R. R. and C. W. Chen, 1964, Physics **1**, 159.
Heisenberg, W., 1928, Z. Phys. **49**, 619.
Henry, W. E. and E. I. Salkovitz, 1959, J. Appl. Phys. **30**, 286S.
Herring, C., 1962, Rev. Mod. Phys. **34**, 631.
Herring, C., 1966, Magnetism, Vol. IIB, Rado, G. T. and H. Suhl, eds. (Academic Press, New York) pp. 1–181.
Heusler, J., 1966, unpublished work.
Hibbard, W. R., Jr., 1958, Trans. Metall. Soc. AIME **212**, 3.
Hibbard, W. R., Jr. and C. G. Dunn, 1956, Acta Met. **4**, 306.
Hogan, C. L., 1953, Rev. Mod. Phys. **25**, 253.
Huber, E. E., Jr., D. O. Smith and J. B. Goodenough, 1958, J Appl. Phys. **29**, 294.
Hume-Rothery, W., 1950, The Structure of Metals and Alloys (The Inst. of Metals, London) p. 57.

Iida, S., H. Sekizawa and Y. Aiyama, 1955, J. Phys. Soc Japan **10**, 907; 1958, **13**, 58.

Ising, E., 1925, Z. Phys. **31**, 253.

Iwata, T., 1958, Sci. Repts. Research Insts., Tôhoku Univ. **A10**, 34.

Jacobs, I. S., 1969, J. Appl. Phys. **40**, 917.

Johnson, R. A. and E. Brown, 1962, Phys. Rev. **127**, 446.

Jones, W. D., 1960, Fundamental Principles of Powder Metallurgy (Arnold, London).

Kanamori, J., 1965, in Magnetism, Vol. 1 (Academic Press, New York).

Kasuya, T., 1956, Progress Theoret. Phys. (Kyoto) **16**, 45.

Kaya, S., 1953, Rev. Mod. Phys. **25**, 49.

Keffer, F., 1966, Handb. Phys. **18-2** (Springer-Verlag, Berlin) p. 1.

Kersten, M., 1938, Probleme der technischen Magnetisierungskurve, Zur Deutung der Koerzitivkraft, Becker, R., ed. (Verlag. Julius Springer) p. 42.

Kersten, M., 1943, Grundlagen einer Theorie der ferromagnetischen Hysterese und der Koerzitivkraft (S. Hirzel, Leipzig).

Kinchin, G. H. and R. S. Pease, 1955, Rept. Progress Phys. **18**, 1.

Kittel, C., 1949, Rev. Mod. Phys. **21**, 541.

Kittel, C., 1950, Phys. Rev. **80**, 918.

Kittel, C. and J. K. Galt, 1956, Solid State Physics 3, Seitz, F. and D. Turnbull, eds. (Academic Press, New York) p. 457.

Kittel, C., 1963, Quantum Theory of Solids (Wiley, New York) p. 346.

Kittel, C., 1976, Int. to Solid State Physics, 5th ed. (Wiley, New York) a. 437, b. 449, c. 444, d. 443 and e. 518.

Klokholm, E. and T. S. Plaskett, 1974, IBM Tech. Disclosure Bull. **16**, 2551.

Koehler, W. C., 1965, J. Appl. Phys. **36**, 1078.

Koehler, W. C., 1972, Magnetic Properties of Rare Earth Metals, Elliott, R. J. ed. (Plenum Press, London) p. 81.

Kondorsky, E., 1937, Physik. Zeit. Sow. **11**, 597.

Kondorsky, E., 1949, Dokl. Akad. Nauk SSSR **68**, 37.

Kooy, C., 1965, Fifth International Symposium Reactivity in Solids (Elsevier, Amsterdam) p. 21.

Kooy, C. and U. Enz, 1960, Phillips Reports **15**, 7.

Köster, W. and J. Heusler, 1966, Z. Metallkunde **57**, 853.

Kouvel, J. S., 1969, Magnetism and Metallurgy, Vol. 2, Berkowitz, A. E. and E. Kneller, eds. (Academic Press, New York) p. 523.

Kramer, H. A., 1934, Physica **1**, 182.

Krén, E. and P. Szabó, 1965, Solid State Comm. **3**, 371.

Kuczynski, G. C., 1961, Powder Metallurgy, Vol. 11, Leszynski, W., ed. (Interscience, New York).

Latsh, V. V., N. G. Minaev, B. Kh. Somin and N. E. Stepina, 1961, Sov. Phys.–Solid State **2**, 1956.

Lawley, A., 1968, Techniques of Materials Preparation and Handling, Vol. 1, Part 2, Bunshan, R. F., ed. (Interscience Publishers, New York).

Lax, B. and K. J. Button, 1962, Microwave Ferrites and Ferrimagnetics (McGraw-Hill, New York).

Lax, M., 1950, Phys. Rev. **80**, 299.

Le Craw, R. C., R. Wolfe, A. H. Bobeck, R. D. Pierce and L. G. Van Uitert, 1971, J. Appl. Phys. **42**, 1641.

Le Craw, R. C., S. C. Blank and G. P. Vella-Coleiro, 1975, Appl. Phys. Lett. **26**, 402.

Leaver, K. D. and B. N. Chapman, 1971, Thin Films (Wykeham, London) p. 804.

Lee, E. W., 1953, Science Progress **1**, 41.

Lee, E. W., 1955, Rept. Prog. Phys. **18**, 184.

Lee, E. W., 1958, Proc. Inst. Electrical Engineers, **105C**, 337.

Lee, E. W., 1963, Magnetism (Penguin, London) p. 106.

Lee, K. and J. C. Suits, 1975, IBM Tech. Disclosure Bull. **17**, 2441.

Legvold, S., 1973, in: Magnetic Properties of Rare Earth Metals, Elliott, R. J., ed. (Plenum Press, London).

Leibfried, G., 1959, J. Appl. Phys. **30**, 1388.

Lihl, F., J. R. Ehold, H. R. Kirchmayr and H. D. Wolf, 1969, Acta Phys. Austr. **30**, 164.

Lilley, B. A., 1950, Phil. Mag. **41**, 792.

Lin, Y. S. and G. E. Keefe, 1973, Appl. Phys. Lett. **22**, 603.

Linares, R. C., 1964, J. Appl. Phys. **35**, 433.

Linden, E. G. and A. L. Long, 1958, U.S. Army Signal R and D Laboratory Tech. Rept. 1965.

Lindhard, J. and M. Scharff, 1961, Phys. Rev. **124**, 128.

Lindhard, J., V. Nielsen, M. Scharff and P. V. Thompsen, 1964, Kgl. Dan. Vid. Selsk., Mat-Fys Medd **33**, No. 10 and No. 13.

Lindhard, J., V. Nielsen, M. Scharff and P. V. Thompsen, 1965, Kgl. Dan. Vid. Selsk., Mat-Fys Medd **34**, No. 14.

Liu, S., R. P. Gupta and S. K. Sinha, 1971, Phys. Rev. **B4**, 1100.

Lomer, W. M., 1967, Theory of Magnetism in Transition Metals, Marshall, W., ed. (Academic Press, New York).

Lomer, W. M., 1973, AIP Conf. Proc. on Magnetism and Magnetic Materials (1972) p. 17.

Low, G. G. E. and M. F. Collins, 1963, J. Appl. Phys. **34**, 1195.

Lucasson, P. G., 1976, in: Fundamental Aspects of Radiation Damage in Metals, (CONF-751006-P1, National Technical Information Service, Springfield VA, USA) p. 42.

Luttinger, J. M., 1951, Phys. Rev. **81**, 1015.

Macklen, E. D., 1965, J. Appl. Phys. **36**, 1022.

Malozemoff, A. P. and J. C. Slonczewski, 1972, Phys. Rev. Letters **29**, 952.

Margolin, H. (ed.), 1966; Recrystallization, Grain Growth and Textures (Am. Soc. for Metals, Metals Park, Ohio, USA).

Marshall, W., ed., 1967, Theory of Magnetism in Transition Metals (Academic Press, New York).

Mason, W. P., 1954, Phys. Rev. **96**, 302.
Matthias, B. T., R. M. Bozorth and J. H. van Vleck, 1961, Phys. Rev. Letters **7**, 161.
McCoy, B. M. and T. C. Wu, 1973, The Two-Dimensional Ising Model (Harvard University Press, Cambridge, Mass., USA).
McGuire, T. R. and P. J. Flanders, 1969, Magnetism and Metallurgy, Vol. 1, Berkowitz, A. E. and E. Kneller, eds. (Academic Press, New York) p. 123.
McGuire, T. R. and M. W. Shafer, 1964, J. Appl. Phys. **35**, 984.
McGuire, T. R., B. E. Argyle, M. W. Shafer and J. S. Smart, 1963, J. Appl. Phys. **34**, 1345.
McGuire, T. R., B. E. Argyle, M. W. Shafer and J. S. Smart, 1964, J. Appl. Phys. **35**, 984.
Mee, J. E., G. R. Pulliam, D. M. Heinz, J. M. Owens and P. J. Besser, 1971, J. Appl. Phys. **42**, 1243.
Menzel, D. H. (ed.), 1960, Fundamental Formulas of Physics (Dover, New York) p. 308.
Metals Handbook, 1961, 8th edition, Vol. 1 (Am. Soc. for Metals, Metals Park, Ohio, USA) p. 785.
Michaelis, P. C., 1967, Proc. International Congress on Magnetism.
Michaelis, P. C. and I. Danylchuk, 1971, IEEE Trans. Magnetics **MAG-7**, 737.
Middelhoek, S., 1971, Magnetic Properties of Materials, Smit, J., ed. (McGraw-Hill, New York) p. 269.
Mitchell, E. N., H. B. Hankaas, H. D. Bale and J. P. Steeper, 1964, J. Appl. Phys. **35**, 2604.
Moon, R. M., 1959, J. Appl. Phys. **30**, 83S.
Moore, W. F., 1968, Techniques of Materials Preparation and Handling, Vol. 1, Part 2, Bunsah, R. F., ed. (Interscience Publishers, New York).
Morrish, A. H., 1965, The Physical Principles of Magnetism (Wiley, New York) a. 279, b. 300 and c.301.
Muto, T. and Y. Takagi, 1955, Solid State Physics, Vol. 3, Seitz, F. and D. Turnbull, eds. (Academic Press, New York).
Nachman, L. F. and W. J. Buehler, 1955, Navord Report 4130.
Nagamiya, T., K. Yosida and R. Kubo, 1955, Adv. Phys. **4**, No. 14, p. 14.
National Bureau of Standards, 1966, Critical Phenomena, NBS Misc. Publication 273 (Green, M. S. and J. V. Sengers, eds.).
Natrita, K., 1974, IEEE Trans. Magnetics **MAG-10**, 104.
Néel, L., 1944, Cahiers de Physique **25**, 21.
Néel, L., 1946, Annales Univ. Grenoble **22**, 299.
Néel, L., 1948a, Ann. Phys. **3**, 137.
Néel, L., 1948b, J. Phys. Radium **9**, 184.
Néel, L., 1952, J. Phys. Radium **12**, 339 and **13**, 249.
Néel, L., 1954a, J. Phys. Radium **15**, 225.
Néel, L., 1954b, Compt. Rend **238**, 305.
Néel, L., 1955, C. r. Acad. Sci. Paris **241**, 533.
Nelson, R. S., 1965, Phil. Mag. **11**, 291.

Nelson, R. S., 1968, The Observation of Atomic Collisions in Crystalline Solids (North-Holland, Amsterdam).
Nelson, T. J., 1974, AIP Conf. Proc. **18**, 95.
Newell, G. F. and E. W. Montroll, 1953, Rev. Mod. Phys. **25**, 353.
Nielsen, J. W., 1960, J. Appl. Phys. **31**, 51S.
Nielsen, J. W., 1971, Metallurgical Trans. **2**, 625.
O'Dell, T. H., 1974, Magnetic Bubbles (MacMillan, London) a. 42, b. 98 and c. 24.
Oen, O. S. and D. K. Holmes, 1959, J. Appl. Phys. **30**, 1289.
Olmen, R. W., 1963, Intermag. 12.1.
Papian, W. N., 1952, Proc. IRE **40**, 475.
Parr, N. L., 1960, Zone Refining and Allied Techniques (George Newnes, London).
Parzefall, F. and B. Lettwin, 1973, IEEE Trans. Magnetics **MAG-2**, 293.
Paulevé, J. D., J. Dautreppe, J. Laugier and L. Néel, 1962, Compt. Rend. **254**, 965.
Pauthenet, R., 1956, C. r. hebd. Séanc. Sci. Paris **243**, 1499.
Pauthenet, R., 1958, Ann. Phys. (Paris) **3**, 424.
Penoyer, R. F. and L. R. Bickford, Jr., 1957, Phys. Rev. **108**, 271.
Perneski, A. J., 1969, IEEE Trans. Magnetics, **MAG-5**, 554.
Pfann, W. G., 1952, Trans. AIME **194**, 861.
Pfann, W. G., 1966, Zone Refining, 2nd ed. (Wiley, New York).
Phillips, J. C., 1967, Theory of Magnetism in Transition Metals, Marshall, W., ed. (Academic Press, New York) p. 22.
Pickering, F. B., 1964, in: Fifty Years Metallurgical Achievements.
Piercy, G. R., 1963, J. Phys. Soc. (Japan) **18**, Supplement III, 169.
Pugh, E. W. and T. O. Mohr, 1964, Thin Films (Am. Soc. for Metals, Metals Park, Ohio, USA) p. 195.
Rado, G. T., R. W. Wright and W. H. Emerson, 1950, Phys. Rev. **80**, 273.
Raffel, J. I., 1959, J. Appl. Phys. **30**, 60S.
Raynor, G. V., 1949, Progress in Metal Physics, Vol. 1 (Interscience, New York).
Read, W. T., Jr., 1953, Dislocations in Crystals (McGraw-Hill, New York) p. 16.
Reed-Hill, R. E., 1973, Physical Metallurgy Principles, 2nd. ed. (van Nostrand, New York) p. 841.
Reijnen, P., 1965, Fifth International Symposium Reactivity in Solids (Elsevier, Amsterdam) p. 562.
Reijnen, P., 1967, Science of Ceramics, Vol. 3, Stewart, A., ed. (Academic Press, New York) p. 245.
Renwick, W. and A. J. Cole, 1971, Digital Storage Systems (Chapman and Hall, London) p. 171.
Rhodes, P. and G. Rowland, 1954, Proc. Leeds Phil. Lit. Soc. **6**, 191.
Rhyne, J. J., 1972, in: Magnetic Properties of Rare Earth Metals, Elliott, R. T., ed. (Plenum Press, London) p. 129.
Richards, P. M., 1976, AIP Conf. Proc. No. 29, 184.
Rieder, G., 1959, Abhandl. Braunschweig. Wiss. Ges. **11**, 20.

Rieger, H., 1964, Z. Angew. Phys. **17**, 166.

Robinson, M. T., 1976, in: Fundamental Aspects of Radiation Damage in Metals (CONF-751006-P1, National Technical Information Service, Springfield VA, USA) p. 15.

Rosencwaig, A., W. J. Tabor and R. D. Pierce, 1971, Phys. Rev. Lett. **26**, 779.

Rossol, F. C. and A. A. Thiele, 1970, J. Appl. Phys. **41**, 1163.

Ruderman, M. A. and Kittel, C., 1954, Phys. Rev. **96**, 99.

Rushton, S., 1963, Thesis, Univ. Nottingham.

Sands, R. L. and C. R. Shakespeare, 1966, Powder Metallurgy, Practice and Applications (George Newnes Ltd., London).

Schiff, L. I., 1955, Quantum Mechanics (McGraw-Hill, New York) p. 103.

Schilling, W. and K. Sonnenberg, 1973, J. Phys. F. **3**, 322.

Schilling, W., P. Ehrhart and K. Sonnenberg, 1976, Fundamental Aspects of Radiation Damage in Metals (CONF-751006-P1, National Technical Information Service, Springfield, VA, USA) p. 470.

Schindler, A. I., E. I. Salkovitz and G. S. Ansell, 1959, J. Appl. Phys. **30**, 282S.

Schumacher, D., 1969, Vacancies and Interstitials in Metals, Seeger, A. D. Schumacher, W. Schilling and J. Diehl, eds. (North-Holland, Amsterdam) p. 889.

Scott, G. G., 1962, Rev. Modern Phys. **34**, 102.

Seeger, A., 1958, Geneva Conf. on Peaceful Uses of Atomic Energy, PUAE **6**, 250.

Seeger, A., 1976, Fundamental Aspects of Radiation Damage in Metals (CONF-751006-P1, National Technical Information Service, Springfield, VA, USA) p. 493.

Seeger, A., H. Kronmüller, H. Rieger and H. Traüble, 1964, J. Appl. Phys. **35**, 740.

Seitz, F., 1949, Disc. Faraday Soc. **5**, 271.

Seitz, F. and J. S. Koehler, 1956, Solid State Physics **2**, Seitz, F. and D. Turnbull, eds. (Academic Press, New York) p. 327.

Shafer, M. W. and T. R. McGuire, 1968, J. Appl. Phys. **39**, 588.

Sherwood, R. C., J. P. Remeika and H. J. Williams, 1959, J. Appl. Phys. **30**, 217.

Shoji, M., 1972, IEEE Trans. on Magnetics **MAG-8**, 240.

Shull, C. G., W. A. Strauser and E. O. Wollan, 1951, Phys. Rev. **83**, 333.

Shull, C. G. and M. K. Wilkinson, 1955, Phys. Rev. **97**, 304.

Sinclair, R. N. and B. N. Brockhouse, 1960, Phys. Rev. **120**, 1638.

Six, W., J. L. Snoek and W. G. Burgers, 1934, Ingenieur (Utrecht) **49**, E195.

Sixtus, K. J., 1935, Physics **6**, 105.

Sixtus, K. J. and L. Tonks, 1933, Phys. Rev. **43**, 931.

Slater, J. C., 1930, Phys. Rev. **35**, 509 and **36**, 57.

Slonczewski, J. C., 1973, J. Appl. Phys. **44**, 1759.

Slonczewski, J. C., J. C. Malozemoff and O. Voegeli, 1972, AIP Conf. Proc. No. 10, 458.

Smallman, R. E., 1970, Modern Physical Metallurgy (Butterworths, London) a. 22 and b. 346.
Smilten, J., 1952, J. Chem. Phys. **20**, 990.
Smit, J. and H. P. J. Wijn, 1954, Adv. Electronics, **6**, 69.
Smit, J., 1966, J. Appl. Phys. **37**, 1455.
Smit, J., 1968, Solid State Communications **6**, 745.
Smit, J. (ed.), 1971, Magnetic Properties of Materials (McGraw-Hill, New York).
Smith, D. V., 1948, Hydrogen in Metals (Univ. of Chicago Press, Chicago) p. 28.
Smith, R. L. and A. A. Hendrickson, Ultra-High-Purity Metals, Smith, R. L., ed. (Am. Soc. for Metals, Metals Park, Ohio, USA).
Smoluchowski, R., 1951, Phys. Rev. **84**, 511.
Snelling, E. C., 1969, Soft Ferrites, Properties and Applications (Iliffe Book Ltd., London).
Snelling, E. C., 1972, IEEE Spectrum **9**, 42.
Snoek, J., 1939, J. Physica **6**, 591.
Soven, P., 1967, Phys. Rev. **156**, 809.
Spedding, F. H. and A. H. Daane, 1961, The Rare Earths (Wiley, New York).
Standley, K. J., 1950, Sci. Prog. (London) **38**, 231.
Standley, K. J., 1972, Oxide Magnetic Materials, 2nd ed. (Clarendon Press, Oxford, England).
Stanley, J. K., 1963, Electrical and Magnetic Properties of Metals (Am. Soc. for Metals, Metals Park, Ohio, USA).
Stein, B. F., 1974, AIP Conf. Proc. No. 18, 48.
Stevens, D. K., 1975, Private communication.
Stoloff, N. S., R. G. Davies and R. C. Ku, 1965, Trans. AIME **233**, 1500.
Stoner, E. C. and E. P. Wohlfarth, 1948, Phil. Trans. Roy. Soc. (London) **A-240**, 599.
Stout, J. W., 1948, Phys. Rev. **74**, 605.
Stuart, R. and W. Marshall, 1960, Phys. Rev. **120**, 353.
Suezawa, M. and H. Kimura, 1973, Nuclear Metallurgy, Vol. 18, Defects and Defect Clusters in BCC Metals and Their Alloys, Arsenault, R. J., ed. (Am. Inst. Met. Eng., New York) p. 440.
Sumino, K., 1958, Sci. Repts. Research Insts. Tôhoku Univ. **A10**, 283.
Tabor, W. J., A. H. Bobeck, G. P. Vella-Coleira and A. Rosencwaig, 1972, J. Bell Syst. Tech. **51**, 1427.
Takahashi, M. and T. Kono, 1960, J. Phys. Soc. Japan **15**, 936.
Tanguchi, S. and A. Sakakura, 1969, J. Appl. Phys. **40**, 1539.
Taniguchi, S. and M. Yamamoto, 1954, Sci. Repts. Res. Inst. Tôhoku Univ. **A6**, 330.
Tao, L. T., R. J. Gambino, S. Kirkpatrick, J. T. Cuomo and H. Lilienthal, 1974, AIP Conf. Proc. No. 18, 641.
Taoka, T. and T. Ohtsuka, 1954, J. Phys. Soc. Japan **9**, 712.
Thiele, A. A., 1969, J. Bell Syst. Tech. **48**, 3287.
Thiele, A. A., 1970, J. Appl. Phys. **41**, 1139.

Thiele, A. A., 1971, J. Bell Syst. Tech. **50**, 725.

Thiele, A. A., 1973, Phys. Rev. Lett. **30**, 230.

Thiele, A. A. and R. H. Morrow, 1967, IEEE Trans. Magnetics **MAG-3**, 458.

Thompson, M. W., 1963, Phys. Lett. **6**, 24.

Thompson, M. W., 1969, Defects and Radiation Damage in Metals (Cambridge University Press, London) a. 89 and b. 37.

Thümmler, F. and W. Thomma, 1968, Europäisches Symposium über Pulvermetallurgie.

Traüble, H., 1969, in: Magnetism and Metallurgy, Berkowitz, A. E. and E. Kneller, eds. (Academic Press, New York) p. 630.

Traüble, H., O. Boser, H. Kronmüller and A. Seeger, 1965, Phys. Stat. Sol. **10**, 283.

Treves, D., 1965, J. Appl. Phys. **36**, 1033.

Vaccari, J. A., 1974, Materials Eng. **80**, No. 2, p. 20.

Van Uitert, L. G., W. A. Bonner, W. H. Grodkiewicz, L. Pictroski and G. J. Zydzik, 1970, Mat. Res. Bull. **5**, 825.

Van Vleck, J. H., 1932, The Theory of Electric and Magnetic Susceptibilities (Oxford Univ. Press, London).

Van Vleck, J. H., 1953, Rev. Modern Phys. **25**, 199.

Varnerin, L. J., 1971, IEEE Trans. Magnetics **MAG-7**, 404.

Vicena, F., 1955, Czech. J. Phys. **5**, 480.

Vella-Coleiro, G. P., A. Rosencwaig and W. J. Tabor, 1972, Phys. Rev. Lett. **29**, 949.

Von Neida, A. R., G. Y. Chin and A. T. English, 1968, J. Appl. Phys. **39**, 610.

Vonsovsky, S. V., 1940, J. Phys. (USSR) **3**, 181.

Vonsovsky, S. V. and E. A. Turov, 1953, J. Exptl. Theoret. Phys. (USSR) **24** 419.

Walker, R. M., 1960, J. Nucl. Mat. **2**, 147.

Walter, J. L., 1965, The Sorby Centennial Symposium on the History of Metallurgy, Smith, C. S., ed. (Gordon and Breach, New York) p. 519.

Webster, P. J. and R. S. Tebble, 1968, J. Appl. Phys. **39**, 471.

Weiss, P., 1907, J. Phys. **6**, 661.

Weiss, P. and R. Forrer, 1929, Ann. Phys. Paris **12**, 297.

Weiss, R. S., 1971, Magnetic Properties of Materials, Smit, J., ed. (McGraw-Hill, New York) p. 209.

Went, J. J. and E. W. Gorter, 1952, Philips Tech. Rev. **13**, 181.

Wiechec, W. and C. M. Kelley, 1962, J. Appl. Phys. **33**, 3054.

Wiener, G. W., P. A. Albert, R. F. Trapp and M. F. Littman, 1958, J. Appl. Phys. **29**, 366.

Wijn, H. P. J., E. W. Gorter, C. J. Esveldt and P. Geldermans, 1954, Phillips Tech. Rev. **16**, 49.

Williams, H. J. and R. C. Sherwood, 1957, J. Appl. Phys. **28**, 548.

Williams, H. J. and R. C. Sherwood, 1958, J. Appl. Phys. **29**, 296.

Williams, H. J. and W. Shockley, 1949, Phys. Rev. **75**, 178.

Williams, H. J., W. Shockley and C. Kittel, 1950, Phys. Rev. **50**, 1090.
Winkler, G., 1971, Magnetic Properties of Materials, Smit, J., ed. (McGraw-Hill, New York) p. 20.
Wohlfarth, E. P., 1949, Nature **163**, 57.
Wohlfarth, E. P. and J. F. Cornwell, 1961, Phys. Rev. Lett. **7**, 342.
Wolfe, R. and J. C. North, 1972, J. Bell Syst. Tech. **51**, 1436.
Wolff, P. A., 1961, Phys. Rev. **124**, 1030.
Woods, J. H., 1962, Phys. Rev. **126**, 517.
Woollons, D. J., 1972, Int. to Digital Computer Design (McGraw-Hill, London).
Yensen, T. D., 1929, Trans. Am. Electrochem. Soc. **56**, 215.
Yosida, K., 1957, Phys. Rev. **106**, 893.
Young, F. W., Jr., T. O. Baldwin and F. A. Sherill, 1967, Lattice Defects and Their Interactions, Hasiguti, R. R., ed. (Gordon and Breach, New York) p. 543.
Zener, C., 1951a, Phys. Rev. **81**, 440.
Zener, C., 1951b, Phys. Rev. **83**, 299.
Zener, C., 1954, Phys. Rev. **96**, 1335.
Zener, C. and R. R. Heikes, 1953, Rev. Mod. Phys. **25**, 191.
Zijlstra, H., 1967, Experimental Methods in Magnetism, Vol. 2 (North-Holland, Amsterdam).
Ziman, J. M., 1964, Principles of the Theory of Solids (Cambridge Univ. Press, Cambridge, England) a. 311, b. 183 and c. 312.

AUTHOR INDEX

SUBJECT INDEX

MATERIALS INDEX

A CATALOG OF SELECTED

DOVER BOOKS

IN ALL FIELDS OF INTEREST

A CATALOG OF SELECTED
DOVER BOOKS
IN SCIENCE AND MATHEMATICS

QUALITATIVE THEORY OF DIFFERENTIAL EQUATIONS, V.V. Nemytskii and V.V. Stepanov. Classic graduate-level text by two prominent Soviet mathematicians covers classical differential equations as well as topological dynamics and ergodic theory. Bibliographies. 523pp. 5⅜ x 8½. 65954-2 Pa. $14.95

MATRICES AND LINEAR ALGEBRA, Hans Schneider and George Phillip Barker. Basic textbook covers theory of matrices and its applications to systems of linear equations and related topics such as determinants, eigenvalues and differential equations. Numerous exercises. 432pp. 5⅜ x 8½. 66014-1 Pa. $10.95

QUANTUM THEORY, David Bohm. This advanced undergraduate-level text presents the quantum theory in terms of qualitative and imaginative concepts, followed by specific applications worked out in mathematical detail. Preface. Index. 655pp. 5⅜ x 8½. 65969-0 Pa. $14.95

ATOMIC PHYSICS (8th edition), Max Born. Nobel laureate's lucid treatment of kinetic theory of gases, elementary particles, nuclear atom, wave-corpuscles, atomic structure and spectral lines, much more. Over 40 appendices, bibliography. 495pp. 5⅜ x 8½. 65984-4 Pa. $13.95

ELECTRONIC STRUCTURE AND THE PROPERTIES OF SOLIDS: The Physics of the Chemical Bond, Walter A. Harrison. Innovative text offers basic understanding of the electronic structure of covalent and ionic solids, simple metals, transition metals and their compounds. Problems. 1980 edition. 582pp. 6⅛ x 9¼. 66021-4 Pa. $16.95

BOUNDARY VALUE PROBLEMS OF HEAT CONDUCTION, M. Necati Özisik. Systematic, comprehensive treatment of modern mathematical methods of solving problems in heat conduction and diffusion. Numerous examples and problems. Selected references. Appendices. 505pp. 5⅜ x 8½. 65990-9 Pa. $12.95

A SHORT HISTORY OF CHEMISTRY (3rd edition), J.R. Partington. Classic exposition explores origins of chemistry, alchemy, early medical chemistry, nature of atmosphere, theory of valency, laws and structure of atomic theory, much more. 428pp. 5⅜ x 8½. (Available in U.S. only) 65977-1 Pa. $11.95

A HISTORY OF ASTRONOMY, A. Pannekoek. Well-balanced, carefully reasoned study covers such topics as Ptolemaic theory, work of Copernicus, Kepler, Newton, Eddington's work on stars, much more. Illustrated. References. 521pp. 5⅜ x 8½. 65994-1 Pa. $12.95

PRINCIPLES OF METEOROLOGICAL ANALYSIS, Walter J. Saucier. Highly respected, abundantly illustrated classic reviews atmospheric variables, hydrostatics, static stability, various analyses (scalar, cross-section, isobaric, isentropic, more). For intermediate meteorology students. 454pp. 6⅛ x 9¼. 65979-8 Pa. $14.95

ASYMPTOTIC METHODS IN ANALYSIS, N.G. de Bruijn. An inexpensive, comprehensive guide to asymptotic methods–the pioneering work that teaches by explaining worked examples in detail. Index. 224pp. 5⅜ x 8½. 64221-6 Pa. $7.95

OPTICAL RESONANCE AND TWO-LEVEL ATOMS, L. Allen and J. H. Eberly. Clear, comprehensive introduction to basic principles behind all quantum optical resonance phenomena. 53 illustrations. Preface. Index. 256pp. 5⅜ x 8½. 65533-4 Pa. $8.95

COMPLEX VARIABLES, Francis J. Flanigan. Unusual approach, delaying complex algebra till harmonic functions have been analyzed from real variable viewpoint. Includes problems with answers. 364pp. 5⅜ x 8½. 61388-7 Pa. $9.95

ATOMIC SPECTRA AND ATOMIC STRUCTURE, Gerhard Herzberg. One of best introductions; especially for specialist in other fields. Treatment is physical rather than mathematical. 80 illustrations. 257pp. 5⅜ x 8½. 60115-3 Pa. $7.95

APPLIED COMPLEX VARIABLES, John W. Dettman. Step-by-step coverage of fundamentals of analytic function theory–plus lucid exposition of five important applications: Potential Theory; Ordinary Differential Equations; Fourier Transforms; Laplace Transforms; Asymptotic Expansions. 66 figures. Exercises at chapter ends. 512pp. 5⅜ x 8½. 64670-X Pa. $12.95

ULTRASONIC ABSORPTION: An Introduction to the Theory of Sound Absorption and Dispersion in Gases, Liquids and Solids, A.B. Bhatia. Standard reference in the field provides a clear, systematically organized introductory review of fundamental concepts for advanced graduate students, research workers. Numerous diagrams. Bibliography. 440pp. 5⅜ x 8½. 64917-2 Pa. $11.95

UNBOUNDED LINEAR OPERATORS: Theory and Applications, Seymour Goldberg. Classic presents systematic treatment of the theory of unbounded linear operators in normed linear spaces with applications to differential equations. Bibliography. I99pp. 5⅜ x 8½. 64830-3 Pa. $7.95

LIGHT SCATTERING BY SMALL PARTICLES, H.C. van de Hulst. Comprehensive treatment including full range of useful approximation methods for researchers in chemistry, meteorology and astronomy. 44 illustrations. 470pp. 5⅜ x 8½. 64228-3 Pa. $12.95

CONFORMAL MAPPING ON RIEMANN SURFACES, Harvey Cohn. Lucid, insightful book presents ideal coverage of subject. 334 exercises make book perfect for self-study. 55 figures. 352pp. 5⅝ x 8¼. 64025-6 Pa. $11.95

OPTICKS, Sir Isaac Newton. Newton's own experiments with spectroscopy, colors, lenses, reflection, refraction, etc., in language the layman can follow. Foreword by Albert Einstein. 532pp. 5⅜ x 8½. 60205-2 Pa. $12.95

GENERALIZED INTEGRAL TRANSFORMATIONS, A.H. Zemanian. Graduate-level study of recent generalizations of the Laplace, Mellin, Hankel, K. Weierstrass, convolution and other simple transformations. Bibliography. 320pp. 5⅜ x 8½. 65375-7 Pa. $8.95

NUMERICAL METHODS FOR SCIENTISTS AND ENGINEERS, Richard Hamming. Classic text stresses frequency approach in coverage of algorithms, polynomial approximation, Fourier approximation, exponential approximation, other topics. Revised and enlarged 2nd edition. 721pp. 5⅜ x 8½. 65241-6 Pa. $15.95

THEORETICAL SOLID STATE PHYSICS, Vol. 1: Perfect Lattices in Equilibrium; Vol. II: Non-Equilibrium and Disorder, William Jones and Norman H. March. Monumental reference work covers fundamental theory of equilibrium properties of perfect crystalline solids, non-equilibrium properties, defects and disordered systems. Appendices. Problems. Preface. Diagrams. Index. Bibliography. Total of 1,301pp. 5⅜ x 8½. Two volumes. Vol. I: 65015-4 Pa. $16.95
Vol. II: 65016-2 Pa. $16.95

OPTIMIZATION THEORY WITH APPLICATIONS, Donald A. Pierre. Broad spectrum approach to important topic. Classical theory of minima and maxima, calculus of variations, simplex technique and linear programming, more. Many problems, examples. 640pp. 5⅜ x 8½. 65205-X Pa. $16.95

THE CONTINUUM: A Critical Examination of the Foundation of Analysis, Hermann Weyl. Classic of 20th-century foundational research deals with the conceptual problem posed by the continuum. 156pp. 5⅜ x 8½. 67982-9 Pa. $6.95

ESSAYS ON THE THEORY OF NUMBERS, Richard Dedekind. Two classic essays by great German mathematician: on the theory of irrational numbers; and on transfinite numbers and properties of natural numbers. 115pp. 5⅜ x 8½.
21010-3 Pa. $5.95

THE FUNCTIONS OF MATHEMATICAL PHYSICS, Harry Hochstadt. Comprehensive treatment of orthogonal polynomials, hypergeometric functions, Hill's equation, much more. Bibliography. Index. 322pp. 5⅜ x 8½. 65214-9 Pa. $9.95

NUMBER THEORY AND ITS HISTORY, Oystein Ore. Unusually clear, accessible introduction covers counting, properties of numbers, prime numbers, much more. Bibliography. 380pp. 5⅜ x 8½. 65620-9 Pa. $10.95

THE VARIATIONAL PRINCIPLES OF MECHANICS, Cornelius Lanczos. Graduate level coverage of calculus of variations, equations of motion, relativistic mechanics, more. First inexpensive paperbound edition of classic treatise. Index. Bibliography. 418pp. 5⅜ x 8½. 65067-7 Pa. $12.95

MATHEMATICAL TABLES AND FORMULAS, Robert D. Carmichael and Edwin R. Smith. Logarithms, sines, tangents, trig functions, powers, roots, reciprocals, exponential and hyperbolic functions, fonnulas and theorems. 269pp. 5⅜ x 8½.
60111-0 Pa. $6.95

THEORETICAL PHYSICS, Georg Joos, with Ira M. Freeman. Classic overview covers essential math, mechanics, electromagnetic theory, thermodynamics, quantum mechanics, nuclear physics, other topics. First paperback edition. xxiii + 885pp. 5⅜ x 8½. 65227-0 Pa. $21.95